ENGINEERING ANALYSIS
A Survey of Numerical Procedures

ENGINEERING SOCIETIES MONOGRAPHS

Bakhmeteff: *Hydraulics of Open Channels*
Bleich: *Buckling Strength of Metal Structures*
Crandall: *Engineering Analysis*
Elevatorski: *Hydraulic Energy Dissipators*
Leontovich: *Frames and Arches*
Nadai: *Theory of Flow and Fracture of Solids*
Timoshenko and Gere: *Theory of Elastic Stability*
Timoshenko and Goodier: *Theory of Elasticity*
Timoshenko and Woinowsky-Krieger: *Theory of Plates and Shells*

Five national engineering societies, the American Society of Civil Engineers, the American Institute of Mining, Metallurgical, and Petroleum Engineers, the American Society of Mechanical Engineers, the American Institute of Electrical Engineers, and the American Institute of Chemical Engineers, have an arrangement with the McGraw-Hill Book Company, Inc., for the production of a series of selected books adjudged to possess usefulness for engineers and industry.

The purposes of this arrangement are: to provide monographs of high technical quality within the field of engineering; to rescue from obscurity important technical manuscripts which might not be published commercially because of too limited sale without special introduction; to develop manuscripts to fill gaps in existing literature; to collect into one volume scattered information of especial timeliness on a given subject.

The societies assume no responsibility for any statements made in these books. Each book before publication has, however, been examined by one or more representatives of the societies competent to express an opinion on the merits of the manuscript.

Ralph H. Phelps, CHAIRMAN
Engineering Societies Library
New York

ENGINEERING SOCIETIES MONOGRAPHS COMMITTEE

A. S. C. E.

Howard T. Critchlow
H. Alden Foster

A. I. M. E.

Nathaniel Arbiter
John F. Elliott

A. S. M. E.

Calvin S. Cronan
Raymond D. Mindlin

A. I. E. E.

F. Malcolm Farmer
Royal W. Sorensen

A. I. Ch. E.

Joseph F. Skelly
Charles E. Reed

ENGINEERING ANALYSIS

A Survey of Numerical Procedures

STEPHEN H. CRANDALL, Ph.D.

Associate Professor of Mechanical Engineering
Massachusetts Institute of Technology

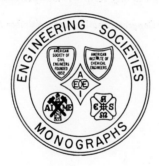

McGRAW-HILL BOOK COMPANY, INC.

New York   Toronto   London

1956

ENGINEERING SOCIETIES MONOGRAPHS

Bakhmeteff: *Hydraulics of Open Channels*
Bleich: *Buckling Strength of Metal Structures*
Crandall: *Engineering Analysis*
Nadai: *Theory of Flow and Fracture of Solids*
Rich: *Hydraulic Transients*
Timoshenko: *Theory of Elastic Stability*
Timoshenko: *Theory of Plates and Shells*
Timoshenko and Goodier: *Theory of Elasticity*

ENGINEERING ANALYSIS

Library of Congress Catalog Card Number 56–6952

v

13430

THE MAPLE PRESS COMPANY, YORK, PA.

PREFACE

The advent of high-speed automatic computing machines is making possible the solution of engineering problems of great complexity. This is a text devoted to the analysis of such problems and the methods for organizing practical programs for their numerical solution. A special feature of the work is the arrangement of the material according to a natural classification of the basic *types of problems* in engineering analysis. A description of this classification and a general outline of the body of the text are given in the Introduction.

The term *engineering analysis* as used in this book means the performance of the following steps within the framework of an engineering problem:

1. Construction of a mathematical model for a physical situation.

2. Reduction of the mathematical problem to a numerical procedure. While the major portion of the text deals with the second step, the author believes that the selection of appropriate numerical procedures cannot be made intelligently without giving due consideration to the first. For this reason, many examples of the first step have been given and considerable discussion is devoted to the structure of the mathematical formulations.

The book is written especially for engineers or engineering mathematicians. The level of mathematical maturity of the reader is assumed to be equivalent to that of an engineering graduate student. Three items which are infrequently included in engineering mathematics courses are developed in the text. They are *matrix notation* in Chap. 2, the *calculus of variations* in Chap. 4, and the *theory of characteristics* of partial differential equations in Chap. 6.

This book originated in a course of lectures given at the Imperial College of Science and Technology, London, in 1949 and at the Massachusetts Institute of Technology since 1951. The author is grateful for the encouragement and cooperation he received at these institutions. He wishes to acknowledge the many useful suggestions that arose out of discussions with his students and colleagues. He would like to express his thanks to Charles E. Thies, who worked out numerical solutions to several of the exercises. Finally, the author would like to pay tribute to his wife, Patricia, who painstakingly typed the first draft of the book and who, by her cheerful good nature, greatly eased the task of getting it ready for publication.

STEPHEN H. CRANDALL

CONTENTS

INTRODUCTION

In this brief introductory section the general outline of the book is described. The division of the text into chapters is according to the type of physical problem dealt with. The first three chapters treat discrete or lumped-parameter systems, i.e., systems with a finite number of degrees of freedom, while the last three deal with continuous systems. The problems in each of these categories are divided into three types: equilibrium problems, eigenvalue problems, and propagation problems. This classification is indicated schematically below.

Lumped-parameter systems				*Continuous systems*
Chap. 1	←	Equilibrium problems	→	Chap. 4
Chap. 2	←	Eigenvalue problems	→	Chap. 5
Chap. 3	←	Propagation problems	→	Chap. 6

Equilibrium problems are problems of steady state. The examples considered here include problems of steady-state temperature distributions, steady subsonic fluid flows, equilibrium stresses in elastic structures, electrostatic fields, and steady flows of direct and alternating current.

Eigenvalue problems may be considered as extensions of equilibrium problems in which critical values of certain parameters are to be determined in addition to the corresponding steady-state configurations. As examples we have natural frequency problems in vibrations, problems of buckling and stability of structures, and problems of resonance in acoustics and in electric circuits.

Propagation problems are initial-value problems. They include transient and unsteady-state phenomena. Problems of this type consist in predicting the subsequent behavior of a system from a knowledge of the present state. Examples treated include the propagation of heat, the propagation of pressure waves in a fluid, and the propagation of stresses and displacements in elastic systems.

It is not claimed that this subdivision is all-inclusive, but a large majority of problems in engineering analysis do fall naturally into one of these three subdivisions.

If the classification of problem types is considered from a strictly mathematical point of view, it is found that the problems of the first two chapters are characterized by sets of simultaneous algebraic equations, the problems of the third chapter are characterized by sets of simultaneous ordinary differential equations with prescribed *initial* conditions, the problems of the fourth and fifth chapters are characterized by ordinary or partial differential equations with closed *boundary* conditions, and the problems of the sixth chapter are characterized by partial differential equations with open boundary conditions. This survey of numerical procedures thus amounts to a catalogue of practical methods for the solution of algebraic, ordinary differential, and partial differential equations. In Chaps. 1, 3, 4, and 6 both linear and nonlinear problems are considered. The discussion of eigenvalue problems in Chaps. 2 and 5 is limited to linear systems.

All six chapters have the same structure. At the beginning of each chapter several representative problems are presented. These serve to identify the class of problems under consideration. The process of formulating a mathematical model is illustrated for each of these problems.

Before leaving these formulations they are each cast into *dimensionless* form. This is an extremely useful organizational tool[1] of the analyst. In connection with numerical calculations it removes all unnecessary symbols, leaving the basic problem in its simplest form.

Then, before surveying numerical procedures applicable to this class of problems, a brief résumé of the classical mathematical theory is given. A complete mathematical development has not been attempted but an effort has been made to describe clearly the properties of the well-behaved or regular system. The possibilities of irregular behavior are hinted at by means of simple counterexamples. Enough theory is presented to provide a background for the explanation of the success (and limitations) of the numerical procedures which follow.

After these preliminaries the actual survey of numerical procedures begins. Illustrative examples are drawn from the problems formulated at the beginning of the chapter. At the end of each section there is a set of exercises for the reader. A few of these are of the nature of drill problems but the majority represent interesting extensions or alternative developments of the text material. Answers or hints for the solution are given in most cases.

The numerical procedures described here are those which in the judgment of the author are of most potential interest to the engineering analyst. Methods for both hand and machine computation are given.

Throughout the text there are references to books and papers having direct bearing on the matter at hand. For the reader's convenience a number of selected general references are grouped together in the Bibliography at the end of the book.

[1] See, for example, H. L. Langhaar, "Dimensional Analysis and Theory of Models," John Wiley & Sons, Inc., New York, 1951.

CHAPTER 1

EQUILIBRIUM PROBLEMS IN SYSTEMS
WITH A FINITE NUMBER OF DEGREES OF FREEDOM

The state of a physical system can often be described with adequate precision by giving the magnitudes of a finite number of state variables. This chapter deals with numerical procedures for determining steady states of such systems. The chapter begins with a preliminary examination of several particular problems. The general problem of this type is then formulated mathematically as a set of simultaneous algebraic equations. There is a review of the classical results from the theory of such systems, including a discussion of the relationship of extremum principles to equilibrium problems. Numerical procedures, both exact and approximate, are then described and illustrated by applying them to the particular problems set up at the beginning of the chapter.

1-1. Particular Examples

We begin with an assortment of examples of how mathematical formulations are set up for particular physical problems. The examples are taken from a variety of fields and, in general, have been chosen for their simplicity despite the fact that the really significant contributions of numerical procedures occur in problems of extended complexity.

It is generally recognized that the most difficult step in the whole process of engineering analysis is that in which a mathematical model is substituted for a real physical system. It is here that judgment, experience, and ingenuity of the highest order are required of the analyst. It is here that the really gross approximations and simplifications are made. In this text the basic structure of the various physical problem types and the corresponding mathematical models is emphasized.

The general equilibrium problem in a lumped-parameter system has the following structure: The given system is made up by interconnecting a number of simple elements. The equilibrium or steady-state requirements for each individual element are known. As examples we have the stress-strain law for elastic elements, Ohm's law for electrical resistances, and the pressure-flow relation for hydraulic resistances. In addition to satisfying the equilibrium requirements of the individual elements it is also necessary to satisfy certain interconnection requirements. Thus in elastic systems we must have geometric fit and balance of forces at all joints; in electric networks we must satisfy both of Kirchhoff's laws; and in hydraulic networks we must have conservation of flow and uniqueness of pressure at every interconnection. The over-all equilibrium problem then consists in finding the state of a system which simultaneously satisfies the equilibrium requirements of the individual elements together with the interconnection requirements.

To serve as concrete illustrations of this general statement and to provide illustrative examples for the numerical procedures which follow, we here consider the following five particular equilibrium problems:

1-1. Elastic spring system.
1-2. D-c network.
1-3. A-c network.
1-4. Continuous beam.
1-5. Hydraulic network.

In each case the problem is cast into nondimensional form, with particular data assumed, in preparation for numerical solution. In most cases complementary forms of the problem are considered. The first four problems are linear, while the fifth represents an example of a practically important nonlinear problem.

Problem 1-1. Elastic Spring System. In Fig. 1-1 a system of four

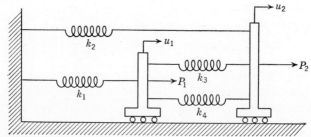

FIG. 1-1. Elastic system of interconnected springs subjected to loads P_1 and P_2.

linear springs is shown. Assume that when P_1 and P_2 are zero then u_1 and u_2 are both zero and that all springs are in their natural positions. The problem here is to find the displacements u_1 and u_2 and the forces $f_1, f_2, f_3,$ and f_4 in the four springs when the loads P_1 and P_2 are applied. The fundamental requirements are:

1. Spring force = k(spring elongation) for each spring.
2. Forces should balance on each movable cart.
3. Spring elongations should be compatible with the displacements of the carts.

A standard method of solution is to choose unknown variables in such a way that requirement 3 above is automatically satisfied. In our problem this is done by taking u_1 and u_2 as unknowns and expressing the spring elongations in terms of them (e.g., elongation of spring 4 is $u_2 - u_1$). Next the spring forces are expressed in terms of u_1 and u_2 by introducing the spring constants. Finally, writing the force-balance conditions for each cart gives us the following equations for u_1 and u_2:

$$\begin{aligned} k_1 u_1 - k_3(u_2 - u_1) - k_4(u_2 - u_1) &= P_1 \\ k_2 u_2 + k_3(u_2 - u_1) + k_4(u_2 - u_1) &= P_2 \end{aligned} \tag{1-1}$$

A complete solution of our problem would require the solution of these simultaneous equations. We stop at this point, however, since we are here concerned only with the formulation of the problem. Summarizing, we limited ourselves to geometrically compatible states as soon as we took u_1 and u_2 as unknowns; requiring that force balance should also hold gave us (1-1).

A *complementary* method of solution for the same problem is to choose unknown variables in such a way that requirement 2 above is automatically satisfied. This may be done by taking the spring forces f_2 and f_3 as unknown and expressing the other spring forces, f_1 and f_4, in terms of them by means of the force-balance conditions.

$$\begin{aligned} f_4 &= P_2 - f_2 - f_3 \\ f_1 &= P_1 + f_3 + f_4 = P_1 + P_2 - f_2 \end{aligned} \tag{1-2}$$

Next the spring elongations are expressed in terms of f_2 and f_3 by introducing the spring constants. Finally we obtain the following equations for f_2 and f_3 by requiring that the spring elongations be compatible with unique displacements of the carts:

$$\begin{aligned} \frac{f_2}{k_2} &= \frac{P_1 + P_2 - f_2}{k_1} + \frac{P_2 - f_2 - f_3}{k_4} \\ \frac{f_3}{k_3} &= \frac{P_2 - f_2 - f_3}{k_4} \end{aligned} \tag{1-3}$$

The second of these expresses the fact that the elongations of springs 3 and 4 should be the same. The first expresses the fact that the elongation of spring 2 must be the same as the sum of the elongations of springs 1 and 4. Again a complete solution would require the simultaneous solution of (1-3), but we stop at this point. Reiterating our logic, we limited ourselves to self-balancing states when we took f_2 and f_3 as unknowns and used (1-2) for the other forces. Among these self-balancing states the true state is selected by (1-3), which requires that the spring elongations should be compatible with the given interconnections of the system.

For future use we now specialize the above problem to the case where

$$
\begin{aligned}
k_1 &= 3k \\
k_2 &= 2k \qquad P_1 = P \\
k_3 &= k \qquad\; P_2 = 2P \\
k_4 &= k
\end{aligned} \tag{1-4}
$$

Substituting these values in (1-1) and (1-3), we obtain

$$
\begin{aligned}
5ku_1 - 2ku_2 &= P \\
-2ku_1 + 4ku_2 &= 2P
\end{aligned} \tag{1-5}
$$

as the equations for the displacements and

$$
\begin{aligned}
\tfrac{11}{6}f_2 + \; f_3 &= 3P \\
f_2 + 2f_3 &= 2P
\end{aligned} \tag{1-6}
$$

as the complementary equations for the forces.   These formulations can be simplified even further by introducing dimensionless variables.   If we define the nondimensional displacements

$$
x_1 = \frac{u_1}{P/k} \qquad x_2 = \frac{u_2}{P/k} \tag{1-7}
$$

the displacement equations (1-5) can be written in the following form:

$$
\begin{aligned}
5x_1 - 2x_2 &= 1 \\
-2x_1 + 4x_2 &= 2
\end{aligned} \tag{1-8}
$$

Similarly, in terms of the nondimensional forces

$$
y_1 = \frac{f_2}{P} \qquad y_2 = \frac{f_3}{P} \tag{1-9}
$$

the force equations (1-6) become

$$
\begin{aligned}
\tfrac{11}{6}y_1 + \; y_2 &= 3 \\
y_1 + 2y_2 &= 2
\end{aligned} \tag{1-10}
$$

Problem 1-2. D-C Network.   We consider the problem of determining the voltages and currents in the network shown in Fig. 1-2.   The resistances and battery emfs are given in the figure in terms of R and E. The equilibrium or steady-state conditions are Ohm's law for each individual resistor plus the interconnection requirements which are the two laws of Kirchhoff.[1]   We can obtain *complementary* formulations of the problem in the following manner: If we represent the state of the sys-

[1] See, for example, C. L. Dawes, "Electrical Engineering," 3d ed., vol. I, McGraw-Hill Book Company, Inc., New York, 1937, p. 72.

tem by a set of independent currents such that Kirchhoff's first law is automatically satisfied, we then obtain equations for determining these currents by requiring that the second law be satisfied. Alternatively if the state of the system is represented by a set of independent voltages such that Kirchhoff's second law is automatically satisfied, equations can then be obtained for determining these voltages by requiring the satisfaction of the first law.

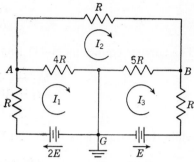

FIG. 1-2. Network of resistors and batteries.

In accordance with the first procedure the state of the system is represented by the three loop currents I_1, I_2, and I_3. The net current flow into any junction is always zero for any values of I_1, I_2, and I_3. Ohm's law together with the requirement that the net voltage drop in any closed loop should vanish yields the following equations:

$$2E - RI_1 - 4R(I_1 - I_2) = 0$$
$$-RI_2 - 5R(I_2 - I_3) - 4R(I_2 - I_1) = 0$$
$$-RI_3 - 5R(I_3 - I_2) - E = 0$$

(1-11)

When the currents which satisfy (1-11) are found, any desired network emf is easily obtained by an elementary application of Ohm's law.

Following the second procedure, the state of the system can be represented by the potentials e_1 and e_2 of the nodes A and B with respect to G. This ensures that the voltage drop around any closed loop vanishes. The requirement that there should be no net current flow into the nodes A and B results in the following equations:

$$\frac{2E - e_1}{R} - \frac{e_1}{4R} + \frac{e_2 - e_1}{R} = 0$$
$$\frac{E - e_2}{R} - \frac{e_2}{5R} + \frac{e_1 - e_2}{R} = 0$$

(1-12)

When the voltages e_1 and e_2 which satisfy (1-12) have been found, any desired network current may be obtained by a simple application of Ohm's law.

The complete solution can thus be obtained by solving either (1-11) or (1-12). Note that here the number of degrees of freedom is not the same for the two analyses. Before leaving this problem, we cast the equations into nondimensional form. Dimensionless currents and voltages are defined as follows:

$$x_1 = \frac{I_1}{E/R} \qquad x_2 = \frac{I_2}{E/R} \qquad x_3 = \frac{I_3}{E/R}$$
$$y_1 = \frac{e_1}{E} \qquad y_2 = \frac{e_2}{E}$$

(1-13)

The current equations (1-11) then become

$$
\begin{aligned}
5x_1 - 4x_2 &= 2 \\
-4x_1 + 10x_2 - 5x_3 &= 0 \\
 - 5x_2 + 6x_3 &= -1
\end{aligned}
$$

(1-14)

while the voltage equations (1-12) take the following form:

$$
\begin{aligned}
2.25y_1 - y_2 &= 2 \\
-y_1 + 2.20y_2 &= 1
\end{aligned}
$$

(1-15)

These last two sets of equations constitute complementary dimensionless formulations of Prob. 1-2.

Problem 1-3. A-C Network. The equilibrium problem here is to determine the steady-state currents in the network of Fig. 1-3. The impedances of the branches at the frequency of the voltage source are indicated in the usual[1] complex notation in terms of R. Complementary formulations of this problem can be obtained in the same manner as in Prob. 1-2. We consider here only the equations for the currents. If we take I_1 and I_2 as the state variables, Kirchhoff's first law is automatically satisfied and the second law yields the following equations:

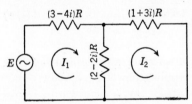

FIG. 1-3. Network of impedances connected to alternating-voltage source.

$$
\begin{aligned}
E - (3 - 4i)RI_1 - (2 - 2i)R(I_1 - I_2) &= 0 \\
-(2 - 2i)R(I_2 - I_1) - (1 + 3i)RI_2 &= 0
\end{aligned}
$$

(1-16)

A nondimensional formulation is obtained by introducing the dimensionless variables

$$I_1' = \frac{I_1}{E/R} \qquad I_2' = \frac{I_2}{E/R}$$

(1-17)

into (1-16) as follows:

$$
\begin{aligned}
(5 - 6i)I_1' - (2 - 2i)I_2' &= 1 \\
-(2 - 2i)I_1' + (3 + i)I_2' &= 0
\end{aligned}
$$

(1-18)

The quantities I_1' and I_2' are expected to be complex. Although procedures exist for the direct solution of sets of equations such as (1-18), it is sometimes useful to trans-

---

[1] See, for example, C. L. Dawes, "A Course in Electrical Engineering," 4th ed., vol. II, McGraw-Hill Book Company, Inc., New York, 1947, p. 70. The symbol i stands for the imaginary unit $\sqrt{-1}$.

form the complex equations into their real equivalents. To illustrate this process for the present example, we define the real quantities $x_1, \ldots, x_4$ as follows:

$$I'_1 = x_1 + ix_2$$
$$I'_2 = x_3 + ix_4$$
(1-19)

When these are substituted in (1-18), each equation can be separated into two: one obtained from the real terms and one from the imaginary terms. We thus obtain the following four real equations, which are equivalent to the two complex equations of (1-18):

$$5x_1 + 6x_2 - 2x_3 - 2x_4 = 1$$
$$6x_1 - 5x_2 - 2x_3 + 2x_4 = 0$$
$$-2x_1 - 2x_2 + 3x_3 - x_4 = 0$$
$$-2x_1 + 2x_2 - x_3 - 3x_4 = 0$$
(1-20)

Problem 1-4. Continuous Beam. In Fig. 1-4 a uniform elastic beam is shown. It is simply supported at A, B, and C and clamped at D. Equilibrium problems for such systems consist in determining the bend-

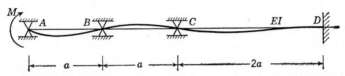

Fig. 1-4. Continuous beam freely supported at A, B, and C, clamped at D, and subjected to external moment M applied at A.

ing moments and deflections resulting from assigned loads. We consider the particular problem of Fig. 1-4, where the load is the single moment M applied at A. The flexural stiffness of the beam is EI, and the span lengths are given in terms of a.

This system may be treated as a lumped parameter system by considering each span as a single element. The total equilibrium problem then involves satisfying the elastic requirements within each span, together with the interconnection requirements at the joints. These interconnection requirements are that adjacent spans should have the same inclination and the same bending moment at their common junction. The internal elastic requirements for a single span are one stage more complicated than the corresponding single-element relations in the foregoing examples. Here each span is itself a two-degree-of-freedom system described by two geometric quantities (the inclinations at the ends) and by two force quantities (the bending moments at the ends). The relations between these which represent the elastic requirements[1] are shown in Fig. 1-5. Clockwise angles have been called positive. Bending moments which tend to stretch the bottom fibers and compress the top fibers have been called positive. A formulation of the equilibrium

[1] See, for example, L. C. Maugh, "Statically Indeterminate Structures," John Wiley & Sons, Inc., New York, 1946, p. 49.

problem may be obtained by using either inclinations or bending moments to represent the state of the system. Thus a set of independent angles which satisfy the compatibility requirements might be chosen. With the aid of the elastic relations bending moments could then be expressed in terms of these angles, and finally, by writing the conditions for moment

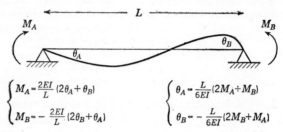

$$\begin{cases} M_A = \dfrac{2EI}{L}(2\theta_A + \theta_B) \\ M_B = -\dfrac{2EI}{L}(2\theta_B + \theta_A) \end{cases} \qquad \begin{cases} \theta_A = \dfrac{L}{6EI}(2M_A + M_B) \\ \theta_B = -\dfrac{L}{6EI}(2M_B + M_A) \end{cases}$$

FIG. 1-5. Elastic relationships for a span whose ends are restrained from translation and which is subjected to end moments.

balance, a set of equations for determining the angles would be obtained. Alternatively a set of independent bending moments which satisfy the requirements of moment balance could be used to represent the state of the system. The compatibility requirements together with the elastic relations would then furnish equations for determining these moments.

Adopting the former procedure, the state of the system of Fig. 1-4 can

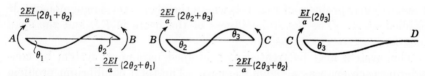

FIG. 1-6. Representation of the beam of Fig. 1-4 in terms of the displacements θ_1, θ_2, and θ_3.

be represented by the clockwise inclinations of the beam at A, B, and C. These angles are denoted by θ_1, θ_2, and θ_3, respectively. Making use of the elastic relations of Fig. 1-5, the terminal bending moments in each span are as indicated in Fig. 1-6.

Governing equations for the angles are now obtained by writing the conditions for moment balance at the supports A, B, and C.

$$M = \frac{2EI}{a}(2\theta_1 + \theta_2)$$

$$-\frac{2EI}{a}(2\theta_2 + \theta_1) = \frac{2EI}{a}(2\theta_2 + \theta_3) \qquad (1\text{-}21)$$

$$-\frac{2EI}{a}(2\theta_3 + \theta_2) = \frac{EI}{a}(2\theta_3)$$

These may be cast into nondimensional form by introducing the following dimensionless inclinations:

$$x_1 = \frac{\theta_1}{Ma/2EI} \qquad x_2 = \frac{\theta_2}{Ma/2EI} \qquad x_3 = \frac{\theta_3}{Ma/2EI} \qquad (1\text{-}22)$$

We thus obtain the following formulation of the equilibrium problem:

$$\begin{aligned}
2x_1 + \ x_2 \qquad\quad &= +1 \\
x_1 + 4x_2 + \ x_3 &= 0 \\
x_2 + 3x_3 &= 0
\end{aligned} \qquad (1\text{-}23)$$

A complementary formulation may be obtained in terms of the bending moments M_1, M_2, and M_3 at B, C, and D, respectively, in the beam of Fig. 1-4. It is left as an exercise for the reader to show that in terms of the dimensionless moments

$$y_1 = \frac{M_1}{M} \qquad y_2 = \frac{M_2}{M} \qquad y_3 = \frac{M_3}{M} \qquad (1\text{-}24)$$

the governing equations are as follows:

$$\begin{aligned}
4y_1 + \ y_2 \qquad\quad &= -1 \\
y_1 + 6y_2 + 2y_3 &= 0 \\
2y_2 + 4y_3 &= 0
\end{aligned} \qquad (1\text{-}25)$$

Problem 1-5. Hydraulic Network. We consider the problem of determining the steady flow of an incompressible fluid in a network of branched pipes under the assumption that the pressure drop in a single

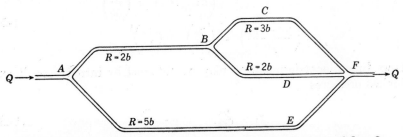

Fig. 1-7. Schematic diagram of hydraulic network passing a total flow Q.

branch is proportional to the square of the rate of flow through that branch. Figure 1-7 shows the plan of a particular pipe network. The total rate of flow, in at A and out at F, is Q. For a single branch the pressure drop in the direction of flow is given[1] by the following resistance law,

$$\Delta p = Rq^2 \tag{1-26}$$

where q is rate of flow through the branch and R is a resistance coefficient. The resistance coefficient of each branch in Fig. 1-7 is given in terms of b.

The equilibrium problem consists in determining the pressure and flow distribution in the steady state. To make the problem definite, we assume that Q is given and that the pressure at F is zero. The governing requirements are that the pressure at each junction should be single-valued, that the rate of flow into any junction should equal the rate of flow out of that junction, and that in each separate branch the resistance law (1-26) should be satisfied. A formulation of the problem can be made in terms of either junction pressures or branch flow rates. Thus the state of the system can be represented by p_1 and p_2, the pressures at A and B, respectively. In terms of these the flow rates in the individual branches are given by (1-26).

$$
\begin{aligned}
q_{AB} &= \left(\frac{p_1 - p_2}{2b}\right)^{\frac{1}{2}} \\
q_{BCF} &= \left(\frac{p_2}{3b}\right)^{\frac{1}{2}} \\
q_{BDF} &= \left(\frac{p_2}{2b}\right)^{\frac{1}{2}} \\
q_{AEF} &= \left(\frac{p_1}{5b}\right)^{\frac{1}{2}}
\end{aligned}
\tag{1-27}
$$

The requirement of continuity of flow at the junctions A and B provides the following governing equations:

$$
\begin{aligned}
Q &= \left(\frac{p_1 - p_2}{2b}\right)^{\frac{1}{2}} + \left(\frac{p_1}{5b}\right)^{\frac{1}{2}} \\
\left(\frac{p_1 - p_2}{2b}\right)^{\frac{1}{2}} &= \left(\frac{p_2}{3b}\right)^{\frac{1}{2}} + \left(\frac{p_2}{2b}\right)^{\frac{1}{2}}
\end{aligned}
\tag{1-28}
$$

A nondimensional formulation may be obtained by introducing dimensionless pressures

$$x_1 = \frac{p_1}{bQ^2} \qquad x_2 = \frac{p_2}{bQ^2} \tag{1-29}$$

---

[1] See, for example, H. W. King, C. O. Wisler, and J. G. Woodburn, "Hydraulics," John Wiley & Sons, Inc., New York, 1948, 5th ed., p. 220. Strictly speaking we should consider Δp and q as directed quantities and write $\Delta p = [\text{sign } (q)]Rq^2$. If we use (1-26), it is incumbent on us to check that all pressure drops are actually in the direction of flow in any proposed solution.

In terms of these (1-28) may be cast into the following form:

$$0.4472x_1^{\frac{1}{2}} + 0.7071(x_1 - x_2)^{\frac{1}{2}} = 1$$
$$0.7071(x_1 - x_2)^{\frac{1}{2}} - 1.2845x_2^{\frac{1}{2}} = 0 \tag{1-30}$$

A *complementary* formulation may be obtained in terms of branch flow rates. Continuity of flow will be preserved in Fig. 1-7 if the flow rates q_1 and q_2 in the branches AB and BCF, respectively, are independent provided the flow rates in the remaining branches are taken as follows:

$$q_{BDF} = q_1 - q_2$$
$$q_{AEF} = Q - q_1 \tag{1-31}$$

With the aid of (1-26) the requirement of single-valued pressures at A and B leads to the following governing equations:

$$2bq_1^2 + 2b(q_1 - q_2)^2 = 5b(Q - q_1)^2$$
$$3bq_2^2 = 2b(q_1 - q_2)^2 \tag{1-32}$$

Introducing the dimensionless flow rates

$$y_1 = \frac{q_1}{Q} \qquad y_2 = \frac{q_2}{Q} \tag{1-33}$$

we obtain a nondimensional formulation as follows:

$$10y_1 - y_1^2 - 4y_1y_2 + 2y_2^2 = 5$$
$$+2y_1^2 - 4y_1y_2 - y_2^2 = 0 \tag{1-34}$$

EXERCISES

1-1. The lengths and cross-sectional areas of the bars of a plane pinned truss are indicated in Fig. 1-8. The bars are joined by frictionless pins, and each one satisfies Hooke's law, $f/A = E\delta/L$, where f is the tensile force and δ is the elongation. The

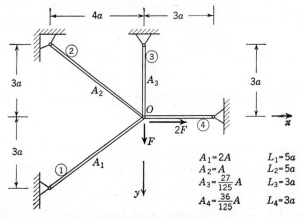

$$A_1 = 2A \qquad L_1 = 5a$$
$$A_2 = A \qquad L_2 = 5a$$
$$A_3 = \frac{27}{125}A \qquad L_3 = 3a$$
$$A_4 = \frac{36}{125}A \qquad L_4 = 3a$$

FIG. 1-8. Exercise 1-1.

modulus of elasticity E is the same for all bars. Set up complementary formulations of the equilibrium problem when the forces F and $2F$ are applied as shown. Show that in terms of nondimensional displacements at the point O the equations are

$$5x_1 - \ \ x_2 = 2$$
$$-x_1 + 3x_2 = 1$$

while in terms of nondimensional forces in bars 1 and 2 the equations are

$$8.5y_1 + \ \ y_2 = 5$$
$$y_1 + 10y_2 = 15$$

1-2. A wall is constructed of two homogeneous slabs in intimate contact. In the steady state the temperatures in the wall are characterized by the external surface

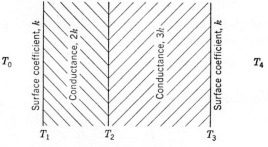

Fig. 1-9. Exercise 1-2.

temperatures T_1 and T_3 and the interface temperature T_2. Formulate the problem of determining these temperatures when the ambient temperatures T_0 and T_4 are known. The conductance[1] per unit area for the individual slabs and the surface coefficients are given in Fig. 1-9. The heat-conduction law is $q/A = K \, \Delta T$, where q is the total heat flux, A is the area, ΔT the temperature drop in the direction of heat flow, and K the conductance or surface coefficient. Show that in terms of dimensionless temperatures the equations are

$$3x_1 - 2x_2 \qquad\quad = 1$$
$$-2x_1 + 5x_2 - 3x_3 = 0$$
$$- 3x_2 + 4x_3 = 0$$

How many degrees of freedom does this system have if the complementary formulation is considered?

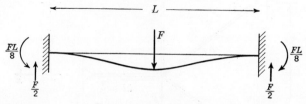

Fig. 1-10. Exercise 1-3.

1-3. Consider the same continuous beam as in illustrated Prob. 1-4 but now with loads F and $2F$ acting vertically downward at the mid-points of spans AB and BC. This may be treated by superposing configurations of the type shown in Fig. 1-10 on

[1] See, for example, W. H. McAdams, "Heat Transmission," 3d ed., McGraw-Hill Book Company, Inc., New York, 1954, p. 15.

the solution of a problem similar to the illustrated example in that the only loads on the system are moments applied at the joints. Carry out this procedure, and show that the equations for the dimensionless inclinations are the same as (1-23) except for the right-hand sides, which become $\frac{1}{8}$, $\frac{1}{8}$, and $-\frac{1}{4}$ in place of 1, 0, and 0.

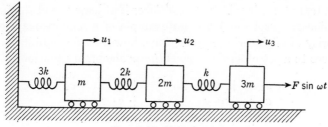

FIG. 1-11. Exercise 1-4.

1-4. The steady-state forced vibration of the system shown in Fig. 1-11 has the form

$$u_1 = x_1 \sin \omega t$$
$$u_2 = x_2 \sin \omega t$$
$$u_3 = x_3 \sin \omega t$$

Show that for the case where $\omega^2 = k/m$ the amplitudes x_1, x_2, and x_3 are given by the solution of the following system:

$$
\begin{aligned}
4x_1 - 2x_2 \quad\quad &= 0 \\
-2x_1 + x_2 - x_3 &= 0 \\
- x_2 - 2x_3 &= \frac{F}{k}
\end{aligned}
$$

1-2. Formulation of the General Problem

In this and the following section we consider the general mathematical structure of the equilibrium problem in lumped parameter systems. As we have seen in the foregoing examples, the problem consists in determining the equilibrium values of the state variables from a set of simultaneous equations. If the state variables are denoted by x_j $(j = 1, \ldots, n)$, we have to solve for them from a set of n equations,

$$
\begin{aligned}
a_1(x_1, \ldots, x_n) &= c_1 \\
a_2(x_1, \ldots, x_n) &= c_2 \\
\cdots\cdots\cdots\cdots\cdots & \\
a_n(x_1, \ldots, x_n) &= c_n
\end{aligned}
\tag{1-35}
$$

where the a_j are known functions of the x_j and the c_j are known constants. The functions a_j describe the intrinsic properties of the passive system, while the constants c_j represent the particular "loading" to which the system is subjected.

The number of degrees of freedom n depends not only on the original physical system but also on the choice of state variables used in the analysis. In the previous section complementary formulations were illustrated for several of the problems.   It was noted in Prob. 1-2 that the number of degrees of freedom is not necessarily the same for both formulations. A marked saving in computing time can sometimes be had by choosing the smaller set of state variables.

The system (1-35) is said to be *nonlinear* if at least one of the functions a_j is nonlinear.   Problem 1-5 is an example of a nonlinear system.   In the majority of engineering applications, however, the equilibrium problem reduces to a set of simultaneous *linear* algebraic equations,

$$
\begin{aligned}
a_{11}x_1 + a_{12}x_2 + \cdots + a_{1n}x_n &= c_1 \\
a_{21}x_1 + a_{22}x_2 + \cdots + a_{2n}x_n &= c_2 \\
&\cdots\cdots\cdots\cdots\cdots\cdots \\
a_{n1}x_1 + a_{n2}x_2 + \cdots + a_{nn}x_n &= c_n
\end{aligned}
\tag{1-36}
$$

where the a_{jk} and c_j are known constants.

Description of Linear Systems.   The first four problems of the previous section all led to one or more formulations which were special cases of (1-36).

The system (1-36) is said to be *real* if all the coefficients a_{jk} and constants c_j are real.   Except for (1-18) all the systems in Sec. 1-1 were real.

The system (1-36) is said to be *symmetric* if, for every j and k, $a_{jk} = a_{kj}$. All the linear systems obtained in Sec. 1-1 were symmetric.   The coefficients a_{11}, a_{22}, . . . , a_{nn} are called the *main diagonal coefficients;* the coefficients not on the main diagonal are called *coupling coefficients.*

The system (1-36) is said to be *positive* if it is real and if the *quadratic form*

$$
\begin{aligned}
Q = \tfrac{1}{2}\{a_{11}x_1{}^2 + a_{12}x_1x_2 &+ \cdots + a_{1n}x_1x_n \\
+ a_{21}x_2x_1 + a_{22}x_2{}^2 &+ \cdots + a_{2n}x_2x_n \\
+ \cdots\cdots\cdots\cdots&\cdots\cdots\cdots \\
+ a_{n1}x_nx_1 + a_{n2}x_nx_2 &+ \cdots + a_{nn}x_n{}^2\}
\end{aligned}
\tag{1-37}
$$

constructed from the coefficients is nonnegative for all possible combinations of real x_j.   The system is said to be *positive definite* if Q is nonnegative and moreover is only zero when every x_j is zero.[1]   Although negative and negative definite systems can also be defined, it is common custom to make a preliminary change of sign throughout (1-36) whenever necessary so that only the positive case need be discussed.   The formulations for the elastic spring system, the d-c network, and the continuous beam are positive definite.   The real system (1-20) obtained by separating the real and imaginary parts of (1-18) is not positive definite or even positive.

---

[1] The basic definition given here does not provide a very direct means of ascertaining whether or not a particular set of equations is positive definite.   For a systematic test see Exercise 1-5.

EXERCISES

1-5. The system (1-36) is positive definite if it is symmetric and the determinant of the coefficients and all its principal minors are positive, i.e., if

$$a_{11} > 0, \quad \begin{vmatrix} a_{11} & a_{12} \\ a_{21} & a_{22} \end{vmatrix} > 0, \quad \begin{vmatrix} a_{11} & a_{12} & a_{13} \\ a_{21} & a_{22} & a_{23} \\ a_{31} & a_{32} & a_{33} \end{vmatrix} > 0, \quad \ldots, \quad \begin{vmatrix} a_{11} & a_{12} & \cdots & a_{1n} \\ a_{21} & a_{22} & \cdots & a_{2n} \\ \vdots & & & \\ a_{n1} & a_{n2} & \cdots & a_{nn} \end{vmatrix} > 0$$

A proof of this theorem appears on page 154 of "The Mathematics of Circuit Analysis" by E. A. Guillemin, John Wiley & Sons, Inc., New York, 1949. Use this theorem to verify that (1-23) and (1-25) are positive definite. Is the system in Exercise 1-4 positive definite?

1-6. Show directly from the text definition that the main diagonal coefficients of a positive definite system must be positive.

1-7. Show that any set of simultaneous linear equations may be transformed into an equivalent *symmetric* set by the following procedure: Multiply the first of (1-36) by a_{11}, the second by a_{21}, . . . , and the last by a_{n1}, and add all these together. This is the first equation of the desired set. Next multiply the first of (1-36) by a_{12}, the second by a_{22}, . . . , and the last by a_{n2}, and add all these to get the second equation, etc.

1-3. Mathematical Properties

No attempt is made here to give a complete treatment of the classical theory of (1-35) and (1-36). Instead we quote the properties of what might be called *well-behaved* systems and describe those classes of problems which can be recognized in advance to possess these properties. As a brief reminder that the general theory is far more complicated, we include a few counterexamples which illustrate common pathological behaviour.

The well-behaved equilibrium problem does have a solution; i.e., there is a set of x_j which satisfies (1-35). Moreover there usually is only one meaningful solution. It can be shown[1] that there exists a single unique solution to the *linear* system (1-36) if the system is positive definite or, more generally, if the determinant of the coefficients,

$$\Delta = \begin{vmatrix} a_{11} & a_{12} & \cdots & a_{1n} \\ a_{21} & a_{22} & \cdots & a_{2n} \\ \vdots & & & \\ a_{n1} & a_{n2} & \cdots & a_{nn} \end{vmatrix} \tag{1-38}$$

does not vanish. For nonlinear systems no such general result exists; however, if the nonlinear system represents a flow network with non-

[1] See, for example, F. B. Hildebrand, "Methods of Applied Mathematics," Prentice-Hall, Inc., New York, 1952, p. 21.

linear resistances a unique solution is guaranteed,[1] provided the flow in any branch is a continuous, strictly increasing function of the potential drop across that branch.

Counterexamples. Despite the practical importance of the well-behaved case it is helpful to keep in mind that the general mathematical problem (1-35) can misbehave in any one of the following ways:

1. *No Solution at All.* There is no set of values (x_1, x_2) which satisfy the positive system

$$\begin{aligned} 4x_1 - 2x_2 &= 3 \\ 2x_1 - x_2 &= 6 \end{aligned} \tag{1-39}$$

The determinant (1-38) for this system is zero.

2. *More than One Solution.* The two sets of values $(1, 0)$ and $(0, 1)$ both satisfy the nonlinear system.

$$\begin{aligned} x_1^2 + x_2^2 &= 1 \\ x_1 + x_2 &= 1 \end{aligned} \tag{1-40}$$

The linear system

$$\begin{aligned} 4x_1 - 2x_2 &= 6 \\ 2x_1 - x_2 &= 3 \end{aligned} \tag{1-41}$$

which has the same coefficients a_{jk} as (1-39), has an infinity of solutions, for example, $(1, -1)$, $(2, 1)$, $(3, 3)$, $(4, 5)$, etc.

3. *Complex Solutions for Real Problems.* The real nonlinear system

$$\begin{aligned} x_1^2 + x_2^2 &= 1 \\ x_1 + x_2 &= 2 \end{aligned} \tag{1-42}$$

has the two complex solutions $(1 + i/\sqrt{2}, 1 - i/\sqrt{2})$ and $(1 - i/\sqrt{2}, 1 + i/\sqrt{2})$.

Ill-conditioned Systems. Consider the two-degree-of-freedom system

$$\begin{aligned} a_{11}x_1 + a_{12}x_2 &= c_1 \\ a_{21}x_1 + a_{22}x_2 &= c_2 \end{aligned} \tag{1-43}$$

A geometrical model of this system consists of two straight lines in the (x_1, x_2) plane. The solution is represented by the point of intersection of the two lines. In Fig. 1-12 two such systems are compared. When the angle between the two lines is small, their intersection is difficult to determine precisely. It is a simple exercise in analytical geometry to show that this angle is given by

$$\tan \theta = \frac{a_{11}a_{22} - a_{12}a_{21}}{a_{11}a_{21} + a_{12}a_{22}} \tag{1-44}$$

The numerator of (1-44) is the determinant of the coefficients of (1-43). When it is small compared with the denominator, it will be difficult to obtain an accurate numerical solution. Such a system is said to be *ill-conditioned.*

[1] G. Birkhoff and J. B. Diaz, Non-linear Network Problems, *Quart. Appl. Math.*, **13**, 431–443 (1956).

This concept may be qualitatively extended to systems of any number of degrees of freedom; i.e., an ill-conditioned system is one in which the solution is relatively weakly indicated by the equations. The equations are very nearly satisfied by values which differ considerably from the exact solution. The quantitative measure (1-44) is not, however, easily generalized to systems with a large number of unknowns. Several proposals[1] have been made in this direction, but the common usage of the term ill-conditioned is still qualitative.

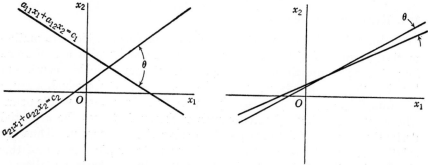

FIG. 1-12. Geometrical models of well-conditioned and ill-conditioned two-degree-of-freedom systems.

Inverse of a Linear System. If the determinant (1-38) of the linear system (1-36) is not zero, it is theoretically possible to manipulate the equations into the form

$$
\begin{aligned}
b_{11}c_1 + b_{12}c_2 + \cdots + b_{1n}c_n &= x_1 \\
b_{21}c_1 + b_{22}c_2 + \cdots + b_{2n}c_n &= x_2 \\
& \\
b_{n1}c_1 + b_{n2}c_2 + \cdots + b_{nn}c_n &= x_n
\end{aligned}
\tag{1-45}
$$

where the coefficients b_{jk} are certain algebraic combinations of the a_{jk}. The system (1-45) is said to be the *inverse* of (1-36). Methods for computing the inverse of a system will be discussed in Sec. 1-5. A theoretical fact of some practical importance is that the inverse system (1-45) will be symmetric if, and only if, the original system (1-36) was symmetric.

The inverse of a linear system is required in certain of the procedures given in Chap. 2 for eigenvalue problems. For equilibrium problems the importance of the inverse is somewhat limited. It is true that once the coefficients b_{jk} of (1-45) are available the solution to (1-36) is readily obtained for any given loading $(c_1, \ldots, c_n)$; however, we shall see that the amount of computation required to obtain the b_{jk} is essentially equivalent to that required to solve (1-36) n times for n different loadings.

[1] See A. M. Turing, Rounding-off Errors in Matrix Processes, *Quart. J. Mech. Appl. Math.*, **1**, 287–308 (1948).

EXERCISE

1-8. The following system is ill-conditioned:

$$9x_1 + 9x_2 + 8x_3 = 26$$
$$9x_1 + 8x_2 + 7x_3 = 24$$
$$8x_1 + 7x_2 + 6x_3 = 21$$

Verify that $(1.3, -0.9, 2.8)$ and $(0.7, 2.9, -0.8)$ both satisfy these equations to within 0.1. The exact solution is $(1, 1, 1)$.

1-4. Extremum Problems

We next describe and illustrate a class of problems which are mathematically equivalent to equilibrium problems and which are of some importance for numerical procedures. An extremum problem consists in locating the set (or sets) of values $(x_1, x_2, \ldots, x_n)$ for which a given function $\Phi(x_1, x_2, \ldots, x_n)$ is a maximum, a minimum, or a saddle point. The classical method of solving such problems is to set the n partial derivatives of Φ equal to zero and to solve simultaneously. Thus the solution of an extremum problem reduces to the basic problem of this chapter, i.e., the solution of a set of simultaneous algebraic equations.

With every extremum problem there is associated a set of simultaneous equations, but the converse is not necessarily true. For a given set of simultaneous equations it may not be possible to find a function Φ whose partial derivatives have the same structure as the given equations. The linear equations

$$
\begin{aligned}
a_{11}x_1 + a_{12}x_2 + &\cdots + a_{1n}x_n = c_1 \\
a_{21}x_1 + a_{22}x_2 + &\cdots + a_{2n}x_n = c_2 \\
&\cdots\cdots\cdots\cdots\cdots \\
a_{n1}x_1 + a_{n2}x_2 + &\cdots + a_{nn}x_n = c_n
\end{aligned}
\tag{1-46}
$$

are equivalent to the conditions for an extremum of

$$
\Phi = \frac{1}{2} \sum_{j=1}^{n} \sum_{k=1}^{n} a_{jk}x_j x_k - \sum_{j=1}^{n} c_j x_j
\tag{1-47}
$$

where the double sum is the quadratic form Q of (1-37) *only* if the system (1-46) is symmetric. The more general nonlinear system of equations

$$
\begin{aligned}
a_1(x_1, \ldots, x_n) &= c_1 \\
a_2(x_1, \ldots, x_n) &= c_2 \\
&\cdots\cdots\cdots\cdots \\
a_n(x_1, \ldots, x_n) &= c_n
\end{aligned}
\tag{1-48}
$$

is equivalent to the conditions for an extremum of

$$\Phi = V - \sum_{j=1}^{n} c_j x_j \tag{1-49}$$

where V is an integral of the a_j only if the a_j satisfy the *integrability relations*

$$\frac{\partial a_j}{\partial x_k} = \frac{\partial a_k}{\partial x_j} \qquad j, k = 1, \ldots, n \tag{1-50}$$

The quadratic function Φ of (1-47) has only one extremum if the corresponding linear system (1-46) has a unique solution. If the linear system is positive definite, then the extremum is a minimum. The more general function Φ of (1-49) will have as many extrema as there are distinct real solutions to (1-48).

There are two principal reasons for studying extremum problems in conjunction with equilibrium problems.   (1) When it is known in advance that a certain extremum problem is equivalent to a given equilibrium problem, the analyst is provided with an alternate method of deriving the equilibrium equations.   It is sometimes easier to construct the function Φ and obtain the equilibrium equations by differentiation than to dissect the physical system and obtain the equilibrium equations by direct analysis.   (2) In studying approximate methods for solving equilibrium problems we sometimes find the procedures easier to understand when they are viewed as methods for obtaining an extremum.   One of the basic ideas here is that the function Φ can be used to measure the goodness of an approximation.   Suppose, for instance, that it is known that the exact equilibrium desired corresponds to a minimum of Φ.   Then if several approximate solutions are available, the corresponding values for Φ may be used as a basis for comparison: the lower the value of Φ, the better the approximation.

Physical Significance.   In several fields of application the extremum problem equivalent to the equilibrium problem has received considerable attention for its own sake.   Rules for constructing Φ directly from the physical system have been given, and the statement that an extremum of Φ is desired is usually called a *stationary principle* or a *minimum principle*. The use of minimum here rather than maximum involves the same arbitrary choice as was involved in the selection of positive definite systems rather than negative definite systems.   The function Φ has traditionally been defined in such a manner that a minimum corresponds to a stable equilibrium configuration.

As an illustration we turn to the principles of minimum potential energy and minimum complementary energy for elastic systems.   A *potential energy* can be assigned to any geometrically compatible state of an elastic system according to the formula

$$\Phi \equiv \text{PE} = U + V \tag{1-51}$$

where U is the potential energy of the passive system (sometimes called

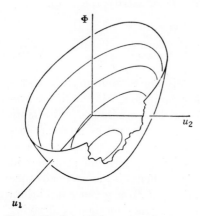

FIG. 1-13. Potential energy of a two-degree-of-freedom linear positive definite system.

strain energy) expressed in terms of displacements and V is the potential energy of the prescribed loads. For example, in the elastic spring system, Prob. 1-1, a geometrically compatible state is represented by the displacement variables u_1 and u_2. The strain energy of a single spring is given by $\frac{1}{2}k\delta^2$, where k is the spring constant and δ is the elongation. The increase in potential energy of a constant force P acting through an in-line displacement δ is $-P\delta$. Expressing the spring elongations and the force displacements of Prob. 1-1 in terms of u_1 and u_2, we obtain the following expression for the total potential energy:

$$\Phi = \tfrac{1}{2}k_1u_1{}^2 + \tfrac{1}{2}k_2u_2{}^2 + \tfrac{1}{2}k_3(u_2 - u_1)^2 + \tfrac{1}{2}k_4(u_2 - u_1)^2 - P_1u_1 - P_2u_2 \tag{1-52}$$

The principle of minimum potential energy[1] states that, *of all geometrically compatible displacement states, those which also satisfy the force-balance conditions give stationary values to the potential energy.* A sketch of Φ, given by (1-52), as a function of u_1 and u_2 is shown in Fig. 1-13. The surface is a paraboloid and has only one stationary value which is a true minimum (this is always the case for positive definite systems). The conditions for locating the minimum

$$\frac{\partial \Phi}{\partial u_1} = k_1u_1 - k_3(u_2 - u_1) - k_4(u_2 - u_1) - P_1 = 0$$

$$\frac{\partial \Phi}{\partial u_2} = k_2u_2 + k_3(u_2 - u_1) + k_4(u_2 - u_1) - P_2 = 0 \tag{1-53}$$

are identical with the force-balance conditions (1-1) obtained directly from the physical system. This amounts to a verification, for this one case, of the principle.

[1] See, for example, R. V. Southwell, "Theory of Elasticity," 2d ed., Oxford University Press, New York, 1941, p. 18.

For elastic systems there is also a complementary minimum principle. Instead of dealing with geometrically compatible displacement states we consider self-balancing force states. A complementary energy can be assigned to any such state as follows

$$\bar{\Phi} \equiv \mathrm{CE} = \bar{U} + \bar{V} \qquad (1\text{-}54)$$

where $\bar{U}$ is the complementary energy of the passive system and $\bar{V}$ is the complementary energy of the prescribed displacements. For example, in the elastic spring system, Prob. 1-1, a self-balancing force state is represented by f_2 and f_3 provided f_1 and f_4 are taken according to (1-2). The complementary energy of a single spring is $f^2/2k$, where k is the spring constant and f is the force in the spring. The complementary energy of a prescribed displacement δ when an in-line force f is required to maintain this displacement is $-\delta f$. In Prob. 1-1 there are no prescribed displacements; so the total complementary energy is simply

$$\bar{\Phi} = \frac{(P_1 + P_2 - f_2)^2}{2k_1} + \frac{f_2^{\,2}}{2k_2} + \frac{f_3^{\,2}}{2k_3} + \frac{(P_2 - f_2 - f_3)^2}{2k_4} \qquad (1\text{-}55)$$

The principle of minimum complementary energy[1] states that *of all self-balancing force states those which also satisfy the requirements of geometric compatibility give stationary values to the complementary energy.* The conditions for locating the extremum of (1-55)

$$\frac{\partial \bar{\Phi}}{\partial f_2} = -\frac{P_1 + P_2 - f_2}{k_1} + \frac{f_2}{k_2} - \frac{P_2 - f_2 - f_3}{k_4} = 0$$

$$\frac{\partial \bar{\Phi}}{\partial f_3} = \frac{f_3}{k_3} - \frac{P_2 - f_2 - f_3}{k_4} = 0 \qquad (1\text{-}56)$$

are identical with the geometric compatibility requirements (1-3) obtained directly from the physical system.

Nonlinear Systems. Both the minimum principles for elastic systems which were described above apply to systems with nonlinear elastic elements provided the strain and complementary strain energies U and $\bar{U}$ are properly defined. For a spring with a nonlinear characteristic as

---

[1] See H. M. Westergaard, On the Method of Complementary Energy, *Trans. ASCE*, **107**, 765–803 (1942), and I. S. Sokolnikoff, "Mathematical Theory of Elasticity," McGraw-Hill Book Company, Inc., New York, 1946, p. 286. The principle represents a slight extension of Castigliano's *theorem of least work* in that prescribed displacements can be handled as well as prescribed force loadings. The realization that the principle applies to nonlinear systems seems to be due to Fr. Engesser, Ueber statisch unbestimmte Träger bei beliebigem Formänderungs-Gesetze und über den Satz von der kleinsten Ergänzungsarbeit, *Z. Arch.- u. Ing.-Ver. Hannover*, **35**, 733–744 (1889).

shown in Fig. 1-14 these are defined as follows:

$$U(\delta) = \int_0^\delta f(\delta)\, d\delta$$
$$\bar{U}(f) = \int_0^f \delta(f)\, df \tag{1-57}$$

Note that for nonlinear systems U and $\bar{U}$ are no longer equal in magnitude as they are for linear systems.

Minimum principles for linear electric circuits have been known[1] for a long time. Recently these principles have been extended[2] to nonlinear systems. In d-c resistance networks W. Millar introduced the terms

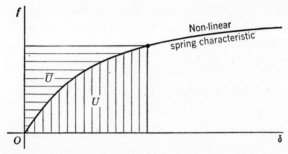

FIG. 1-14. Definition of strain energy U and complementary strain energy $\bar{U}$ for a nonlinear spring.

content and *co-content*, which play the same role as strain energy and complementary strain energy in elastic problems. In heat-transfer and hydraulic problems analogous minimum principles exist, but names have not yet been assigned to the corresponding quantities.

We turn to Prob. 1-5 to illustrate the extremum problems corresponding to a nonlinear system. In Fig. 1-15 the hydraulic resistance law (1-26) for a single branch is sketched. Borrowing the terminology used by Millar for d-c circuits, we call the area under the curve, expressed in terms of the flow rate, the content C of the branch, while the area above the curve expressed in terms of the pressure drop is called the co-content $\bar{C}$ of the branch. For simplicity we shall consider only ideal pumps. These are of two kinds: (1) *pressure sources*, which maintain a given pressure rise P independently of the flow rate; (2) *flow sources*, which pump a given rate of flow Q independently of the pressure rise. We define the *active content* of a pressure source maintaining a pressure rise P when the flow rate is q as $-Pq$. We define the *active co-content* of a flow

[1] J. C. Maxwell, "A Treatise on Electricity and Magnetism," 3d ed., 1892, p. 407.

[2] See W. Millar, Some General Theorems for Non-linear Systems Possessing Resistance, *Phil. Mag.*, (7) **42**, 1150–1160 (1951), and E. C. Cherry, Some General Theorems for Non-linear Systems Possessing Reactance, *Phil. Mag.*, (7) **42**, 1161–1177 (1951).

source which delivers a flow rate Q against a pressure rise p as $-Qp$. Finally the total content of a system is defined as the sum of all the internal or passive contents plus all the active contents. A similar statement applies for the total co-content.

In Fig. 1-7 the hydraulic network is shown with only passive elements; however, a fixed flow rate Q is given so that the complete system con-

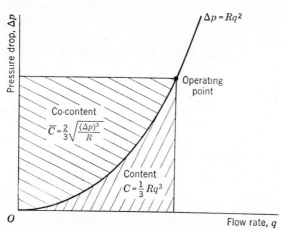

FIG. 1-15. Nonlinear resistance law for a single branch of a hydraulic network.

sists of the resistances shown plus an ideal flow source. Any flow state which meets the continuity requirements can be represented by arbitrary values of q_1 and q_2 provided the remaining flow rates are taken according to (1-31). The total content is then as follows:

$$\Phi \equiv C_{tot} = \tfrac{1}{3}\{2bq_1{}^3 + 3bq_2{}^3 + 2b(q_1 - q_2)^3 + 5b(Q - q_1)^3\} \quad (1\text{-}58)$$

To determine the true flow state, we enunciate the principle of minimum total content: *Of all flow states which satisfy the requirements of continuity those which also provide single-valued pressures give stationary values to the total content.* The reader can readily show that equating to zero the partial derivatives of (1-58) with respect to q_1 and q_2 yields the same governing equation of the flow rates as previously given in (1-32).

The complementary development for pressure states is exactly analogous. A single-valued pressure state for Prob. 1-5 is represented by arbitrary junction pressures p_1 and p_2. In terms of these the total co-content is as follows:

$$\bar{\Phi} \equiv \bar{C}_{tot} = \frac{2}{3}\left\{\sqrt{\frac{(p_1 - p_2)^3}{2b}} + \sqrt{\frac{p_2{}^3}{3b}} + \sqrt{\frac{p_2{}^3}{2b}} + \sqrt{\frac{p_1{}^3}{5b}}\right\} - p_1 Q \quad (1\text{-}59)$$

The true pressure state is selected by means of the principle of minimum co-content: *Of all single-valued pressure states those which also satisfy the continuity requirements give stationary values to the total co-content.* Setting the partial derivatives of (1-59) with respect to p_1 and p_2 equal to zero yields the same pressure equation (1-28) previously obtained by the direct application of the continuity requirements.

Upper and Lower Bounds for System Properties. In linear systems there are certain over-all system properties such as the *over-all stiffness* of an interconnected set of springs or the *over-all impedance* of a passive electrical network which are proportional to the quantities Φ and $\bar{\Phi}$ employed in the extremum principles. For definiteness let us call such an over-all property K. Using an extremum principle to compute K

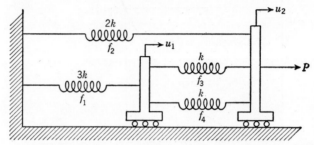

Fig. 1-16. Elastic system for which over-all stiffness is to be obtained.

usually involves more labor than direct calculation and hence cannot be recommended when an exact solution is contemplated. If, however, one must deal with an approximate solution, the computation of K by means of an extremum principle has the advantage of giving greater accuracy since in the neighborhood of the true solution the error in Φ (or $\bar{\Phi}$) will be of second order in comparison with an error in the solution itself. In positive definite systems the true value of Φ is known to be a minimum, and hence the direction of any error in K can be forecast. It is often possible to relate K to *both* Φ and $\bar{\Phi}$ in such a manner that approximations to the first will always give values of K which are too high while approximations to the second will always give values which are too low. We will illustrate this in terms of a simple elastic example related to the elastic spring system of Prob. 1-1.

Figure 1-16 shows the elastic system subjected to the single load P. The desired system property is the over-all system *stiffness*

$$K = \frac{P}{u_2} \tag{1-60}$$

The total potential energy Φ of any displacement state is the sum of the internal strain energy U and the potential energy of the load V when these are expressed in terms of geometrically compatible displacement variables (for example, u_1 and u_2 in the present case). The principle of minimum potential energy tells us that the state which has equilibrium of forces has minimum potential energy. In this case we also know something else about the true equilibrium state. Since the only active load is P, if we consider a succession of equilibrium states as P is gradually increased from zero to its prescribed value, we obtain the result that the work done by P in this process is equal to the internal strain energy U. Since the system is linear, this work is one-half the product of the final values of P and u_2. We can therefore write

$$\Phi_{\min} = U + V$$
$$= \tfrac{1}{2}Pu_2 - Pu_2 = -\tfrac{1}{2}Pu_2 \qquad (1\text{-}61)$$
$$= -\frac{P^2}{2K} \qquad \text{or} \qquad K = \frac{P^2}{-2\Phi_{\min}}$$

Thus our desired stiffness is inversely proportional to $\Phi_{\min}$. If we have only an approximate solution and hence a value of Φ which is algebraically greater than $\Phi_{\min}$ (note that $\Phi_{\min}$ is negative), we have

$$K \leq \frac{P^2}{-2\Phi} \qquad (1\text{-}62)$$

For example, the true displacement state in Fig. 1-16 is $u_1 = 0.1250P/k$, $u_2 = 0.3125P/k$, and the true value of K is $3.2000k$. Let us suppose that instead of the true solution we have the approximate solution $u_1 = 0.1P/k$, $u_2 = 0.3P/k$. We can obtain an approximation to K from this by going directly to (1-60), which gives us $K = 3.3333k$ (4.2 per cent too high). Alternatively we can use these values of u_1 and u_2 to evaluate

$$2\Phi = \frac{P^2}{k}\{(3)(0.1)^2 + (2)(0.3)^2 + 2(0.2)^2 - (2)(0.3)\} = -\frac{0.310P^2}{k}$$

which when inserted in (1-62) gives $K \leq 3.2258k$ (0.8 per cent too high).

Turning to the complementary approach, we remember that the complementary energy $\bar{\Phi} = \bar{U} + \bar{V}$ may be computed for any self-balancing force state as in (1-55). The principle of minimum complementary energy states that $\bar{\Phi}$ obtains its minimum for that state which is also geometrically compatible. In the present instance $\bar{V} = 0$ (no prescribed displacements), and in the true equilibrium state we have, because of the linearity, $\bar{U} = U$ and hence

$$\bar{\Phi}_{\min} = \frac{1}{2}\frac{P^2}{K} \qquad \text{or} \qquad K = \frac{P^2}{2\bar{\Phi}_{\min}} \qquad (1\text{-}63)$$

If we have an approximation to the true minimizing state, then $\bar{\Phi}$ will be too large and thus

$$\frac{P^2}{2\bar{\Phi}} \leq K \qquad (1\text{-}64)$$

For example, proceeding as in (1-2), it is possible to represent a self-balancing force state for Fig. 1-16 in terms of f_2 and f_3. An exact solution for them gives $f_2 = 0.6250P$, $f_3 = 0.1875P$. If instead we consider the approximate state $f_2 = 0.6P$, $f_3 = 0.2P$, we can obtain approximations to K in two ways. We can go directly to Fig. 1-16 and note that if $f_2 = 0.6P$ then u_2 must be $0.3P/k$, and then from (1-60) we obtain $K = 3.3333k$ (4.16 per cent too high); or we can use our values of f_2 and f_3 to evaluate $\bar{\Phi}$ as in (1-55). In this way we find $2\bar{\Phi} = 0.3133P^2/k$, and hence from (1-64) $K \geq 3.1915k$ (0.27 per cent too low).

EXERCISES

1-9. By analogy with the text discussion for elastic and hydraulic systems formulate the rules for setting up the content and co-content of a d-c network, and state the corresponding minimum principles. Apply these to the d-c network of Prob. 1-2, and verify that the extremum conditions are equivalent to (1-11) and (1-12).

1-10. Show that the internal strain energy of the single-span beam of Fig. 1-5 is

$$U = \frac{2EI}{L} (\theta_A{}^2 + \theta_A\theta_B + \theta_B{}^2)$$

What is $\bar{U}$, the complementary strain energy, in terms of M_A and M_B for the same span? Obtain the total potential energy for the continuous beam, Prob. 1-4. Verify that the principle of minimum potential energy leads to (1-21). Construct the total complementary energy, and use the principle of minimum complementary energy to verify (1-25).

1-11. Show that the total potential energy of the pinned truss of Exercise 1-1 is

$$\Phi = \frac{125}{12} \frac{aF^2}{EA} \left\{ \frac{1}{2} (5x_1{}^2 - 2x_1x_2 + 3x_2{}^2) - 2x_1 - x_2 \right\}$$

where x_1 and x_2 are the same dimensionless variables used in Exercise 1-1. Do the same for the total complementary energy

$$\bar{\Phi} = \frac{5}{6} \frac{aF^2}{EA} \left\{ \frac{17}{2} y_1{}^2 + 2y_1y_2 + 10y_2{}^2 - 10y_1 - 30y_2 + \frac{100}{3} \right\}$$

Use the minimum principles to verify the equations given in Exercise 1-1.

1-12. For steady-state heat-transfer problems of the type illustrated by Exercise 1-2 formulate a minimum principle for states having single-valued temperatures at each junction. Verify that when this principle is applied to Exercise 1-2 the equations given there are obtained. Relate the value of Φ in your minimum principle to the over-all conductance of the wall defined as

$$C \equiv \frac{q}{A(T_4 - T_0)}$$

1-13. If the over-all stiffness of the continuous beam, Prob. 1-4, is defined as

$$K \equiv \frac{M}{\theta_A}$$

find the relations which exist between K and the total energies Φ and $\bar{\Phi}$ of Exercise 1-10.

1-5. Elimination Method for Linear Systems

We turn now to procedures for obtaining numerical solutions for the type of problems formulated and discussed in the foregoing sections.

The available methods may be broadly classified into *direct*, or *exact*, procedures and *indirect*, or *approximate*, procedures. The direct methods prescribe finite routines of elementary arithmetical operations which if carried out exactly would provide mathematically exact solutions. The indirect methods prescribe routines for obtaining successive approximations; any specified degree of accuracy can be obtained by carrying out enough steps, but mathematically exact solutions would theoretically require infinite processes. In most computational work even the direct methods provide approximate numerical solutions because of the practical necessity of restricting the number of decimal places carried in the working.

In this section we consider direct methods for solving simultaneous linear equations. We discuss two different methods. There is *Cramer's rule*,[1] which gives the solution in the form of quotients of determinants, and there is the *successive elimination of unknowns* by algebraic manipulation. The elimination method is usually presented first in elementary algebra, while the determinant method is reserved for treatment at a higher level. Because of its elegance the determinant method appears in most mathematical studies of simultaneous equations. For actual computation when the number of equations is large, the elimination method proves to be superior because, as we shall see, it is considerably shorter.

In applying the elimination method to systems with a large number of unknowns it is desirable to have a systematic arrangement of the work. Such a systematic program was thoroughly examined by C. F. Gauss[2] in 1826. The basic algorithm of Gauss has been "engineered" into efficient concise routines by several authors.[3] We shall describe a recent variation of the basic algorithm which has been designed[4] to exploit the

---

[1] G. Cramer, "Introduction à l'analyse des lignes courbes algébriques," Geneva, 1750, pp. 656–659. The rule is given in most modern treatments of simultaneous equations. See, for example, L. A. Pipes, "Applied Mathematics for Engineers and Physicists," McGraw-Hill Book Company, Inc., New York, 1946, p. 86.

[2] See "Carl Friedrich Gauss Werke," vol. 4, Göttingen, 1873, pp. 55–93.

[3] For example, M. H. Doolittle, Method Employed in the Solution of Normal Equations and the Adjustment of a Triangulation, *U.S. Coast Geodetic Survey Rept.*, pp. 115–120 (1878). A. L. Cholesky, ca. 1916. Cholesky was killed in World War I. His procedure is described in a paper by Benoit, Note sur une méthode, etc. (Procédé du Commandant Cholesky), *Bull. géodésique*, pp. 67–77 (1924). A. C. Aitken, On the Evaluation of Determinants, the Formation of their Adjugates and the Practical Solution of Simultaneous Linear Equations, *Proc. Edinburgh Math. Soc.*, [2]**3**, 207–219 (1932). For an extensive treatment of these and other methods see P. S. Dwyer, "Linear Computations," John Wiley & Sons, Inc., New York, 1951.

[4] The method described here was developed independently by T. Banachiewicz, Méthode de résolution numérique des équations linéaires, du calcul des déterminants et des inverses et de réduction des formes quadratiques, *Bull. inter. acad. polon. sci., Sér. A*, pp. 393–404 (1938), and P. D. Crout, A Short Method for Evaluating Determinants and Solving Systems of Linear Equations with Real or Complex Coefficients, *Trans. AIEE*, **60**, 1235–1240 (1941).

special properties of the desk calculating machine. The computation is so arranged that partial results are mostly carried on the machine itself. To solve the general linear system

$$a_{11}x_1 + a_{12}x_2 + \cdots + a_{1n}x_n = c_1$$
$$a_{21}x_1 + a_{22}x_2 + \cdots + a_{2n}x_n = c_2 \qquad (1\text{-}65)$$
$$\cdots \cdots \cdots \cdots \cdots \cdots \cdots \cdots \cdots$$
$$a_{n1}x_1 + a_{n2}x_2 + \cdots + a_{nn}x_n = c_n$$

we compute $n^2 + n$ auxiliary quantities in a systematic fashion. Each of these quantities is obtained by a computation of the form

$$\frac{p_0 + p_1q_1 + p_2q_2 + \cdots + p_mq_m}{d} \qquad (1\text{-}66)$$

which can be carried out on a modern desk calculating machine in a single sequence of operations. Then from these auxiliary quantities the unknowns are computed by means of n more calculations of the form (1-66). This computing routine represents a considerable abbreviation of the usual solution by successive elimination. The essential steps are still[1] there, but all duplication and redundant notation have been deleted.

The description of the actual procedure is clarified if we illustrate for a particular value of n. There is an underlying pattern in the sequence of operations which can be readily grasped from studying a single example. This same pattern can be shown, by induction,[2] to extend to successively larger values of n. Thus we consider the system (1-65) for $n = 4$. The given coefficients a_{jk} and the given constants c_j are written in the following array:

a_{11}	a_{12}	a_{13}	a_{14}	c_1
a_{21}	a_{22}	a_{23}	a_{24}	c_2
a_{31}	a_{32}	a_{33}	a_{34}	c_3
a_{41}	a_{42}	a_{43}	a_{44}	c_4

$$(1\text{-}67)$$

Immediately below this we leave room for a similar array of the auxiliary quantities. When this has been filled in, it will appear as follows:

g_{11}	h_{12}	h_{13}	h_{14}	C_1
g_{21}	g_{22}	h_{23}	h_{24}	C_2
g_{31}	g_{32}	g_{23}	h_{34}	C_3
g_{41}	g_{42}	g_{43}	g_{44}	C_4

$$(1\text{-}68)$$

[1] See Exercise 1-14.   For a general proof using matrix notation see Exercise 2-42.
[2] See Crout, *op. cit.*

We now describe how the auxiliary quantities (1-68) are obtained from (1-67).

First of all, there is a simple pattern for the order in which the auxiliary quantities are computed. The quantities in the first column, g_{11} to g_{41}, are obtained first, then the quantities in the first row, h_{12}, h_{13}, h_{14}, and C_1, are filled in. The second column (g_{22} to g_{42}) is obtained next, and then the second row (h_{23}, h_{24}, and C_2) is filled in. Continuing in this fashion, we work diagonally downward, first filling in the g_{ij} ($i \geq j$) in the jth column and then filling in the h_{jk} ($k > j$) and C_j in the jth row.

Second, the g_{ij} are all computed according to one fixed pattern, while the h_{jk} and C_j are all computed from a similar but slightly different fixed pattern. The explicit formulas follow.

$$
\begin{aligned}
g_{11} &= a_{11} \\
g_{21} &= a_{21} & g_{22} &= a_{22} - g_{21}h_{12} \\
g_{31} &= a_{31} & g_{32} &= a_{32} - g_{31}h_{12} & g_{33} &= a_{33} - g_{31}h_{13} - g_{32}h_{23} \\
g_{41} &= a_{41} & g_{42} &= a_{42} - g_{41}h_{12} & g_{43} &= a_{43} - g_{41}h_{13} - g_{42}h_{23} & g_{44} &= a_{44} - g_{41}h_{14} - g_{42}h_{24} - g_{43}h_{34}
\end{aligned}
\tag{1-69}
$$

$$
\begin{aligned}
h_{12} &= \frac{a_{12}}{g_{11}} & h_{13} &= \frac{a_{13}}{g_{11}} & h_{14} &= \frac{a_{14}}{g_{11}} & C_1 &= \frac{c_1}{g_{11}} \\[2mm]
& & h_{23} &= \frac{a_{23} - g_{21}h_{13}}{g_{22}} & h_{24} &= \frac{a_{24} - g_{21}h_{14}}{g_{22}} & C_2 &= \frac{c_2 - g_{21}C_1}{g_{22}} \\[2mm]
& & & & h_{34} &= \frac{a_{34} - g_{31}h_{14} - g_{32}h_{24}}{g_{33}} & C_3 &= \frac{c_3 - g_{31}C_1 - g_{32}C_2}{g_{33}} \\[2mm]
& & & & & & C_4 &= \frac{c_4 - g_{41}C_1 - g_{42}C_2 - g_{43}C_3}{g_{44}}
\end{aligned}
\tag{1-70}
$$

The reader is advised to trace out several of these operations until the basic pattern is clearly evident. Note that each operation begins with a quantity from the original array (1-67) which corresponds in position with the auxiliary quantity being computed. From this are subtracted products of the auxiliary quantities already obtained which lie in the same row and column. The operations of (1-70) differ only from those of (1-69) in requiring a final division by the diagonal quantity g_{jj}.

When the auxiliary array (1-68) is complete, the solution values for the x_j are obtained by the following process of *back substitution*:

$$
\begin{aligned}
x_4 &= C_4 \\
x_3 &= C_3 - h_{34}x_4 \\
x_2 &= C_2 - h_{24}x_4 - h_{23}x_3 \\
x_1 &= C_1 - h_{14}x_4 - h_{13}x_3 - h_{12}x_2
\end{aligned}
\tag{1-71}
$$

The reader is again advised to trace out these operations and note the pattern that is established.

Example. As an illustration we show the solution of the system (1-20) obtained in Prob. 1-3, the a-c network. The array of given quantities is first entered in the top of Fig. 1-17. The auxiliary quantities are then computed according to (1-69) and

Given Quantities

5	6	-2	-2	1
6	-5	-2	2	0
-2	-2	3	-1	0
-2	2	-1	-3	0

Auxiliary Quantities

5	$\dfrac{6}{5}$	$-\dfrac{2}{5}$	$-\dfrac{2}{5}$	$\dfrac{1}{5}$
6	$-\dfrac{61}{5}$	$-\dfrac{2}{61}$	$-\dfrac{22}{61}$	$\dfrac{6}{61}$
-2	$\dfrac{2}{5}$	$\dfrac{135}{61}$	$-\dfrac{101}{135}$	$\dfrac{22}{135}$
-2	$\dfrac{22}{5}$	$-\dfrac{101}{61}$	$-\dfrac{466}{135}$	$-\dfrac{16}{233}$

$$\text{Solution}\begin{cases} x_4 = -\dfrac{16}{233} = -0.06867 \\[2mm] x_3 = \dfrac{26}{233} = 0.11159 \\[2mm] x_2 = \dfrac{18}{233} = 0.07725 \\[2mm] x_1 = \dfrac{29}{233} = 0.12446 \end{cases}$$

FIG. 1-17. Solution of (1-20) by elimination.

(1-70). These values are recorded in fractional form so that the reader can follow the steps without a computing machine. Thus, for example, we have

$$h_{23} = \frac{-2 - (6)(-\frac{2}{5})}{-\frac{61}{5}} = -\frac{2}{61} \tag{1-72}$$

The solution values of the x_j are obtained at the end by means of (1-71), for example

$$x_2 = \tfrac{6}{61} - (-\tfrac{22}{61})(-\tfrac{16}{233}) - (-\tfrac{2}{61})(\tfrac{26}{233}) = \tfrac{18}{233} \tag{1-73}$$

Abnormal Case: Zero Diagonal Element. The procedure described above breaks down if one of the g_{jj} terms vanishes because the subsequent divisions required for the h_{jk} and C_j become impossible. Suppose that the computation has been carried to the row $j - 1$ without mishap and then on beginning the jth column it is found that $g_{jj} = 0$. The rest of the jth column, g_{ij} $(i > j)$, should be computed exactly as before. Now if at least one of these g_{ij} is different from zero, it will be possible to continue. Suppose g_{pj} is different from zero. Then interchanging the pth and jth rows will put the former g_{pj} in the diagonal position, and the computation can be continued. A simple way to effect this change without disturbing the previous work is to cover both arrays from row j down to row n with blank sheets and recopy the previous entries in rows j to n, interchanging only the entries in the pth and jth rows.[1]

If $g_{jj} = 0$ and all the g_{ij} $(i > j)$ are also zero, then the solution cannot be continued. It can be shown that in this case the determinant of the system coefficients is zero and that a unique solution does not exist. This case presumably will not arise in meaningful equilibrium problems.

Even when g_{jj} does not vanish, it is good practice to introduce the row-interchange device whenever g_{jj} is very much smaller than the g_{ij} $(i > j)$. This is because of the relatively larger effect of round-off errors on a small quantity that has been obtained as the difference between two large, but nearly equal, quantities and because the effect of a small error in g_{jj} upon subsequent divisions is minimized when g_{jj} is as large as possible.

Symmetric Systems. The basic algorithm described above applies to the general linear system. For symmetric systems there exists a simplification which essentially cuts the work in half. The basis for this short cut is the identity

$$h_{jk} = \frac{g_{kj}}{g_{jj}} \tag{1-74}$$

which may be easily verified[2] to hold when the original system is symmetric. The careful reader may have already noted this property in the foregoing example. Thus for symmetrical systems both the jth column and jth row of the auxiliary array can be filled in at the same time. When g_{kj} has been obtained, it is left on the machine and a single division by g_{jj} provides h_{jk}.

Systems with Complex Coefficients. The basic algorithm applies equally well to systems with complex coefficients. Each entry in the computation scheme will in general be of the form $a + ib$. Each multiplication will take the form

$$(a + ib)(c + id) = (ac - bd) + i(bc + ad)$$

that is, a complex product requires four real products and two real additions. In computing the g_{jk} using formulas (1-69) the real parts may be obtained from one single sequence of machine operations and the imaginary parts from another single sequence. Crout[3] suggests that the division by g_{jj} required for the h_{jk} and C_j in (1-70) is most easily effected by multiplying by $1/g_{jj}$. This requires that the reciprocals

---

[1] See Exercise 1-16 for a simple example in which this interchange is necessary.
[2] See Exercise 1-17.
[3] *Op. cit.*

of the g_{ii} be obtained in supplementary computations of the following type:

$$\frac{1}{a+ib} = \frac{a}{a^2+b^2} - \frac{ib}{a^2+b^2} \tag{1-75}$$

Furthermore it is necessary to provide space to store the numerators in (1-70) in preparation for the final multiplication by $1/g_{ii}$.

These points are illustrated in Fig. 1-18, where the solution of the complex equations (1-18) for Prob. 1-3, the a-c network, is displayed. The given quantities are entered

Given Quantities

5	-2	1
$-i6$	$i2$	$i0$
-2	3	0
$i2$	$i1$	$i0$

a^2+b^2	$\frac{1}{g_{jj}}$	Auxiliary Quantities			$g_{jj}C_j$
61	$\frac{5}{61}$	5	$-\frac{22}{61}$	$\frac{5}{61}$	1
	$i\frac{6}{61}$	$-i6$	$-i\frac{2}{61}$	$i\frac{6}{61}$	$i0$
$\frac{466}{61}$	$\frac{135}{466}$	-2	$\frac{135}{61}$	$\frac{26}{233}$	$\frac{22}{61}$
	$-i\frac{101}{466}$	$i2$	$i\frac{101}{61}$	$-i\frac{16}{233}$	$i\frac{2}{61}$

$$I_2' = \frac{26}{233} - i\frac{16}{233} = 0.11159 - i0.06867$$

$$I_1' = \frac{29}{233} + i\frac{18}{233} = 0.12446 + i0.07725$$

Fig. 1-18. Solution of (1-18) by elimination.

in the top array. The imaginary part of each coefficient is written directly under the real part. Underneath the given array is the auxiliary array. Two supplementary columns appear at the left, and one supplementary column appears at the right. All entries have been left in fractional form so that the reader can easily verify each step. The computation is begun by entering g_{11} and g_{21} in the auxiliary array. Next h_{12} is obtained. Since the system is symmetric, the numerator of h_{12} already appears as g_{21}; no supplementary column is required to hold it. We therefore form $1/g_{11}$ in the supplementary columns at the left. The sum of the squares of the real and imaginary parts of g_{11} ($5^2 + 6^2 = 61$) is entered in the column labeled $a^2 + b^2$, and $1/g_{11}$, computed according to (1-75), is placed in the column labeled $1/g_{ii}$. Then h_{12} is obtained by complex multiplication of g_{21} and $1/g_{11}$. Proceeding to C_1, the numerator is placed in the column $g_{ii}C_j$ and then multiplied by $1/g_{11}$. The real and imaginary parts of g_{22} are now computed according to the basic algorithm.

$$\begin{aligned} \tfrac{135}{61} &= 3 - \{(-2)(-\tfrac{22}{61}) - (2)(-\tfrac{2}{61})\} \\ \tfrac{101}{61} &= 1 - \{(-2)(-\tfrac{2}{61}) + (2)(-\tfrac{22}{61})\} \end{aligned} \tag{1-76}$$

The reciprocal $1/g_{22}$ is then computed at the left as before. Finally C_2 is obtained: the numerator $c_2 - g_{21}C_1$ is calculated and stored in the $g_{ii}C_j$ column and then multiplied by $1/g_{22}$, which completes the auxiliary array. The actual solution values follow, with the aid of (1-71). This solution is equivalent to that obtained in Fig. 1-11. A discussion of the relative merits of the two procedures appears on page 36.

Other Applications. If it is required to solve the same system of equations for several different right-hand sides, the major portion of the auxiliary array remains unchanged. Only the column of C_j is altered. Hence, once the system has been solved for one particular right-hand side, it is necessary only to compute the C_j column and to reevaluate the solution according to (1-71). This fact is the basis of an efficient means for obtaining the *inverse* of a given system. We briefly illustrate by means of an example.

Consider the system

$$\begin{aligned} 5x_1 - \ 4x_2 \quad\quad &= y_1 \\ -4x_1 + 10x_2 - 5x_3 &= y_2 \\ - \ 5x_2 + 6x_3 &= y_3 \end{aligned} \quad\quad (1\text{-}77)$$

which is closely related to (1-14) of Prob. 1-2, the d-c network. To invert, we must solve for the x_j in terms of the y_j. The complete solution by the algorithm of this section is displayed in Fig. 1-19. This may be viewed as consisting of three separate "unit problems" in which the right-hand sides are (1, 0, 0), (0, 1, 0), and (0, 0, 1). The final solution is obtained by superposing the three unit solutions with the multipliers y_1, y_2, and y_3. Note that this interpretation indicates that the computational labor involved in obtaining an inverse is essentially the same as solving the given system for n different particular loadings.

In carrying out the computation the same column followed by row order is used as before, the only difference being that in completing each row there are three C_j calculations to be made instead of one. When the auxiliary array is complete, (1-71) is used three times over to obtain the three unit solutions. The entries in Fig. 1-19 are in fractional form to facilitate verification by the reader. This example illustrates the general principle that the inverse of a symmetric system is also symmetric. This principle may be used to shorten the computation of symmetric inverses. Thus, in the above example, the parts of the unit solutions below the slanting dotted line can be written down by symmetry as soon as the parts above the line have been computed.

The same basic algorithm of this section may be used to evaluate *determinants.* Let the given determinant be $\Delta = |a_{jk}|$. Then if we construct the auxiliary array (1-68) corresponding to the given a_{jk}, we obtain Δ from the following identity:[1]

$$\Delta = (g_{11})(g_{22}) \cdots (g_{nn}) \quad\quad (1\text{-}78)$$

[1] See Exercise 1-18.

Given coefficients   Three Unit Problems

5	-4	0	1	0	0
-4	10	-5	0	1	0
0	-5	6	0	0	1

Auxiliary Quantities

5	$-\frac{4}{5}$	0	$\frac{1}{5}$	0	0
-4	$\frac{34}{5}$	$-\frac{25}{34}$	$\frac{4}{34}$	$\frac{5}{34}$	0
0	-5	$\frac{79}{34}$	$\frac{20}{79}$	$\frac{25}{79}$	$\frac{34}{79}$

Solutions to
Unit Problems

$$
\begin{array}{l}
x_3 \rightarrow \quad \dfrac{20}{79} \quad\quad \dfrac{25}{79} \quad\quad \dfrac{34}{79} \\[2mm]
x_2 \rightarrow \quad \dfrac{24}{79} \quad\quad \dfrac{30}{79} \quad\quad \dfrac{25}{79} \\[2mm]
x_1 \rightarrow \quad \dfrac{35}{79} \quad\quad \dfrac{24}{79} \quad\quad \dfrac{20}{79}
\end{array}
$$

Inverse of
Given System

$$
\begin{cases}
x_1 = \dfrac{35}{79}\,y_1 + \dfrac{24}{79}\,y_2 + \dfrac{20}{79}\,y_3 \\[2mm]
x_2 = \dfrac{24}{79}\,y_1 + \dfrac{30}{79}\,y_2 + \dfrac{25}{79}\,y_3 \\[2mm]
x_3 = \dfrac{20}{79}\,y_1 + \dfrac{25}{79}\,y_2 + \dfrac{34}{79}\,y_3
\end{cases}
$$

FIG. 1-19. Computation of the inverse of (1-77).

For instance, if we wanted to know the value of the determinant of the coefficients of (1-77), we would multiply the elements g_{11}, g_{22}, and g_{33} of the auxiliary array in Fig. 1-19 as follows:

$$\Delta = (5)\left(\tfrac{34}{5}\right)\left(\tfrac{79}{34}\right) = 79 \tag{1-79}$$

This method of evaluating determinants is essentially the same as the method of systematic reduction[1] of order, sometimes called *pivotal condensation*.

Length of Computation. When a numerical procedure has been systematized to the extent of the algorithm of this section, it is possible to count the number of operations involved and thereby to estimate the total computing time required for any particular problem. We take as

[1] The first application of the underlying method of this section to the reduction of order of determinants is given by F. Chio, "Mémoire sur les fonctions connues sous le nom de résultantes ou de déterminants," Turin, 1853, p. 11.

the basic operational unit the multiplication of two real numbers. Since the partial sums of such operations are automatically held on the computing machine, we do not trouble to count the additions and subtractions. We count a division of two real numbers as one unit. As an illustration of our counting procedure, the calculation (1-66) is counted as $m + 2$ units: $m + 1$ units in the numerator and one unit in the final division. Transferring the factor p_0 to the machine is counted as a unit operation even though no multiplication is required.

TABLE 1-1. UNSYMMETRIC REAL SYSTEM
Number of unit operations required for:

Number of unknowns	Single solution	Inversion of system	Evaluation of determinant
2	12	15	6
3	29	43	17
4	56	93	36
5	95	171	65
6	148	283	106
n	$\frac{1}{3}n^3 + 2n^2 + \frac{2}{3}n$	$n^3 + 2n^2 - n + 1$	$\frac{1}{3}n^3 + n^2 - \frac{n}{3}$

With this convention it is a simple, if tedious, matter to count the number of unit operations contained in a single solution of a system with $n = 4$ as given by (1-69) to (1-71). We find 56. The results of similar counts for several values of n are given in Table 1-1. The numbers of unit operations involved in inverting systems and evaluating determinants[1] by the methods of this section are also tabulated.

TABLE 1-2. SYMMETRIC REAL SYSTEM
Number of unit operations required for:

Number of unknowns	Single solution	Inversion of system	Evaluation of determinant
2	12	13	6
3	27	33	15
4	49	66	29
5	79	115	49
6	118	183	76
n	$\frac{1}{6}n^3 + 2n^2 + \frac{11}{6}n - 1$	$\frac{1}{2}n^3 + 2n^2 + \frac{1}{2}n$	$\frac{1}{6}n^3 + n^2 + \frac{5}{6}n - 1$

Table 1-2 shows the corresponding numbers of operations required for symmetric systems. Note that for large n the amount of computation increases essentially in proportion to n^3 in all cases but that the labor in

[1] The calculation (1-78) is counted as $n - 1$ unit operations.

symmetric systems approaches half that required for nonsymmetric systems. Note also that the amount of computation required for solving a system of equations becomes of the same order of magnitude as evaluating its determinant but that inverting the system becomes three times as much work.

If we compare the above results for a single solution with Cramer's rule, which requires the evaluation of $n + 1$ determinants, we see the computational advantage of the elimination method. If the determinants are evaluated by the *Laplace expansion*,[1] the number of unit operations required for each is greater than *factorial n*. If the determinants are evaluated by the above algorithm (or by any process in which the labor is proportional to n^3), the total computational labor for a single solution would be proportional to n^4.

Although Cramer's rule is unsuited for computational work when n is large, there is one case in which it provides the quickest solution, i.e., when $n = 2$. The complete solution of a two-degree-of-freedom system by Cramer's rule is given below. It requires eight unit operations.

$$
\begin{aligned}
a_{11}x_1 + a_{12}x_2 &= c_1 \\
a_{21}x_1 + a_{22}x_2 &= c_2
\end{aligned}
\qquad
\begin{cases}
\Delta = a_{11}a_{22} - a_{21}a_{12} \\[2mm]
x_1 = \dfrac{a_{22}c_1 - a_{12}c_2}{\Delta} \\[4mm]
x_2 = \dfrac{a_{11}c_2 - a_{21}c_1}{\Delta}
\end{cases}
\qquad (1\text{-}80)
$$

It may similarly be shown that determinant methods[2] for inverting systems are much more laborious than the above elimination method except for the cases $n = 2$ and $n = 3$.

Our counting technique may also be applied to complex systems to compare the labor involved in solving as an n-degree-of-freedom complex system with that involved in solving as a $2n$-degree-of-freedom real system. The algorithm for complex systems is essentially the same as that for real systems except that each unit operation of the real system is replaced by 4 unit operations in the complex system. There are some supplementary operations required in the complex solution, but their number is only proportional to n for symmetric systems (n^2 for unsymmetric systems). Thus for large n the number of unit operations required for the solution of n symmetric[3] complex equations is of order $4(1/6n^3)$. If the n equations are separated into real and imaginary parts to obtain $2n$ real equations, the number of unit operations is of order $1/6(2n)^3 = 8(1/6n^3)$. Therefore it is worthwhile (by a factor of 2, for large n) to use the slightly more cumbersome form of the algorithm directly on the complex equations. Even for small n the complex treatment comes out slightly ahead. For $n = 2$ the complex solution as exemplified by Fig. 1-12 requires 46 unit operations, whereas when the same system is separated into real and imaginary parts, as in Fig. 1-11, 49 unit operations are required. For $n = 3$ the count is 100 for the complex treatment against 118 for the real.

---

[1] See, for example, Pipes, *op. cit.*, p. 71.

[2] See, for example, Pipes, *op. cit.*, p. 80. See also Exercise 1-22.

[3] A similar result holds for unsymmetrical systems with $1/6n^3$ replaced by $1/3n^3$.

Practical Suggestions. It makes for less confusion regarding decimal points and less trouble with round-off errors if all the given quantities in (1-65) are of roughly the same order of magnitude. If there are great discrepancies in the magnitudes of the various elements, it is worthwhile to try to reduce the spread by means of *preliminary scaling operations*. All the elements in any row of (1-67) can be scaled up by a common factor without changing the solution. Any column of a_{jk} can be scaled up by a common factor provided the corresponding unknown, x_k, is scaled down by the same factor. The column of c_j can be scaled up by a common factor provided all of the unknowns are scaled up by the same factor.

To compensate for possible loss of accuracy due to round-off errors, it is common to introduce a fictitious accuracy into the calculation by adding so-called *guarding places* to the given data. Thus if the given quantities contain four significant figures, one might add 2 zero decimal places to each quantity and treat all intermediate results as if they had six significant figures. Then at the end the final answers would be rounded off to 4 places again. It is not possible to say in advance how many decimal places of accuracy will be lost during the solution of any one particular system. By considering a statistical distribution of all possible linear systems J. von Neumann and H. H. Goldstine[1] have shown that the number of decimal places of accuracy lost increases logarithmically with the number of unknowns. They estimate that the probability of losing 6 decimal places in a 4-degree-of-freedom system is the same as that of losing 9 decimal places in a 40-degree-of-freedom system or of losing 12 decimal places in a 400-degree-of-freedom system. In their view this probability is quite remote.

Sad experience indicates that a simple *checking device* is an almost necessary addition to an extended computational scheme. For the algorithm of this section it is possible to check the computation of the auxiliary array, row by row, by making use of a check column of *row sums*. This is illustrated in Fig. 1-20, where the computation of the auxiliary array of Fig. 1-17 is repeated with the addition of a check column. After entering the given quantities the sums of the elements in each row are computed and placed in the check column as shown. This check column is then treated just as if it were another column of c_j. A corresponding column in the auxiliary array is computed in the usual manner of the basic algorithm, e.g.,

$$\frac{43}{61} = \frac{1 - (6)(\frac{8}{5})}{-\frac{61}{5}} \tag{1-81}$$

---

[1] Numerical Inverting of Matrices of High Order, II, *Proc. Am. Math. Soc.*, **2**, 188–202 (1951). The estimates given apply to inversion of systems by an elimination method very similar to the one described in this section.

A check on each row of the auxiliary array is then provided by the fact[1] that the *corresponding element in the check column should equal* 1 *plus the sum of the elements in that row which are to the right of the heavy staircase*

						Check Column
5	6	−2	−2	1		8
6	−5	−2	2	0		1
−2	−2	3	−1	0		−2
−2	2	−1	−3	0		−4

5	$\frac{6}{5}$	$-\frac{2}{5}$	$-\frac{2}{5}$	$\frac{1}{5}$	$\frac{8}{5}$
6	$-\frac{61}{5}$	$-\frac{2}{61}$	$-\frac{22}{61}$	$\frac{6}{61}$	$\frac{43}{61}$
−2	$\frac{2}{5}$	$\frac{135}{61}$	$-\frac{101}{135}$	$\frac{22}{135}$	$\frac{56}{135}$
−2	$\frac{22}{5}$	$-\frac{101}{61}$	$-\frac{466}{135}$	$-\frac{16}{233}$	$\frac{217}{233}$

Fig. 1-20. Adding a row-sum check to the computation of Fig. 1-17.

line. Thus from the second row we compute

$$1 + \{(-\tfrac{2}{61}) + (-\tfrac{22}{61}) + \tfrac{6}{61}\} = \tfrac{43}{61} \tag{1-82}$$

which should check with (1-81).

When the complete solution has been obtained, it should be subjected to a final check by substituting the values obtained back into the original equations.

EXERCISES

1-14. Solve the following system by the Banachiewicz-Crout algorithm:

$$\begin{aligned}
x_1 + 2x_2 - 2x_3 + x_4 &= 4 \\
2x_1 + 5x_2 - 2x_3 + 3x_4 &= 7 \\
-2x_1 - 2x_2 + 5x_3 + 3x_4 &= -1 \\
x_1 + 3x_2 + 3x_3 + 2x_4 &= 0
\end{aligned}$$

This system has been specially constructed so that small whole numbers are obtained at every stage of the process.

1-15. Solve a general set of three simultaneous linear algebraic equations with literal coefficients a_{jk} in the following manner: Write the three equations, and divide the first equation through by a_{11}, the coefficient of x_1. Next write two simultaneous equations for x_2 and x_3 obtained from subtracting appropriate multiples of the first equation from the second and third. Treat these two new equations in the same manner; i.e., divide the first equation through by the coefficient of x_2, and then sub-

[1] See Exercise 1-20.

tract an appropriate multiple of it from the second equation to obtain a single equation for x_3. Solve for x_3, and back-substitute for x_2 and x_1. Now apply the Banachiewicz-Crout algorithm to the same system, and identify the auxiliary quantities with coefficients obtained in the above elimination.

1-16. Apply the algorithm of this section to the following system. This is a case in which a diagonal element of the auxiliary array vanishes.

$$\begin{aligned} x_1 + 2x_2 + 3x_3 &= 2 \\ 3x_1 + 6x_2 + x_3 &= 14 \\ x_1 + x_2 + x_3 &= 2 \end{aligned}$$

1-17. Verify the identity (1-74) for a symmetrical system of order 4. The result can be seen from a study of (1-69) and (1-70).

1-18. Verify the identity (1-78) for a general determinant of order 3 by making use of the results of Exercise 1-15.

1-19. Let the right-hand sides of (1-20) be y_1, y_2, y_3, and y_4, respectively. Invert this system to find the x's in terms of the y's. Make use of the results in Fig. 1-17.

1-20. Verify the row-sum check described on page 38 for the general three-degree-of-freedom system of Exercise 1-15.

1-21. In addition to the row-sum check there is a *column-sum check*, which may be described as follows: The sum of each column of the original array is entered into a check row directly beneath the nth row of given quantities. The basic algorithm is then applied to obtain the auxiliary array, the check row being treated the same as any other row. This produces a check row in the auxiliary array. A check on the computation of a particular column in the auxiliary array is then provided by the fact that *the element in the check row of that column should equal the sum of the elements in that column which are below the heavy staircase line.* Verify this for the general three-degree-of-freedom system of Exercise 1-15. Apply the test to the computation of Fig. 1-17.

1-22. The coefficients b_{jk} in the inverse (1-45) of the system (1-36) may be computed as the ratio of determinants as follows:

$$b_{jk} = (-1)^{j+k} \frac{A_{kj}}{\Delta}$$

where Δ is the determinant (1-38) of order n and A_{kj} is the determinant of order $n-1$ obtained by deleting the kth row and jth column from Δ. Count the number of operations required to form the inverse by this method when $n = 2$ and $n = 3$. Show that for large n the number of operations increases in proportion to $n^5/3$ if all determinants are evaluated according to (1-78).

1-23. Solve the system (1-8) by the Banachiewicz-Crout algorithm and by Cramer's rule (1-78). Compare. Obtain the inverse for the set of coefficients of (1-8) by the elimination algorithm and by the method of Exercise 1-22. Compare.

1-6. Iteration

We turn now to indirect, or approximate, procedures for solving the linear system

$$\begin{aligned} a_{11}x_1 + a_{12}x_2 + \cdots + a_{1n}x_n &= c_1 \\ a_{21}x_1 + a_{22}x_2 + \cdots + a_{2n}x_n &= c_2 \\ \cdots\cdots\cdots\cdots\cdots\cdots\cdots\cdots\cdots \\ a_{n1}x_1 + a_{n2}x_2 + \cdots + a_{nn}x_n &= c_n \end{aligned} \qquad (1\text{-}83)$$

The basic idea underlying the iteration process may be described as follows: We solve the first of (1-83) for x_1, the second for x_2, etc., in the following manner:

$$x_1 = \frac{1}{a_{11}} (c_1 \qquad\quad - a_{12}x_2 - \cdots - a_{1n}x_n)$$

$$x_2 = \frac{1}{a_{22}} (c_2 - a_{21}x_1 \qquad - \cdots - a_{2n}x_n) \qquad\qquad (1\text{-}84)$$

$$\cdots\cdots\cdots\cdots\cdots\cdots\cdots\cdots\cdots\cdots$$

$$x_n = \frac{1}{a_{nn}} (c_n - a_{n1}x_1 - a_{n2}x_2 - \cdots \qquad\qquad)$$

In the above derivation the x_j on the left are the same as the x_j on the right. Let us temporarily drop this requirement. Then (1-84) may be interpreted as a rule for transforming a given set of values $(x_1, \ldots, x_n)$ on the right into a new set on the left. The new set so obtained will be different from the given set unless the given values are actually the solution to (1-83). Thus solving (1-83) may be viewed as the problem of finding that set of values $(x, \ldots, x_n)$ which under this transformation reproduces itself.

A possible procedure for doing this would be simply to guess a set of values, insert them in the right of (1-84), and see whether the values obtained on the left were identical with the guessed set. If they were not, they would be discarded and another guess made.

This guessing procedure becomes an *iteration procedure* when we stop making random guesses for each new set of values and use the set obtained on the left from the first trial set as the next trial set to be inserted on the right. We thus construct a repetitive process which, starting from a guessed initial set, transforms the guessed set according to (1-84) into a second set, which is in turn transformed by the same rule into a third set, etc. If ever a stage is reached where a new set is exactly the same as its predecessor, then we have the solution to (1-83). The mathematically exact solution is seldom obtained in a finite number of steps, but the process does converge rapidly toward the correct solution when the main diagonal coefficients in (1-83) are large in comparison with the coupling coefficients. We shall return to the question of convergence, but first we go on to describe some of the variations on the basic scheme.

Iteration by Total Steps and by Single Steps. More explicitly the iteration procedure outlined above consists of the construction of a sequence of sets of values

$$(x_1^{(0)}, \ldots, x_n^{(0)}), (x_1^{(1)}, \ldots, x_n^{(1)}), \ldots, (x_1^{(r)}, \ldots, x_n^{(r)}), \ldots$$
$$(1\text{-}85)$$

according to the following rule:

$$x_1^{(r+1)} = \frac{1}{a_{11}} (c_1 \qquad\qquad - a_{12}x_2^{(r)} - \cdots - a_{1n}x_n^{(r)})$$

$$x_2^{(r+1)} = \frac{1}{a_{22}} (c_2 - a_{21}x_1^{(r)} \qquad\qquad - \cdots - a_{2n}x_n^{(r)}) \qquad (1\text{-}86)$$

$$x_n^{(r+1)} = \frac{1}{a_{nn}} (c_n - a_{n1}x_1^{(r)} - a_{n2}x_2^{(r)} - \cdots \qquad\qquad)$$

This process is called *iteration by total steps*[1] since all n variables are treated at once.

In contrast to this there is *iteration by single steps*, which treats one variable at a time. The basic step in this case consists in replacing the variable x_p in the set $(x_1, \ldots, x_p, \ldots, x_n)$ by the improved value

$$x_p' = \frac{1}{a_{pp}} (c_p - a_{p1}x_1 - \cdots - a_{p(p-1)}x_{p-1}$$
$$-a_{p(p+1)}x_{p+1} - \cdots - a_{pn}x_n) \qquad (1\text{-}87)$$

This has the same form as one of the single equations in (1-86), but the idea here is that one variable at a time is improved and that this improved value is used immediately in computing the improvement for the next variable. The order in which the variables are improved is not necessarily specified. For automatic machine computation a cyclic order is desirable. If p in (1-87) is allowed to run from 1 to n, the resulting equations can be cast in a form similar to (1-86).

$$x_1^{(r+1)} = \frac{1}{a_{11}} (c_1 \qquad\qquad - a_{12}x_2^{(r)} - \cdots - a_{1n}x_n^{(r)})$$

$$x_2^{(r+1)} = \frac{1}{a_{22}} (c_2 - a_{21}x_1^{(r+1)} \qquad\qquad - \cdots - a_{2n}x_n^{(r)}) \qquad (1\text{-}88)$$

$$x_n^{(r+1)} = \frac{1}{a_{nn}} (c_n - a_{n1}x_1^{(r+1)} - a_{n2}x_2^{(r+1)} - \cdots \qquad\qquad)$$

Iteration according to this scheme is commonly called the *Gauss-Seidel procedure*.[2]

Successive Approximations vs. Successive Corrections. The iteration procedures so far described have all dealt directly with the variables them-

[1] It was employed by C. G. J. Jacobi in 1844. See his "Gessamelte Werke," vol. 3, Berlin, 1884, pp. 467–478.

[2] Iteration by single steps was used by C. F. Gauss in his own calculations. A description of his method appears in C. L. Gerling, "Die Ausgleichs-Rechnungen der practischen Geometrie," Hamburg and Gotha, 1843. The method was independently described by L. Seidel, Ueber ein Verfahren die Gleichungen auf welche die Methode der Kleinsten Quadrate führt, so wie lineäre Gleichungen überhaupt durch successive Annäherung aufzulösen, *Abhandl. bayer. Akad. Wiss.*, **11**, 81–108 (1874). In neither case was a fixed cyclic order recommended.

selves.   As such they may be called methods of *successive approximation.*

A popular variation is to arrange the computation so that at each stage one deals only with corrections to the variables.   Such procedures are called methods of *successive corrections.*

To show how the Gauss-Seidel process, for instance, can be transformed into a method of successive corrections, we write the identity

$$x_p^{(s)} = x_p^{(0)} + (x_p^{(1)} - x_p^{(0)}) + \cdots + (x_p^{(s)} - x_p^{(s-1)})$$
$$p = 1, \ldots, n \quad (1\text{-}89)$$

and then define the corrections $\varphi_p^{(r)}$ as

$$\varphi_p^{(r)} = x_p^{(r)} - x_p^{(r-1)} \qquad p = 1, \ldots, n \qquad (1\text{-}90)$$

If for simplicity we take the $x_p^{(0)}$ to be zero, we have from (1-89)

$$x_p^{(s)} = \sum_{r=1}^{s} \varphi_p^{(r)} \qquad p = 1, \ldots, n \qquad (1\text{-}91)$$

i.e., the sth approximation is the sum of the first s corrections.

Iteration formulas for the $\varphi_p^{(r)}$ are obtained by substituting (1-88) into (1-90) to get

$$\varphi_1^{(r+1)} = \frac{1}{a_{11}} (\qquad\qquad - a_{12}\varphi_2^{(r)} - \cdots - a_{1n}\varphi_n^{(r)})$$
$$\varphi_2^{(r+1)} = \frac{1}{a_{22}} (-a_{21}\varphi_1^{(r+1)} \qquad\qquad - \cdots - a_{2n}\varphi_n^{(r)})$$
$$\cdots\cdots\cdots\cdots\cdots\cdots\cdots\cdots\cdots\cdots\cdots\cdots\cdots\cdots$$
$$\varphi_n^{(r+1)} = \frac{1}{a_{nn}} (-a_{n1}\varphi_1^{(r+1)} - a_{n2}\varphi_2^{(r+1)} - \cdots \qquad\qquad)$$

(1-92)

for $r \geq 1$.   The set $\varphi_p^{(1)} = x_p^{(1)}$ must be obtained directly from (1-88) and the initial set $x_p^{(0)} = 0$, $p = 1, \ldots, n$.

The arguments in favor of successive corrections are that there is one less arithmetical operation in every step of (1-92) as compared with (1-88) and that as the process converges the size of the correction gets smaller and hence presumably easier to handle.   On the other hand the argument in favor of successive approximations is that it is self-checking. No matter how many mistakes have been made during an iteration, when a set of values finally does reproduce itself, that is the solution.   In a process of successive correction an external check is required after the correcting process is completed,[1] and if a mistake has been made, the entire calculation is invalidated.

[1] It is possible to set up a running check on the arithmetic of each cycle (see Exercise 1-29).

Example. We take the system (1-23) obtained for the continuous beam, Prob. 1-4, which when written in the form of (1-84) is

$$x_1 = 0.5000 \qquad\qquad - 0.5000x_2$$
$$x_2 = \qquad - 0.2500x_1 \qquad\qquad - 0.2500x_3 \qquad (1\text{-}93)$$
$$x_3 = \qquad\qquad - 0.3333x_2$$

If we use this in the manner of (1-86) for iteration by total steps, we obtain, starting from the (arbitrary) initial trial (0, 0, 0), the successive iterates shown in Table 1-3.

TABLE 1-3. SUCCESSIVE ITERATES OF (1-93) USING ITERATION BY TOTAL STEPS

Cycle	x_1	x_2	x_3
(0)	0	0	0
(1)	0.5000	0	0
(2)	0.5000	−0.1250	0
(3)	0.5625	−0.1250	0.0417
(4)	0.5625	−0.1510	0.0417
(5)	0.5755	−0.1510	0.0503
(6)	0.5755	−0.1564	0.0503
(7)	0.5782	−0.1564	0.0521
(8)	0.5782	−0.1576	0.0521
(9)	0.5788	−0.1576	0.0525
(10)	0.5788	−0.1578	0.0525
(11)	0.5789	−0.1578	0.0526
(12)	0.5789	−0.1579	0.0526

At the end of 12 cycles the values agree with the correct solution to four decimal places.

If we apply the Gauss-Seidel process of (1-88) to (1-93), we get, again starting from (0, 0, 0), the successive iterates shown in Table 1-4. We note in this case that the

TABLE 1-4. SUCCESSIVE ITERATES OF (1-93) USING THE GAUSS-SEIDEL PROCESS

Cycle	x_1	x_2	x_3
(0)	0	0	0
(1)	0.5000	−0.1250	0.0417
(2)	0.5625	−0.1510	0.0503
(3)	0.5755	−0.1564	0.0521
(4)	0.5782	−0.1576	0.0525
(5)	0.5788	−0.1578	0.0526
(6)	0.5789	−0.1579	0.0526

same stage as in Table 1-3 is reached in half as many cycles.

To illustrate successive corrections, we repeat the Gauss-Seidel process, but using corrections instead. We first write according to (1-92) the iteration formula for the corrections which corresponds to (1-93).

$$\varphi_1^{(r+1)} = \qquad\qquad - 0.5000\varphi_2^{(r)}$$
$$\varphi_2^{(r+1)} = -0.2500\varphi_1^{(r+1)} \qquad\qquad - 0.2500\varphi_3^{(r)} \qquad (1\text{-}94)$$
$$\varphi_3^{(r+1)} = \qquad\qquad - 0.3333\varphi_2^{(r+1)}$$

Then, starting with the first corrections taken from the first cycle of Table 1-4, we obtain from (1-94) the successive corrections shown in Table 1-5. The slight differ-

TABLE 1-5. SUCCESSIVE-CORRECTION SOLUTION OF (1-23) USING THE GAUSS-SEIDEL PROCESS

Cycle	φ_1	φ_2	φ_3
(1)	0.5000	−0.1250	0.0417
(2)	0.0625	−0.0260	0.0087
(3)	0.0130	−0.0054	0.0018
(4)	0.0027	−0.0011	0.0004
(5)	0.0006	−0.0002	0.0001
(6)	0.0001		
Sums	0.5789	−0.1577	0.0527

ences between the results of Table 1-5 and Table 1-4 are due to the fact that in working to exactly four decimal places we have made rounding errors. These errors remain uncompensated in the successive-correction process.

Convergence. Whether an iteration procedure converges or not and whether the convergence is slow or fast depend on the properties of the system (1-83). Speaking broadly, iterative methods are useful only for those systems whose main diagonal coefficients are large in comparison with the coupling coefficients. We give first a convergence proof which emphasizes this point. A second proof is then given which makes use of the extremum problem corresponding to the given equilibrium system.

In the system (1-83) we define the quantities

$$\alpha_j = \frac{\sum_{k=1}^{n}{}' |a_{jk}|}{|a_{jj}|} \qquad j = 1, \ldots, n \qquad (1\text{-}95)$$

where the prime on the summation sign indicates that $k = j$ has been omitted, as measures of the predominance of the main diagonal coefficients. The α_j are called the *row-sum criteria*. It has been shown[1] that *if the largest of the α_j satisfies*

$$\alpha_{\max} < 1 \qquad (1\text{-}96)$$

then iteration either by total or by single steps will converge. We sketch the proof in the case of iteration by total steps.[2]

Let the true solution to (1-83) be $(x_1, \ldots, x_n)$, and let the iterated set after r

---

[1] H. Geiringer, "On the Solution of Linear Equations by Certain Iteration Methods," Reissner Anniversary Volume, J. W. Edwards, Publisher, Inc., Ann Arbor, Mich., 1949, pp. 365–393.

[2] See Exercise 1-28 for iteration by the Gauss-Seidel process.

cycles be $(x_1^{(r)}, \ldots, x_n^{(r)})$. The set of errors $(e_1^{(r)}, \ldots, e_n^{(r)})$ is defined as

$$e_j^{(r)} = x_j^{(r)} - x_j \qquad j = 1, \ldots, n \tag{1-97}$$

We next obtain the iteration formula for $e_j^{(r)}$ by substituting $x_j^{(r)}$ from (1-97) into (1-86),

$$e_1^{(r+1)} = \frac{1}{a_{11}} (\qquad - a_{12}e_2^{(r)} - \cdots - a_{1n}e_n^{(r)})$$

$$e_2^{(r+1)} = \frac{1}{a_{22}} (-a_{21}e_1^{(r)} \qquad - \cdots - a_{2n}e_n^{(r)}) \tag{1-98}$$

$$\cdots\cdots\cdots\cdots\cdots\cdots\cdots\cdots\cdots\cdots\cdots$$

$$e_n^{(r+1)} = \frac{1}{a_{nn}} (-a_{n1}e_1^{(r)} - a_{n2}e_2^{(r)} - \cdots \qquad)$$

where we have made use of the fact that the true solution satisfies (1-84). If $|e_{max}^{(r)}|$ is the largest of the $|e_j^{(r)}|$, we then have from (1-98)

$$|e_{max}^{(r+1)}| \le \alpha_{max}|e_{max}^{(r)}| \tag{1-99}$$

by using (1-95) and the fact that the absolute value of a sum is less than (or equal to) the sum of the individual absolute values. The convergence is proved by (1-99), which states that each cycle of iteration accomplishes at least a fixed percentage decrease of the maximum error. This result indicates that, the smaller α_{max} is, the faster the convergence will be. In the example worked in Tables 1-3 to 1-5 α_{max} was 0.50.

We next prove[1] that *iteration by single steps always converges when applied to symmetric positive definite systems.* When the system (1-83) is symmetric and positive definite, then the function

$$\Phi = \frac{1}{2} \sum_{j=1}^{n} \sum_{k=1}^{n} a_{jk}x_jx_k - \sum_{j=1}^{n} c_jx_j \tag{1-100}$$

has a single extremum which is a true minimum for the set of values which are the solution to (1-83). The main part of the proof consists in showing that each single step of iteration involves a decrease in Φ.

We take as our basic step the act of replacing x_p by x_p' according to (1-87). Let the values of (1-100) before and after be denoted by Φ and Φ', respectively. We then evaluate $\Phi' - \Phi$ by using the symmetry of the coefficients together with the relation

$$x_p' - x_p = \frac{1}{a_{pp}} \left(c_p - \sum_{k=1}^{n} a_{pk}x_k \right) \tag{1-101}$$

which is obtained directly from (1-87). In the following the prime on the summation

[1] The proof given here is essentially the same as that given by Seidel, *op. cit.*

sign indicates that the term with $k = p$ has been omitted:

$$
\begin{aligned}
\Phi' - \Phi &= \sideset{}{'}\sum_{k=1}^{n} a_{pk}x_k(x_p' - x_p) + \frac{1}{2} a_{pp}(x_p'^2 - x_p^2) - c_p(x_p' - x_p) \\
&= (x_p' - x_p) \left\{ \sideset{}{'}\sum_{k=1}^{n} a_{pk}x_k + \frac{1}{2} a_{pp}(x_p' + x_p) - c_p \right\} \\
&= (x_p' - x_p) \left\{ \frac{1}{2} a_{pp}(x_p' - x_p) + \sum_{k=1}^{n} a_{pk}x_k - c_p \right\} \\
&= -\frac{a_{pp}}{2} (x_p' - x_p)^2
\end{aligned}
\tag{1-102}
$$

Now the main diagonal coefficients of a positive definite system are positive (see Exercise 1-6) so that (1-102) indicates a decrease in Φ when x_p is replaced by x_p'. If iteration by single steps is continued in any order (just so long as each variable is eventually treated an unlimited number of times), the process must converge since Φ, having a minimum, cannot decrease indefinitely. Furthermore Φ cannot stop short of its true minimum since (1-102) states that Φ can stop decreasing only when every $x_p' = x_p$, that is, when the iteration process reproduces the variables without change. But this means that the solution of the original system has been obtained. Note that this proof gives no information regarding the rate of convergence.

The two theorems just proved give *sufficient* conditions for convergence. The conditions are not *necessary*. The question as to exactly where the border line between convergence and nonconvergence lies is mathematically very interesting and is still far from being completely answered. It is possible to construct systems[1] for which iteration by total steps converges while iteration by single steps diverges. It is possible to construct other systems[2] for which the reverse is true. From a practical point of view *rapid* convergence is usually desired, and this can be assured only if the main diagonal coefficients are large compared with the coupling coefficients. The Gauss-Seidel procedure is generally favored under these circumstances.

Length of Computation. It is difficult to estimate in advance how much computation will be required to obtain a solution by iteration. The number of cycles required depends on the accuracy desired and on the particular system to be solved. It is generally true that once an iteration is well under way a roughly fixed number of cycles is required for each additional correct decimal place in the answer. This number of cycles varies widely from system to system. Note that the number of multiplications required for a cycle of iteration is essentially n^2 except in special systems where a majority of the coefficients a_{jk} are zero. For instance, in systems where each variable is coupled to only one or two

---

[1] See Exercise 1-26.
[2] See Exercise 1-27.

other variables the number of operations required for a cycle of iteration may be less than $3n$.

Other Iterative Procedures. The iteration methods described here are only the simplest in an extensive family[1] of procedures that have been studied mathematically. Within this family one of the most complicated yet at the same time most interesting procedures is the *n-step iteration* of Hestenes and Stiefel.[2] This is actually a direct, or exact, method since the iterated operations are so chosen that at the end of exactly n steps the true solution is obtained; i.e., the true solution would be obtained if it were possible to carry out all the operations without any round-offs. The iteration rules for the n-step procedure cannot be easily described without making liberal use of matrix notation. For this reason we defer a description until Sec. 2-9.

An entirely different method for solving linear equations is based on transforming the coefficient matrix into diagonal form by an iterative process. The process requires much more computation than the methods we have already described but may be useful when a diagonalizing program is immediately available on a high-speed computing machine. The procedure is described[3] in Sec. 2-12.

EXERCISES

1-24. Solve the system (1-25) by iteration according to the Gauss-Seidel procedure. Evaluate the row-sum criteria of (1-95).

1-25. What are the row-sum criteria for the system (1-14)? Solve (1-14) by the Gauss-Seidel procedure, and compare the rate of convergence with that in Exercise 1-24.

1-26. Verify that iteration by total steps converges when applied to the system[4]

$$x_1 + 2x_2 - 2x_3 = 1$$
$$x_1 + x_2 + x_3 = 3$$
$$2x_1 + 2x_2 + x_3 = 5$$

Demonstrate that the Gauss-Seidel process diverges.

1-27. Show that the system

$$5x_1 + 4x_2 + 3x_3 = 12$$
$$4x_1 + 7x_2 + 4x_3 = 15$$
$$3x_1 + 4x_2 + 4x_3 = 11$$

is symmetric and positive definite and hence that the Gauss-Seidel process would converge. Demonstrate that iteration by total steps diverges.

[1] See A. S. Householder, "Principles of Numerical Analysis," McGraw-Hill Book Company, Inc., New York, 1953, chap. 2.

[2] M. R. Hestenes and E. Stiefel, Method of Conjugate Gradients for Solving Linear Systems, *Natl. Bur. Standards Rept.* 1659 (1952).

[3] See, in particular, Exercise 2-84.

[4] This example was given by L. Collatz, Fehlerabschatzüng für das Iterations verfahren zur Auflösung linearen Gleichungs-systeme, *Z. angew. Math. Mech.*, **22,** 357–361 (1942).

1-28. Prove that if $\alpha_{max} < 1$ the Gauss-Seidel process will converge. The proof is very similar to that given in the text for iteration by whole steps.

1-29. Write out the iteration formulas corresponding to (1-92) for the successive corrections in iteration by total steps. Show that a check on the arithmetic of each cycle is obtained by the following device: At the end of the cycle compute

$$S^{(r+1)} = -s_1\varphi_1^{(r)} - s_2\varphi_2^{(r)} - \cdots - s_n\varphi_n^{(r)}$$

where the coefficients s_k are computed in advance as the column sums

$$s_k = \sum_{j=1}^{n}{}' \frac{a_{jk}}{a_{jj}}$$

The check consists in comparing $S^{(r+1)}$ with the sum

$$\sum_{j=1}^{n} \varphi_j^{(r+1)}$$

1-30. Any system (1-83) of simultaneous linear equations can always be rearranged so that the Gauss-Seidel process will converge. Let

$$r_j = c_j - \sum_{k=1}^{n} a_{jk}x_k \qquad j = 1, \ldots, n$$

Show that the function

$$\Phi = \frac{1}{2} \sum_{j=1}^{n} (r_j{}^2 - c_j{}^2)$$

has a single extremum, a minimum, at the solution of (1-83). Show that the equations obtained by setting the partial derivatives of Φ with respect to the x_j equal to zero are identical with the set obtained in Exercise 1-7. These equations are symmetric and positive definite, and hence the Gauss-Seidel process will converge for them. Note that the number of multiplications required to obtain these equations from (1-83) is in excess of n^3.

1-31. Show that the maximum change in any variable in an iteration by total steps

$$\max|x_j^{(r+1)} - x_j^{(r)}| = d_{max}^{(r+1)}$$

is greater than (or equal to) $|e_{max}^{(r)}| - |e_{max}^{(r+1)}|$ and hence, using (1-99), that we can give an upper bound[1] for the absolute error at any stage as follows:

$$|e_{max}^{(r+1)}| \leq \frac{\alpha_{max}}{1 - \alpha_{max}} d_{max}^{(r+1)}$$

1-7. Relaxation

Actually carrying out the iterative procedures described in the foregoing section is a very routine matter and hence suited to automatic computation. We turn in this section to a more flexible procedure which has several advantages when the computations must be done by hand by an intelligent computer.

[1] This result was given by Collatz, *loc. cit.*

We begin by describing the relaxation method[1] for solving the linear system

$$\sum_{k=1}^{n} a_{jk}x_k = c_j \qquad j = 1, \ldots, n \qquad (1\text{-}103)$$

Our first step is to define the *residuals*

$$r_j = c_j - \sum_{k=1}^{n} a_{jk}x_k \qquad j = 1, \ldots, n \qquad (1\text{-}104)$$

as the amounts by which the n equations of (1-103) are *not satisfied* by a given set of values $(x_1, \ldots, x_n)$. Solving the system (1-103) may then be viewed as the problem of finding that set of values $(x_1, \ldots, x_n)$ which makes all the residuals of (1-104) vanish.

The relaxation process consists in making a sequence of alterations to an initial trial set $(x_1, \ldots, x_n)$ in such a manner that the residuals are continually reduced. The bookkeeping in this process has been efficiently engineered so that it is easy to ascertain the effect on the residuals of a proposed change in the x_j. The basic difference between iteration and relaxation lies in the fact that once this bookkeeping machinery has been set up the particular path by which the residuals are reduced and the true solution approached is not dictated but left instead to the initiative of the computer.

TABLE 1-6. UNIT OPERATIONS FOR THE RESIDUALS (1-104)

Operations				Changes in residuals			
Δx_1	Δx_2	$\cdots$	Δx_n	Δr_1	Δr_2	$\cdots$	Δr_n
1	0	$\cdots$	0	$-a_{11}$	$-a_{21}$	$\cdots$	$-a_{n1}$
0	1	$\cdots$	0	$-a_{12}$	$-a_{22}$	$\cdots$	$-a_{n2}$
$\cdot$	$\cdot$		$\cdot$	$\cdots$	$\cdots$	$\cdots$	$\cdots$
0	0	$\cdots$	1	$-a_{1n}$	$-a_{2n}$	$\cdots$	$-a_{nn}$

Our second step is to set up a *unit-operations table* which gives us the increment to each residual occasioned by a *unit* increase in each of the x_j. This is shown in Table 1-6, where each row contains a single unit operation and its effect on the residuals. By using linear superposition we

---

[1] R. V. Southwell, "Relaxation Methods in Engineering Science," Oxford University Press, New York, 1940. For a simple introduction see also F. S. Shaw, "An Introduction to Relaxation Methods," Dover Publications, New York, 1953, and D. N. de G. Allen, "Relaxation Methods," McGraw-Hill Book Company, Inc., New York, 1954.

can obtain from the table the changes in the residuals occasioned by any combination of increments to the x_j.

The actual computation is carried out in a *relaxation table* in which we record at each step the alteration made in the x_j and the new values of the residuals which result after the alteration has been made. The further details of the process are best illustrated in terms of a particular example.

Example. We take the system (1-23) obtained for the continuous beam, Prob. 1-4. This was the example we solved in Tables 1-3 to 1-5 by iteration. The residuals according to (1-104) are

$$
\begin{aligned}
r_1 &= 1 - 2x_1 - x_2 \\
r_2 &= \quad - x_1 - 4x_2 - x_3 \\
r_3 &= \quad\quad - x_2 - 3x_3
\end{aligned}
\tag{1-105}
$$

We next fill in the unit-operations table at the top of Table 1-7 in lines (1), (2), and (3).

The relaxation table is started in line (4) by selecting an initial trial set (x_1, x_2, x_3). We have taken (arbitrarily) the set $(0, 0, 0)$. From (1-105) we compute the residuals $(1, 0, 0)$, which are placed on the right of line (4). Our next step in line (5) is simply a scaling operation occasioned by the desire to avoid decimals in the mental arithmetic soon to follow. We multiply both sides of (1-105) by 100, and hence in line (5) we multiply each entry of line (4) by 100.

We are now ready to begin relaxation. Our aim is to reduce all residuals to zero; we are given the three unit operations, but we are not told in what order or in what amounts they are to be used. As a first step, feeling our way, we might try 50 units of operation (1). We record this by writing 50 under x_1 in line (6). The entries in the residual columns are obtained by doing a little mental arithmetic as follows: For each unit of operation (1) r_1 is decreased by 2; we have 50 units and hence a total decrease of 100, which when subtracted from the 100 in line (5) gives the entry 0 for r_1 in line (6). Similarly 50 times the unit decrease of r_2 gives us the entry -50 for r_2 in line (6). The entry for r_3 remains zero because operation (1) has no effect on r_3.

We might pause at this point to compare the relaxation process with the Gauss-Seidel process of Table 1-4. There in the first single step of iteration x_1 was changed from 0 to 0.5000 by temporarily satisfying the first of the given system of equations. Note that this is exactly what we have done in our first step of relaxation. Since $r_1 = 0$, the first equation is temporarily satisfied. We could continue following the Gauss-Seidel process by simply using operations (1), (2), (3) in strict order and by each time using just enough units so that the corresponding residual was brought exactly to zero. The advantage of the relaxation method is that we are not tied to this or to any other fixed pattern. Instead of bringing a particular residual exactly to zero, we are free to *overrelax* or *underrelax*. We are encouraged to experiment. Any operation is a good one if it leads shortly to a substantial reduction of the residuals.

Returning to Table 1-7, the reader should follow the operations of lines (7) to (9). Note that, while only the *increments* of x_j are listed, the new *total residual values* r_j are recorded; e.g., in line (7) we mentally multiply -15 by the unit increments of line (2), obtaining (15, 60, 15), which are mentally added to the existing total residuals $(0, -50, 0)$ of line (6) to obtain finally the new total residuals (15, 10, 15) recorded in line (7). At the end of line (9) the residuals have been reduced so far that fractional or decimal entries would be required for further progress. We then decide to introduce another scale factor of 100.

At this juncture it is convenient to institute a check of our work so far. The calculations of lines (6) to (9) are of the nature of successive corrections and hence are not self-checking. The check is performed in line (10), where we add the increments in the x_1 columns to obtain the current total values (57, -15, 5). We then compute the residuals for these from the basic definition (1-105). The residuals so obtained are listed on the right of line (10). They should agree with those of line (9). If a mistake has been made, it is not necessary to find it. We merely start again from line (10) with the correct residuals just computed.

In line (11) we have scaled up all the entries of line (10) by the factor 100. We return to the actual relaxation process in lines (12) to (15), reducing the residuals until once more a scaling operation would be required if decimal entries are to be avoided. The increments to the x_j have been summed in line (16) and the residuals checked by computing them from (1-105).

The results of the calculation at this stage give us as the approximate solution to (1-23)

$$x_1 = 0.5790$$
$$x_2 = -0.1580$$
$$x_3 = 0.0527$$

These are the entries of line (16) after they have been scaled back down by the factor 10,000. We know, further, that these values satisfy (1-23) to within a maximum residual of 0.0003.

TABLE 1-7. SOLUTION OF (1-23) BY RELAXATION

	Line	Δx_1	Δx_2	Δx_3	Δr_1	Δr_2	Δr_3
Operations table	(1)	1	0	0	-2	-1	0
	(2)	0	1	0	-1	-4	-1
	(3)	0	0	1	0	-1	-3
Relaxation table	Line	x_1	x_2	x_3	r_1	r_2	r_3
	(4)	0	0	0	1	0	0
$\times$ 100	(5)	0	0	0	100	0	0
	(6)	50			0	-50	0
	(7)		-15		15	10	15
	(8)			5	15	5	0
	(9)	7			1	-2	0
Check	(10)	57	-15	5	1	-2	0
$\times$ 100	(11)	5,700	$-1,500$	500	100	-200	0
	(12)		-60		160	40	60
	(13)	90			-20	-50	60
	(14)			27	-20	-77	-21
	(15)		-20		0	3	-1
Check	(16)	5,790	$-1,580$	527	0	3	-1

Technique of Relaxation. Once the details of the bookkeeping have been mastered, the computer can devote his attention to the problem of selecting an effective sequence of operations. A considerable amount of skill can be attained with practice. No two computers will follow exactly the same path, but an experienced computer will be able to bring his residuals down much faster than a neophyte, and even the beginner will usually do better than by a fixed iterative procedure. For the intelligent computer relaxation becomes a game on a par with a crossword puzzle. This introduction of an intellectual challenge into what would otherwise be a dull, tedious hand calculation has been a major contribution of the relaxation method.

In acquiring skill the computer will learn to study the effect of each operation, feeling out, as it were, the response of the system. He will learn to look ahead as in chess, but he will also learn where to draw the line between studying out optimum operations and plunging quickly ahead with not quite such optimum operations.

An important device for speeding up relaxation is an extended operations table in which are placed *group operations*. We illustrate this for the system (1-14) obtained in the d-c network, Prob. 1-2. In the first three lines of Table 1-8 we enter the basic unit operations according to Table 1-6. These operations could be used exactly as in Table 1-7 to effect a solution. From a relaxational point of view operations such as (1) and (3) are not the most desirable because they are not selective, i.e., because they affect two residuals nearly equally. It requires a certain finesse to bring all residuals down simultaneously by working only with such operators.

TABLE 1-8. EXTENDED-OPERATIONS TABLE FOR SYSTEM OF (1-14)

Line	Operations			Changes in residuals		
	Δx_1	Δx_2	Δx_3	Δr_1	Δr_2	Δr_3
(1)	1	0	0	-5	4	0
(2)	0	1	0	4	-10	5
(3)	0	0	1	0	5	-6
(4)	1	1	1	-1	-1	-1
(5)	3	2	2	-7	2	-2
(6)	4	5	4	0	-14	1
(7)	2	3	4	2	-2	-9

In such cases it is worthwhile to take the time to construct a few group operations such as those shown in lines (4) to (7) of Table 1-8. The operation of line (4) consists in simultaneously adding unity to all three variables. The net change in the residuals is obtained by simply adding together the effects of lines (1) to (3). This simple group operator would be employed to advantage whenever all three residuals were nearly equal. The operations of lines (5) to (7) were similarly obtained by superposition of the indicated units of operations (1) to (3). Note that they are quite selective in their action against r_1, r_2, and r_3, respectively. Constructing such selective operators may be considered as equivalent to transforming the given system of equations into an equivalent system in which the main diagonal coefficients predominate. The reader may wish to take up a pencil and obtain by trial some selective group operations of his own. If he spends much time, he can easily find operations even more selective than those of Table 1-8. The ultimate in this direction would be operations which would reduce one residual without affecting any of the others. Having a set of these would amount, however, to having the inverse of the given system. Here again the experienced relaxation operator must decide where to draw the line between groping for more selective operations and actually getting down to the relaxation with the moderately selective operations which have come easily.

The bookkeeping of relaxation with group operations is no more difficult than with the basic unit operations. By looking at the residual patterns the computer decides how many units of the group operation he wants to use. Thus, if he wants four units of operation (7) of Table 1-8, he would record (8, 12, 16) as the increments to (x_1, x_2, x_3) and mentally add to the existing residuals the residual increments $(8, -8, -36)$.

Relaxation Method for Nonlinear Systems. The basic ideas of the method just described can be extended to nonlinear systems, although some of the simplifications in detail are unavoidably lost. If we consider the equilibrium problem

$$a_1(x_1, \ldots, x_n) = c_1$$
$$a_2(x_1, \ldots, x_2) = c_2 \qquad (1\text{-}106)$$
$$\ldots \ldots \ldots \ldots$$
$$a_n(x_1, \ldots, x_n) = c_n$$

where the a_j are nonlinear functions of $x_1, x_2, \ldots, x_n$, we can still define *residuals* as follows:

$$r_j = c_j - a_j \qquad j = 1, \ldots, n \qquad (1\text{-}107)$$

Since these residuals are nonlinear functions of the x_k, it will not be possible to obtain a simple operations table that will be valid throughout all the ranges of the x_k. In an infinitesimal neighborhood of a point where all the a_j are continuously differentiable, we can easily obtain an operations table for *infinitesimal alterations* in the x_k. This table would be exactly the same as Table 1-6 but with

$$-a_{jk} = -\frac{\partial a_j}{\partial x_k} = \frac{\partial r_j}{\partial x_k} \tag{1-108}$$

In the linear system a_{jk} was constant; in the nonlinear system a_{jk} is a function of the trial solution under consideration.

The relaxation method handles this difficulty as follows: We select an initial trial solution (using as much insight as possible) and compute the residuals according to (1-107). We also construct an operations table with the elements (1-108). Now this table is strictly valid only in an infinitesimal neighborhood of the initial trial. In spite of this we proceed to operate with it, exactly as in a linear system, until the initial residuals have been considerably reduced. We then total the increments to the x_j and recompute the residuals according to (1-107) just as in the checking operation for linear systems. Here we do not expect a check, but we hope that the true residuals at this stage are at least somewhat smaller than the original initial residuals. If this is the case, we would repeat the process with a new operations table valid in the neighborhood of our present trial. The closer we get to the true solution, the more accurately does our linearized operations table represent the nonlinear system and the more confidence we can have in our relaxation.

For a large class of nonlinear network problems it has been proved[1] that the process just outlined will lead to the true solution. For more general nonlinear systems the process may not "pull in," or it may begin to converge toward a solution which is physically uninteresting even though mathematically correct. Several initial trials may be necessary to locate the neighborhood of the desired solution.

Example. We take the nonlinear system (1-30) for the hydraulic network, Prob. 1-5. The residuals, according to (1-107), are

$$r_1 = 1 - 0.4472 \ x_1^{\frac{1}{2}} \quad - 0.7071 \ (x_1 - x_2)^{\frac{1}{2}}$$
$$r_2 = \quad + 0.7071 \ (x_1 - x_2)^{\frac{1}{2}} - 1.2845 \ x_2^{\frac{1}{2}} \tag{1-109}$$

and the entries for an operations table in the neighborhood of (x_1, x_2) would be obtained by evaluating the following:

	$\dfrac{\partial r_1}{\partial x_k}$	$\dfrac{\partial r_2}{\partial x_k}$	
$k = 1$	$-0.2236x_1^{-\frac{1}{2}} - 0.3536(x_1 - x_2)^{-\frac{1}{2}}$	$0.3536(x_1 - x_2)^{-\frac{1}{2}}$	(1-110)
$k = 2$	$0.3536(x_1 - x_2)^{-\frac{1}{2}}$	$-0.3536(x_1 - x_2)^{-\frac{1}{2}} - 0.6422x_2^{-\frac{1}{2}}$	

We begin by selecting an initial trial based on rather rough estimates of how the flow divides in Fig. 1-7: $x_1 = 1.0$, $x_2 = 0.4$. We construct the operations table in lines (1) and (2) of Table 1-9 by substituting these values in (1-110). In line (3) the trial solution is inserted with the residuals computed according to (1-109). Three steps

[1] Birkhoff and Diaz, *op. cit.*

of relaxation are then performed in lines (4) to (6), using operations (1) and (2) exactly as in a linear system.

We are usually forced to work with decimals in nonlinear systems because the introduction of scale factors is awkward, if not impossible. In line (7) the increments to x_1 and x_2 are summed and the corresponding residuals are obtained from (1-109). In this case we are fortunate in that the true residuals of line (7) are actually smaller than those of line (6).

To carry the computation one step further, we have recomputed the operations table in lines (8) and (9) from (1-110), using $x_1 = 0.88$ and $x_2 = 0.20$. Line (10) is line (7) repeated. In lines (11), (12), and (13) there are three relaxations using

TABLE 1-9. RELAXATION SOLUTION OF NONLINEAR SYSTEM (1-30)

Line	Δx_1	Δx_2	Δr_1	Δr_2
(1)	1	0	-0.68	0.46
(2)	0	1	0.46	-1.47

	x_1	x_2	r_1	r_2
(3)	1.00	0.40	0.0051	-0.2646
(4)		-0.20	-0.0869	$+0.0294$
(5)	-0.10		-0.0189	-0.0166
(6)	-0.02		-0.0053	-0.0258
(7)	0.88	0.20	-0.00262	$+0.00867$

	Δx_1	Δx_2	Δr_1	Δr_2
(8)	1	0	-0.667	0.429
(9)	0	1	0.429	-1.865

	x_1	x_2	r_1	r_2
(10)	0.8800	0.2000	-0.00262	$+0.00867$
(11)		0.0050	-0.00047	-0.00066
(12)		-0.0005	-0.00069	$+0.00028$
(13)	-0.0010		-0.00002	-0.00015
(14)	0.8790	0.2045	-0.000017	-0.000121

operations (8) and (9). In line (14) the increments to x_1 and x_2 are again summed and the true residuals computed from (1-109). The largest residual is now 2,000 times smaller than that for the initial trial.

If greater accuracy were required, the relaxation could be continued. In subsequent steps it would be satisfactory to continue using the operations (8) and (9) provided the correct residuals were occasionally recomputed from (1-109). Note that the discrepancy between the residuals of lines (13) and (14) is already small.

EXERCISES

1-32. Solve the system (1-25) by relaxation.

1-33. Solve the system (1-20) by relaxation. Here rapid progress will be difficult if only the basic unit operations are employed. Construct a few selective group operations.

1-34. Show that a *running check* on the bookkeeping of the relaxation process can be obtained with the following scheme: Add an extra column to the operations table. Enter in this the sum of the entries in each row of the operations table. Add to the relaxation table a column labeled "Sum residual." The initial value to be placed in this column is the sum of the initial residuals. Now, as the relaxation proceeds, the alteration to the sum residual is computed from the additional column in the unit-operations table exactly the same as for the other residuals. At each stage a check is obtained by comparing the sum of the actual residuals with the entry in the sum-residual column.

1-35. Solve the nonlinear system (1-34) by relaxation.

1-36. Outline the relaxation method for solving a nonlinear system of one degree of freedom.

$$a(x) = c$$

where $a(x)$ is a nonlinear function of x and c is a constant. Compare this procedure with *Newton's method*, which consists in constructing a sequence of approximations according to the following rule:

$$x^{(r+1)} = x^{(r)} - \frac{a(x^{(r)}) - c}{(da/dx)_{x=x^{(r)}}}$$

In terms of this example what is the basis for the suggestion of the text that once a rough solution has been obtained further accuracy can be obtained by using the same operations table provided the true residual is occasionally recomputed directly from its definition?

1-8. Iteration Combined with Elimination

When an ill-conditioned linear system is solved by elimination, the round-off errors may accumulate in such a fashion as to produce considerable error in the solution. This can be remedied by carrying several extra decimal places throughout the computation or by combining iteration with an elimination solution of limited accuracy. We illustrate this latter process for the ill-conditioned system of Exercise 1-8.

In Fig. 1-21 the complete solution by the Banachiewicz-Crout algorithm appears to the left of the double line. To accentuate the role of round-off error, we have deliberately worked to only two decimal places. The solution labeled $x_j^{(0)}$ differs from the true solution (1, 1, 1) by as much as 16 per cent.

In columns (1) and (2) on the right of the double line in Fig. 1-21 we show the calculations for two steps of iteration. We begin by computing the residuals

$$r_j^{(1)} = c_j - \sum_{k=1}^{n} a_{jk}x_k^{(0)} \tag{1-111}$$

which are entered at the top of column (1). (The fact that these residuals are small in comparison with the solution errors indicates that the system is ill-conditioned.) We then treat the residuals as new right-hand sides of the original system and ask for the corresponding solution, $\varphi_j^{(1)}$,

Given Array				Residuals	
a_{jk}			c_j	$r_j^{(1)}$	$r_j^{(2)}$
9	9	8	26	0.02	0.006
9	8	7	24	0.02	0.006
8	7	6	21	0.00	0.004

Auxiliary Array				Auxiliary Columns for Residuals	
9.00	1.00	0.89	2.88	0.002	0.0007
9.00	−1.00	1.00	2.00	0.000	0.0000
8.00	−1.00	−0.12	0.84	0.148	0.0113

Solution				Corrections	
			$x_j^{(0)}$	$\varphi_j^{(1)}$	$\varphi_j^{(2)}$
$x_1 \rightarrow$			0.98	0.018	0.0019
$x_2 \rightarrow$			1.16	−0.148	−0.0113
$x_3 \rightarrow$			0.84	0.148	0.0113

			Iterated Solutions	
			$x_j^{(1)}$	$x_j^{(2)}$
$x_1 \rightarrow$			0.998	0.9999
$x_2 \rightarrow$			1.012	1.0007
$x_3 \rightarrow$			0.988	0.9993

Fig. 1-21. Solution of Exercise 1-8 by elimination combined with iteration.

which we shall consider as a correction to our initial solution. To obtain the solution $\varphi_j^{(1)}$ from the $r_j^{(1)}$, we use the auxiliary array previously obtained in computing $x_j^{(0)}$ so that it is necessary to compute only a single new column in the auxiliary array according to the Banachiewicz-Crout algorithm. When the $\varphi_j^{(1)}$ have been obtained, they are added to the $x_j^{(0)}$ to obtain the first iterated solution $x_j^{(1)}$.

The process is repeated in column (2). The $r_j^{(2)}$ are the residuals corresponding to the $x_j^{(1)}$, and the $\varphi_j^{(2)}$ are the solutions to the original system (according to the elimination algorithm) when the $r_j^{(2)}$ are the right-hand members. When the $\varphi_j{}^{(2)}$ are added to the $x_j{}^{(1)}$, we obtain the second iterated solution $x_j{}^{(2)}$. Note that in two steps of iteration we have reduced the maximum solution error more than two-hundred-fold.

1-9. Procedures Applicable Directly to Physical Systems

We have thus far considered the formulation of an equilibrium problem in terms of a system of algebraic equations as being a necessary step in the over-all process of obtaining a solution. In some situations the problems can be so systematized that this step can be omitted; i.e., fixed numerical routines can be set up to handle the problem data without ever explicitly writing the equations which connect these data to the solution.

As an example, we may cite the *tabular methods*[1] of solving electrical-network problems. Tables are set up directly on the diagrams of networks, and by filling in these tables in a systematic manner the solutions are obtained. The routine has been carefully engineered to fit the capabilities of a hand computer using a slide rule. The basis for the tables is the solution of the network equations by elimination.

As another example, there is the method of *moment distribution*[2] for solving statically indeterminate beams and frames. The computation is performed directly on a sketch of a structure. After certain preliminary calculations the process consists in "balancing up" one joint at a time and "distributing" the unbalance to the adjacent joints. The procedure is mathematically equivalent to iteration by single steps applied to the equations for the deflection angles, although these equations are never written.

The relaxation method may also be applied directly to a physical system without formulating the governing equations. All that is needed for a relaxation process is a definition for residuals and an operations table. Obtaining these directly from the physical system is often as easy as formulating the governing equations. We illustrate this for the elastic spring system, Prob. 1-1.

Following Southwell,[3] let us imagine a pair of fictitious screw jacks, J_1 and J_2, attached to the moving carts as shown in Fig. 1-22. These screw jacks are called *constraints*. To visualize the relaxation procedure, imagine that before the application of the external loads, P_1 and P_2, the constraints are locked in the position $u_1 = u_2 = 0$, which leaves the elastic

[1] D. E. Richardson, "Electrical Network Calculations," D. Van Nostrand Company, Inc., New York, 1946.

[2] H. Cross, Analysis of Continuous Frames by Distributing Fixed-end Moments, *Trans. ASCE*, **96**, 1–10 (1932).

[3] It was this type of problem for which the relaxation method was first devised and which suggested most of the nomenclature of relaxation. See R. V. Southwell, Stress-calculation in Frameworks by the Method of "Systematic Relaxation of Constraints," pts. I and II, *Proc. Roy. Soc. (London)*, **A151**, 56–95 (1935).

system unstrained. Thus when the P_1 and P_2 are applied, they are taken entirely by the constraints J_1 and J_2. Now if we begin to crank down the jack J_2, increasing u_2, the springs 2, 3, and 4 will exert tensions which tend to relieve the force on the constraint. This part of the process is called *relaxing the constraint*. After J_2 has been relaxed a convenient amount, we lock it and turn to J_1. By turning the crank we alter u_1 and

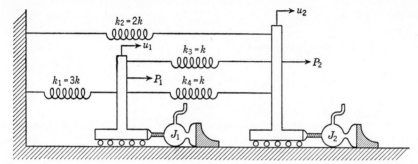

Fig. 1-22. Introduction of a pair of screw jacks as constraints into the system of Fig. 1-1.

transfer some of the load from the constraint to the system of springs. We continue in this manner easing off on the constraints until finally all the load is carried by the system and none by the constraints.

At any stage the *residuals* are the forces carried by the constraints. A *unit-operations table* can be constructed by considering the effect on these constraint forces of a unit displacement of one jack. For instance, if u_1 is increased by unity, the springs 1, 3, and 4 will exert a net force equal to $5k$ to the left on J_1 and the springs 3 and 4 will exert a force $2k$ to the right on J_2. This is the first line of Table 1-10. The reader should verify the second line. He should also verify that this table is the same as that derivable from the governing equations (1-5).

TABLE 1-10. UNIT-OPERATIONS TABLE OBTAINED FROM FIG. 1-22

Δu_1	Δu_2	Δr_1	Δr_2
1	0	$-5k$	$2k$
0	1	$2k$	$-4k$

EXERCISES

1-37. Give a physical basis for relaxation applied to the complementary treatment of the elastic spring system, Prob. 1-1. The system is cut in two places as shown in Fig. 1-23. The force pairs f_2 and f_3 are applied at the cuts. The residuals are the separation distances at the cuts. Obtain the initial residuals and the unit-operations table directly from the figure.   Compare with corresponding results obtained from (1-6).

1-38. Show how the residuals could be represented physically if we were to apply relaxation directly to the d-c network of Fig. 1-2.   Obtain the unit-operations table.

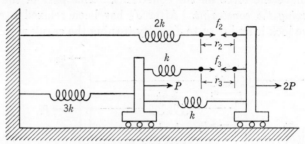

Fig. 1-23. Exercise 1-37.

Do this for the two complementary analyses, and compare the direct physical approach with results obtained from (1-11) and (1-12).

1-39. Repeat Exercise 1-38 for the hydraulic network, Prob. 1-5, comparing the physical approach with results obtained from (1-28) and (1-32).

1-40. Repeat Exercise 1-38 for the continuous beam, Prob. 1-4, comparing the physical approach with results obtained from (1-21) and (1-25).

CHAPTER 2

EIGENVALUE PROBLEMS FOR SYSTEMS
WITH A FINITE NUMBER OF DEGREES OF FREEDOM

Equilibrium problems involve the determination of system configurations under prescribed loading conditions. An eigenvalue problem may also involve the determination of system configurations, but of greater importance is the determination of the critical loading conditions under which these configurations are possible. A parameter which describes such a critical condition is called an eigenvalue. As examples we have the *natural frequencies* in oscillating systems and the *buckling loads* in elastic-stability problems.

We consider only *linear* eigenvalue problems. Matrix notation is used because it facilitates the theoretical discussion and because it provides a useful system for laying out the actual computations. The necessary rules are briefly reviewed in Sec. 2-2.

2-1. Particular Examples

Two examples are used to illustrate the formulation of eigenvalue problems from physical systems:

2-1. Three-mass vibrating system.

2-2. Buckling of a structure.

In both cases the formulations are left in ordinary algebraic form. Matrix formulations will be given in Sec. 2-2.

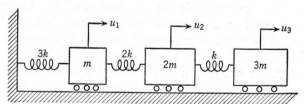

Fig. 2-1. Vibrational system with three degrees of freedom.

Problem 2-1. Three-mass Vibrating System. The system is shown in Fig. 2-1. The displacements of the three masses from the unstrained configuration are measured by u_1, u_2, and u_3. The equations of motion may be written by imagining the system disturbed from equilibrium and

61

applying Newton's second law to each mass. Neglecting friction, we obtain

$$-3ku_1 + 2k(u_2 - u_1) \qquad\qquad = m\frac{d^2u_1}{dt^2}$$

$$- 2k(u_2 - u_1) + k(u_3 - u_2) = 2m\frac{d^2u_2}{dt^2} \qquad (2\text{-}1)$$

$$- k(u_3 - u_2) = 3m\frac{d^2u_3}{dt^2}$$

For a *natural* vibration we would have

$$\begin{aligned} u_1 &= x_1 \sin{(\omega t + \varphi)} \\ u_2 &= x_2 \sin{(\omega t + \varphi)} \\ u_3 &= x_3 \sin{(\omega t + \varphi)} \end{aligned} \qquad (2\text{-}2)$$

where x_1, x_2, and x_3 represent the amplitudes of vibration, ω is the natural frequency, and φ is a phase angle. If we substitute (2-2) into (2-1) and set

$$\frac{m\omega^2}{k} = \lambda \qquad (2\text{-}3)$$

we find the following equations as the conditions for determining the amplitudes and frequency:

$$\begin{aligned} 5x_1 - 2x_2 \qquad &= \lambda(x_1) \\ -2x_1 + 3x_2 - x_3 &= \lambda(2x_2) \\ - x_2 + x_3 &= \lambda(3x_3) \end{aligned} \qquad (2\text{-}4)$$

The parameter λ is a dimensionless measure of the frequency. An *eigenvalue* is a value of λ for which there are nonzero amplitudes which satisfy (2-4). A configuration of amplitudes which meets these requirements is called a *natural mode*. The corresponding frequency is called a *natural frequency*. A complete solution would involve finding all the natural frequencies and their associated natural modes. In technical problems it may not be of interest to obtain the complete solution. Sometimes only the lowest natural frequency is desired; sometimes just the lowest frequency and the corresponding mode or just the two lowest frequencies are desired.

Problem 2-2. Buckling of a Structure. A system of rigid weightless links hinged together and supported by springs is shown in Fig. 2-2a. In this position all three links are exactly vertical, and there is no force in any of the springs. We consider the stability of this system when subjected to a vertical load P applied at B. For small loads the three links will remain vertical, moving down as a unit against the springs k_1. For large loads the links will buckle; that is, B and C will undergo transverse displacements as shown in Fig. 2-2b. Our problem is to determine the stability limit for the vertical position. We want to know the value of P for which an equilibrium position with transverse displacements first becomes possible.

To obtain a quantitative analysis, we assume that the desired critical buckling load is holding the system in equilibrium and find the equi-

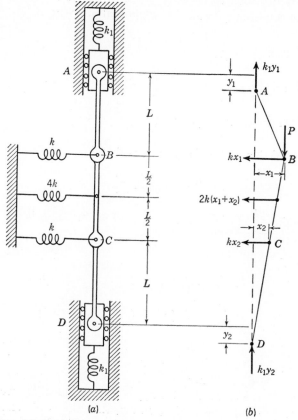

FIG. 2-2. Buckling of a system of spring-supported rigid links.

librium conditions by applying the principle of minimum potential energy. A geometrically compatible state can be represented by arbitrary (small) values of y_1, x_1, and x_2 if we take y_2 as

$$y_2 = y_1 - \frac{x_1{}^2}{2L} - \frac{(x_1 - x_2)^2}{2L} - \frac{x_2{}^2}{2L} \tag{2-5}$$

The usual small-angle approximations, $1 - \cos \theta \approx \frac{1}{2}\theta^2$ and $\sin \theta \approx \theta$, have been made here. By adding the strain energy of the springs to the potential energy of the load P we have the total potential energy

$$\Phi = \tfrac{1}{2}k_1y_1^2 + \tfrac{1}{2}kx_1^2 + \tfrac{1}{2}k(x_1 + x_2)^2 + \tfrac{1}{2}kx_2^2 + \tfrac{1}{2}k_1y_2^2 - P\left(y_1 - \frac{x_1^2}{2L}\right)$$
$$(2\text{-}6)$$

where y_2 is understood to take the value (2-5). The equilibrium equations are the conditions for stationary potential energy.

$$\frac{\partial \Phi}{\partial y_1} = k_1y_1 + k_1y_2 - P = 0$$

$$\frac{\partial \Phi}{\partial x_1} = kx_1 + k(x_1 + x_2) + k_1y_2\left(-\frac{x_1}{L} - \frac{x_1 - x_2}{L}\right) + \frac{Px_1}{L} = 0 \quad (2\text{-}7)$$

$$\frac{\partial \Phi}{\partial x_2} = k(x_1 + x_2) + kx_2 + k_1y_2\left(\frac{x_1 - x_2}{L} - \frac{x_2}{L}\right) = 0$$

One solution of this system is $x_1 = x_2 = 0$ and

$$y_2 = y_1 = \frac{P}{2k_1} \qquad (2\text{-}8)$$

which is obtained from (2-5) and the first of (2-7). This is the unbuckled equilibrium position.

If x_1 and x_2 do not vanish, we obtain

$$y_1 = \frac{P}{2k_1} + \frac{1}{4L}[x_1^2 + (x_1 - x_2)^2 + x_2^2]$$
$$(2\text{-}9)$$
$$y_2 = \frac{P}{2k_1} - \frac{1}{4L}[x_1^2 + (x_1 - x_2)^2 + x_2^2]$$

by solving (2-5) and the first of (2-7). We next insert the second of (2-9) into the last two relations of (2-7) to get a pair of simultaneous equations in x_1 and x_2. These equations contain linear and cubic terms. Since we are interested in the first appearance of buckling, we need consider only such small values of x_1 and x_2 that the cubic terms may be neglected in comparison with the linear terms. The linearized equations for x_1 and x_2 then appear as follows:

$$kx_1 + k(x_1 + x_2) + \frac{P}{2L}(-2x_1 + x_2) + \frac{Px_1}{L} = 0$$
$$(2\text{-}10)$$
$$k(x_1 + x_2) + kx_2 + \frac{P}{2L}(x_1 - 2x_2) = 0$$

By introducing the dimensionless parameter

$$\lambda = \frac{2kL}{P} \qquad (2\text{-}11)$$

we obtain

$$- \quad x_2 = \lambda(2x_1 + \quad x_2)$$
$$-x_1 + 2x_2 = \lambda(\quad x_1 + 2x_2)$$

(2-12)

as our formulation of the eigenvalue problem.

An *eigenvalue* is a value of λ for which the equations permit nonvanishing displacements. Such a configuration of displacements is called a *buckling mode*.

A complete solution of an eigenvalue problem involves finding all possible eigenvalues with their associated modes. In technical buckling problems a complete solution is not of interest. Very often the magnitude of the smallest buckling load is all that is required. Sometimes the corresponding buckling mode is of interest in order to assist in the design of stiffening reinforcement.

The present system has the interesting feature that if the load P is reversed (i.e., applied vertically upward) there is still the possibility of buckling. In such cases both the smallest positive and smallest negative buckling loads are of practical interest.

EXERCISES

2-1. Show that the eigenvalue problem for determining the natural frequencies and modes of torsional vibration of the system of Fig. 2-3 may be formulated as follows:

$$\begin{aligned}
x_1 - \quad x_2 \qquad\qquad\qquad &= \lambda x_1 \\
-x_1 + 2x_2 - \quad x_3 \qquad\qquad &= \lambda x_2 \\
- \quad x_2 + 2x_3 - \quad x_4 \qquad &= \lambda x_3 \\
- \quad x_3 + \tfrac{3}{2}x_4 - \tfrac{1}{2}x_5 &= \lambda x_4 \\
- \tfrac{1}{2}x_4 + \tfrac{1}{2}x_5 &= 4\lambda x_5
\end{aligned}$$

where $\lambda = \omega^2 J/k$ and k is the torsional stiffness of a shaft and J is the moment of

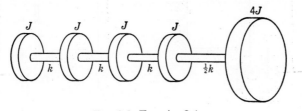

FIG. 2-3. Exercise 2-1.

inertia of a disk. The system is so supported on frictionless bearings that it is free to rotate without any bending of the shafts.

2-2. At resonance let the currents in Fig. 2-4 be

$$I_1 = x_1 \sin (\omega t + \varphi)$$
$$I_2 = x_2 \sin (\omega t + \varphi)$$

Show that the eigenvalue problem for determining the resonant frequencies and modes may be formulated as follows:

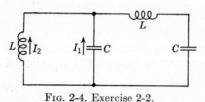

FIG. 2-4. Exercise 2-2.

$$2x_1 + x_2 = \lambda(x_1 + x_2)$$
$$x_1 + x_2 = \lambda(x_1 + 2x_2)$$

where $\lambda = \omega^2 LC$.

2-3. The system shown in Fig. 2-5 consists of a rocker arm hinged to rigid links which are spring-supported. Show that the eigenvalue problem for determining the buckling load P may be formulated as follows:

$$3x_1 - \tfrac{3}{2}x_2 \qquad = \lambda x_1$$
$$-\tfrac{3}{2}x_1 + 3x_2 \qquad = \lambda x_2$$
$$\qquad\qquad - 2x_3 = \lambda x_3$$

where $\lambda = kL/P$ and x_1, x_2, and x_3 are the transverse displacements at A, B, and C.

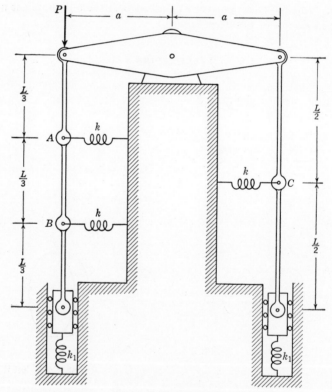

FIG. 2-5. Exercise 2-3.

2-4. The elastic properties of the system of Fig. 2-1 may be described by its *influence coefficients*. The influence coefficient α_{jk} is defined as the static deflection at j due to a unit force applied at k. Show that

$$\alpha_{11} = \frac{1}{3k} \qquad \alpha_{12} = \frac{1}{3k} \qquad \alpha_{13} = \frac{1}{3k}$$

$$\alpha_{21} = \frac{1}{3k} \qquad \alpha_{22} = \frac{5}{6k} \qquad \alpha_{23} = \frac{5}{6k}$$

$$\alpha_{31} = \frac{1}{3k} \qquad \alpha_{32} = \frac{5}{6k} \qquad \alpha_{33} = \frac{11}{6k}$$

Obtain the equations of motion by writing the displacement at j due to the *inertia forces* acting at 1, 2, and 3. In this way show that an alternative formulation of the eigenvalue problem (2-4) is

$$x_1 = \lambda(\tfrac{1}{3}x_1 + \tfrac{2}{3}x_2 + x_3)$$
$$x_2 = \lambda(\tfrac{1}{3}x_1 + \tfrac{5}{3}x_2 + \tfrac{5}{2}x_3)$$
$$x_3 = \lambda(\tfrac{1}{3}x_1 + \tfrac{5}{3}x_2 + \tfrac{11}{2}x_3)$$

2-5. Derive the equilibrium equations (2-7) by directly balancing forces in Fig. 2-2.

2-6. Consider the one-degree-of-freedom systems in Fig. 2-6. Show that the eigenvalue problem of determining the natural frequency in (a) is completely analogous

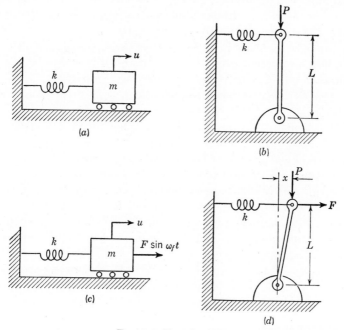

Fig. 2-6. Exercise 2-6.

to the eigenvalue problem of determining the buckling load in (b). Show that the equilibrium problem of determining the steady-state amplitude in (c) is completely analogous to the equilibrium problem of determining the (small) transverse displacement under F in (d). The two systems are not completely analogous when it comes to examining the stability of (c) and (d); i.e., there are stable steady motions in (c) when $\omega_f{}^2 > k/m$, but there is no stable equilibrium position (for small x) in (d) when $P > kL$. Where does the analogy break down?

2-2. Matrix Notation

Discussion of the eigenvalue problem in systems with a finite number of degrees of freedom is considerably simplified by the introduction of matrix notation.

We begin[1] with some definitions. A *matrix* is an array of *elements* arranged in rows and columns. For instance, a matrix of mn elements e_{jk} having m *rows* and n *columns* is shown below.

$$\begin{bmatrix} e_{11} & e_{12} & \cdots & e_{1n} \\ e_{21} & e_{22} & \cdots & e_{2n} \\ \cdot & \cdot & \cdots & \cdot \\ e_{m1} & e_{m2} & \cdots & e_{mn} \end{bmatrix} \qquad (2\text{-}13)$$

The elements may be a wide variety of things in general. The e_{jk} may represent numbers, functions, coded instructions for a computing machine, or even other matrices. In our applications, however, the elements will always be *real numbers*. Furthermore we shall only use *square* matrices $(m = n)$, *column* matrices $(n = 1)$, and *row* matrices $(m = 1)$.

If a square matrix has n rows and n columns, it is said to be of order n. The following special square matrices are often employed:

1. A *symmetric matrix* has elements which satisfy the condition that $e_{jk} = e_{kj}$.

2. If the elements of a matrix are the coefficients of a positive definite form, the matrix is said to be *positive definite*.

3. A *diagonal matrix* is one whose elements e_{jk} vanish except when $j = k$. These nonvanishing elements constitute the *main diagonal* of the matrix.

4. The *unit matrix* is a diagonal matrix whose main diagonal elements are ones. It will be denoted by the symbol I.

A matrix with one column and n rows is called a column matrix, or column *vector, of order* n. A matrix with one row and n columns is called a row matrix, or *row vector, of order* n.

The *transpose* of a matrix A is a matrix A' whose rows are the same as the columns of A. The transpose of a square matrix may be visualized as the matrix obtained by turning over the given matrix about its main diagonal as an axis. Note that a symmetric matrix is its own transpose. Our use of the transpose will be mainly limited to column vectors. The transpose of a column vector is a row vector.

[1] For more extensive treatment see R. A. Frazer, W. J. Duncan, and A. R. Collar, "Elementary Matrices and Some Applications to Dynamics and Differential Equations," Cambridge University Press, New York, 1938.

As illustrations we give in (2-14) some matrices of order 3 associated with the eigenvalue problem (2-4). The square matrix A contains the coefficients on the left side of (2-4). Note that A is symmetric (it is also positive definite; see Exercise 1-5).

$$A = \begin{bmatrix} 5 & -2 & 0 \\ -2 & 3 & -1 \\ 0 & -1 & 1 \end{bmatrix} \quad B = \begin{bmatrix} 1 & 0 & 0 \\ 0 & 2 & 0 \\ 0 & 0 & 3 \end{bmatrix}$$

$$I = \begin{bmatrix} 1 & 0 & 0 \\ 0 & 1 & 0 \\ 0 & 0 & 1 \end{bmatrix} \quad X = \begin{bmatrix} x_1 \\ x_2 \\ x_3 \end{bmatrix} \quad X' = [x_1 \ x_2 \ x_3] \tag{2-14}$$

The matrix B contains the coefficients of the right side of (2-4). It is a diagonal matrix. I is the unit matrix of order 3. X is a column matrix containing the three amplitudes of (2-4), and X' is its transpose.

We consider next the rules for manipulating these arrays of elements. Let two matrices which have the same shape and size be denoted by A and B with corresponding elements a_{jk} and b_{jk}. A and B are said to be *equal*, and we write $A = B$ if $a_{jk} = b_{jk}$ for every pair of corresponding elements. The *sum* of A and B is defined to be a matrix $(A + B)$ whose elements are $(a_{jk} + b_{jk})$. *The difference* of A and B is defined to be a matrix $(A - B)$ whose elements are $(a_{jk} - b_{jk})$. A matrix B is said to be the *product of a number c and a matrix A*, and we write $B = cA$ if $b_{jk} = ca_{jk}$. As an illustration of these operations the reader should verify the following matrix equation:

$$\begin{bmatrix} 3 & 2 \\ -1 & 4 \end{bmatrix} + \begin{bmatrix} -1 & 2 \\ 5 & 6 \end{bmatrix} = 2 \begin{bmatrix} 1 & 2 \\ 2 & 5 \end{bmatrix} \tag{2-15}$$

If A is a matrix of elements a_{jk} with m rows and p columns and B is a matrix of p rows and n columns, then their *matrix product* $C = AB$ is defined to be a matrix of elements c_{jk} with m rows and n columns, where

$$c_{jk} = \sum_{i=1}^{p} a_{ji} b_{ik} \tag{2-16}$$

It is important to note that matrix multiplication has been defined only when the number of columns of the first matrix is equal to the number of rows of the second matrix. There are four cases of matrix multiplication that we have to deal with.

1. *Product of a square matrix into a column vector.* If A is a square matrix of order n with elements a_{jk} and X is a column vector of order n with elements x_j, the matrix product AX is a column vector with ele-

ments as shown,

$$
\begin{bmatrix}
a_{11} & a_{12} & \cdots & a_{1n} \\
a_{21} & a_{22} & \cdots & a_{2n} \\
\cdots & \cdots & \cdots & \cdots \\
a_{n1} & a_{n2} & \cdots & a_{nn}
\end{bmatrix}
\begin{bmatrix}
x_1 \\ x_2 \\ \cdots \\ x_n
\end{bmatrix}
=
\begin{bmatrix}
a_{11}x_1 + a_{12}x_2 + \cdots + a_{1n}x_n \\
a_{21}x_1 + a_{22}x_2 + \cdots + a_{2n}x_n \\
\cdots \cdots \cdots \cdots \cdots \cdots \cdots \\
a_{n1}x_1 + a_{n2}x_2 + \cdots + a_{nn}x_n
\end{bmatrix}
$$

$$(2\text{-}17)$$

e.g.,

$$
\begin{bmatrix} 3 & 4 \\ -1 & 2 \end{bmatrix}
\begin{bmatrix} 5 \\ 3 \end{bmatrix}
=
\begin{bmatrix} 27 \\ 1 \end{bmatrix}
\qquad
\begin{bmatrix} 5 & -2 & 0 \\ -2 & 3 & -1 \\ 0 & -1 & 1 \end{bmatrix}
\begin{bmatrix} 3 \\ -2 \\ 1 \end{bmatrix}
=
\begin{bmatrix} 19 \\ -13 \\ 3 \end{bmatrix}
\qquad (2\text{-}18)
$$

This particular matrix product is a useful symbolism in dealing with simultaneous linear algebraic equations. We put the coefficients in a square matrix and the variables in a column matrix. The matrix product is then a convenient shorthand for writing out the combinations of the coefficients and the variables which customarily appear in the simultaneous equations.

2. *Product of a row vector into a square matrix.* The result in this case is a row vector, e.g.,

$$
[1, 2, 3]
\begin{bmatrix}
2 & -1 & 0 \\
3 & 1 & -2 \\
1 & 2 & 1
\end{bmatrix}
= [11, 7, -1]
\qquad (2\text{-}19)
$$

3. *Product of two square matrices.* The result is a square matrix, e.g.,

$$
\begin{bmatrix} 4 & 5 \\ -2 & 0 \end{bmatrix}
\begin{bmatrix} 1 & 2 \\ 3 & -2 \end{bmatrix}
=
\begin{bmatrix} 19 & -2 \\ -2 & -4 \end{bmatrix}
$$

$$
\begin{bmatrix} 1 & 2 \\ 3 & -2 \end{bmatrix}
\begin{bmatrix} 4 & 5 \\ -2 & 0 \end{bmatrix}
=
\begin{bmatrix} 0 & 5 \\ 16 & 15 \end{bmatrix}
$$

$$(2\text{-}20)$$

We note here that where AB and BA are both defined they are not equal. Matrix multiplication is not *commutative* in general.

4. *Product of a row vector into a column vector.* This kind of product of two vectors is called the *scalar product* because the result is a matrix of only one element which is taken to be a scalar; e.g.,

$$
[1, 2, 3]
\begin{bmatrix} 4 \\ -1 \\ 1 \end{bmatrix}
= [5] = 5
\qquad (2\text{-}21)
$$

The scalar product of a column vector by its transpose is equal to the sum of the squares of the elements, e.g.,

$$[x_1, \ x_2, \ x_3] \begin{bmatrix} x_1 \\ x_2 \\ x_3 \end{bmatrix} = x_1{}^2 + x_2{}^2 + x_3{}^2 \tag{2-22}$$

It can be shown that continued matrix products obey the associative law, i.e.,

$$(AB)C = A(BC) \tag{2-23}$$

We shall make considerable use of the triple matrix product where A is a row vector, B is a square matrix, and C is a column vector. The resulting product is a scalar, e.g.,

$$\begin{aligned} [1, \ -3] \begin{bmatrix} 5 & 2 \\ 2 & 1 \end{bmatrix} \begin{bmatrix} 3 \\ 1 \end{bmatrix} &= [1, \ -3] \begin{bmatrix} 17 \\ 7 \end{bmatrix} = -4 \\ &= [-1, \ -1] \begin{bmatrix} 3 \\ 1 \end{bmatrix} = -4 \end{aligned} \tag{2-24}$$

If the row vector is the transpose of the column vector and the square matrix is the matrix of a positive definite quadratic form, then the triple matrix product *is* the quadratic form, and hence necessarily positive; e.g.,

$$[x_1, \ x_2] \begin{bmatrix} 2 & 1 \\ 1 & 2 \end{bmatrix} \begin{bmatrix} x_1 \\ x_2 \end{bmatrix} = 2x_1{}^2 + 2x_1x_2 + 2x_2{}^2 = x_1{}^2 + (x_1 + x_2)^2 + x_2{}^2 \tag{2-25}$$

If X and Y are column vectors and B is a symmetric square matrix, then

$$X'BY = Y'BX \tag{2-26}$$

as the reader can easily show.[1]

A square matrix, A, multiplied by itself yields a square matrix which is denoted by A^2. If A^2 is in turn multiplied by A, the resulting matrix is $A^3 = A(A^2) = A^2(A)$. In this way positive integer powers of a square matrix are defined.

If, for a given square matrix A, a square matrix B can be found such that $BA = I$, where I is the unit matrix, then B is called the *inverse* of A and is denoted by A^{-1}. We discussed the problem of computing the inverse in Chap. 1. Matrix notation does not make the task of actually obtaining a numerical inverse any easier, but it does give an effective symbolism which is useful in general or theoretical discussions. For instance, the linear equilibrium problem of Chap. 1 may be briefly represented as

$$AX = C \tag{2-27}$$

[1] See Exercise 2-12.

where A is the square matrix of the coefficients of (1-36), X is the column matrix of the x_j, and C is the column matrix of the c_j. If we know the square matrix B which is the inverse of A, we can premultiply both sides of (2-27) by B to obtain

$$BAX = IX = X = BC = A^{-1}C \qquad (2\text{-}28)$$

which is the matrix equivalent of (1-45).

It was noted in Chap. 1 that a system can be inverted if the determinant of the coefficients is not zero. A square matrix of coefficients whose determinant is zero has no inverse and is said to be *singular*. If a symmetric matrix is nonsingular, it has an inverse which is also symmetric. A positive definite matrix can always be inverted, and the inverse matrix is itself positive definite.[1]

We conclude this section by recasting the formulation of Probs. 2-1 and 2-2 into matrix notation. Equations (2-4) for the three-mass vibrating system may be written as

$$\begin{bmatrix} 5 & -2 & 0 \\ -2 & 3 & -1 \\ 0 & -1 & 1 \end{bmatrix} \begin{bmatrix} x_1 \\ x_2 \\ x_3 \end{bmatrix} = \lambda \begin{bmatrix} 1 & 0 & 0 \\ 0 & 2 & 0 \\ 0 & 0 & 3 \end{bmatrix} \begin{bmatrix} x_1 \\ x_2 \\ x_3 \end{bmatrix} \qquad (2\text{-}29)$$

and Eqs. (2-12) for the buckling problem become

$$\begin{bmatrix} 0 & -1 \\ -1 & 2 \end{bmatrix} \begin{bmatrix} x_1 \\ x_2 \end{bmatrix} = \lambda \begin{bmatrix} 2 & 1 \\ 1 & 2 \end{bmatrix} \begin{bmatrix} x_1 \\ x_2 \end{bmatrix} \qquad (2\text{-}30)$$

EXERCISES

2-7. Write the formulations of Exercises 2-2 and 2-3 in matrix notation

2-8. Write the formulation of Exercise 2-1 in matrix notation. Show that the square matrix on the left is *singular*.

2-9. Show that the inverse of a diagonal matrix is also a diagonal matrix whose elements are the reciprocals of the elements of the given matrix.

2-10. The procedure of *iteration by total steps* given by (1-86) may be concisely described in matrix notation as follows: Let A be the square matrix of the coefficients, let D be the diagonal matrix whose elements are $a_{11}, a_{22}, \ldots, a_{nn}$, let C be the column matrix of the right-hand sides of the given equations, and let $X^{(r)}$ be the column matrix containing the rth iterates. Show that

$$X^{(r+1)} = D^{-1}C - (D^{-1}A - I)X^{(r)}$$

is the matrix representation of (1-86).

2-11. If X is a column matrix and B is a square matrix, show that $X'B'$ is a row matrix whose elements are the same as the column matrix BX, that is,

$$(X'B')' = BX$$

[1] See Exercise 2-13.

2-12. By using Exercise 2-11 show that if B is a symmetric matrix and X and Y are arbitrary column matrices then

$$X'BY = Y'BX$$

2-13. The matrix A is positive definite if $X'AX \geq 0$ for all possible columns X and the equality holds only when all the elements of X are zero. By considering the column Y obtained from $AX = Y$ show that $Y'A^{-1}Y \geq 0$ for arbitrary Y (equality holding only when all the elements of Y vanish) and hence that the inverse of a positive definite matrix is also positive definite.

2-14. Construct a square matrix from the influence coefficients given in Exercise 2-4. Show that this matrix (except for the factor k) is the inverse of the square matrix on the left of (2-29).

2-15. Count the number of simple multiplications required to multiply a square matrix of order n into a column matrix of order n. Show that n^3 operations are required to multiply two square matrices of order n and that a triple matrix product such as (2-26) requires $n^2 + n$ operations.

2-16. The following device is used to obtain a check on the arithmetic when the matrix product AB of two square matrices is computed: An additional column is placed on the right of B. The elements in this column are the sums of the elements in the corresponding rows of B. The matrix A is now multiplied into this column and the results placed in a column on the right of the matrix product AB. The elements in this column *should* equal the sums of the elements in the rows of AB. Show the basis for this.

2-3. Formulation of the General Problem

The formulations of (2-29) and (2-30) are both special cases of the following: Given square matrices A and B of order n, to find scalars λ and associated column matrices X of order n which satisfy

$$AX = \lambda BX \qquad (2\text{-}31)$$

Throughout the chapter we consider this problem under the restrictions that *both* A and B are *symmetric* and that at least one of them is *positive definite*. For the sake of uniformity we shall always consider B to be positive definite [if A is positive definite and B is not, we simply divide (2-31) by the scalar λ, call $1/\lambda$ the eigenvalue, and interchange the names of the square matrices].

The vast majority of eigenvalue problems which arise from discrete physical systems and from the application of approximate methods to continuous physical systems as in Chap. 5 meet these restrictive requirements. From the mathematical point of view these restrictions assure us that we have a well-behaved problem with certain fairly simple characteristic properties. We shall describe these properties in the following section.

We shall call (2-31) a *special* eigenvalue problem if B is a *diagonal matrix of positive elements*. One[1] of the numerical procedures to be described applies only to special eigenvalue problems.

[1] See Sec. 2-10.

A common preliminary step in many of the discussions which follow is to combine the two square matrices A and B into a single square matrix of coefficients. Since we have agreed that B is positive definite, it can always be inverted and (2-31) can be transformed into

$$
\begin{aligned}
B^{-1}AX &= \lambda B^{-1}BX = \lambda IX \\
HX &= \lambda X
\end{aligned}
\tag{2-32}
$$

where $H = B^{-1}A$. We may note that while this process is easily described in matrix notation it represents a considerable calculation when n is large and the matrices A and B do not have any special simplifying characteristics. Inverting B requires more than $\frac{1}{2}n^3$ multiplication operations by the algorithm of Sec. 1-5, and the matrix product $B^{-1}A$ requires n^3 operations. A combined algorithm[1] can be used to form H, which requires only somewhat more than $\frac{7}{6}n^3$ operations. If, however, B is a diagonal matrix, the inversion becomes trivial and only about n^2 operations are required to form H.

When A is nonsingular, it may be inverted and (2-31) can be transformed into

$$
\begin{aligned}
A^{-1}AX &= \lambda A^{-1}BX \\
X &= \lambda GX
\end{aligned}
\tag{2-33}
$$

where $G = A^{-1}B$. The matrices H and G in (2-32) and (2-33) will not usually be symmetric. Note that H is the inverse of G.

As an illustration of these transformations we consider the system (2-29) for the three-mass vibrating system. The matrix B is diagonal and can be inverted by inspection. We thus obtain

$$
\begin{bmatrix} 5 & -2 & 0 \\ -1 & \frac{3}{2} & -\frac{1}{2} \\ 0 & -\frac{1}{3} & \frac{1}{3} \end{bmatrix}
\begin{bmatrix} x_1 \\ x_2 \\ x_3 \end{bmatrix}
= \lambda
\begin{bmatrix} x_1 \\ x_2 \\ x_3 \end{bmatrix}
\tag{2-34}
$$

according to (2-32). On the other hand the matrix A is not immediately inverted, and it requires some calculation to find

$$
\begin{bmatrix} x_1 \\ x_2 \\ x_3 \end{bmatrix}
= \lambda
\begin{bmatrix} \frac{1}{3} & \frac{2}{3} & 1 \\ \frac{1}{3} & \frac{5}{3} & \frac{5}{2} \\ \frac{1}{3} & \frac{5}{3} & \frac{11}{2} \end{bmatrix}
\begin{bmatrix} x_1 \\ x_2 \\ x_3 \end{bmatrix}
\tag{2-35}
$$

according to (2-33). This formulation can be obtained with less trouble directly from the physical system by using the *influence coefficients*[2] to describe the elastic properties of the system. The reader may verify that the matrix H of (2-34) is the inverse of the matrix G of (2-35).

---

[1] See Exercise 2-36.

[2] See Exercise 2-4.

2-17. Show that the matrices H and G corresponding to the formulation (2-30) for the buckling problem are

$$H = \begin{bmatrix} \frac{1}{3} & -\frac{4}{3} \\ -\frac{2}{3} & \frac{5}{3} \end{bmatrix} \qquad G = \begin{bmatrix} -5 & -4 \\ -2 & -1 \end{bmatrix}$$

and verify that $H = G^{-1}$.

2-18. Find the matrix H for the system of Exercise 2-8. Show that the matrix G does not exist.

2-4. Mathematical Properties

The eigenvalue problem

$$AX = \lambda BX \qquad (2\text{-}36)$$

where A and B are symmetric matrices of order n and B is positive definite has a number of properties which can be guaranteed in advance. We proceed to review these properties, mostly without proof.[1]

First of all, solutions do *exist*. There is at least one value of λ for which a nonzero X satisfies (2-36). Note that if X satisfies (2-36) so does any scalar multiple of X; that is, the eigenvalue problem does not fix the absolute magnitude of the elements of X, but only their relative ratios. We call X an *eigenvector*, or *modal column*, or simply a *mode*, and any vector proportional to X is considered to be essentially the same mode. In numerical work it is often convenient to scale the vector X up until its largest element is unity. In theoretical work it is convenient to consider X to be so scaled that

$$X'BX = 1 \qquad (2\text{-}37)$$

A vector which satisfies (2-37) is said to be *normalized* with respect to B.

Second, the eigenvalue problem possesses exactly n distinct solutions,

$$\lambda_1, X_1; \qquad \lambda_2, X_2; \qquad \dots ; \qquad \lambda_n, X_n \qquad (2\text{-}38)$$

where the X_j are essentially different from one another. A quantitative measure of this differentness is the *orthogonality condition*,[2]

$$X'_j BX_k = 0 \qquad \text{if } j \neq k \qquad (2\text{-}39)$$

The corresponding eigenvalues λ_j are usually different from one another (the case of repeated eigenvalues is discussed below). The requirement

[1] See, for example, T. von Kármán and M. A. Biot, "Mathematical Methods in Engineering," McGraw-Hill Book Company, Inc., New York, 1940, pp. 162–210. Counterexamples showing behavior when the conditions on A and B are not met are given in Exercises 2-25 and 2-27.

[2] For a proof of this see Exercise 2-20.

that B be positive definite assures us that all the eigenvalues are *real*. If A is also positive definite, then all the eigenvalues are *positive*.

These general statements may be illustrated in terms of the eigenvalue problem (2-29) for the three-mass vibrating system. Here $n = 3$, and both A and B are positive definite. The complete solution (which we shall later obtain by various procedures) is

$$\lambda_1 = 0.1546 \qquad \lambda_2 = 1.1751 \qquad \lambda_3 = 5.5036$$

$$X_1 = \begin{bmatrix} 0.2213 \\ 0.5361 \\ 1.0000 \end{bmatrix} \qquad X_2 = \begin{bmatrix} 0.5229 \\ 1.0000 \\ -0.3960 \end{bmatrix} \qquad X_3 = \begin{bmatrix} 1.0000 \\ -0.2518 \\ 0.0162 \end{bmatrix} \qquad (2\text{-}40)$$

The eigenvalues are all positive and different from one another. Each of the modes has been scaled so that the largest element is unity. The modes are essentially different, as may be noted qualitatively by the sign patterns and by the positions of the largest elements. Quantitatively we have the orthogonality condition (2-39); e.g., when $j = 3$ and $k = 2$,

$$[1.0000, \; -0.2518, \; 0.0162] \begin{bmatrix} 1 & 0 & 0 \\ 0 & 2 & 0 \\ 0 & 0 & 3 \end{bmatrix} \begin{bmatrix} 0.5229 \\ 1.0000 \\ -0.3960 \end{bmatrix} = 0 \qquad (2\text{-}41)$$

The reader is urged to visualize the solution (2-40) in terms of the physical system of Fig. 2-1; e.g., the first mode associated with the lowest natural frequency $\omega_1 = \sqrt{0.1546k/m}$ is a mode in which all three masses swing simultaneously in the same direction, the right-hand mass moving through the largest excursion and the other two moving with excursions 22.13 and 53.61 per cent as large.

Exceptional Case: Repeated Eigenvalues. The description above must be modified slightly when not all n eigenvalues are different from one another. Let us first examine a simple physical example. Consider a shallow elliptical bowl. If a mass particle is allowed to slide back and forth without friction in the bottom of the bowl, the general motion has a complicated nonrepeating two-dimensional path. The eigenvalue problem for this system consists in finding the paths and the frequencies of back-and-forth motion for the natural modes of motion in which each back-and-forth excursion of the particle is always on the same path.

The answer to this problem is simple. The paths of the natural modes are along the major and minor axes of the ellipse. The natural frequencies depend on the radii of curvature of these paths, the lower frequency being associated with the larger radius, i.e., the flattest path. Here the eigenvalues (as measures of the natural frequencies) are different, and with each is associated a unique mode. The orthogonality of the modes shows up in the perpendicularity of the two paths.

Now imagine that the elliptical bowl is gradually transformed into a circular bowl. The two principal radii of curvature, and hence the eigenvalues, will approach one another. The directions of the paths of the natural modes will remain unchanged in the limiting process; however, in the circular bowl *any* straight-line path through the bottom of the bowl is equally well the path of a natural mode. Thus, when two eigenvalues coalesce, there is introduced a whole infinity of modes. We may still consider that there are two orthogonal modes with the understanding that when the eigenvalues are equal any linear combination of these modes is also a mode.

Returning to the general solution of the eigenvalue problem (2-38), with this convention we can still consider that there are n modes all orthogonal to one another and that with each mode is associated an eigenvalue. If the eigenvalues are all distinct, that is all that need be said. If, however, p of the eigenvalues are alike, we must admit that any linear combination[1] of the p corresponding modes is equally well a mode for this eigenvalue.

Expansion Theorem. The fact that the n eigenvectors or modal columns of an eigenvalue problem are orthogonal implies that they may be used as a base set of vectors for the decomposition of any arbitrary vector V of order of n. Thus if $X_1, \ldots, X_n$ are the modes of an eigenvalue problem, any vector V with elements $(v_1, \ldots, v_n)$ can be expanded in the form

$$V = c_1 X_1 + c_2 X_2 + \cdots + c_n X_n \tag{2-42}$$

It is left as an exercise[2] for the reader to show that the coefficients may be obtained from the following formula:

$$c_j = \frac{X_j' B \hat{V}}{X_j' B X_j} \tag{2-43}$$

Enclosure Theorem. Let the eigenvalue problem (2-31) be transformed into the form

$$HX = \lambda X \tag{2-44}$$

as indicated in (2-32). For an arbitrary vector V of order n we define a set of n ratios

$$l_j = \frac{j\text{th element of } HV}{j\text{th element of } V} \qquad j = 1, \ldots, n \tag{2-45}$$

where HV is the column matrix obtained after multiplying H into V. Notice that, if V were actually an eigenvector X_p, then by definition all the l_j would equal λ_p. For an arbitrary vector V let l_{max} and l_{min} be

---

[1] See Exercise 2-28.
[2] See Exercise 2-22.

the algebraically largest and smallest of the l_j. Then the statement that there is always a true eigenvalue λ satisfying

$$l_{min} \leq \lambda \leq l_{max} \tag{2-46}$$

is called an *enclosure theorem*. Such a theorem does *not* hold[1] for all systems of the form

$$AX = \lambda BX \tag{2-47}$$

but it is valid[2] when (2-47) is a *special* eigenvalue problem, i.e., when B is a diagonal matrix of positive elements.

Useful Relations. When the eigenvalue problems (2-47) have been reduced to the form (2-44), there are two useful relations connecting the eigenvalues and the elements of the matrix H. If the elements of H are denoted by h_{jk}, then[3]

$$\sum_{j=1}^{n} h_{jj} = \sum_{j=1}^{n} \lambda_j \tag{2-48}$$

$$\sum_{j=1}^{n} \sum_{k=1}^{n} h_{jk}h_{kj} = \sum_{j=1}^{n} \lambda_j{}^2 \tag{2-49}$$

The left side of (2-48) is the sum of the main diagonal elements of H, while the left side of (2-49) is the sum of the main diagonal elements of the matrix H^2. These relations are sometimes employed in checking. For instance, we may verify that the eigenvalues (2-40) and the matrix H of (2-34) satisfy (2-48) and (2-49) as follows:

$$
\begin{aligned}
5 + \tfrac{3}{2} + \tfrac{1}{3} &= 0.1546 + 1.1751 + 5.5036 = 6.833 \\
25 + (2)(2) + \tfrac{9}{4} + (2)(\tfrac{1}{6}) + \tfrac{1}{9} & \\
&= (0.1546)^2 + (1.1751)^2 + (5.5036)^2 = 31.6944
\end{aligned} \tag{2-50}
$$

The relation (2-48) is helpful in making a preliminary order-of-magnitude estimate of the eigenvalue locations. Thus for the three-mass vibrating system before making any computations it would be known that there were three positive eigenvalues whose sum was 6.833.

When the matrix A can be inverted and the formulation (2-33) obtained, then relations similar to (2-48) and (2-49) can be written[4] connecting the elements of G with the reciprocals of the eigenvalues.

---

[1] For a counterexample see Exercise 2-23.

[2] L. Collatz "Eigenwertaufgaben mit technischen Anwendung," Akademische Verlagsgesellschaft m.b.H., Leipzig, 1949, p. 289.

[3] For proof of these see C. B. Biezeno and R. Grammel, "Technische Dynamik," Springer-Verlag OHG, Berlin, 1939, p. 153.

[4] See Exercise 2-24.

EXERCISES

2-19. Show that the triple scalar product $X_1'BX_1$ for the vector X_1 of (2-40) has the value 3.6238 and hence that

$$X_1 = \begin{bmatrix} 0.1163 \\ 0.2816 \\ 0.5253 \end{bmatrix}$$

is the same mode *normalized* with respect to B.

2-20. Consider two different modes X_j and X_k of the eigenvalue problem (2-36), i.e.,

$$AX_j = \lambda_j BX_j$$
$$AX_k = \lambda_k BX_k$$

Premultiply the first of these by X_k' and the second by X_j'. Finally use (2-26) to prove the orthogonality condition (2-39).

2-21. Verify the following solution to the buckling problem (2-30):

$$\lambda_1 = 2.1547 \qquad\qquad \lambda_2 = -0.1547$$

$$X_1 = \begin{bmatrix} -0.7321 \\ 1.0000 \end{bmatrix} \qquad X_2 = \begin{bmatrix} 1.0000 \\ 0.3660 \end{bmatrix}$$

What is the significance of the negative eigenvalue? Show that X_1 and X_2 satisfy the orthogonality requirements.

2-22. Derive the formula (2-43) by premultiplying both sides of (2-42) by the row matrix $X_j'B$ and using the orthogonality condition. The reader familiar with Fourier series will note the similarity here to the technique used to evaluate the Fourier coefficients. Show that the coefficients for expanding the vector (2, 2, 1) in terms of the modes of (2-40) are 1.5416, 1.4060, and 0.9236.

2-23. Consider the eigenvalue problem

$$\begin{bmatrix} 2 & 1 \\ 1 & 1 \end{bmatrix}\begin{bmatrix} x_1 \\ x_2 \end{bmatrix} = \lambda \begin{bmatrix} 5 & 2 \\ 2 & 1 \end{bmatrix}\begin{bmatrix} x_1 \\ x_2 \end{bmatrix}$$

The matrices A and B are both positive definite. Verify that

$$H = \begin{bmatrix} 0 & -1 \\ 1 & 3 \end{bmatrix}$$

according to (2-32). Show that $l_1 = 1$, $l_2 = 2$ for the vector $(1, -1)$ according to (2-45) but that there is no eigenvalue *enclosed* by these values since the solution to this system is

$$\lambda_1 = 0.382 \qquad\qquad \lambda_2 = 2.618$$

$$X_1 = \begin{bmatrix} 1.000 \\ -0.382 \end{bmatrix} \qquad X_2 = \begin{bmatrix} -0.382 \\ 1.000 \end{bmatrix}$$

2-24. Verify the relations (2-48) and (2-49) for the buckling problem formulated in (2-30) (see Exercise 2-21 for the solution and Exercise 2-17 for the matrix H). Predict that

$$\frac{1}{\lambda_1} + \frac{1}{\lambda_2} = -6$$

by using the matrix G.

2-25. In the eigenvalue problem

$$\begin{bmatrix} 1 & -4 \\ 1 & 1 \end{bmatrix} \begin{bmatrix} x_1 \\ x_2 \end{bmatrix} = \lambda \begin{bmatrix} 1 & 0 \\ 0 & 1 \end{bmatrix} \begin{bmatrix} x_1 \\ x_2 \end{bmatrix}$$

the matrix B is symmetric and positive definite, but the matrix A is neither. Verify the following solution:

$$\lambda_1 = 1 + 2i \qquad \lambda_2 = 1 - 2i$$

$$X_1 = \begin{bmatrix} 1.0 \\ -0.5i \end{bmatrix} \qquad X_2 = \begin{bmatrix} 1.0 \\ 0.5i \end{bmatrix}$$

2-26. In the matrix formulation of Exercise 2-1 the matrix B is symmetric and positive definite, while the matrix A is symmetric and positive but *not* definite. A is singular. Verify that $X = (1, 1, 1, 1, 1)$ is a mode and that the corresponding eigenvalue is zero.

2-27. In the two-degree-of-freedom eigenvalue problem

$$\begin{bmatrix} 2 & 1 \\ -1 & 4 \end{bmatrix} \begin{bmatrix} x_1 \\ x_2 \end{bmatrix} = \lambda \begin{bmatrix} 1 & 0 \\ 0 & 1 \end{bmatrix} \begin{bmatrix} x_1 \\ x_2 \end{bmatrix}$$

the matrix A is positive definite but not symmetric. There is only a *single* eigenvalue $\lambda = 3$ and a *single* mode $X = (1, 1)$. This is an example of a system with a *quadratic divisor*, whereas the systems treated in this chapter all have linear divisors. For an introduction to this theory see, for example, H. W. Turnbull and A. C. Aitken, "Theory of Canonical Matrices," Blackie & Son, Ltd., Glasgow, 1932.

2-28. The eigenvalue problem

$$\begin{bmatrix} 3 & 0 & 2 \\ 0 & 5 & 0 \\ 2 & 0 & 3 \end{bmatrix} \begin{bmatrix} x_1 \\ x_2 \\ x_3 \end{bmatrix} = \lambda \begin{bmatrix} x_1 \\ x_2 \\ x_3 \end{bmatrix}$$

has the solution

$$\lambda_1 = 1 \qquad \lambda_2 = 5 \qquad \lambda_3 = 5$$

$$X_1 = \begin{bmatrix} 1 \\ 0 \\ -1 \end{bmatrix} \qquad X_2 = \begin{bmatrix} 1 \\ 0 \\ 1 \end{bmatrix} \qquad X_3 = \begin{bmatrix} 0 \\ 1 \\ 0 \end{bmatrix}$$

Verify that these modes are orthogonal with respect to $B = I$. Show that $c_2 X_2 + c_3 X_3$ for arbitrary values of c_2 and c_3 is also an eigenvector with $\lambda = 5$.

2-29. An alternate expansion theorem for the eigenvalue problem (2-36) states that an arbitrary vector V can be expanded in the form

$$V = c_1 B X_1 + c_2 B X_2 + \cdots + c_n B X_n$$

Derive a formula for evaluating the coefficients in this expansion.

2-5. An Extremum Principle for Eigenvalues

We first consider Probs. 2-1 and 2-2 and illustrate how an extremum principle for the eigenvalues can be deduced in these particular cases. This is followed by a general discussion of the principle and its applications.

Let us take the buckling problem first. The potential energy (2-6) from which we originally derived the eigenvalue problem was a function of y_1 and y_2 as well as x_1 and x_2. If we restrict ourselves to the neighborhood of very small x_1 and x_2, we can eliminate y_1 and y_2 from (2-6) by using (2-9) and neglecting fourth powers in comparison with second powers. In this way we obtain

$$\Phi = -\frac{P^2}{4k_1} + \tfrac{1}{2}k(2x_1{}^2 + 2x_1x_2 + 2x_2{}^2) - \frac{P}{4L}(-2x_1x_2 + 2x_2{}^2) \quad (2\text{-}51)$$

as the total potential energy. The reader may verify that Eqs. (2-12) follow from this on setting the partial derivatives of Φ equal to zero. Let us instead see what can be learned directly from (2-51). Introducing the dimensionless parameter λ of (2-11), we rewrite (2-51) as follows:

$$\lambda(2x_1{}^2 + 2x_1x_2 + 2x_2{}^2) - (-2x_1x_2 + 2x_2{}^2) = \frac{4L}{P}\left(\Phi + \frac{P^2}{4k_1}\right) \quad (2\text{-}52)$$

We are now ready to make two deductions. Using only the fact that Φ is stationary for equilibrium positions, we first evaluate the left side of (2-52) as follows: If (x_1, x_2) is a buckling mode and P is the corresponding buckling force, then the right side of (2-52) is stationary for small variations in x_1 and x_2. Consider now the special type of variation in which we move from (x_1, x_2) to (cx_1, cx_2), where c is a parameter which is to be thought of as gradually decreasing from unity to zero. For every value of c we have an equilibrium position, and hence Φ remains stationary for an infinitesimal displacement. But if we think of the total displacement between the positions $c = 1$ and $c = 0$ as the integral of such infinitesimal displacements, we conclude that the right side of (2-52) has the same value for a buckling mode (x_1, x_2) as it does for $(0, 0)$. Now the left side of (2-52) is clearly zero when $x_1 = x_2 = 0$ and hence also zero for a buckling mode. We then have

$$\lambda = \frac{-2x_1x_2 + 2x_2{}^2}{2x_1{}^2 + 2x_1x_2 + 2x_2{}^2} \quad (2\text{-}53)$$

by solving for λ. Thus our first deduction gives us a formula for obtaining the eigenvalue when the buckling mode is known.

Our second deduction is that since (2-52) is stationary for small displacements from equilibrium so also is (2-53). When x_1 and x_2 differ from a buckling mode by infinitesimals of first order, (2-52) differs from zero by an infinitesimal of second order and hence on solving for λ under the assumption that (2-52) is exactly zero we make an error in (2-53) of second order. Note that both the above deductions are based on the stationary property of Φ. In the first case we consider variations from one equilibrium position to another, while in the second we consider variations from an equilibrium position to neighboring nonequilibrium positions.

By using the matrices defined in (2-30) we can write the quotient (2-53) in the following form:

$$\lambda = \frac{[x_1 x_2] \begin{bmatrix} 0 & -1 \\ -1 & 2 \end{bmatrix} \begin{bmatrix} x_1 \\ x_2 \end{bmatrix}}{[x_1 x_2] \begin{bmatrix} 2 & 1 \\ 1 & 2 \end{bmatrix} \begin{bmatrix} x_1 \\ x_2 \end{bmatrix}} = \frac{X'AX}{X'BX} \tag{2-54}$$

Vibratory System. We shall briefly apply Hamilton's[1] principle to the three-mass system of Prob. 2-1. Hamilton's principle states that of all geometrically compatible motions of a conservative dynamical system between fixed positions at the beginning and end of an interval the motions which also satisfy Newton's laws give a stationary value to

$$\Phi = \int_{t_1}^{t_2} (\text{KE} - \text{PE}) \, dt \tag{2-55}$$

We consider the system of Fig. 2-1 making a natural vibration (2-2). Then, if we take the time interval $t_2 - t_1$ to be $2\pi/\omega$, the position of the system at the beginning and end of this interval can be $u_1 = u_2 = u_3 = 0$. Now

$$\begin{aligned} \text{KE} &= \tfrac{1}{2} m\omega^2 [x_1^2 + 2x_2^2 + 3x_3^2] \cos^2 (\omega t + \varphi) \\ \text{PE} &= \tfrac{1}{2} k [3x_1^2 + 2(x_2 - x_1)^2 + (x_3 - x_2)^2] \sin^2 (\omega t + \varphi) \end{aligned} \tag{2-56}$$

and by carrying out the integration of (2-55) we obtain

$$\Phi = \frac{K\pi}{2\omega} \{\lambda[x_1^2 + 2x_2^2 + 3x_3^2] - [3x_1^2 + 2(x_2 - x_1)^2 + (x_3 - x_2)^2]\} \tag{2-57}$$

where $\lambda = m\omega^2/k$ as before. The same argument as used in the foregoing example leads to the result that

$$\lambda = \frac{3x_1^2 + 2(x_2 - x_1)^2 + (x_3 - x_2)^2}{x_1^2 + 2x_2^2 + 3x_3^2} \tag{2-58}$$

when (x_1, x_2, x_3) is a natural mode and that (2-58) is stationary with respect to small variations in x_1, x_2, and x_3 in the neighborhood of a

---

[1] See, for example, S. Timoshenko and D. H. Young, "Advanced Dynamics," McGraw-Hill Book Company, Inc., New York, 1948, p. 230.

natural mode.   This quotient may be written as

$$\lambda = \frac{[x_1,\ x_2,\ x_3]\begin{bmatrix} 5 & -2 & 0 \\ -2 & 3 & -1 \\ 0 & -1 & 1 \end{bmatrix}\begin{bmatrix} x_1 \\ x_2 \\ x_3 \end{bmatrix}}{[x_1 x_2 x_3]\begin{bmatrix} 1 & 0 & 0 \\ 0 & 2 & 0 \\ 0 & 0 & 3 \end{bmatrix}\begin{bmatrix} x_1 \\ x_2 \\ x_3 \end{bmatrix}} = \frac{X'AX}{X'BX} \tag{2-59}$$

by using the matrices defined in (2-29).

Rayleigh's Quotient.   In the general eigenvalue problem of this chapter

$$AX = \lambda BX \tag{2-60}$$

we can premultiply both sides by X' and divide to obtain

$$\lambda = \frac{X'AX}{X'BX} \tag{2-61}$$

If V is an arbitrary nonvanishing vector, we call the corresponding expression *Rayleigh's quotient*[1] and denote it by the symbol λ_R,

$$\lambda_R = \frac{V'AV}{V'BV} \tag{2-62}$$

Let us consider the properties of λ_R as V ranges over all possible vectors with n elements.   From (2-61) we see that if V is actually an eigenvector then λ_R is the corresponding eigenvalue.   We next show that λ_R is stationary when V varies in the neighborhood of an eigenvector.   To obtain this result without appealing to a physical extremum principle (as we did in the foregoing examples), we use the expansion theorem and the orthogonality property.

Let X_j ($j = 1, \ldots, n$) be the eigenvectors of (2-60) *normalized* with respect to B; then according to (2-42) we can expand an arbitrary V in the form

$$V = \sum_{j=1}^{n} c_j X_j \tag{2-63}$$

We then evaluate

$$\lambda_R = \frac{(\Sigma c_j X_j') A (\Sigma c_j X_j)}{(\Sigma c_j X_j') B (\Sigma c_j X_j)} \tag{2-64}$$

[1] Rayleigh obtained the quotient by equating time integrals of energies such as those of (2-56). He demonstrated the stationary property in J. W. Strutt, Some General Theorems Relating to Vibrations, *Proc. London Math. Soc.*, **4**, 357–368 (1873). See also J. W. Strutt, 3d Baron Rayleigh, "The Theory of Sound," 2d ed., vol. **1**, Dover Publications, New York, 1945, p. 109.

by multiplying out numerator and denominator and using orthogonality and the fact that

$$X_j'AX_j = X_j'(\lambda_j BX_j) = \lambda_j X_j'BX_j = \lambda_j \qquad (2\text{-}65)$$

to obtain finally

$$\lambda_R = \frac{c_1{}^2\lambda_1 + c_2{}^2\lambda_2 + \cdots + c_n{}^2\lambda_n}{c_1{}^2 + c_2{}^2 + \cdots + c_n{}^2} \qquad (2\text{-}66)$$

This states that λ_R is a weighted average of the true eigenvalues where the weighting factors are the squares of the coefficients in the expansion (2-63). If V differs only slightly from a mode X_p, then c_p will be much larger than the other c_j. If

$$\left|\frac{c_j}{c_p}\right| < \epsilon \ll 1 \qquad j \neq p \qquad (2\text{-}67)$$

we may say that V is an approximation to X_p with error of order ϵ. Substituting in (2-66), we have

$$\lambda_R = \lambda_p \frac{1 + (c_1/c_p)^2(\lambda_1/\lambda_p) + \cdots + (c_n/c_p)^2(\lambda_n/\lambda_p)}{1 + (c_1/c_p)^2 + \cdots + (c_n/c_p)^2}$$
$$= \lambda_p[1 + O(\epsilon^2)] \qquad (2\text{-}68)$$

where $O(\epsilon^2)$ stands for an expression whose order of magnitude is ϵ^2 in the sense that as $\epsilon \to 0$ the expression remains smaller than a fixed multiple of ϵ^2. Thus λ_R approximates λ_p with error of order ϵ^2, or alternatively λ_R changes only by an infinitesimal of second order when V moves in an infinitesimal neighborhood of a true mode. This constitutes a proof of the extremum property of Rayleigh's quotient.

Applications. Several of the procedures for solving eigenvalue problems are essentially methods of successive improvement of approximate *modes*. If the corresponding *eigenvalue* is required, Rayleigh's quotient may be used to obtain an approximation of high accuracy. If the elements of an approximate mode are correct to k decimal places, then λ_R can be expected to provide about $2k$ correct decimal places in the corresponding eigenvalue. Note that $2n(n + 1)$ multiplications and one division are required to construct λ_R from a given V.

A common engineering application of Rayleigh's quotient involves direct calculation of λ_R from a trial mode based on physical insight. As an illustration we might guess that the first mode of the three-mass vibrating system of Prob. 2-1 was $(\frac{1}{4}, \frac{1}{2}, 1)$. Inserting this trial in (2-59), we compute $\lambda_R = 0.1579$, which is only about 2 per cent in error, whereas the elements of our trial mode differ from those of X_1 of (2-40) by 13 and 7 per cent.

According to (2-66) λ_R is a weighted average of the n true eigenvalues. This means that λ_R can never be smaller than the smallest eigenvalue or larger than the largest (speaking algebraically). This fact is often used to obtain a one-sided bound for the smallest or largest eigenvalue.

EXERCISES

2-30. Consider the surface defined by (2-52) when $4L/P(\Phi + P^2/4k_1)$ is plotted against x_1 and x_2. Show that, (a) when λ is large and positive (P is small and positive), the surface is an elliptic paraboloid opening upward with vertex at the origin; (b) when $\lambda = \lambda_1 = 2.1547$, the surface is a parabolic cylinder opening upward with the line $x_1 + 0.7321x_2 = 0$ as the generator through the vertex; (c) when $\lambda_2 < \lambda < \lambda_1$, the surface is a hyperbolic paraboloid (saddle surface); (d) when $\lambda = \lambda_2 = -0.1547$, the surface is a parabolic cylinder opening downward with the line $-0.3660x_1 + x_2 = 0$ as the generator through the vertex. Sketch a few contour lines in each case, and correlate these results with the elastic-stability problem.

2-31. Show that the quotient (2-58) can be obtained directly by equating the maximum kinetic energy in a principal vibration with the maximum potential energy.

2-32. Estimate the natural mode of vibration for the lowest nonzero natural frequency of the system of Exercise 2-1. Compute λ_R. The true value is $\lambda = 0.1585$. Note that since $\lambda = 0$ is an eigenvalue, λ_R for this mode may be either too high or too low.

2-33. If the matrix B is a diagonal matrix of positive elements show that λ_R may be considered as a weighted average of the ratios l_j of (2-45) where the weighting factors are $b_{jj}v_j^2$.

$$\lambda_R = \frac{l_1b_{11}v_1^2 + l_2b_{22}v_2^2 + \cdots + l_nb_{nn}v_n^2}{b_{11}v_1^2 + b_{22}v_2^2 + \cdots + b_{nn}v_n^2}$$

2-34. When the matrix A can be inverted and the matrix G constructed according to (2-33), demonstrate that the quotient

$$\frac{V'B(GV)}{V'BV}$$

can be put in a form similar to (2-66) to show that it is a weighted average of the reciprocals of the λ_j. Verify that, when V is $(\frac{1}{4}, \frac{1}{2}, 1)$ for the three-mass vibrating system of Prob. 2-1, this quotient produces the approximation 0.15475 for λ_1. Why is this more accurate than λ_R? Which quotient would give the better approximation to λ_3 if V were an approximation to X_3?

2-6. Direct Methods of Solution

The classical method of solving the eigenvalue problem

$$AX = \lambda BX \tag{2-69}$$

is to reason that the system of simultaneous equations

$$(A - \lambda B)X = 0 \tag{2-70}$$

can have solutions for $(x_1, \ldots, x_n)$ other than zero only if the determinant of the coefficients vanishes.

$$\begin{vmatrix} a_{11} - \lambda b_{11} & a_{12} - \lambda b_{12} & \cdots & a_{1n} - \lambda b_{1n} \\ a_{21} - \lambda b_{21} & a_{22} - \lambda b_{22} & \cdots & a_{2n} - \lambda b_{2n} \\ \vdots & & & \vdots \\ a_{n1} - \lambda b_{n1} & a_{n2} - \lambda b_{n2} & \cdots & a_{nn} - \lambda b_{nn} \end{vmatrix} = 0 \tag{2-71}$$

When the determinant in (2-71) is expanded, we have an algebraic equation of nth degree for λ. This equation is called the *characteristic equation* of the system. The n roots of this equation are the n eigenvalues of (2-69). The mode corresponding to λ_j can be found by inserting λ_j in (2-70) and solving for the ratios of the elements in X_j. A practical way to do this is to set x_n, say, equal to unity and solve the first $n - 1$ equations for $x_1, \ldots, x_{n-1}$. The last equation may be used as a check.

We call the process just outlined a *direct* method even though there is one step, finding the roots of an nth-degree polynomial, which cannot be performed directly. When n is greater than 3, there is considerable labor involved, not only in finding the roots but in expanding the determinant in (2-71) and in solving for the modes from (2-70) after the roots have been found. Nevertheless, when a complete solution for all modes is required, it appears that this process is probably as short as any, particularly if high accuracy is desired.

A good deal of effort has been expended in "engineering" the above procedure into an efficient computational algorithm.[1] In many of the methods proposed the number of multiplication operations required (exclusive of finding the roots of the characteristic equation) is proportional to n^4. The method about to be described is one of the few schemes in which the number of operations is proportional to n^3. It is especially designed to make use of a desk calculating machine which can construct a sum of products followed by a final division as in (1-66), and it requires a minimum of recording of intermediate results.

A preliminary step before applying this algorithm is to transform the problem (2-69) into the form

$$HX = \lambda X \tag{2-72}$$

according to (2-32). This requires somewhat more than $\frac{7}{6} n^3$ operations (see Exercise 2-36), but the total computation appears to be shorter[2] this way than by any method which begins to expand (2-71) immediately.

Hessenberg Algorithm. Only a description of the computational rules[3] and an example are given here. The first step is to obtain the equation whose roots are the eigenvalues. The computation is shown schematically in Fig. 2-7 for $n = 4$. The pattern of operations can be easily recognized by tracing through the steps required to construct this table. Exactly the same pattern is followed for any other value of n.

The coefficients h_{jk} are entered at the upper left in Fig. 2-7, and the

[1] For a survey of some of the methods see H. Wayland, Expansion of Determinantal Equations into Polynomial Form, *Quart. Appl. Math.*, **2**, 277–306 (1945), and A. S. Householder, "Principles of Numerical Analysis," McGraw-Hill Book Company, Inc., New York, 1953, p. 166.

[2] See Wayland, *op. cit.*

[3] The algorithm was given by K. Hessenberg in his doctoral dissertation at Darmstadt Technische Hochschule in 1941. For a complete description and proof see R. Zurmühl, "Praktische Mathematik für Ingenieure und Physiker," Springer-Verlag OHG, Berlin, 1953, p. 136. See also Exercises 2-37 to 2-39.

several zeros and ones are entered as shown, blocking in the spaces for the entries p_{jk}, z_{jk}, and f_{jk}, which are to be filled in as the computation proceeds. The sequence that we follow is indicated by the arrows.

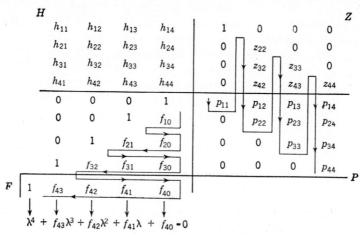

Fig. 2-7. Construction of the equation for the eigenvalues by the Hessenberg algorithm.

Starting with p_{11}, we fill in the columns at the right and then go to the rows of f_{jk} until finally, with the entries in the bottom row, we have the desired coefficients in the equation for the eigenvalues. The computation for the p_{jk} and z_{jk} is shown below.

$$p_{11} = -(h_{11} \cdot 1 + h_{12} \cdot 0 + h_{13} \cdot 0 + h_{14} \cdot 0)/1$$

$$z_{22} = h_{21} \cdot 1 + h_{22} \cdot 0 + h_{23} \cdot 0 + h_{24} \cdot 0 + 0 \cdot p_{11}$$
$$z_{32} = h_{31} \cdot 1 + h_{32} \cdot 0 + h_{33} \cdot 0 + h_{34} \cdot 0 + 0 \cdot p_{11}$$
$$z_{42} = h_{41} \cdot 1 + h_{42} \cdot 0 + h_{43} \cdot 0 + h_{44} \cdot 0 + 0 \cdot p_{11}$$
$$p_{12} = -(h_{11} \cdot 0 + h_{12}z_{22} + h_{13}z_{32} + h_{14}z_{42})/1$$
$$p_{22} = -(h_{21} \cdot 0 + h_{22}z_{22} + h_{23}z_{32} + h_{24}z_{42} + 0 \cdot p_{12})/z_{22}$$

$$z_{33} = h_{31} \cdot 0 + h_{32}z_{22} + h_{33}z_{32} + h_{34}z_{42} + 0 \cdot p_{12} + z_{32}p_{22}$$
$$z_{43} = h_{41} \cdot 0 + h_{42}z_{22} + h_{43}z_{32} + h_{44}z_{42} + 0 \cdot p_{12} + z_{42}p_{22} \qquad (2\text{-}73)$$
$$p_{13} = -(h_{11} \cdot 0 + h_{12} \cdot 0 + h_{13}z_{33} + h_{14}z_{43})/1$$
$$p_{23} = -(h_{21} \cdot 0 + h_{22} \cdot 0 + h_{23}z_{33} + h_{24}z_{43} + 0 \cdot p_{13})/z_{22}$$
$$p_{33} = -(h_{31} \cdot 0 + h_{32} \cdot 0 + h_{33}z_{33} + h_{34}z_{43} + 0 \cdot p_{13} + z_{32}p_{23})/z_{33}$$

$$z_{44} = h_{41} \cdot 0 + h_{42} \cdot 0 + h_{43}z_{33} + h_{44}z_{43} + 0 \cdot p_{13} + z_{42}p_{23} + z_{43}p_{33}$$
$$p_{14} = -(h_{11} \cdot 0 + h_{12} \cdot 0 + h_{13} \cdot 0 + h_{14}z_{44})/1$$
$$p_{24} = -(h_{21} \cdot 0 + h_{22} \cdot 0 + h_{23} \cdot 0 + h_{24}z_{44} + 0 \cdot p_{14})/z_{22}$$
$$p_{34} = -(h_{31} \cdot 0 + h_{32} \cdot 0 + h_{33} \cdot 0 + h_{34}z_{44} + 0 \cdot p_{14} + z_{32}p_{24})/z_{33}$$
$$p_{44} = -(h_{41} \cdot 0 + h_{42} \cdot 0 + h_{43} \cdot 0 + h_{44}z_{44} + 0 \cdot p_{14} + z_{42}p_{24} + z_{43}p_{34})/z_{44}$$

Note the sign pattern and the final divisions for the p_{jk}. The above operations may be visualized as row by column matrix multiplication where the rows extend across the h_{jk} and the z_{jk} and the columns contain the z_{jk} and the p_{jk}. The zeros and ones have been deliberately left in (2-73) so as to emphasize this. Note that in computing p_{jk} the j row and k column are multiplied but that in computing z_{jk} the j row and $k - 1$ column are used. The reader should follow this carefully until he recognizes the pattern.

The f_{jk} in Fig. 2-7 are computed as follows:

$$f_{10} = 1 \cdot p_{11}$$

$$f_{20} = 1 \cdot p_{12} + f_{10}p_{22}$$
$$f_{21} = (0 \cdot p_{12} + 1 \cdot p_{22}) + f_{10}$$

$$f_{30} = 1 \cdot p_{13} + f_{10}p_{23} + f_{20}p_{33}$$
$$f_{31} = (0 \cdot p_{13} + 1 \cdot p_{23} + f_{21}p_{33}) + f_{20} \qquad (2\text{-}74)$$
$$f_{32} = (0 \cdot p_{13} + 0 \cdot p_{23} + 1 \cdot p_{33}) + f_{21}$$

$$f_{40} = 1 \cdot p_{14} + f_{10}p_{24} + f_{20}p_{34} + f_{30}p_{44}$$
$$f_{41} = (0 \cdot p_{14} + 1 \cdot p_{24} + f_{21}p_{34} + f_{31}p_{44}) + f_{30}$$
$$f_{42} = (0 \cdot p_{14} + 0 \cdot p_{24} + 1 \cdot p_{34} + f_{32}p_{44}) + f_{31}$$
$$f_{43} = (0 \cdot p_{14} + 0 \cdot p_{24} + 0 \cdot p_{34} + 1 \cdot p_{44}) + f_{32}$$

Here again there is a simple pattern. Each f_{jk} contains a sum of the products of corresponding elements in the jth column of p's and the kth column of f's. When $k \neq 0$, the quantity $f_{j-1,k-1}$ is also added in.

The characteristic equation whose roots are the eigenvalues is then constructed as shown in Fig. 2-7. The process of actually obtaining the roots[1] is not considered a part of the algorithm. Once a root λ has been found, the corresponding mode X is found in two steps. We first solve for the vector Y from

$$\begin{bmatrix} p_{11} + \lambda & p_{12} & p_{13} & p_{14} \\ -1 & p_{22} + \lambda & p_{23} & p_{24} \\ 0 & -1 & p_{33} + \lambda & p_{34} \\ 0 & 0 & -1 & p_{44} + \lambda \end{bmatrix} \begin{bmatrix} y_1 \\ y_2 \\ y_3 \\ y_4 \end{bmatrix} = 0 \qquad (2\text{-}75)$$

by setting $y_4 = 1$ and solving for y_3 from the bottom row of (2-75), for y_2 from the next row, and for y_1 from the second row. The top row provides a check. Finally the mode X is obtained from Y by a matrix multiplication.

---

[1] For a survey of methods see Householder, *op. cit.*, p. 106. A common procedure is to obtain the roots one at a time by *Newton's method* (see Exercise 1-36).

$$
\begin{bmatrix} x_1 \\ x_2 \\ x_3 \\ x_4 \end{bmatrix} = \begin{bmatrix} 1 & 0 & 0 & 0 \\ 0 & z_{22} & 0 & 0 \\ 0 & z_{32} & z_{33} & 0 \\ 0 & z_{42} & z_{43} & z_{44} \end{bmatrix} \begin{bmatrix} y_1 \\ y_2 \\ y_3 \\ y_4 \end{bmatrix} \tag{2-76}
$$

The relation of the square matrices in (2-75) and (2-76) to the table in Fig. 2-7 should be noted.

Using the conventions of Sec. 1-5 to count operations, there are in general $n^3 + n$ operations required to obtain the characteristic equation as in Fig. 2-7. The computation of each mode according to (2-75) and (2-76) requires $n^2 - n + 1$ operations, and thus the complete solution for all modes (exclusive of obtaining the roots of the nth-degree polynomial) requires $2n^3 - n^2 + 2n$ operations.

Example. In Fig. 2-8 we show the solution of the three-mass vibrating system as formulated in (2-34). The polynomial for the roots is obtained from the table at the

$$
\begin{array}{ccc|ccc}
\multicolumn{3}{c}{H} & & Z & \\
5 & -2 & 0 & 1 & 0 & 0 \\
-1 & \frac{3}{2} & -\frac{1}{2} & 0 & -1 & 0 \\
0 & -\frac{1}{3} & \frac{1}{3} & 0 & 0 & \frac{1}{3} \\
\hline
0 & 0 & 1 & -5 & -2 & 0 \\
0 & 1 & -5 & 0 & -\frac{3}{2} & -\frac{1}{6} \\
1 & -\frac{13}{2} & \frac{11}{2} & 0 & 0 & -\frac{1}{3} \\
\end{array}
$$

$$
F \begin{array}{|ccc|} 1 & -\frac{41}{6} & \frac{15}{2} & -1 \end{array} \qquad P
$$

$$
\lambda^3 - \tfrac{41}{6}\lambda^2 + \tfrac{15}{2}\lambda - 1 = 0
$$

$$
\text{roots} \begin{cases} \lambda_1 = 0.1546 \\ \lambda_2 = 1.1751 \\ \lambda_3 = 5.5036 \end{cases}
$$

For $\lambda_1 = 0.1546$

$$
\begin{bmatrix} -4.8454 & -2 & 0 \\ -1 & -1.3454 & -0.1667 \\ 0 & -1 & -0.1787 \end{bmatrix} \begin{bmatrix} y_1 \\ y_2 \\ y_3 \end{bmatrix} = 0 \qquad \begin{array}{l} y_1 = 0.0737 \\ y_2 = -0.1787 \\ y_3 = 1.0000 \end{array}
$$

$$
X_1 = \begin{bmatrix} x_1 \\ x_2 \\ x_3 \end{bmatrix} = \begin{bmatrix} 1 & 0 & 0 \\ 0 & -1 & 0 \\ 0 & 0 & \frac{1}{3} \end{bmatrix} \begin{bmatrix} 0.0737 \\ -0.1787 \\ 1.0000 \end{bmatrix} = \begin{bmatrix} 0.0737 \\ 0.1787 \\ 0.3333 \end{bmatrix}
$$

FIG. 2-8. Solution of (2-34) for eigenvalues and first mode by Hessenberg algorithm.

top, using the formulas of (2-73) and (2-74). For the root λ_1 the construction of X_1 is shown according to (2-75) and (2-76). Except for the scale factor of $\frac{1}{3}$ this is the same mode as in (2-40).

Abnormal Case: Zero Diagonal Element. In the above algorithm the elements $(z_{11}, z_{21}, z_{31}, z_{41})$ in the first column of Z were taken as $(1, 0, 0, 0)$. The choice is arbitrary in that the method gives the same results for any set $(z_{11},\ z_{21},\ z_{31},\ z_{41})$ if the same row by column multiplication scheme as (2-73) is followed. The set $(1, 0, 0, 0)$ is chosen so as to reduce the number of multiplications. It may happen that starting from this set one of the subsequent main diagonal elements z_{jj} comes out to be zero. The procedure then fails. A simple way[1] out of this difficulty is to start all over again with a new first column such as $(1, 1, 0, 0)$. Alternatively there are ways of rearranging the computation which do not require discarding any of the work already done. The interested reader is referred to the account in Zurmühl.[2]

EXERCISES

2-35. Continue the computation of Fig. 2-8 to obtain X_2 and X_3.

2-36. The algorithm of Sec. 1-5 may be used to obtain $H = B^{-1}A$. The elements of B and A make up the given array. The auxiliary array is obtained in the usual way, treating B as the coefficient matrix and the columns of A as n different loadings (see Fig. 1-19). Show that the n different solutions provide the elements of the matrix H. The number of operations required is $(7n^3 + 3n^2 - 7n)/6$.

2-37. Let Y be a mode and λ the corresponding eigenvalue of the system

$$-PY = \lambda Y \tag{1}$$

where $-P$ is a given square matrix. If $X = ZY$, where Z is a nonsingular square matrix, show that the eigenvalue problem

$$HX = \lambda X \tag{2}$$

has the same eigenvalues as (1) if the matrix H satisfies

$$HZ + ZP = 0 \tag{3}$$

2-38. If

$$
Z = \begin{bmatrix} z_{11} & 0 & 0 & 0 \\ z_{21} & z_{22} & 0 & 0 \\ z_{31} & z_{32} & z_{33} & 0 \\ z_{41} & z_{42} & z_{43} & z_{44} \end{bmatrix}
\qquad
P = \begin{bmatrix} p_{11} & p_{12} & p_{13} & p_{14} \\ -1 & p_{22} & p_{23} & p_{24} \\ 0 & -1 & p_{33} & p_{34} \\ 0 & 0 & -1 & p_{44} \end{bmatrix}
$$

show that the relations (2-73) are equivalent to $HZ + ZP = 0$. Note that from a given matrix H and an arbitrary first column in Z it is possible to work out, element by element, the matrices Z and P.

2-39. The eigenvalues of the system $-PY = \lambda Y$ are the values for which the determinant

$$
\begin{vmatrix} p_{11} + \lambda & p_{12} & p_{13} & p_{14} \\ -1 & p_{22} + \lambda & p_{23} & p_{24} \\ 0 & -1 & p_{33} + \lambda & p_{34} \\ 0 & 0 & -1 & p_{44} + \lambda \end{vmatrix}
$$

vanishes. Show that this determinant equals

$$
(p_{44} + \lambda) \begin{vmatrix} p_{11} + \lambda & p_{12} & p_{13} \\ -1 & p_{22} + \lambda & p_{23} \\ 0 & -1 & p_{33} + \lambda \end{vmatrix} + p_{34} \begin{vmatrix} p_{11} + \lambda & p_{12} \\ -1 & p_{22} + \lambda \end{vmatrix}
$$
$$
+ p_{24}|p_{11} + \lambda| + p_{14}
$$

---

[1] See Exercise 2-41.

[2] Zurmühl, *op. cit.*, p. 144.

and hence equals F_4 in the following recursion scheme:

$$F_1 = \lambda + p_{11}$$
$$F_2 = (\lambda + p_{22})F_1 + p_{12}$$
$$F_3 = (\lambda + p_{33})F_2 + p_{23}F_1 + p_{13}$$
$$F_4 = (\lambda + p_{44})F_3 + p_{34}F_2 + p_{24}F_1 + p_{14}$$

Verify that the relations (2-74) are equivalent to evaluating this sequence.

2-40. Show that, if a double-length row whose elements are the *sums* of the columns in H and in Z is multiplied into a double-length column made up of a column of Z and the corresponding column of P, the result is zero if $HZ + ZP = 0$. Show how this can be used to provide a check on the arithmetic of computing Z and P in the Hessenberg algorithm.

2-41. When the Hessenberg algorithm is applied to

$$H = \begin{bmatrix} 1 & 2 & 3 \\ 2 & 1 & 2 \\ 3 & 3 & 2 \end{bmatrix}$$

show that the method fails ($z_{33} = 0$) when the first column of Z is taken as $(1, 0, 0)$. Show that the procedure works when this column is taken as $(1, 0, 1)$. The correct polynomial here is $\lambda^3 - 4\lambda^2 - 14\lambda - 9$. Verify that the procedure fails when $(1, 1, 0)$ is used.

2-42. The linear equilibrium problem (1-36) can be written in matrix form as

$$AX = C \tag{1}$$

If G is a nonsingular square matrix and $GY = C$, where Y is a column matrix, show that the unknown column X also satisfies

$$HX = Y \tag{2}$$

where H is a square matrix satisfying $GH = A$. Show that (1-69) and (1-70) are equivalent to requiring $GH = A$ when G and H have the following form,

$$G = \begin{bmatrix} g_{11} & 0 & 0 & 0 \\ g_{21} & g_{22} & 0 & 0 \\ g_{31} & g_{32} & g_{33} & 0 \\ g_{41} & g_{42} & g_{43} & g_{44} \end{bmatrix} \qquad H = \begin{bmatrix} 1 & h_{12} & h_{13} & h_{14} \\ 0 & 1 & h_{23} & h_{24} \\ 0 & 0 & 1 & h_{34} \\ 0 & 0 & 0 & 1 \end{bmatrix}$$

and that the capital C_j of (1-70) are the elements of the column Y defined above. Verify that the back substitution (1-71) is the solution of (2) above.

2-7. Iteration

Two fundamentally different iterative procedures for obtaining approximate solutions to the eigenvalue problem are described in this chapter. In the first, which is described in this section, the basic iterated operation involves the replacement of a trial *vector* by an improved vector. The process leads in general to only one mode of the system. Modifications of the procedure which permit other modes to be obtained are described in Secs. 2-8 and 2-9.

In the second type of iteration the basic operation consists in the replacement of a *square matrix* by an improved square matrix. The process is described in Sec. 2-12.   It leads simultaneously to all the modes and their corresponding eigenvalues.   We call this second procedure *diagonalization by successive rotations*.   The first procedure we refer to simply as *iteration*.

Iteration was first applied to technical eigenvalue problems in 1898 by L. Vianello[1] in a study of buckling problems.   The process was applied by A. Stodola[2] in 1904 to the problem of critical speeds of rotating shafts. The theory of the method had already been presented by H. A. Schwarz[3] in 1885 and had been developed by E. Picard.[4]   All these early works dealt with eigenvalue problems in continua.   The iteration process is, however, completely analogous for systems with a finite number of degrees of freedom.   In 1921 E. Pohlhausen[5] applied the method to the specific problem of this chapter.   Matrix notation was used in discussion of the method by R. von Mises and H. Geiringer[6] in 1929.   In 1934 W. J. Duncan and A. R. Collar[7] gave a number of practical applications of iteration, using matrix notation for the calculations.

Iteration for eigenvalue problems is similar in principle to iteration for equilibrium problems.   The system equations are arranged to provide a test for a proposed trial vector such that, if the trial vector is actually an eigenvector, it will reproduce itself.   For other vectors this same operation becomes a means of transforming one trial vector into another. The iteration procedure consists in continually transforming successive transforms until finally a vector is obtained which transforms into itself.

If we take the eigenvalue problem of this chapter,

$$AX = \lambda BX \tag{2-77}$$

in the form [see (2-32)]

$$HX = \lambda X \tag{2-78}$$

we have the following test for a trial vector V: Compute the vector HV. If V is actually an eigenvector, HV will be proportional to V; that is, each element of HV will be a scalar multiple of the corresponding ele-

[1] Graphische Untersuchung der Knickfestigkeit gerader Stäbe, *Z. Ver. deut. Ing.*, **42**, 1436–1443 (1898).

[2] See A. Stodola, "Steam and Gas Turbines," 2d German ed., transl. by A. Loewenstein, D. Van Nostrand Company, New York, 1905, p. 185.

[3] See "Gesammelte Mathematische Abhandlungen von H. A. Schwarz," vol. I, Berlin, 1890, pp. 241–265.

[4] "Traité d'analyse," vol. III, Gauthier-Villars & Cie, Paris, 1896, chap. 6.

[5] Eigenschwingungen Statisch-bestimmter Fachwerk, *Z. angew. Math. u. Mech.*, **1**, 28–42 (1921).

[6] Praktische Verfahren der Gleichungsauflosung, *Z. angew. Math. u. Mech.*, **9**, 152–164 (1929).

[7] A Method for the Solution of Oscillation Problems by Matrices, *Phil. Mag.*, (7)**17**, 865–909 (1934).

ment of V. The scalar multiple is the eigenvalue. If V is not an eigen-vector, HV will not be proportional to V; there will be no single ratio between corresponding elements of V and HV. As a convenient means of ascertaining whether the vectors V and HV are proportional, we adopt the following procedure: We shall take our trial vector V so that one of its elements, usually the largest, is unity. Then, after computing HV, we shall first factor out the necessary scalar so that the same element is unity. Then, if the remaining elements of V and HV are identical, V is an eigenvector and the scalar just factored out is the eigenvalue. When V is not an eigenvector, this process becomes a transformation from one trial vector to another. Continued repetition of this transformation constitutes the iteration process.

To illustrate the technique, we return to the vibratory system of Prob. 2-1, for which we already have the matrix H given in (2-34). Let us begin with an initial trial vector having elements $(1, 0, 0)$. We show below the computation of HV and the scaling down required to make the first element unity.

$$\begin{bmatrix} 5 & -2 & 0 \\ -1 & \frac{3}{2} & -\frac{1}{2} \\ 0 & -\frac{1}{3} & \frac{1}{3} \end{bmatrix} \begin{bmatrix} 1 \\ 0 \\ 0 \end{bmatrix} = \begin{bmatrix} 5 \\ -1 \\ 0 \end{bmatrix} = 5 \begin{bmatrix} 1.0 \\ -0.2 \\ 0.0 \end{bmatrix} \qquad (2\text{-}79)$$

We see that the vector $(1.0, -0.2, 0.0)$ is not identical with our initial trial. We next repeat the previous step, but now using $(1.0, -0.2, 0.0)$ instead of $(1, 0, 0)$. The results of the first four steps of iteration are shown below. We give the succeeding vectors together with the scalar multiple factored out at each stage.

$$\begin{bmatrix} 1 \\ 0 \\ 0 \end{bmatrix} \qquad \begin{bmatrix} 1.0 \\ -0.2 \\ 0.0 \end{bmatrix} \qquad \begin{bmatrix} 1.00 \\ -0.24 \\ 0.01 \end{bmatrix} \qquad \begin{bmatrix} 1.000 \\ -0.249 \\ 0.015 \end{bmatrix} \qquad \begin{bmatrix} 1.000 \\ -0.252 \\ 0.016 \end{bmatrix} \qquad (2\text{-}80)$$
$$ 5.0 \qquad\quad 5.40 \qquad\quad 5.48 \qquad\quad 5.49$$

This sequence of vectors gives evidence of approaching the mode X_3 with elements $(1.0000, -0.2518, 0.0162)$, and the sequence of scalars which represent the ratios of successive first elements is apparently approaching the eigenvalue $\lambda_3 = 5.5036$.

If iteration is to be useful, we must know when and to what it will converge. We turn to a brief examination of this question. The pro-cedure for the problem defined by (2-78) may be described as follows: From an initial vector V_0, the sequence of vectors

$$V_0, V_1, V_2, \ldots, V_r, \ldots \qquad (2\text{-}81)$$

is constructed by making

$$V_{r+1} = HV_r \qquad (2\text{-}82)$$

The scaling operation, while helpful for comparison, is not an essential part of the process and will be omitted in this discussion. To investi-gate the behavior of the sequence (2-81), let us suppose that V_0 has been expanded in terms of eigenvectors, i.e.,

$$V_0 = \sum_{j=1}^{n} c_j X_j \tag{2-83}$$

Then
$$V_1 = HV_0 = \sum_{j=1}^{n} c_j \lambda_j X_j$$

$$V_2 = HV_1 = \sum_{j=1}^{n} c_j(\lambda_j)^2 X_j \tag{2-84}$$

$$V_r = HV_{r-1} = \sum_{j=1}^{n} c_j(\lambda_j)^r X_j$$

If now λ_n is the eigenvalue with largest absolute value and $c_n \neq 0$, we have finally

$$V_r = c_n(\lambda_n)^r \left[X_n + \sum_{j=1}^{n-1} \frac{c_j}{c_n} \left(\frac{\lambda_j}{\lambda_n}\right)^r X_j \right] \tag{2-85}$$

The characteristics of the iteration process can be readily deduced from (2-85). Since

$$\left| \frac{\lambda_j}{\lambda_n} \right| < 1 \qquad j \neq n \tag{2-86}$$

the summation in (2-85) becomes negligible compared with X_n, as r gets large and hence V_r approaches a multiple of X_n. We also see that the scalar ratio between corresponding elements of V_{r+1} and V_r approaches λ_n. Thus the iteration process defined by (2-82) yields convergence to the mode corresponding to the *eigenvalue of largest absolute value.*

Theoretically it would be possible to choose an initial vector V_0 so that $c_n = 0$, in which case a slight modification of (2-85) would show that iteration should converge to the mode corresponding to λ_p, where λ_p was the eigenvalue of largest absolute value for which $c_p \neq 0$. This possibility is of negligible importance for numerical work, for even if $c_n = 0$ in V_0, the inevitable round-off errors will introduce a minute component of X_n which will steadily be increased by iteration. We can see from (2-85) that the number of iterations required to achieve a certain accuracy will be decreased if

(a) the $\left| \dfrac{c_j}{c_n} \right|$ $(j \neq n)$ are small quantities

and

(b) the $\left| \dfrac{\lambda_j}{\lambda_n} \right|$ $(j \neq n)$ are small quantities

The condition (a) is somewhat under the control of the computer, who should utilize any auxiliary knowledge or intuition in choosing an appropriate initial vector. The condition (b) depends on the properties of the system. If the eigenvalues are well separated, convergence will be fairly rapid, but if one or more eigenvalues are close to λ_n in magnitude, convergence may be very slow.[1] If the largest of the quantities (2-86) is $\frac{1}{10}$, then each iteration adds substantially one more correct decimal place to the elements of V. If the largest of the quantities (2-86) is $\frac{8}{10}$, then each additional correct decimal place in the elements of V will require more than 10 iterations.[2] Note that each iteration requires n^2 multiplications.

The eigenvalue of largest absolute value is usually of primary interest in buckling problems when formulated as in Prob. 2-2 since the largest λ means the smallest critical load; however, for natural-frequency problems when formulated as in Prob. 2-1 the largest λ corresponds to the highest natural frequency, which is seldom of interest. In such cases iteration may be used to obtain the smallest λ if the problem (2-77) can be put in the form [see (2-33)[3]]

$$X = \lambda G X \tag{2-87}$$

and iteration performed with G instead of H. A development similar to (2-85) shows that iteration with G converges to the mode corresponding to the *eigenvalue of smallest absolute value*.

For example, the first four iterates obtained from the trial vector $(0, 0, 1)$ for Prob. 2-1, using the matrix G of (2-35), are

$$\begin{bmatrix} 0 \\ 0 \\ 1 \end{bmatrix} \quad \begin{bmatrix} 0.18 \\ 0.45 \\ 1.00 \end{bmatrix} \quad \begin{bmatrix} 0.217 \\ 0.527 \\ 1.000 \end{bmatrix} \quad \begin{bmatrix} 0.221 \\ 0.535 \\ 1.000 \end{bmatrix} \quad \begin{bmatrix} 0.2212 \\ 0.5361 \\ 1.0000 \end{bmatrix} \tag{2-88}$$

$$ 5.5 6.31 6.46 6.465$$

where the scalar multiple factored out at each stage is given below each vector. These sequences are clearly approaching the mode X_1 (0.2213, 0.5361, 1.0000) and the reciprocal of the eigenvalue λ_1 ($1/\lambda_1 = 6.473$) as given in (2-40).

Schwarz Quotients. Iteration is essentially a process of successive improvement of an eigen*vector*. We consider here methods of obtaining approximations to the corresponding eigen*value*.

[1] See Exercises 2-52 and 2-53 for behavior in case of repeated eigenvalues and eigenvalues of equal absolute value.

[2] See Exercise 2-51.

[3] We have also indicated (see Exercise 2-4) how this form can be obtained directly in vibrating systems by using *influence coefficients*. In systems for which $B = I$ iteration with G can be performed without actually obtaining G by using a modification of the Banachiewicz-Crout algorithm (see Exercise 2-54).

We have already seen that the ratio of the largest elements of $V^{(r+1)}$ and $V^{(r)}$ may be used as an approximation to the eigenvalue. An extension of this which is useful in systems for which the *enclosure theorem* is valid is to consider all the l_j [see (2-45)], thereby obtaining upper and lower bounds for the eigenvalue.

The error involved in such estimates of λ are of the same order as the error in the elements of V. An approximation to λ with second-order error can be obtained by evaluating Rayleigh's quotient for V. H. A. Schwarz[1] showed that there is a whole family of similar quotients associated with the sequence of vectors (2-81). These may be defined as follows

$$\nu_0 = \frac{V_1'BV_0}{V_0'BV_0}$$

$$\nu_1 = \frac{V_1'BV_1}{V_1'BV_0}$$

$$\nu_2 = \frac{V_2'BV_1}{V_1'BV_1} \tag{2-89}$$

$$\nu_3 = \frac{V_2'BV_2}{V_2'BV_1}$$

$$. \quad . \quad . \quad . \quad . \quad .$$

although a large number of alternative forms are possible. For instance, it is a simple exercise[2] to show that

$$\nu_1 = \frac{V_1'BV_1}{V_0'BV_1} = \frac{V_1'AV_0}{V_1'BV_0} = \frac{V_0'AV_1}{V_0'AV_0} = \frac{V_1'BV_1}{V_0'AV_0} \tag{2-90}$$

The Schwarz quotients with even subscripts are Rayleigh quotients, e.g.,

$$\nu_2 = \frac{V_2'BV_1}{V_1'BV_1} = \frac{V_1'AV_1}{V_1'BV_1} = \lambda_R(V_1) \tag{2-91}$$

The Schwarz quotients have the following properties: As the sequence of vectors (2-81) approach a multiple of X_n, the ν_k approach the corresponding eigenvalue λ_n; the larger the value of k, the better the approximation. The errors in ν_{2r-1} and ν_{2r} are of second order in comparison with the errors in the elements of V_r. These properties can be easily deduced[3] from the following expression, which is obtained by substituting the vectors[4] of (2-84) into the definitions of (2-89):

$$\nu_k = \frac{c_1^2\lambda_1^{k+1} + c_2^2\lambda_2^{k+1} + \cdots + c_n^2\lambda_n^{k+1}}{c_1^2\lambda_1^k + c_2^2\lambda_2^k + \cdots + c_n^2\lambda_n^k} \tag{2-92}$$

---

[1] *Loc. cit.*

[2] See Exercise 2-46.

[3] For a more complete discussion see Collatz, *op. cit.*, p. 298.

[4] The eigenvectors of (2-83) and (2-84) are here considered to be *normalized* with respect to B.

The discussion follows the same pattern as for (2-66) and is left as an exercise for the reader.

To illustrate, we have computed the first five Schwarz quotients from the sequence of vectors in (2-80). It should be noted that in the definitions of (2-89) the iterated vectors are not scaled down and therefore it is necessary to reintroduce the scalar multiples factored out in (2-80). We obtain

$$\begin{aligned}
\nu_0 &= 5.0000 = \lambda_R(V_0) \\
\nu_1 &= 5.4000 \\
\nu_2 &= 5.4815 = \lambda_R(V_1) \\
\nu_3 &= 5.4989 \\
\nu_4 &= 5.5026 = \lambda_R(V_2)
\end{aligned} \tag{2-93}$$

using the matrices A and B of (2-29). The value of λ_3 is 5.5036.

δ^2 **Extrapolation.** We have seen in (2-85) that iteration is a process in which the undesired components (that is, X_j for $j \neq n$) in a vector are reduced by a fixed percentage (that is, λ_j/λ_n) with each iteration. Those modes for which λ_j/λ_n is very small will die out rapidly. The mode for which λ_j/λ_n is largest will be the most resistant.

A. C. Aitken[1] has devised a simple procedure which may be used for obtaining an improved approximation by extrapolation from a sequence of iterates. This extrapolation may be applied to the elements of a sequence of vectors V_r or to a sequence of Schwarz quotients[2] or a sequence of Rayleigh quotients. The extrapolation is based on the assumption that the major contribution to the error comes from a single undesired mode.

Let v_r be a sequence of values obtained in an iterative process which are made up of a true value t plus a single error component, i.e.,

$$v_r = t + c(b)^r \tag{2-94}$$

where b and c are constants. With each iteration the error is multiplied by b. If $|b| < 1$, the process is convergent. In performing the actual iteration one works only with the v_r; the quantities t, b, and c are unknown. If we obtain three successive iterates v_{r-1}, v_r, and v_{r+1}, we can substitute in (2-94) to obtain three simultaneous equations which are easily solved for t.

$$t = \frac{v_{r-1}v_{r+1} - v_r^2}{v_{r-1} - 2v_r + v_{r+1}} \tag{2-95}$$

Thus if an iterative sequence is known to contain only a single error component as in (2-94) the true limit of the sequence is given by (2-95). When we use iteration in an eigenvalue problem with n degrees of free-

---

[1] A. C. Aitken, The Evaluation of the Latent Roots and Latent Vectors of a Matrix, *Proc. Roy. Soc. Edinburgh*, **62**, 269–304 (1937). The designation δ^2 arises from the fact that in the notation of finite differences the denominator of (2-95) is denoted by δ^2.

[2] See Exercise 5-46.

dom, there are $n - 1$ error components and the extrapolation (2-95) will not give the true limit; however, if the iteration has been carried to the point where one of the error modes dominates the others, a substantial advance toward the limit can be had by using (2-95). We seldom know with certainty when such a stage has been reached, and hence there is an element of risk in accepting the result of an extrapolation without some auxiliary confirmation.

As an illustration we show in (2-96) the second, third, and fourth vectors of the sequence (2-80) and the vector obtained by extrapolating the elements according to (2-95).

$$\begin{bmatrix} 1.00 \\ -0.20 \\ 0.00 \end{bmatrix} \quad \begin{bmatrix} 1.000000 \\ -0.240741 \\ 0.012346 \end{bmatrix} \quad \begin{bmatrix} 1.000000 \\ -0.249437 \\ 0.015390 \end{bmatrix} \rightarrow \begin{bmatrix} 1.0000 \\ -0.2518 \\ 0.0164 \end{bmatrix} \qquad (2\text{-}96)$$

The computation of the second element, for example, is

$$-\frac{(0.200000)(0.249437) - (0.240741)^2}{0.200000 - (2)(0.240741) + 0.249437} = -0.2518 \qquad (2\text{-}97)$$

Note that, since both numerator and denominator of (2-95) involve subtractions of nearly equal quantities, it is necessary to carry several more decimal places in the iterates than is required from the extrapolated result. This result should be compared with the true mode $X_3 = (1.0000, -0.2518, 0.0162)$.

As a further illustration the extrapolation (2-95) applied to the first three Schwarz quotients of (2-93) yields the value 5.5023 as an approximation to $\lambda_3 = 5.5036$. The accuracy of this result is remarkable considering that it is based only on the two vectors $(1, 0, 0)$ and $(5, -1, 0)$, whose elements have errors of 25 per cent and 5 per cent, respectively. The second-order property of the Schwarz quotients coupled with the near cancellation of the dominant error component by the δ^2 extrapolation gives a result with an error of less than 0.03 per cent.

EXERCISES

2-43. Use iteration to find the mode corresponding to the largest eigenvalue of

$$\begin{bmatrix} 2 & 2 & 2 \\ 2 & 5 & 5 \\ 2 & 5 & 11 \end{bmatrix} \begin{bmatrix} x_1 \\ x_2 \\ x_3 \end{bmatrix} = \lambda \begin{bmatrix} x_1 \\ x_2 \\ x_3 \end{bmatrix}$$

Use the enclosure theorem as you go along to obtain upper and lower bounds for λ_3.
Ans. $\lambda_3 = 14.43$.

2-44. Obtain the mode corresponding to the largest eigenvalue of the system in Fig. 2-3 (see Exercises 2-1 and 2-8).

2-45. Obtain both modes for Prob. 2-2 by iteration (see Exercise 2-17 for the matrices H and G). Check the result by applying the orthogonality condition.

2-46. Show that for the sequence (2-81) the triple scalar products

$$\begin{array}{ll} V_i'BV_{k-i} & i = 0, 1, \ldots, k \\ V_i'AV_{k-i-1} & i = 0, 1, \ldots, k-1 \end{array}$$

are independent of i and also equal to each other. Verify the relations (2-90) and (2-91).

2-47. If for an eigenvalue problem in which $B = I$, $A = H$ the iterates V_0, V_1, and V_2 are available, what would be the easiest forms from which to compute ν_3 and ν_4? Show that ν_4 requires $n^2 + 2n + 1$ operations while ν_3 requires only $2n + 1$ operations.

2-48. Show that if the quotient of Exercise 2-34 is computed for the vector V_1 of the sequence (2-81) that result is the reciprocal of the Schwarz quotient ν_1.

2-49. Schwarz quotients may also be defined for the sequence V_r obtained from

$$V_{r+1} = GV_r$$

Show that the triple scalar products

$$a_k = V_i' B V_{k-i} = V_i' A V_{k+1-i} \qquad i = 0, \ldots, k$$

are independent of i and equal to each other. The Schwarz quotients

$$\mu_k = \frac{a_{k-1}}{a_k}$$

then have the same properties as the ν_k except that they approach the eigenvalue of smallest absolute value. Show that μ_0, μ_2, μ_4, ... are Rayleigh quotients. Compute μ_0 and μ_1 for the sequence (2-88).

2-50. Use the expression (2-92) to show that if all the $\lambda_j \geq 0$ then $\nu_{k+1} \geq \nu_k$.

2-51. Use the expression (2-94) to show that the number of iterations required to reduce the error magnitude by the factor 10^p is

$$\frac{p}{\log_{10}(1/b)}$$

2-52. Apply iteration to the system

$$\begin{bmatrix} 3 & 0 & 2 \\ 0 & 5 & 0 \\ 2 & 0 & 3 \end{bmatrix} \begin{bmatrix} x_1 \\ x_2 \\ x_3 \end{bmatrix} = \lambda \begin{bmatrix} x_1 \\ x_2 \\ x_3 \end{bmatrix}$$

starting first from $(0, 0, 1)$ and then again from $(1, 2, 3)$. Discuss the results in the light of Exercise 2-28.

2-53. Suppose that λ_n is the largest positive eigenvalue but that the largest negative eigenvalue is $\lambda_1 = -\lambda_n$. Alter the analysis of (2-85) to take this into account. What will happen to the sequence (2-81)? How can X_1 and X_n be extracted from the sequence? Apply this theory to obtain the modes X_1 and X_3 for the system

$$\begin{bmatrix} 3 & -4 & 6 \\ 2 & 1 & 2 \\ 2 & 6 & -3 \end{bmatrix} \begin{bmatrix} x_1 \\ x_2 \\ x_3 \end{bmatrix} = \lambda \begin{bmatrix} x_1 \\ x_2 \\ x_3 \end{bmatrix}$$

by iteration starting from $(2, 1, 0)$. *Ans.* $\lambda_1 = -5$, $\lambda_3 = 5$.

2-54. If the matrix A is nonsingular in the eigenvalue problem $AX = \lambda X$, iteration with $G = A^{-1}$ may be performed by using the Banachiewicz-Crout algorithm as follows: The auxiliary array of g_{jk} and h_{jk} in (1-68) are computed from the elements of A. This required $\frac{1}{6}n^3 + n^2 - \frac{1}{6}n$ operations. An initial trial vector V_0 is now entered in place of the c_j in (1-67) and the auxiliary quantities C_j computed therefrom. The "solution" obtained by back-substituting with these C_j is the first iterate V_1, which is entered as a new right-hand side and the process repeated to obtain V_2, etc. Show that each subsequent step of iteration requires $n^2 + 2n - 1$ operations.

2-8. Intermediate Eigenvalues

The basic iteration procedure of the previous section converges to the mode with largest eigenvalue (or smallest eigenvalue if the inverse form is used). There are several ways in which the procedure can be altered so as to obtain other modes. We shall describe two approaches which illustrate the principles underlying most of the possible methods.

In the first approach ordinary iteration is used to obtain the mode X_n, say, and then some device, usually based on the orthogonality principle, is used to suppress X_n in subsequent iterations so that the procedure converges to X_{n-1}. When X_{n-1} has been obtained, both X_n and X_{n-1} may be suppressed so that further iteration will converge to X_{n-2}.

In the second approach the given system is transformed into a related one which has the same modes, but the corresponding eigenvalues are relocated at the computer's option. If the mode X_p, which originally had an intermediate eigenvalue, now has the largest eigenvalue, then iteration will converge to X_p.

Use of Orthogonality. When the mode X_n has been obtained, any other mode X_j must satisfy the orthogonality condition

$$X_n'BX_j = 0 \qquad j \neq n \tag{2-98}$$

as well as the governing equation

$$HX_j = \lambda X_j \tag{2-99}$$

Equation (2-98) may be solved for one of the elements of X_j, x_p say, in terms of the others. Introducing this into (2-99) enables us to eliminate x_p, thereby obtaining n algebraic equations in $n - 1$ unknowns. If we merely lay aside the pth equation, the remaining equations make up an eigenvalue problem with $n - 1$ degrees of freedom. The eigenvalues of this reduced system are the $n - 1$ eigenvalues λ_j ($j \neq n$) of the original system, and the modes are vectors whose elements x_j ($j \neq p$) are the elements in the corresponding modes of the original system. Iteration in the reduced system will converge to the mode X_{n-1}, which now has the eigenvalue of greatest magnitude.

To illustrate, we take the mode X_3 for the three-mass system of Prob. 2-1, which was obtained by iteration in (2-80). The orthogonality condition (2-98) when written out is

$$x_1 - 0.504x_2 + 0.048x_3 = 0 \tag{2-100}$$

where x_1, x_2, and x_3 are the elements of either X_1 or X_2. We solve (2-100) for x_1 and substitute in (2-34) as follows:

$$
\begin{aligned}
5(0.504x_2 - 0.048x_3) - 2x_2 &= \lambda x_1 \\
-(0.504x_2 - 0.048x_3) + \tfrac{3}{2}x_2 - \tfrac{1}{2}x_3 &= \lambda x_2 \\
-\tfrac{1}{3}x_2 + \tfrac{1}{3}x_3 &= \lambda x_3
\end{aligned}
\tag{2-101}
$$

to get

$$
\begin{aligned}
0.520x_2 - 0.240x_3 &= \lambda x_1 \\
0.996x_2 - 0.452x_3 &= \lambda x_2 \\
-0.333x_2 + 0.333x_3 &= \lambda x_3
\end{aligned}
\tag{2-102}
$$

The last two equations of (2-102) constitute our reduced system. If we iterate with the matrix

$$
H^* = \begin{bmatrix} 0.996 & -0.452 \\ -0.333 & 0.333 \end{bmatrix}
\tag{2-103}
$$

starting with the initial trial (1, 0), we obtain the following sequence:

$$
\begin{bmatrix} 1 \\ 0 \end{bmatrix}
\quad
\begin{matrix} \begin{bmatrix} 1.000 \\ -0.334 \end{bmatrix} \\ 0.996 \end{matrix}
\quad
\begin{matrix} \begin{bmatrix} 1.000 \\ -0.387 \end{bmatrix} \\ 1.147 \end{matrix}
\quad
\begin{matrix} \begin{bmatrix} 1.000 \\ -0.394 \end{bmatrix} \\ 1.171 \end{matrix}
\quad
\begin{matrix} \begin{bmatrix} 1.000 \\ -0.395 \end{bmatrix} \\ 1.174 \end{matrix}
\tag{2-104}
$$

If we stop here, we have the approximation 1.174 to $\lambda_2 = 1.1751$. We also have approximations to the elements x_2 and x_3 of X_2. To obtain the element x_1, we return to (2-100) to find

$$
x_1 = (0.504)(1.000) - (0.048)(-0.395) = 0.524
\tag{2-105}
$$

This result should be compared with (2-40).

We may continue this process. For example, in the above illustration the mode X_1 must satisfy not only the governing equation (2-99) but the two orthogonality conditions

$$
\begin{aligned}
x_1 - 0.504x_2 + 0.048x_3 &= 0 \\
0.524x_1 + 2.000x_2 - 1.185x_3 &= 0
\end{aligned}
\tag{2-106}
$$

We solve here for x_1 and x_2 in terms of x_3, obtaining $x_1 = 0.221x_3$ and $x_2 = 0.534x_3$, that is, $X_1 = (0.221, 0.534, 1.000)$.

The above procedure was systematically laid out with the aid of matrix notation by Duncan and Collar.[1] A somewhat different method for reducing the order of a system when one mode is known was given by Wielandt.[2] An alternative method in which the order of the system remains n after suppressing X_n was presented by Aitken.[3] Another way of utilizing orthogonality, initially proposed for continuous systems, was

[1] Duncan and Collar, *loc. cit.* See Exercise 2-56.

[2] H. Wielandt, Das Iterations verfahren bei nicht-selbst-adjungierten linearen Eigenwertaufgaben, *Math. Z.*, **50**, 93–143 (1944).

[3] Aitken, *loc. cit.* See Exercise 2-57. Recent experience with automatic machine computation by this method indicates that unexpectedly high accuracy is retained even after many modes have been extracted. See J. H. Wilkinson, The Calculation of the Latent Roots and Vectors of Matrices on the Pilot Model of the A. C. E., *Proc. Camb. Philos. Soc.*, **50**, 536–566 (1954).

given by Koch.[1]   In all these methods there is a progressive loss of accuracy as more and more modes are obtained.   This may be overcome by computing the early modes with greater precision than necessary.   A rule of thumb which has been used in this connection advises carrying an additional decimal place for each additional mode.   These methods are probably most useful when only the two or three modes with largest eigenvalues (or smallest eigenvalues if inverse iteration is used) are required.

Polynomial Operators.   Let X be a mode and λ be the corresponding eigenvalue of a system with the matrix $H = B^{-1}A$, that is,

$$HX = \lambda X \tag{2-107}$$

Let us multiply both sides by the square matrix H.

$$H^2X = \lambda HX = \lambda^2 X \tag{2-108}$$

This means that X is also an eigenvector for a system having the matrix H^2.   The eigenvalue is λ^2.   In the same way we show that

$$H^mX = \lambda^m X \qquad m = 1, 2, 3, \ldots \tag{2-109}$$

Finally, if we construct a square matrix by adding together various multiples of powers of H, we find that X is an eigenvector of the resulting system.

$$(a_mH^m + \cdots + a_1H + a_0I)X = (a_m\lambda^m + \cdots + a_1\lambda + a_0)X \tag{2-110}$$

The square matrix on the left is a *matrix polynomial*, or *polynomial operator*.   We have just shown that a polynomial operator constructed from a matrix H has the same modes as H.   The corresponding eigenvalues are the same polynomial functions of the original eigenvalues.

The idea of systematically using polynomial operators in conjunction with iteration to obtain intermediate modes was developed by Kincaid[2] in 1947 and independently by Richardson[3] in 1950.   The use of the linear polynomial $H + a_0I$ had been suggested by Aitken[4] in 1937.

To give a simple introduction, we show along the abcissa of Fig. 2-9 the eigenvalues of the three-mass vibrating system of (2-34).   These eigenvalues are the eigenvalues of H.   This is indicated by running ordinates up to the 45° line through the origin.

[1] J. J. Koch, Bestimmung höherer kritischer Drehzahlen schnell laufender Wellen, *Proc. 2d Intern. Congr. Appl. Mech.*, Zurich, 1926, pp. 213–218.   See Exercise 2-58.

[2] W. M. Kincaid, Numerical Methods for Finding Characteristic Roots and Vectors of Matrices, *Quart. Appl. Math.*, **5**, 320–346 (1947).

[3] L. F. Richardson, A Purification Method for Computing the Latent Columns of Numerical Matrices and Some Integrals of Differential Equations, *Trans. Roy. Soc.* *(London)*, **A242**, 439–491 (1950).

[4] Aitken, *loc. cit.*

As we have seen, iteration converges to the mode with the eigenvalue of largest absolute value, i.e., the greatest ordinate in this figure, namely, 5.5036.

Now consider the matrix $H - 3.500I$.

$$
\begin{bmatrix} 5 & -2 & 0 \\ -1 & 1.500 & -0.500 \\ 0 & -0.333 & 0.333 \end{bmatrix} - \begin{bmatrix} 3.500 & 0 & 0 \\ 0 & 3.500 & 0 \\ 0 & 0 & 3.500 \end{bmatrix} = \begin{bmatrix} 1.500 & -2 & 0 \\ -1 & -2 & -0.500 \\ 0 & -0.333 & -3.167 \end{bmatrix}
$$

$$(2\text{-}111)$$

Its eigenvalues according to (2-110) are $\lambda - 3.500$, which are shown in Fig. 2-9 as the ordinates running up to the curve $\lambda - 3.500$. Of these the greatest in absolute value is -3.3454; so iteration with the matrix of (2-111) will yield convergence to the mode X_1.

As another example suppose that approximations to λ_1 and λ_3 of the same system are available. (The corresponding modes are not required.) We shall construct a

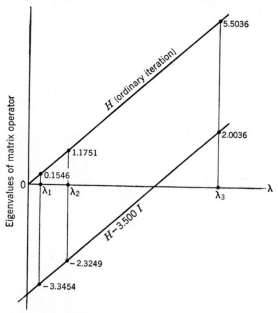

Fig. 2-9. The linear polynomial $H - 3.500I$ has eigenvalues which are 3.500 units smaller than the eigenvalues of H.

polynomial operator which will give rapid convergence to the intermediate mode X_2.

If we take 0.155 and 5.50 as approximations to $\lambda_1 = 0.1546$ and $\lambda_3 = 5.5036$, then

$$(\lambda - 0.155)(\lambda - 5.50) = \lambda^2 - 5.655\lambda + 0.8525 \qquad (2\text{-}112)$$

is a polynomial with small ordinates at λ_1 and λ_3 but with a relatively large ordinate at λ_2. The situation is sketched in Fig. 2-10. We must next compute the matrix polynomial corresponding to (2-112). To avoid round-off error here, we have changed the coefficients in (2-112) as follows:

$$\lambda^2 - 5\tfrac{2}{3}\lambda + \tfrac{31}{36} = P(\lambda) \qquad (2\text{-}113)$$

This has the effect of slightly changing the curve in Fig. 2-10, but the ordinates at

λ_1 and λ_3 are still small in comparison with the ordinate at λ_2. Then we construct

$$\begin{bmatrix} 5 & -2 & 0 \\ -1 & \frac{3}{2} & -\frac{1}{2} \\ 0 & -\frac{1}{3} & \frac{1}{3} \end{bmatrix} \begin{bmatrix} 5 & -2 & 0 \\ -1 & \frac{3}{2} & -\frac{1}{2} \\ 0 & -\frac{1}{3} & \frac{1}{3} \end{bmatrix} - 5\frac{2}{3}\begin{bmatrix} 5 & -2 & 0 \\ -1 & \frac{3}{2} & -\frac{1}{2} \\ 0 & -\frac{1}{3} & \frac{1}{3} \end{bmatrix} + \frac{31}{36}\begin{bmatrix} 1 & 0 & 0 \\ 0 & 1 & 0 \\ 0 & 0 & 1 \end{bmatrix}$$

$$= \frac{1}{36}\begin{bmatrix} -17 & -60 & 36 \\ -30 & -116 & 69 \\ 12 & 46 & -27 \end{bmatrix} = P(H) \quad (2\text{-}114)$$

To illustrate the sort of convergence that may be expected from an operator constructed in this manner, we show two successive iterates obtained by using (2-114), starting from the initial trial vector $(0, 1, 0)$.

$$\begin{bmatrix} 0 \\ 1 \\ 0 \end{bmatrix} \qquad \begin{bmatrix} 0.517 \\ 1.000 \\ -0.397 \end{bmatrix} \qquad \begin{bmatrix} 0.5229 \\ 1.0000 \\ -0.3960 \end{bmatrix} \qquad (2\text{-}115)$$

$$-116 \qquad\qquad -158.9$$

The elements of the second iterate are already correct to four decimal places. Note that the corresponding eigenvalue is not immediately given. While it would be possible to equate the polynomial (2-113) to the scalar multiple -158.9 and solve for λ, it will generally be simpler to multiply the mode obtained in (2-115) by H and use one of the element ratios l_j [see (2-45)] as an approximation to λ_2. Accuracy to about twice as many places can be had from about twice as much calculation by computing Rayleigh's quotient for the mode obtained in (2-115).

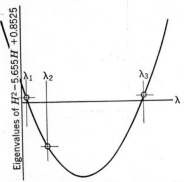

FIG. 2-10. Polynomial operator with small eigenvalues for first and third modes and a large eigenvalue for the second mode.

For further applications the reader is referred to the papers of Kincaid and Richardson. The principal advantage of the method lies in the possibility of obtaining a particular mode with great accuracy when only rough approximations to the eigenvalues of the other modes are known. These approximate eigenvalues might be furnished, for example, by the procedure described in the beginning of Sec. 2-10. A reasonably high level of skill and ingenuity is required on the part of the computer in order to take maximum advantage of polynomial operators. They are not readily incorporated into fixed routines.

EXERCISES

2-55. Apply the method of suppression of X_n given in the beginning of the section to obtain the second mode for the system of Exercise 2-43.   *Ans.* $\lambda_2 = 2.615$.

2-56. Let the mode X_n corresponding to the eigenvalue of greatest magnitude be known. Let BX_n be scaled so that its largest element, the pth, is unity. Let the elements of BX_n then be $(\xi_1, \ldots, \xi_{p-1}, 1, \xi_{p+1}, \ldots, \xi_n)$. Show that if V is any vector S_nV is orthogonal to X_n [that is, $X'_nB(S_nV) = 0$], where S_n is the square matrix

$$
S_n = \begin{bmatrix}
1 & 0 & \cdots & 0 & 0 & 0 & \cdots & 0 \\
0 & 1 & \cdots & 0 & 0 & 0 & \cdots & 0 \\
\cdots & \cdots & \cdots & \cdots & \cdots & \cdots & \cdots \\
-\xi_1 & -\xi_2 & \cdots & -\xi_{p-1} & 0 & -\xi_{p+1} & \cdots & -\xi_n \\
\cdots & \cdots & \cdots & \cdots & \cdots & \cdots & \cdots \\
0 & 0 & \cdots & 0 & 0 & 0 & \cdots & 1
\end{bmatrix}
$$

Compare this with (2-100). Show that iteration with the square matrix $H^* = HS_n$ will converge to the mode X_{n-1}. Construct the matrix H^* according to this rule for the system of (2-34), and compare with (2-103). When X_{n-1} has been obtained, how should a matrix S_{n-1} be constructed so that iteration with H^*S_{n-1} will converge to X_{n-2}?

2-57. Show that if X_j is a mode of the eigenvalue problem $HX = \lambda X$ the *square* matrix

$$H_j = \frac{X_jX'_jB}{X'_jBX_j}$$

satisfies the relations $H_jX_j = X_j$ and $H_jX_k = 0$ for $j \neq k$. Then the given matrix may be decomposed as follows:

$$H = \sum_{j=1}^{n} \lambda_j H_j$$

Once X_n and λ_n have been obtained, X_n may be suppressed in further iterations by iterating with

$$H^* = H - \lambda_n H_n$$

Apply this to obtain X_2 for the three-mass system of (2-34).

2-58. Show that for any vector V_r the vector

$$V_r^* = V_r - \frac{X'_nBV_r}{X'_nBX_n} X_n$$

is orthogonal to X_n and hence that iteration according to the rule

$$V_{r+1} = HV_r^*$$

will converge to the mode X_{n-1}. How can this be modified when both X_n and X_{n-1} are known so as to obtain convergence to X_{n-2}?

2-59. Sketch the eigenvalues of the operator G of (2-35) on the same diagram with Fig. 2-9. Compare the processes of obtaining X_1 by iteration with G [see (2-88)] and by iteration with $H - 3.500I$. Use Exercise 2-51 to estimate the relative lengths of computation.

2-60. Starting from the matrix G of (2-35) and the mode X_1 obtained in (2-88), construct polynomial functions of G which upon iteration will give convergence to X_3 and X_2.

2-9. n-step Iteration

The iteration procedures described in the two preceding sections were indirect, or approximate, methods. Mathematical convergence could not be obtained in a finite number of iterations. In this section we examine procedures which although having the form of iteration processes are actually direct methods in that exact solutions are provided at the end of just n iterations if all operations are carried out without round-off.

The n-step iteration for the eigenvalue problem of this chapter was given by Lanczos,[1] while the related procedure for equilibrium problems in linear symmetric positive definite systems is due to Hestenes and Stiefel.[2] The usefulness of these methods for actual calculation is still being evaluated. When considered as direct methods, they require more computation than the direct methods already given. When considered as iterations, it is not clear whether the increased complication of each step is compensated for by the accelerated convergence obtained. There is, however, no denying the mathematical elegance of the methods. An understanding of the principles involved will give a deeper appreciation of the structure of linear systems.

n-step Iteration for Eigenvalue Problems. We consider the system $AX = \lambda BX$ with $H = B^{-1}A$. Starting from an arbitrary trial vector V_1, we construct

$$V_2^* = HV_1 \tag{2-116}$$

as in ordinary iteration. Instead of using V_2^* as our iterate we first make it *orthogonal* to V_1; that is, we use

$$V_2 = V_2^* - \alpha_1 V_1 \tag{2-117}$$

where

$$\alpha_1 = \frac{V_1' B V_2^*}{V_1' B V_1} \tag{2-118}$$

The reader should verify that $V_1' B V_2 = 0$. We continue by constructing

$$V_3^* = HV_2 \tag{2-119}$$

and then making this orthogonal to *both* V_1 and V_2 as follows,

$$V_3 = V_3^* - \alpha_2 V_2 - \beta_1 V_1 \tag{2-120}$$

where

$$\alpha_2 = \frac{V_2' B V_3^*}{V_2' B V_2} \qquad \beta_1 = \frac{V_1' B V_3^*}{V_1' B V_1} \tag{2-121}$$

[1] C. Lanczos, An Iteration Method for the Solution of the Eigenvalue Problem of Linear Differential and Integral Operators, *J. Research Natl. Bur. Standards,* **45,** 255–282 (1950).

[2] M. H. Hestenes and E. Stiefel, Method of Conjugate Gradients for Solving Linear Systems, *Natl. Bur. Standards (U.S.) Rept.* 1659 (1952).

In this way we build up a set[1] of n vectors $V_1, \ldots, V_n$ which are orthogonal[2] to one another. The general formulas are

$$V_k^* = HV_{k-1}$$
$$V_k = V_k^* - \alpha_{k-1}V_{k-1} - \beta_{k-2}V_{k-2} \qquad (2\text{-}122)$$

where
$$\alpha_{k-1} = \frac{V_{k-1}'BV_k^*}{V_{k-1}'BV_{k-1}} \qquad \beta_{k-2} = \frac{V_{k-2}'BV_k^*}{V_{k-2}'BV_{k-2}}$$

Since only n vectors can be mutually orthogonal,[3] the vector V_{n+1} constructed in this way should be identically zero. This provides a check. The total number of operations required is $4n^3 + 5n^2 - n$.

Another way of describing this sequence is in terms of the following polynomials:

$$p_1(\lambda) = \lambda - \alpha_1$$
$$p_2(\lambda) = (\lambda - \alpha_2)p_1(\lambda) \quad - \beta_1$$
$$p_3(\lambda) = (\lambda - \alpha_3)p_2(\lambda) \quad - \beta_1 p_1(\lambda)$$
$$\cdots\cdots\cdots\cdots\cdots\cdots\cdots\cdots \qquad (2\text{-}123)$$
$$p_n(\lambda) = (\lambda - \alpha_n)p_{n-1}(\lambda) - \beta_{n-1}p_{n-2}(\lambda)$$

If we put the matrix H in place of λ, we have matrix polynomials which satisfy

$$p_1(H)V_1 = (H - \alpha_1 I)V_1 = V_2$$
$$p_2(H)V_1 = \qquad\qquad = V_3$$
$$\cdots\cdots\cdots\cdots\cdots\cdots\cdots \qquad (2\text{-}124)$$
$$p_n(H)V_1 = \qquad\qquad = V_{n+1} = 0$$

which the reader should verify. The last of these tells us that the arbitrary vector V_1 is an eigenvector of $p_n(H)$ with eigenvalue zero. The eigenvalue of $p_n(H)$ is the nth-order polynomial $p_n(\lambda)$. Therefore setting

$$p_n(\lambda) = 0 \qquad (2\text{-}125)$$

gives the characteristic equation for the eigenvalues of our system. Using the sequence (2-123) to write out $p_n(\lambda)$ requires $\frac{3}{2}n^2 + n/2$ operations. Thus a total of $4n^3 + \frac{13}{2}n^2 - \frac{1}{2}n$ operations is required to obtain the characteristic equation. This is to be compared with $n^3 + n$ operations in the Hessenberg algorithm.

So far we have shown how to construct a set of n orthogonal vectors and how to obtain the characteristic equation. Let us now consider how to extract *approximate* modes and eigenvalues when only m ($m < n$) steps of iteration have been made. We assemble our m vectors and ask what linear combinations

$$c_1 V_1 + c_2 V_2 + \cdots + c_m V_m \qquad (2\text{-}126)$$

give good approximations to eigenvectors. As a criterion we use the Rayleigh quotient λ_R. Since λ_R is stationary in the neighborhood of an eigenvector, we look

---

[1] If the initial trial vector happens to be chosen orthogonal to one or more eigenvectors, then the sequence will terminate with less than n vectors. For a treatment of this exceptional case see Householder, *op. cit.*, p. 173.

[2] See Exercise 2-63.

[3] See Exercise 2-64.

for those combinations (2-126) for which λ_R is stationary with respect to variations in the c_j. Now

$$\lambda_R = \frac{\displaystyle\sum_{j=1}^{m}\sum_{k=1}^{m} c_j c_k V'_j A V_k}{\displaystyle\sum_{j=1}^{m}\sum_{k=1}^{m} c_j c_k V'_j B V_k} = \frac{N}{D} \qquad (2\text{-}127)$$

say, and,

$$\frac{\partial \lambda_R}{\partial c_j} = 0$$

if

$$D\frac{\partial N}{\partial c_j} - N\frac{\partial D}{\partial c_j} = 0$$

or if

$$\frac{\partial N}{\partial c_j} - \lambda_R \frac{\partial D}{\partial c_j} = 0 \qquad (2\text{-}128)$$

Writing out this last gives

$$\sum_k c_k V'_j A V_k - \lambda_R \sum_k c_k V'_j B V_k = 0 \qquad (2\text{-}129)$$

and upon introducing[1] the orthogonality of the V_j and the definitions (2-122) we obtain the following eigenvalue problem within an eigenvalue problem:

$$\begin{bmatrix} \alpha_1 & \beta_1 & 0 & 0 & \cdots & 0 & 0 \\ 1 & \alpha_2 & \beta_2 & 0 & \cdots & 0 & 0 \\ 0 & 1 & \alpha_3 & \beta_3 & \cdots & 0 & 0 \\ \multicolumn{7}{c}{\cdots\cdots\cdots\cdots\cdots\cdots\cdots} \\ 0 & 0 & 0 & 0 & \cdots & 1 & \alpha_m \end{bmatrix} \begin{bmatrix} c_1 \\ c_2 \\ c_3 \\ \cdots \\ c_m \end{bmatrix} = \lambda_R \begin{bmatrix} c_1 \\ c_2 \\ c_3 \\ \cdots \\ c_m \end{bmatrix} \qquad (2\text{-}130)$$

The modes of this m-degree-of-freedom system give the sets of coefficients to be used in (2-126) to provide approximate modes of our original system. The eigenvalues of (2-130) are the Rayleigh quotients of these approximate modes.

The solution of (2-130) is already at hand. If we obtain its characteristic equation by expanding its determinant, we find simply[2]

$$p_m(\lambda_R) = 0 \qquad (2\text{-}131)$$

where p_m is defined in (2-123). If λ^* is a root of (2-131), the corresponding mode of (2-130) is obtained by setting $c_1 = 1$ and solving the rows of (2-130) in sequence for c_2, c_3, etc. In this way we obtain

$$V_1 + \frac{p_1(\lambda^*)}{\beta_1} V_2 + \frac{p_2(\lambda^*)}{\beta_1 \beta_2} V_3 + \cdots + \frac{p_{m-1}(\lambda^*)}{\beta_1 \beta_2 \cdots \beta_{m-1}} V_m \qquad (2\text{-}132)$$

as the approximate mode (2-126) corresponding to λ^*. When $m = n$, then (2-131) becomes the true characteristic equation and the vectors (2-132) become the true eigenvectors of our system.

We note that the roots of a polynomial must be obtained at any stage where we wish to stop and obtain approximate eigenvalues and modes. If the roots of each

[1] See Exercise 2-65.
[2] See Exercise 2-66.

polynomial (2-123) are found in sequence, there is the advantage that the roots of p_{m-1} are approximations to $m-1$ of the roots of p_m and hence provide good starting places for *Newton's method*.[1]

n-step Iteration for Equilibrium Problems. We consider the system of simultaneous equations

$$AX = C \qquad (2\text{-}133)$$

where A is a symmetric positive definite square matrix of order n. For a trial vector V_k we define the *residual vector* R_k as

$$R_k = C - AV_k \qquad (2\text{-}134)$$

In n-step iteration we start with an initial trial V_1 and build up successive corrections from a family of vectors P_k which are constructed in step fashion to be *orthogonal with respect to* A. After n such corrections the true solution is achieved.

The first member of the family P_k is taken as $P_1 = R_1$, and the first improvement of V_1 is taken as

$$V_2 = V_1 + \lambda_1 P_1 \qquad (2\text{-}135)$$

where $$\lambda_1 = \frac{R_1' R_1}{P_1' A P_1} \qquad (2\text{-}136)$$

The process is continued by calculating the new residual R_2. This may be done by returning to the definition (2-134), or we have more briefly in the form of a correction

$$R_2 = R_1 - \lambda_1 A P_1 \qquad (2\text{-}137)$$

The choice of λ_1 in (2-136) was such as to make R_2 orthogonal[2] to R_1 with respect to the unit matrix, that is, $R_1' R_2 = 0$.

In order to make P_2 orthogonal to P_1 with respect to A, we take

$$P_2 = R_2 + \mu_1 P_1 \qquad (2\text{-}138)$$

where $$\mu_1 = \frac{R_2' R_2}{R_1' R_1} \qquad (2\text{-}139)$$

The reader should check[3] that $P_1' A P_2 = 0$.

We continue in this fashion, using the following formulas:

$$\begin{aligned} R_{k+1} &= R_k - \lambda_k A P_k \\ P_{k+1} &= R_{k+1} + \mu_k P_k \end{aligned} \qquad (2\text{-}140)$$

$$\lambda_k = \frac{R_k' R_k}{P_k' A P_k} \qquad \mu_k = \frac{R_{k+1}' R_{k+1}}{R_k' R_k} \qquad (2\text{-}141)$$

At any stage the approximate solution is

$$V_{k+1} = V_1 + \lambda_1 P_1 + \lambda_2 P_2 + \cdots + \lambda_k P_k \qquad (2\text{-}142)$$

If no round-offs occur, R_{n+1} will vanish and V_{n+1} will be the desired solution X in (2-133). If round-offs (or mistakes) have occurred, the iteration process (2-140) may be continued without alteration until the residual is as small as desired.

[1] See Exercise 1-36.
[2] See Exercise 2-67.
[3] See Exercise 2-67.

Each cycle (2-140) requires $n^2 + 7n + 2$ operations so that a complete n-step solution requires something more than n^3 operations. This is to be compared with $\frac{1}{6}n^3$ operations by the Banachiewcz-Crout algorithm (see Table 1-2).

EXERCISES

2-61. Apply the Lanczos method to the system

$$\begin{bmatrix} 1 & 1 & 1 & 1 \\ 1 & 2 & 2 & 2 \\ 1 & 2 & 3 & 3 \\ 1 & 2 & 3 & 4 \end{bmatrix} \begin{bmatrix} x_1 \\ x_2 \\ x_3 \\ x_4 \end{bmatrix} = \lambda \begin{bmatrix} x_1 \\ x_2 \\ x_3 \\ x_4 \end{bmatrix}$$

starting from $V_1 = (0, 0, 0, 1)$.

Ans. $V_2 = (1, 2, 3, 0)$, $V_3 = (1, 1, -1, 0)$, $V_4 = (\frac{5}{42}, -\frac{4}{42}, \frac{1}{42}, 0)$. From $p_1 = 0$, $\lambda = 4$; from $p_2 = 0$, $\lambda = 8.2749$, 0.7251; from $p_3 = 0$, $\lambda = 8.2908$, 0.9955, 0.3804; from $p_4 = 0$, $\lambda = 8.29086$, 1.00000, 0.42602, 0.28312.

2-62. Show that α_k defined by (2-122) is the Rayleigh quotient for V_k.

2-63. Show that V_3 of (2-120) is orthogonal with respect to B to both V_1 and V_2. Verify the identities

$$V_j' B V_{k+1}^* = V_j' A V_k = V_k' A V_j = V_k' B V_{j+1}^*$$

and use them to show that V_k of (2-122) is orthogonal to every V_j ($j = 1, \ldots, k - 1$) provided the V_j are already mutually orthogonal. Then use induction to show that all the V_j must be orthogonal.

2-64. Use the expansion theorem (2-42) to show that if the vector V_{n+1} is orthogonal to every one of n nonzero vectors V_j ($j = 1, \ldots, n$) which are mutually orthogonal then the elements of V_{n+1} must all vanish.

2-65. Show that for the sequence defined by (2-122) the quotient

$$\frac{V_j' A V_k}{V_j' B V_j}$$

equals 1, α_j, and β_j when k equals $j - 1$, j, and $j + 1$, respectively, and is zero for all other values at k. The identities of Exercise 2-63 are helpful here.

2-66. Write out the determinant (2-79) for the system (2-130), and show that for increasing m the characteristic equations are $p_m(\lambda_R) = 0$ ($m = 1, \ldots, n$) where the p_m are defined by (2-123).

2-67. Show that $R_1' R_2 = 0$ when R_2 is constructed according to (2-137). Show also that $P_1' A P_2 = 0$ when P_2 is constructed according to (2-138). Then show that R_{k+1} of (2-140) is orthogonal with respect to the unit matrix to all the previous R_j, provided that all these previous R_j are mutually orthogonal with respect to I and that all the P_j ($j = 1, \ldots, k$) are mutually orthogonal with respect to A. Similarly show that P_{k+1} of (2-140) is orthogonal with respect to A to all the previous P_j provided that these are already mutually orthogonal with respect to A and that the R_j ($j = 1, \ldots, k + 1$) are mutually orthogonal with respect to I.

2-68. Apply the Hestenes-Stiefel method to solve the system (1-23) for the continuous beam, starting with the initial trial $V_1 = (0, 0, 0)$.

$$Ans. \quad V_2 = (\tfrac{1}{2}, 0, 0), \quad V_3 = (\tfrac{4}{7}, -\tfrac{1}{7}, 0), \quad V_4 = (\tfrac{11}{19}, -\tfrac{3}{19}, \tfrac{1}{19}).$$

2-10. Relaxation Methods

By a relaxation method we mean an informal type of iteration in which a large measure of initiative is left to the computer. These methods are primarily suited to hand computation. For a skilled computer they are more interesting and faster than the routine fixed procedures previously described. We describe two types of relaxation methods designed for eigenvalue problems. One method is based on the enclosure theorem, while the other makes use of Rayleigh's quotient.

Use of the Enclosure Theorem. A method of this type was proposed by Vazsonyi[1] in 1944. In the eigenvalue problem $AX = \lambda BX$ let B be a diagonal matrix of positive elements. Then the enclosure theorem (2-46) holds[2] for

$$HX = \lambda X \qquad (2\text{-}143)$$

with $H = B^{-1}A$. For a trial vector V we compute HV and the n ratios $l_j(j = 1, \ldots, n)$ as in (2-45). The relaxation procedure then consists in altering V with intent to decrease the spread between l_{max} and l_{min}. The method is particularly useful when approximate locations of several eigenvalues are desired. For each mode an approximating vector and upper and lower bounds for the eigenvalue are obtained.

We illustrate the method by applying it to the formulation (2-34) for the three-mass vibration system. The l_j are here defined as follows:

$$\begin{aligned}
5v_1 - 2v_2 \qquad &= l_1 v_1 \\
-v_1 + \tfrac{3}{2}v_2 - \tfrac{1}{2}v_3 &= l_2 v_2 \\
- \tfrac{1}{3}v_2 + \tfrac{1}{3}v_3 &= l_3 v_3
\end{aligned} \qquad (2\text{-}144)$$

For the initial trial vector the left sides of (2-144) are evaluated and divided in turn by v_1, v_2, and v_3 to obtain l_1, l_2, l_3. For subsequent trials the *changes* in the left sides of (2-144) can be obtained from the *changes* in V by using the unit-operations table shown in the first three lines of Table 2-1.

The actual relaxation solution begins in line (4) of Table 2-1. The trial vector (25, 50, 100) is recorded in the v_j columns and the left sides of (2-144) entered in the $l_j v_j$ columns. The quotients of these (obtained to two places on a slide rule) are entered in the l_j columns. The results of line (4) tell us that there is at least one eigenvalue between 0 and 1. In lines (5) to (7) the trial vector is altered one element at a time. In each case the element altered is underlined. The changes in $l_j v_j$ are obtained

---

[1] A. Vazsonyi, A Numerical Method in the Theory of Vibrating Bodies, *J. Appl. Phys.*, **15**, 598–606 (1944). See also Collatz, *op. cit.*, p. 291, and S. H. Crandall, On a Relaxation Method for Eigenvalue Problems, *J. Math. and Phys.*, **30**, 140–145 (1951).

[2] The enclosure theorem is probably valid for a much wider class of matrices B, but criteria for deciding in advance whether or not a system will have the enclosure property are not yet available.

TABLE 2-1. RELAXATION SOLUTION OF (2-144)

Line	Δv_1	$\Delta(l_1v_1)$	Δv_2	$\Delta(l_2v_2)$	Δv_3	$\Delta(l_3v_3)$
(1)	1	5	0	-1	0	0
(2)	0	-2	1	1.5	0	-0.333
(3)	0	0	0	-0.5	1	0.333

	v_1	l_1v_1	l_1	v_2	l_2v_2	l_2	v_3	l_3v_3	l_3
(4)	25	25	1.00	50	0	0	100	16.7	0.17
(5)	25	17	0.68	54	6	0.11	100	15.3	0.15
(6)	22	2	0.09	54	9	0.17	100	15.3	0.15
(7)	22.3	3.5	0.16	54	8.7	0.16	100	15.3	0.15
(8)	0	-200	$-\infty$	100	150	1.5	0	-33.3	$-\infty$
(9)	50	50	1.0	100	100	1.0	0	-33.3	$-\infty$
(10)	50	50	1.0	100	120	1.2	-40	-46.7	1.16
(11)	52	60	1.15	100	118	1.18	-40	-46.7	1.16
(12)	100	500	5.0	0	-100	$-\infty$	0	0	....
(13)	100	560	5.6	-30	-145	4.8	0	10	∞
(14)	100	560	5.6	-30	-146	4.9	2	10.7	5.4

from the unit operations of lines (1) to (3). Only the final values are recorded in the table. At the end of line (7) we have located an eigenvalue between 0.15 and 0.16.

In lines (8) to (11) and (12) to (14) the other two eigenvalues are roughly located. Note that the same operations table and the same procedure are followed for all modes. No explicit rules are given for choosing alterations. In general it is wise to concentrate on equalizing the l_j corresponding to the v_j of greatest magnitude. The l_j corresponding to small v_j can then be readily brought into line without changing the others very much. See lines (7) and (14) of Table 2-1 for examples.

When one mode has been isolated, the computer must explore to locate another. An aid here is the orthogonality condition. It is unnecessary to construct a new trial vector which is *exactly* orthogonal to the known mode, but the new trial should at least be *different*. See lines (8) and (12) of Table 2-1.

In principle the above procedure could be used to obtain unlimited accuracy. In practice the method becomes too cumbersome as soon as the divisions required for the l_j require greater accuracy than can be obtained from a slide rule. We can always get greater accuracy for the eigenvalues if we construct Rayleigh's quotient for the approximate mode. For instance, the vector (22.3, 54, 100) of line (7), Table 2-1, yields

$\lambda_R = 0.15463$, which differs from λ_1 by only one unit in the fifth decimal place.

Use of Rayleigh's Quotient. For the eigenvalue problem $AX = \lambda BX$, where B is *any* symmetric positive definite matrix, let V be an approximate mode and Λ be an approximate eigenvalue. We then define a corresponding *residual vector* R as follows:

$$AV - \Lambda BV = R \qquad (2\text{-}145)$$

If not all the elements of V are zero, then all the elements of R can vanish only if V is actually an eigenvector X_p and Λ is the corresponding eigenvalue λ_p. If $V = X_p$ but $\Lambda \neq \lambda_p$, then R is the nonzero vector $(\lambda_p - \Lambda)BX_p$. Several procedures have been proposed[1] which amount to altering V and Λ until R is made as small as desired. A step common to many of these procedures is the replacement of Λ by an improved value which is the Rayleigh quotient of the current vector V.

A method of this type was suggested by Rayleigh,[2] whose description (with minor changes in the notation) follows: "Beginning with assumed rough approximations to the modal elements $v_1, v_2, \ldots, v_n$ we may calculate a first approximation Λ from λ_R. With this value of Λ recalculate the v_j from any $n - 1$ of the n equations $AV - \Lambda BV = 0$, then again by application of λ_R determine an improved value of Λ, and so on."

This amounts to requiring that $n - 1$ of the elements of the residual R in (2-145) should be zero. The remaining element is not zero until the true mode has been obtained.

A relaxation method using the same general principle was developed by Southwell.[3] In this procedure the shape of the residual R in (2-145) is not prescribed. With Λ fixed V is altered with intent to decrease the elements of R. The operations table described in Sec. 1-7 is employed in this process. After some decrease has been obtained, Λ is recalculated as the Rayleigh quotient of the altered V. In selecting operations the computer should try not only to reduce the elements of R but to make them proportional to the elements of BV since V is an eigenvector if the elements of R and BV are exactly proportional. Here again the same procedure can be followed for all modes. It is up to the computer to select new trials and operate on them in such a manner that convergence back to modes already known is avoided.

[1] See S. H. Crandall, Iterative Procedures Related to Relaxation Methods for Eigenvalue Problems, *Proc. Roy. Soc.* (*London*), **A207**, 416–423 (1951).

[2] J. W. Strutt, 3d Baron Rayleigh, "The Theory of Sound," 2d ed., vol. 1, Dover Publications, New York, 1945, p. 110. See also W. Kohn, A Variational Iteration Method for Solving Secular Equations, *J. Chem. Phys.*, **17**, 670 (1949).

[3] R. V. Southwell, "Relaxation Methods in Engineering Science," Oxford University Press, New York, 1940, pp. 131–161.

In all these methods it should be remembered that modes are determined only to within a multiplicative factor. All residuals can be reduced by simply scaling down the trial vector. Such false indications of progress can be avoided by fixing the magnitude of the largest element of V at some arbitrary level, e.g., unity.

EXERCISES

2-69. Apply the relaxation method based on the enclosure theorem to isolate the eigenvalues of the system of Exercise 2-43. See Exercise 2-75 for the true eigenvalues.

2-70. If the matrix B in the eigenvalue problem $AX = \lambda BX$ is a diagonal matrix of positive elements, show that orthogonal modes cannot have the same algebraic sign patterns; e.g., $(+, +, -)$ and $(+, -, +)$ are different sign patterns for third-order vectors, but $(+, +, +)$ and $(-, -, -)$ are essentially the same.

2-71. For torsional vibration systems like that of Fig. 2-3 the vector V in (2-145) may be obtained, when Λ is given and R is to have zero for every element but the last, by using a *Holzer*[1] *table*. For Fig. 2-3 the table would have the following form:

No.	(1) J	(2) $J\omega^2$	(3) v	(4) $J\omega^2 v$	(5) $\Sigma J\omega^2 v$	(6) k	(7) $(1/k)\Sigma J\omega^2 v$
1	J	$\cdots$	1.00	$\cdots$	$\cdots$	k	$\cdots$
2	J	$\cdots$	v_2	$\cdots$	$\cdots$	k	$\cdots$
3	J	$\cdots$	v_3	$\cdots$	$\cdots$	k	$\cdots$
4	J	$\cdots$	v_4	$\cdots$	$\cdots$	$\frac{1}{2}k$	$\cdots$
5	$4J$	$\cdots$	v_5	$\cdots$	r_5		

When ω has been selected, the second column is filled in by multiplying each entry in column (1) by ω^2. The computation then proceeds across each line from (3) to (7). The first element of V is arbitrarily taken as unity. The succeeding elements are obtained in sequence by subtracting the entry in column (7), line p, from v_p to give v_{p+1}. The entry in column (5) at any stage is the sum of all the entries in column (4) up to this point. The final entry in column (5) is the single nonzero element of the residual vector. Complete this table when $\omega^2 = 0.20k/J$. *Ans.* $r_5 = -0.274k$.

2-72. Show that completing the Holzer table of Exercise 2-71 is equivalent to requiring that the equations of Exercise 2-1 when put in the form of (2-145) have the residual $(0, 0, 0, 0, -r_5)$. Interpret this as a forced vibration with excitation supplied only at the last disk. Show that when $0 < \omega < \omega_2$, where ω_2 is the first nonzero natural frequency, the sign of r_5 is positive.

2-73. Show that the entries in column (7) of Exercise 2-71 are the amplitudes of the angles of twist in between each disk. Show that the sum of the products of corresponding elements in columns (5) and (7) is proportional to the maximum potential energy during the vibration. Show that the sum of the products of corresponding elements in columns (3) and (4) is proportional to the maximum kinetic energy during the vibration. Use Exercise 2-31 to obtain a simple rule for computing the Rayleigh quotient of the vector V standing in column (3) of the Holzer table. Verify that $\lambda_R = \omega^2 J/k = 0.1513$ for Exercise 2-71.

[1] H. Holzer, "Die Berechnung der Drehschwingungen," Berlin, 1921.

2-74. Show that if $\Lambda = 1.5$ the operations table for the three-mass system of (2-29), using Southwell's relaxation method, is:

Δv_1	Δv_2	Δv_2	Δr_1	Δr_2	Δr_2
1	0	0	3.5	-2	0
0	1	0	-2	0	-1
0	0	1	0	-1	-3.5

Use this table to improve a trial vector. Try to make the residual elements proportional to the elements of BV. After a few steps compute λ_R. Set $\Lambda = \lambda_R$, and recompute the operations table. Toward which mode does the process seem to be converging?

2-11. Upper and Lower Bounds for Eigenvalues

In using procedures which yield approximate eigenvalues it is sometimes necessary to know bounds on the possible error involved. Some of the procedures already discussed give information of this type. Whenever the *enclosure theorem* can be used, it gives upper and lower bounds for eigenvalues. If the eigenvalues are ordered as follows,

$$\lambda_1 \leq \lambda_2 \leq \cdots \leq \lambda_n \qquad (2\text{-}146)$$

then Rayleigh's quotient or any of the quotients of Schwarz will give an upper bound for λ_1 and a lower bound for λ_n.

We now describe a method[1] for obtaining upper *and* lower bounds for any eigenvalue both of which possess the second-order error property of the Rayleigh and Schwarz quotients. Let V be an approximation to the eigenvector X_p of the eigenvalue problem $AX = \lambda BX$. Construct the vector HV, where $H = B^{-1}A$, and the following quotients:

$$\lambda_R = \frac{V'AV}{V'BV} = \frac{V'B(HV)}{V'BV}$$
$$Q = \frac{(HV)'B(HV)}{V'BV} \qquad (2\text{-}147)$$

If λ_R is sufficiently close to λ_p so that $\lambda_{p-1} < \lambda_R < \lambda_{p+1}$, we then have the upper and lower bounds for λ_p,

$$\lambda_R - \frac{Q - \lambda_R^2}{\lambda_{p+1} - \lambda_R} \leq \lambda_p \leq \lambda_R + \frac{Q - \lambda_R^2}{\lambda_R - \lambda_{p-1}} \qquad (2\text{-}148)$$

[1] This method was given independently by W. Kohn, A Note on Weinstein's Variational Method, *Phys. Rev.*, (2)**71**, 902–904 (1947), and T. Kato, On the Upper and Lower Bounds of Eigenvalues, *J. Phys. Soc. Japan*, **4**, 334 (1949). The proof given here is that of Kato. It represents an extension of a method for obtaining a lower bound for λ_1 given by G. Temple, The Computation of Characteristic Numbers and Characteristic Functions, *Proc. London Math. Soc.*, (2)**29**, 257–280 (1928).

in which the numerators $Q - \lambda_R^2$ are small quantities of second order in comparison to the error components in V. We shall first prove this result and then give an illustration of its application.

The foundation of the demonstration lies in the obvious remark that if λ_j is *any* one of the eigenvalues (2-146) then the expression

$$(\lambda_j - \lambda_p)(\lambda_j - \lambda_{p-1}) = \lambda_j^2 - \lambda_j(\lambda_p + \lambda_{p-1}) + \lambda_p\lambda_{p-1} \quad (2\text{-}149)$$

can never be negative when λ_{p-1} and λ_p are adjacent eigenvalues. Now let the vector V be expanded in terms of eigenvectors which have been *normalized* with respect to B.

$$V = \sum_{j=1}^{n} c_j X_j \quad (2\text{-}150)$$

We then can evaluate the following triple matrix products:

$$V'BV = \sum_{j=1}^{n} c_j^2$$

$$V'B(HV) = \sum_{j=1}^{n} \lambda_j c_j^2 \quad (2\text{-}151)$$

$$(HV)'B(HV) = \sum_{j=1}^{n} \lambda_j^2 c_j^2$$

To introduce these into (2-149), we multiply (2-149) by c_j^2 and then sum over j. The sum of nonnegative expressions is also nonnegative, and hence we obtain

$$(HV)'B(HV) - (\lambda_p + \lambda_{p-1}) V'B(HV) + \lambda_p\lambda_{p-1} V'BV \geq 0 \quad (2\text{-}152)$$

Dividing through by $V'BV$ and using the definitions (2-147), we have

$$Q - (\lambda_p + \lambda_{p-1})\lambda_R + \lambda_p\lambda_{p-1} \geq 0 \quad (2\text{-}153)$$

Finally, if we add and subtract λ_R^2 on the left of (2-153), we obtain

$$Q - \lambda_R^2 + (\lambda_R - \lambda_p)(\lambda_R - \lambda_{p-1}) \geq 0 \quad (2\text{-}154)$$

which when λ_R is greater than λ_{p-1} yields

$$\frac{Q - \lambda_R^2}{\lambda_R - \lambda_{p-1}} + \lambda_R \geq \lambda_p \quad (2\text{-}155)$$

This completes the proof of the right-hand inequality in (2-148). The left-hand inequality is obtained in the same manner.[1]

In order to use these bounds for λ_p, it is necessary to have estimates of λ_{p-1} and λ_{p+1}. These estimates can, however, be very crude, as illus-

[1] See Exercise 2-78 for the proof that the bounds have second-order error.

trated in the following example. Note that when $p = 1$ in (2-155) the denominator of the fraction is not defined, but this is exactly the case for which we already know that λ_R itself is an upper bound. A similar remark applies to the lower bound for λ_n.

To illustrate the use of (2-148), we choose the vector $(0.52, 1.00, -0.40)$ obtained in line (11) of Table 2-1 for the three-mass vibrating system. The matrices A, B, and H are given in (2-29) and (2-34). Evaluating the quotients (2-147), we find

$$\lambda_R = 1.175102$$
$$Q = 1.380938$$
$$\lambda_R{}^2 = 1.380864 \tag{2-156}$$

and hence (2-148) becomes

$$1.175102 - \frac{0.000074}{\lambda_3 - \lambda_R} \le \lambda_2 \le 1.175102 + \frac{0.000074}{\lambda_R - \lambda_1} \tag{2-157}$$

We now need estimates for λ_1 and λ_3. From Table 2-1 we know that these are about 0.16 and 5, respectively, and therefore that $\lambda_R - \lambda_1$ and $\lambda_3 - \lambda_R$ are about 1 and 4,

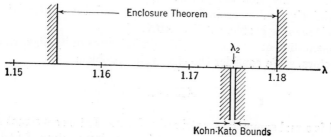

FIG. 2-11. Upper and lower bounds for λ_2 obtained from the approximate mode $(0.52, 1.00, -0.40)$

respectively. Our error bound will be conservative if we use instead 0.7 and 2 for $\lambda_R - \lambda_1$ and $\lambda_3 - \lambda_R$. When this is done, (2-157) becomes

$$1.175065 \le \lambda_2 \le 1.175208 \tag{2-158}$$

A comparison of these bounds with those furnished by the enclosure theorem is shown in Fig. 2-11.

EXERCISES

2-75. Apply the Kohn-Kato bounds to the approximate modes for the system of Exercise 2-43, which were obtained in Exercise 2-69.

$$Ans.\ \lambda_1 = 0.9539,\ \lambda_2 = 2.6152,\ \lambda_3 = 14.4309.$$

2-76. Verify the identity

$$\frac{(HV - \lambda_R V)'B(HV - \lambda_R V)}{V'BV} = Q - \lambda_R{}^2$$

and hence show that $Q - \lambda_R{}^2$ must always be positive.

2-77. Show that when the Schwarz quotients are properly identified with V the expression $Q - \lambda_R{}^2$ may be written as $\nu_0(\nu_1 - \nu_0)$.

2-78. Use either of the identities of the two foregoing exercises to show that if in (2-150) $|c_j/c_p| \leq \epsilon$ for $j \neq p$ then $Q - \lambda_R^2$ is less than a constant multiple of ϵ^2 when $\epsilon \to 0$.

2-12. Diagonalization of Matrices by Successive Rotations

In Sec. 2-7 we considered a type of iteration in which the basic step consisted in multiplying a square matrix into a column matrix to obtain an improved column. In this section we consider a type of iteration in which the basic step consists in multiplying a square matrix by a pair of square matrices to obtain an improved square matrix. Whereas the earlier iteration led only to a single mode, the present procedure leads simultaneously to all the modes and their eigenvalues. The method is an old one[1] but has not been used much until recently because of the excessive amount of arithmetic required. It is, however, quite well suited[2] to automatic computation because of its simple standardized program. To describe the procedure, we first develop the theory for the special eigenvalue problem $AX = \lambda X$ and then extend it to the general eigenvalue problem $AX = \lambda BX$.

Representation of a Symmetric Matrix in Terms of its Eigenvalues and Eigenvectors. We consider the eigenvalue problem

$$AX = \lambda X \tag{2-159}$$

where A is a real symmetric matrix of order n. Let us construct a square *modal matrix*, M, whose columns are the true eigenvectors of (2-159) and a diagonal matrix $D(\lambda)$ whose elements are the corresponding eigenvalues.

$$M = \begin{bmatrix} \uparrow & \uparrow & & \uparrow \\ X_1 & X_2 & \cdots & X_n \\ \downarrow & \downarrow & & \downarrow \end{bmatrix} \qquad D(\lambda) = \begin{bmatrix} \lambda_1 & 0 & \cdots & 0 \\ 0 & \lambda_2 & \cdots & 0 \\ \cdot & \cdot & \cdots & \cdot \\ 0 & 0 & \cdots & \lambda_n \end{bmatrix} \tag{2-160}$$

The eigenvectors are assumed to be *normalized* ($X_j' X_j = 1$) as well as *orthogonal* ($X_j' X_k = 0$, $j \neq k$). According to the definition of matrix multiplication, we have

$$AM = MD(\lambda) \tag{2-161}$$

To solve this for A, we study the properties of the modal matrix M.

[1] C. G. J. Jacobi, Ein leichtes Verfahren die in der Theorie der Sakularstörungen vorkommenden Gleichungen numerisch aufzulösen, *J. reine u. angew. Math.*, **30**, 51–94 (1846).

[2] R. T. Gregory, Computing Eigenvalues and Eigenvectors of a Symmetric Matrix on the Illiac, *Math. Tables, Other Aids to Comp.*, **7**, 215–220 (1953).

Since the eigenvectors are orthonormal, we see that

$$M'M = \begin{bmatrix} \leftarrow X_1 \rightarrow \\ \leftarrow X_2 \rightarrow \\ \cdots \cdots \\ \leftarrow X_n \rightarrow \end{bmatrix} \begin{bmatrix} \uparrow & \uparrow & & \uparrow \\ X_1 & X_2 & \cdots & X_n \\ \downarrow & \downarrow & & \downarrow \end{bmatrix} = \begin{bmatrix} 1 & 0 & \cdots & 0 \\ 0 & 1 & \cdots & 0 \\ \cdots\cdots\cdots\cdots \\ 0 & 0 & \cdots & 1 \end{bmatrix} \quad (2\text{-}162)$$

Thus the *transpose* of M is also its *inverse*. Multiplying both sides of (2-161) into M' yields

$$A = MD(\lambda)M' \quad (2\text{-}163)$$

This is a useful representation of a symmetric matrix. It puts the eigenvalues and eigenvectors clearly in evidence. When this representation is available, we can easily derive from it matrices which are functions of A; for example, A^2 is given by

$$A^2 = MD(\lambda^2)M' \quad (2\text{-}164)$$

where $D(\lambda^2)$ is a diagonal matrix whose elements are the squares of the corresponding elements in $D(\lambda)$. The proof of (2-164) is given by simply multiplying the right side of (2-163) by itself and using (2-162). Similarly, if A is nonsingular (if all its eigenvalues are nonzero), its inverse is given by

$$A^{-1} = MD\left(\frac{1}{\lambda}\right)M' \quad (2\text{-}165)$$

In the development for the general eigenvalue problem we will need the square root of a symmetrical matrix.

$$A^{\frac{1}{2}} = MD(\lambda^{\frac{1}{2}})M' \quad (2\text{-}166)$$

Returning to (2-161), we can multiply M' into both sides to obtain

$$M'AM = D(\lambda) \quad (2\text{-}167)$$

This expression forms the basis of our iteration. In stepwise fashion we build up the square matrices M' and M, which reduce A to the diagonal matrix $D(\lambda)$.

Diagonalization of a Symmetric Matrix. Starting with the given matrix A, we select a sequence of square matrices T_r and operate as follows,

$$\begin{aligned} A &= A_0 \\ T'_1 A_0 \ T_1 &= A_1 \\ T'_2 A_1 \ T_2 &= A_2 \\ &\cdots\cdots\cdots \\ T'_r A_{r-1} T_r &= A_r \\ &\cdots\cdots\cdots \end{aligned} \quad (2\text{-}168)$$

so that

$$T'_r \cdots T'_2 T'_1 A T_1 T_2 \cdots T_r = A_r \quad (2\text{-}169)$$

It remains to select the T_r so that the continued product $T_1 T_2 \cdots T_r$ converges toward M and A_r converges toward $D(\lambda)$.

Note that the matrices A_r will all be symmetrical[1] no matter what T_r are used. If each of the T_r satisfies the condition that its *transpose* is identical with its *inverse*, then the continued product of the T_r will also satisfy[2] this condition.

The most important property of the T_r actually employed is that each T_r causes a particular off-diagonal element in A_r to vanish. For simplicity of notation, let the elements of A_{r-1} be a_{jk} and the elements of A_r be b_{jk}. Then if $a_{pq} \neq 0$, we set

$$
T_r = \begin{bmatrix}
1 & 0 & \cdots & 0 & 0 & 0 & 0 & \cdots & 0 \\
0 & 1 & \cdots & 0 & 0 & 0 & 0 & \cdots & 0 \\
 & & \cdots & & & & & \cdots & \\
0 & 0 & \cdots & c & 0 & 0 & -s & \cdots & 0 \\
0 & 0 & \cdots & 0 & 1 & 0 & 0 & \cdots & 0 \\
0 & 0 & \cdots & 0 & 0 & 1 & 0 & \cdots & 0 \\
0 & 0 & \cdots & s & 0 & 0 & c & \cdots & 0 \\
 & & \cdots & & & & & \cdots & \\
0 & 0 & \cdots & 0 & 0 & 0 & 0 & \cdots & 1
\end{bmatrix}
\qquad (2\text{-}170)
$$

where $c = \cos \theta$ and $s = \sin \theta$. T_r may be considered[3] as a *rotation* through the angle θ in the (p, q) plane. It has the property that its transpose is also its inverse. A simple exercise in matrix multiplication shows that $b_{jk} = a_{jk}$ except in the p row and column and the q row and column, where we have

$$
\begin{aligned}
b_{pp} &= a_{pp} \cos^2 \theta + 2a_{pq} \sin \theta \cos \theta + a_{qq} \sin^2 \theta \\
b_{pq} &= a_{pq}(\cos^2 \theta - \sin^2 \theta) - (a_{pp} - a_{qq}) \sin \theta \cos \theta \\
b_{qq} &= a_{pp} \sin^2 \theta - 2a_{pq} \sin \theta \cos \theta + a_{qq} \cos^2 \theta \\
b_{pj} &= a_{pj} \cos \theta + a_{qj} \sin \theta \\
b_{qj} &= -a_{pj} \sin \theta + a_{qj} \cos \theta
\end{aligned} \right\} \quad j \neq p, q
\qquad (2\text{-}171)
$$

If now θ is taken so that

$$
\tan 2\theta = \frac{2a_{pq}}{a_{pp} - a_{qq}}
\qquad (2\text{-}172)
$$

we shall have $b_{pq} = 0$. This is obtained at the expense of altering all

---

[1] See Exercise 2-82.
[2] See Exercise 2-83.
[3] See Exercise 2-85.

the elements in the two rows and two columns involved. This represents, however, a definite step toward the diagonal form because[1] the sum of the squares of the off-diagonal elements in A_r is less than that for A_{r-1} by the amount $2a_{pq}^2$. The process (2-169) therefore converges to (2-167) provided that every off-diagonal element (such as a_{pq}) is treated an unlimited number of times in the sequence of T_r.

In practice some programs select T_r according to the largest off-diagonal element a_{pq} in A_{r-1}, while others[2] simply take the off-diagonal elements in systematic order row by row above the main diagonal. When the latter process was applied to a number of random symmetric matrices of orders 4 to 20, it was found[3] that seven complete sweeps through a matrix were always sufficient for convergence.

For automatic computation the values c and s in (2-170) can be obtained[4] by direct algebraic operations on the elements of A_{r-1} as follows:

$$t = \frac{2a_{pq}}{a_{pp} - a_{qq} - \sqrt{(a_{pp} - a_{qq})^2 + 4a_{pq}^2}}$$
$$c = (1 + t^2)^{-\frac{1}{2}} \qquad (2\text{-}173)$$
$$s = tc$$

The time required to compute c and s from (2-173) is independent of the order of the system being treated. Beyond this the number of multiplication operations involved in obtaining A_r from A_{r-1} according to (2-171) is essentially $4n$ if we make use of the symmetry of A_r. The number of multiplications involved in multiplying $T_1T_2 \cdots T_{r-1}$ into T_r is also $4n$. Thus each step of the iteration process involves somewhat more than $8n$ operations. To make 7 complete sweeps through a system of order n would require more than $28n^3$ operations.

Application to the General Eigenvalue Problem. The general eigenvalue problem

$$AX = \lambda BX \qquad (2\text{-}174)$$

where A and B are symmetric and B is positive definite, can be solved by making two successive applications of the diagonalization process just

---

[1] See Exercise 2-86.

[2] The relative advantage of the two methods depends among other things on how the speed of the machine in searching for a largest element compares with its speed in performing arithmetic (see Gregory, *op. cit.*).

[3] See Gregory, *op. cit.*

[4] See Exercise 2-87.

outlined. We begin by introducing the matrices M and $D(\lambda)$ of (2-160). Here the columns of the modal matrix M are the true eigenvectors of (2-174) normalized with respect to B, and $D(\lambda)$ is the diagonal matrix of the corresponding eigenvalues. The modal matrix then satisfies

$$AM = BMD(\lambda) \qquad (2\text{-}175)$$

To reduce this to the problem of diagonalizing a symmetric matrix, we first diagonalize the matrix B, obtaining the representation corresponding to (2-163). This permits us to construct $B^{\frac{1}{2}}$ according to (2-166) and $B^{-\frac{1}{2}}$. The positive definiteness of B assures us that $B^{-\frac{1}{2}}$ exists. We then manipulate (2-175) as follows:

$$
\begin{aligned}
A(B^{-\frac{1}{2}}B^{\frac{1}{2}})M &= (B^{\frac{1}{2}}B^{\frac{1}{2}})MD(\lambda) \\
AB^{-\frac{1}{2}}(B^{\frac{1}{2}}M) &= B^{\frac{1}{2}}(B^{\frac{1}{2}}M)D(\lambda) \\
(B^{-\frac{1}{2}}AB^{-\frac{1}{2}})(B^{\frac{1}{2}}M) &= (B^{\frac{1}{2}}M)D(\lambda)
\end{aligned}
\qquad (2\text{-}176)
$$

If we now define

$$
\begin{aligned}
\bar{A} &= B^{-\frac{1}{2}}AB^{-\frac{1}{2}} \\
\bar{M} &= B^{\frac{1}{2}}M
\end{aligned}
\qquad (2\text{-}177)
$$

our last result becomes

$$\bar{A}\bar{M} = \bar{M}D(\lambda) \qquad (2\text{-}178)$$

which is precisely the form of (2-161). Furthermore $\bar{A}$ is symmetric,[1] and $\bar{M}$ satisfies the condition that its transpose is also its inverse.[2] The eigenvalues and modal matrix of $\bar{A}$ can thus be obtained by the iterative diagonalization process just described. These eigenvalues are the same as those of the original system (2-174). The modal matrix of the original system requires an additional matrix multiplication.

$$M = B^{-\frac{1}{2}}\bar{M} \qquad (2\text{-}179)$$

The diagonalizations of B and $\bar{A}$ are the principal computations in this process. Forming $B^{-\frac{1}{2}}$, $\bar{A}$ and obtaining M from (2-179) require somewhat more than $3n^3$ additional multiplication operations if we make use of symmetry wherever possible.

EXERCISES

2-79. Apply the diagonalization process of this section to the system of Exercise 2-28. This is an unrepresentative example in that convergence to the exact solution is obtained in just one step.

2-80. Show that if A, B, and C are matrices for which the triple product ABC is

[1] See Exercise 2-88.
[2] See Exercise 2-89.

defined then the transpose of the triple product is given by the triple product of the transposes taken in reverse order,

$$(ABC)' = C'B'A'$$

Suggestion. Prove the corresponding result for a double product by actual evaluation of sample elements of the products $(AB)'$ and $B'A'$, and then use induction.

2-81. Show that if A, B, and C are nonsingular square matrices, then

$$(ABC)^{-1} = C^{-1}B^{-1}A^{-1}$$

2-82. Use the result of Exercise 2-80 to show that, if A is a symmetric matrix, then $T'AT$ is also symmetric for any square matrix T.

2-83. If $T_r'T_r = I$ for all values of r, where I is the identity matrix, show that the continued product $U = T_1 T_2 \cdots T_r$ also satisfies $U'U = I$. *Suggestion.* Use Exercise 2-80 to expand U'.

2-84. When the program for diagonalizing a symmetric matrix A is available, show that the inverse of A can be constructed by using somewhat more than $n^3/2$ operations to evaluate (2-165). This provides a possible way to solve[1] the *equilibrium* problem $AX = C$, where X and C are unknown and known matrices. Jacobi[2] suggested that diagonalization could be joined with iteration by total steps to solve the equilibrium problem. The idea was that diagonalization should be carried out only far enough to give A_r a strongly predominating main diagonal, for then iteration by whole steps would converge very rapidly. Show that, if the diagonalization (2-169) has been carried to the point $U'AU = A_r$, the solution to $AX = C$ is given by $X = U\bar{X}$, where $\bar{X}$ is the solution to the strongly diagonal equilibrium problem $A_r\bar{X} = \bar{C} = U'C$.

2-85. Consider a point P in a plane whose coordinates are (x_1, x_2) with respect to rectangular axes through the origin. Let these axes be rotated through the angle θ. The same point P now has the coordinates $(\bar{x}_1, \bar{x}_2)$. Let X be the column matrix with elements x_1 and x_2, and similarly let $\bar{X}$ be the column with elements $\bar{x}_1$ and $\bar{x}_2$. Show that $X = T\bar{X}$ and $\bar{X} = T'X$, where

$$T = \begin{bmatrix} \cos\theta & -\sin\theta \\ \sin\theta & \cos\theta \end{bmatrix}$$

Verify that $T'T = I$. Consider the quadratic form $Q = X'AX$, where A is a symmetric matrix. Verify that in the rotated coordinates $Q = \bar{X}'B\bar{X}$, where $B = T'AT$. Evaluate the elements of B, and compare with (2-171).

2-86. Show that the sum of the squares of *all* the elements a_{jk} of A_{r-1} equals the corresponding sum of the squares of all the elements b_{jk} of A_r no matter what angle θ is used in (2-171). If θ is chosen to satisfy (2-172), show that

$$\sum_{j=1}^{n} b_{jj}^2 = 2a_{pq}^2 + \sum_{j=1}^{n} a_{jj}^2$$

and hence that the sum of the squares of the off-diagonal elements is decreased by $2a_{pq}^2$.

---

[1] This idea can be extended to nonsymmetrical matrices A. See E. G. Kogbetliantz, Solution of Linear Equations by Diagonalization of Coefficients Matrix, *Quart. Appl. Math.*, **13**, 123–132 (1955).

[2] G. G. J. Jacobi, Über eine neue Auflösungsart der bei der Methode der kleinsten Quadrate vorkommenden linearen Gleichungen, *Schumacher Astron. Nochrichten*, **22** (523) (1844). See his "Gesammelte Werke," vol. 3, Berlin, 1884, pp. 467–478.

2-87. Verify that the relations (2-173) are equivalent to (2-172). Note that there are two acute angles satisfying (2-172). The choice in (2-173) makes $b_{qq} > b_{pp}$, which tends to put the eigenvalues in $D(\lambda)$ in algebraic order.

2-88. If A is a symmetric square matrix, show that $T'AT$ is also symmetric for any symmetric square matrix T of the same order.

2-89. Verify that in the terminology of (2-177) $\bar{M}'\bar{M} = I$ implies $M'BM = I$, and vice versa. The latter is a statement that the modes of (2-174) are normalized with respect to B.

CHAPTER 3

PROPAGATION PROBLEMS IN SYSTEMS
WITH A FINITE NUMBER OF DEGREES OF FREEDOM

The diffusion of heat and the transmission of disturbances by wave motion are familiar physical illustrations of propagation phenomena. Engineering examples include the problems of cooling and heating of systems with known initial temperature distributions, and the transient response of mechanical and electrical systems when disturbed from known configurations. The fundamental problem in the analysis of these phenomena is that of predicting the future behavior of a system in terms of initial data taken at the starting instant. In the present chapter we consider numerical procedures for the solution of propagation problems in systems whose instantaneous state can be described by a finite number of quantities.

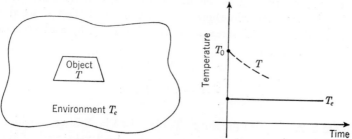

FIG. 3-1. Cooling of an object subject to Newton's law of cooling.

3-1. Particular Examples

Mathematical formulation of propagation problems from physical systems is illustrated by means of three examples.

3-1. Cooling of an object.

3-2. Hydraulic surge tank.

3-3. Response of a nonlinear spring-mass system.

In each case particular data are assumed and the formulations reduced to nondimensional terms in preparation for numerical solution.

Problem 3-1. Cooling of an Object. We consider in Fig. 3-1 an object with temperature T in an environment of temperature T_e. Newton's law of cooling, which states that the time rate of cooling is proportional to the difference between the temperatures of the object and the environment, is assumed, i.e.,

$$-\frac{dT}{dt} = k(T - T_e) \tag{3-1}$$

The propagation problem for this system consists in predicting the temperature history of the object from a knowledge of the environment's temperature history. Let us consider the almost trivial case in which T_e remains constant for $t > 0$. If the object temperature is T_0 at $t = 0$, the problem is to integrate (3-1) for $t > 0$ starting from the initial value $T = T_0$.

This may be cast into dimensionless form by introducing the following variables:

$$x = \frac{T - T_e}{T_0 - T_e} \qquad t' = kt \tag{3-2}$$

We then have the problem of determining $x(t')$ for $t' > 0$ from the conditions

$$
\begin{aligned}
x &= 1 \qquad \text{at } t' = 0 \\
\frac{dx}{dt'} &= -x \qquad \text{for } t' > 0
\end{aligned}
\tag{3-3}
$$

When we return to this problem, we shall always consider this nondimensional form. It will therefore cause no confusion if we drop the prime from the dimensionless time variable in (3-3) and consider the basic formulation to be

$$
\begin{aligned}
x &= 1 \qquad t = 0 \\
\frac{dx}{dt} &= -x \qquad t > 0
\end{aligned}
\tag{3-4}
$$

Problem 3-2. Hydraulic Surge Tank. In Fig. 3-2 we consider a conduit supplying a hydraulic turbine. In the steady state when the valve is open the flow velocity in the conduit is V_0. Because of fluid friction in the long conduit the pressure p_2 is less than p_1, and the water level in the surge tank does not rise the full height H. There is a head loss given[1] by

$$H_L = f\frac{L}{D}\frac{V_0^2}{2g} \tag{3-5}$$

where f is the friction factor and g is the acceleration of gravity.

[1] See, for example, J. C. Hunsaker and B. G. Rightmire, "Engineering Applications of Fluid Mechanics," McGraw-Hill Book Company, Inc., New York, 1947, p. 128. For an extended treatment of the surging problem see G. R. Rich, "Hydraulic Transients," McGraw-Hill Book Company, Inc., New York, 1951.

We shall study the propagation problem that ensues when the steady state just described is disrupted by suddenly closing the valve at $t = 0$. The level in the surge tank will surge up and down until all the excess kinetic energy is dissipated by fluid friction. Of practical interest is the maximum value of y reached in the first surge.

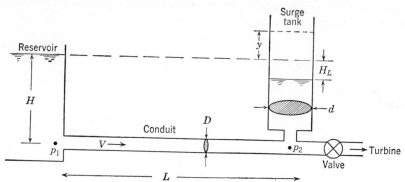

FIG. 3-2. Surging takes place when the valve is suddenly closed.

The physical laws governing the surging process are continuity of flow and Newton's second law of motion. If the instantaneous level in the surge tank is y, the velocity is dy/dt and continuity requires that

$$\frac{\pi d^2}{4} \frac{dy}{dt} = \frac{\pi D^2}{4} V \tag{3-6}$$

where V is the instantaneous velocity in the conduit. The deceleration of the liquid column in the conduit is due to the pressure difference $p_2 - p_1$ and the friction drag. If ρ is the mass density of the liquid, we have Newton's law

$$-\rho L \frac{\pi D^2}{4} \frac{dV}{dt} = (p_2 - p_1) \frac{\pi D^2}{4} + \tau \pi D L \tag{3-7}$$

where τ is shear stress at the conduit wall. Because of the slowness of the surge $p_2 - p_1$ may be approximated by the static pressure difference $\rho g y$ due to the head difference y. The wall shear stress is[1] $\tau = f\rho V^2/8$. Inserting these in (3-7), we obtain

$$L \frac{dV}{dt} = -gy - f \frac{L}{D} \frac{V^2}{2} \tag{3-8}$$

Equations (3-6) and (3-8) are simultaneous ordinary differential equations for the variables y and V. The propagation problem consists in

[1] See, for example, Hunsaker and Rightmire, *op. cit.*, p. 125.

finding solutions which also satisfy the initial conditions

$$V = V_0$$
$$y = -f\frac{L}{D}\frac{V_0^2}{2g} \tag{3-9}$$

at $t = 0$.

A dimensionless formulation can be obtained by introducing the following nondimensional variables:

$$x_1 = \frac{y}{f(L/D)(V_0^2/2g)} \qquad x_2 = \frac{V}{V_0} \qquad t' = \frac{t}{2D/fV_0} \tag{3-10}$$

The initial conditions (3-9) and the governing equations (3-6) and (3-8) then become

$$\left.\begin{array}{l} x_1 = -1 \\ x_2 = 1 \end{array}\right\} \quad t = 0$$

$$\left.\begin{array}{l} \dfrac{dx_1}{dt} = \theta x_2 \\[2mm] \dfrac{dx_2}{dt} = -x_1 - x_2^2 \end{array}\right\} \quad t > 0 \tag{3-11}$$

where the dimensionless parameter θ is defined as follows:

$$\theta = \frac{g}{L}\left(\frac{2D^2}{f\,dV_0}\right)^2 \tag{3-12}$$

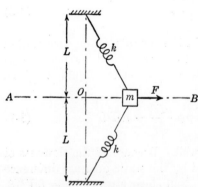

FIG. 3-3. Mass m moves along AOB, restrained by linear springs with stiffness k and initial tension T_0.

Note that in (3-11) we have already dropped the prime from the nondimensional time variable.

Problem 3-3. Response of a Nonlinear Spring-mass System. In Fig. 3-3 the mass m is constrained to remain on the line AOB by the springs which have initial tension T_0 when the mass is at O. We consider the propagation problem of predicting the displacement history of the mass when a constant force F is suddenly applied at $t = 0$. We assume that the system has been at rest prior to this time.

During the motion the governing equation is Newton's second law. If we denote the distance of the mass m from O as u, we have

$$F - 2\frac{u}{\sqrt{L^2 + u^2}}[T_0 + k(\sqrt{L^2 + u^2} - L)] = m\frac{d^2u}{dt^2} \tag{3-13}$$

The initial conditions are

$$u = 0$$

$$\frac{du}{dt} = 0 \qquad \text{at } t = 0 \tag{3-14}$$

Although it is perfectly possible to treat (3-13) by numerical procedures, we shall first introduce a simplification which is often employed in the analysis of such systems. We expand the nonlinear term in (3-13) in powers of u/L (using the binomial theorem) and then retain only the first two terms of the expansion. In this way we obtain

$$F - 2T_0 \left[\frac{u}{L} + \frac{kL - T_0}{2T_0} \left(\frac{u}{L} \right)^3 \right] = m \frac{d^2u}{dt^2} \tag{3-15}$$

in place of (3-13). The solution of the propagation problem then requires the integration of (3-15) subject to the initial conditions (3-14).

To formulate an explicit problem, we shall assume $F = 2T_0$. Then if the dimensionless variables

$$x = \frac{u}{L} \qquad t' = \frac{t}{\sqrt{mL/2T_0}} \tag{3-16}$$

are introduced into (3-14) and (3-15), we have

$$x = \frac{dx}{dt} = 0 \qquad\qquad t = 0$$

$$\frac{d^2x}{dt^2} = 1 - x - \epsilon x^3 \qquad t > 0 \tag{3-17}$$

as a formulation of the propagation problem where the dimensionless parameter ϵ is defined as follows:

$$\epsilon = \frac{kL - T_0}{2T_0} \tag{3-18}$$

Note that the prime has already been dropped from the nondimensional time variable in (3-17).

In the surge-tank example our formulation (3-11) contained a pair of *first*-order differential equations, whereas here we have a single *second*-order equation. This may easily be reduced to a pair of first-order equations as follows: Let

$$x_1 = x$$

$$x_2 = \frac{dx}{dt} \tag{3-19}$$

be the dimensionless displacement and velocity in (3-17). We then obtain the following formulation, which is equivalent to (3-17):

$$\left. \begin{array}{l} x_1 = 0 \\ x_2 = 0 \end{array} \right\} \quad t = 0$$

$$\left. \begin{array}{l} \dfrac{dx_1}{dt} = x_2 \\[2mm] \dfrac{dx_2}{dt} = 1 - x_1 - \epsilon x_1{}^3 \end{array} \right\} \quad t > 0 \qquad (3\text{-}20)$$

EXERCISES

3-1. Fig. 3-4 shows a projectile of mass m in flight subject to gravity (assumed constant) and a drag force R which is a known function of $\sqrt{u^2 + v^2}$ and of y, the elevation above sea level. The projectile is released at $y = 0$ with $u = u_0$, $v = v_0$. Formulate the propagation problem in the form of three first-order differential equations for $u(t)$, $v(t)$, and $y(t)$ with suitable initial conditions. If the solution of this system were available, how could the horizontal displacement $x(t)$ be obtained?

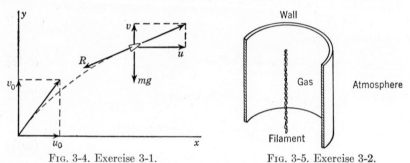

FIG. 3-4. Exercise 3-1.   FIG. 3-5. Exercise 3-2.

3-2. An idealized case of transient heat flow in an electron tube is shown in Fig. 3-5. A filament is heated to a temperature T_f by an electric current. Heat is convected from the filament to the surrounding gas and also radiated to the wall. The wall receives heat by convection from the gas and by radiation from the filament. Finally, there is convective heat transfer from the wall to the surrounding atmosphere at temperature T_a. Assume that the thermal state of the system can be adequately described in terms of a single gas temperature x_1 and a single wall temperature x_2. The whole system is at temperature T_a when at $t = 0$ the filament temperature is suddenly raised to T_f. Formulate the propagation problem for x_1 and x_2 in the form of a pair of first-order differential equations with suitable initial conditions. The heat capacities of the gas and wall, C_1 and C_2, and the necessary conductance and radiation coefficients K_{f1}, K_{12}, K_{2a}, and K_{f2} are known.   *Ans.* See Exercise 3-19.

3-3. A rocket plus fuel initially weighs W_0. When the power is on, exhaust gases are discharged with velocity E relative to the rocket at a rate of w units of weight per unit time. The rocket starts from rest under full power, moving vertically up against gravity and a drag force kV^2, where V is the rocket velocity. Show that the propagation problem for the first stage of the flight (gravity and k considered constant) takes the dimensionless form

$$x = 0 \qquad\qquad t = 0$$

$$\frac{dx}{dt} = \frac{1}{1 - t} - \theta \left(1 + \frac{\epsilon x^2}{1 - t} \right) \qquad t > 0$$

where x is a dimensionless velocity and t is a dimensionless time. The parameters θ and ϵ are dimensionless combinations of the given constants.

3-4. Formulate the propagation problem for the heating of an object similar to that in Prob. 3-1 if the environment has its temperature raised uniformly from T_0 to T_1 during the interval $0 < t < t_0$. The environment then remains at T_1 for $t > t_0$. Assume the object to be at temperature T_0 at $t = 0$.

3-5. The formulation (3-11) is valid during upward surges in the tank. How does it have to be altered before it can be applied during the time when the level in the surge tank is receding?

3-6. Radium B decays at a rate of 0.026 moles per minute per mole present. For each mole of radium B which is decomposed, 1 mole of radium C is produced. Radium C decays at a rate of 0.036 moles per minute per mole present. Formulate the propagation problem for the amounts of radium B and C present in a system which starts off initially with an amount M of radium B and no radium C.

3-2. Formulation of the General Problem

The particular examples of the previous section all led to ordinary differential equations with initial conditions. The most general problem of this type that we shall consider takes the following standard form:

The state of a system is specified by n quantities $x_1(t)$, $x_2(t)$, . . . , $x_n(t)$. The problem is to predict the behavior of these variables from a knowledge of the initial values

$$x_1(0), \ x_2(0), \ . \ . \ . \ , x_n(0) \tag{3-21}$$

prescribed at $t = 0$ and a knowledge of the governing laws

$$\frac{dx_1}{dt} = f_1(x_1, x_2, \ . \ . \ . \ , x_n, t)$$

$$\frac{dx_2}{dt} = f_2(x_1, x_2, \ . \ . \ . \ , x_n, t)$$

$$. \ . \ . \ . \ . \ . \ . \ . \ . \ . \ . \ . \ . \ . \tag{3-22}$$

$$\frac{dx_n}{dt} = f_n(x_1, x_2, \ . \ . \ . \ , x_n, t)$$

which hold for $t > 0$.

This standard form represents a fairly general problem in the theory of differential equations. It is not the *most* general initial-value problem that can be posed, but it is sufficiently general for most practical applications. Note that we take as the basic standard form the system of n first-order differential equations. There are other alternative formulations which are easily reduced to this standard form. We have already seen in (3-20) how a single second-order system was transformed into a system with two first-order equations. This same method can be applied to transform an nth-order equation into n first-order equations.

Most of this chapter deals with the standard form of (3-21) and (3-22). There is, however, one class of systems which has such simplifying properties that the systems, although reducible to the standard form, are more easily treated in their original form. These systems are the undamped mechanical and electrical systems whose governing equations contain second derivatives but no first derivatives. We shall call such systems *special second-order systems.* Their propagation problems may be formulated as follows:

The behavior of the n variables $x_1(t)$, $x_2(t)$, . . . , $x_n(t)$ is to be predicted from a knowledge of the initial values of the x_j *and* the dx_j/dt $(j = 1, \ldots, n)$ prescribed at $t = 0$ and a knowledge of the governing laws

$$\frac{d^2x_j}{dt^2} = f_j(x_1, x_2, \ldots, x_n, t) \qquad j = 1, \ldots, n \qquad (3\text{-}23)$$

which hold for $t > 0$.

EXERCISES

3-7. Reduce the following propagation problem to the standard form of (3-21) and (3-22):

$$\left.\begin{array}{lll} y_1 = 1 & \dfrac{dy_1}{dt} = 0 & \dfrac{d^2y_1}{dt^2} = 0 \\[2ex] y_2 = 0 & \dfrac{dy_2}{dt} = 1 & \end{array}\right\} \quad t = 0$$

$$\left.\begin{array}{l} \dfrac{d^3y_1}{dt^3} + y_2\dfrac{dy_1}{dt} - y_1{}^2\dfrac{dy_2}{dt} = \sin t \\[2ex] \dfrac{d^2y_2}{dt^2} - y_1\dfrac{dy_2}{dt} + t^2y_2 = 0 \end{array}\right\} \quad t > 0$$

3-8. Show that the special second-order system (3-23) can be reduced to the standard form of (3-21) and (3-22).

3-9. Show that (3-22) can be written in matrix notation as

$$\frac{d}{dt}X = F$$

where X and F are column matrices of order n.

3-3. Mathematical Properties

The solution of the propagation problem formulated in the previous section entails marching forth a set of variables $x_j(t)$ $(j = 1, \ldots, n)$ from given initial values subject to a set of governing differential equations

$$\frac{dx_j}{dt} = f_j(x_1, x_2, \ldots, x_n, t) \qquad j = 1, \ldots, n \qquad (3\text{-}24)$$

In a well-behaved system the course of a typical propagating variable might be as sketched in Fig. 3-6a. In general the $x_j(t)$ can be expected to be better-behaved functions than the f_j since integration is a smoothing process. Some possible types of misbehavior where the solution does not exist or is not unique are sketched in Fig. 3-6b to d. The presence of such singularities can be detected by applying the tests given in the following theorems.

Existence and Uniqueness Theorems. We give without proof[1] two basic theorems. The first guarantees that a unique solution exists in the neighborhood of a trial position. The second guarantees that the solution may be expanded in a Taylor's series centered about the trial

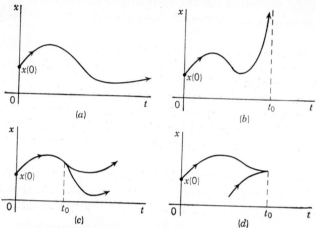

FIG. 3-6. (a) Well-behaved propagation problem. (b) Solution does not exist at $t = t_0$. (c) Solution is not unique at $t = t_0$. (d) Solution is not unique at $t = t_0$, and real solution does not exist for $t > t_0$.

position. In the theorems a set of values $(x_1, x_2, \ldots, x_n, t)$ is referred to as a point in $(n + 1)$-dimensional space.

1. There is one and only one solution of

$$\frac{dx_j}{dt} = f_j(x_1, x_2, \ldots, x_n, t) \qquad j = 1, \ldots, n \qquad (3\text{-}25)$$

which passes through each point of a certain domain of $(x_1, x_2, \ldots, x_n, t)$ space provided that within this domain the functions f_j and their partial derivatives[2] $\partial f_j / \partial x_k (j, k = 1, \ldots, n)$ are finite continuous functions of all their arguments.

2. Let the functions f_j in (3-25) be expandable in $n + 1$ dimensional Taylor's series about the point $P(\xi_1, \xi_2, \ldots, \xi_n, t_0)$. If these are all convergent within the domain

$$|x_j - \xi_j| < a$$
$$|t - t_0| < b \qquad (3\text{-}26)$$

[1] For proof see, for example, E. L. Ince, "Ordinary Differential Equations," Longmans, Green & Co., Inc., New York, 1927, chap. 3, p. 284.

[2] For most engineering problems the theorem as given is sufficiently powerful. It is possible, however, to substitute a somewhat weaker requirement than the existence of partial derivatives and still guarantee a unique solution (see Exercise 3-10).

and if within this domain the maximum $|f_j|$ is less than M, then the solution to (3-25) which passes through P consists of n functions $x_j(t)$ which may be expanded in Taylor's series in powers of $t - t_0$. Furthermore, these series will all be convergent for

$$|t - t_0| < m \tag{3-27}$$

where m is the smaller of the two quantities b and a/M.

Simple examples of the use of these theorems are provided for the reader.[1]

Exact Solutions. Although the foregoing theorems may guarantee the existence of a well-behaved solution to a propagation problem, they do not provide a recipe for writing down the solution in a neat mathematical form. Closed-form solutions to differential equations are available only for a very limited number of special types of equations.[2] If the functions f_j in the standard formulation (3-22) are at all complicated, it is very probable that a procedure for obtaining an exact solution is unknown.

There is one class of systems which is of great engineering importance and which does permit a systematic exact solution. These are the systems in which the $f_j(x_1, x_2, \ldots, x_n, t)$ have the special form

$$f_j = a_{j1}x_1 + a_{j2}x_2 + \cdots + a_{jn}x_n + g_j(t) \tag{3-28}$$

where the a_{jk} are constants, i.e., when the f_j are *linear* with *constant coefficients*. Propagation problems in such systems may be treated in a variety of ways. Perhaps the most elegant procedure is the Heaviside operational or Laplace-transform method in which the algebraic manipulations are minimized and the prescribed initial conditions are automatically worked into the solution. For descriptions of the method together with numerous engineering applications the reader is referred to the texts of Gardner and Barnes,[3] McLachlan,[4] and Carslaw and Jaeger.[5]

Approximate Solutions. All the procedures to be described in this chapter are approximate methods. Many of them have the merit of being applicable to *any* system having the standard form of (3-21) and (3-22). This is in contrast with the very specialized techniques for

---

[1] See Exercises 3-11 to 3-13.

[2] A comprehensive catalogue of equations and techniques is given by E. Kamke, "Differentialgleichungen, Lösungsmethoden und Lösungen," Akademische Verlagsgesellschaft m.b.H., Leipzig, 1943.

[3] M. F. Gardner and J. L. Barnes, "Transients in Linear Systems," John Wiley & Sons, Inc., New York, 1942.

[4] N. W. McLachlan, "Complex Variable and Operational Calculus with Technical Applications," Cambridge University Press, New York, 1942.

[5] H. S. Carslaw and J. C. Jaeger, "Operational Methods in Applied Mathematics," Oxford University Press, New York, 1941.

obtaining exact solutions; however, since the procedures are approximate, there is always a question regarding the accuracy of approximation.

In most cases there are theorems to the effect that if an iteration is carried on without limit, or if the number of adjustable parameters is increased without limit, or if the size of an increment is decreased without limit, then the process converges to the true solution. Such theorems are not without value in encouraging an analyst to adopt a certain procedure, and they will be quoted where known. However, it is usually possible to take only the first 2 or 5 or 100 steps (if a high-speed automatic computing machine is used) of an infinite process, and hence a realistic error bound applicable at any stage of a calculation is far more valuable than a statement of eventual convergence.

Realistic error bounds are unfortunately still relatively scarce. Where none exists, a common method of demonstrating the power of an approximate method has been to apply the method to a problem whose solution is already known. The error can then be ascertained exactly. The presumption then is that in similar problems the method will produce errors of the same order of magnitude. While this is somewhat less than an ideal procedure, it has been resorted to on several occasions in the following pages.

EXERCISES

3-10. Sketch the graph of $|x|$ as a function of x. Note that $|x|$ has no derivative at the origin; $d|x|/dx$ is $+1$ for $x > 0$ and -1 for $x < 0$ and is undefined at $x = 0$. This function does, however, satisfy the *Lipschitz condition*

$$|f(\bar{x}) - f(x)| \leq M|\bar{x} - x|$$

where M is a constant for all values of $\bar{x}$ and x. A weaker form of theorem 1 states that unique solutions exist within any domain where the f_j are continuous and satisfy Lipschitz conditions

$$|f_j(\bar{x}_1, \bar{x}_2, \ldots, \bar{x}_n, t) - f_j(x_1, x_2, \ldots, x_n, t)| \leq M \sum_{k=1}^{n} |\bar{x}_k - x_k|$$

Study the system

$$\frac{dx}{dt} = 1 + |x|$$

Show that the solution through the origin is $x = 1 - e^{-t}$ for $t \leq 0$ and $x = e^t - 1$ for $t \geq 0$. What is the solution of the propagation problem defined by this equation and the initial condition $x = -1$ at $t = 0$?

3-11. Consider the system

$$\frac{dx}{dt} = \frac{x}{t}$$

Note that $f(x, t)$ here is not finite at $t = 0$. Show that the general solution has the form $x = ct$, where c is a constant at integration. Sketch this family of solutions, and discuss existence and uniqueness of solutions at $t = 0$.

3-12. Consider the propagation problem

$$x = 0 \qquad\qquad t = 0$$
$$\frac{dx}{dt} = + \sqrt{|x|} \qquad t > 0$$

Note that $f(x, t)$ is continuous but $\partial f/\partial x$ is unbounded at $x = 0$. [This $f(x, t)$ does not satisfy the Lipschitz condition either.] Show that $x = t^2/4$ is a solution. There are, however, an infinity of other solutions. One of these is $x = 0$, for $0 \leq t \leq 1$, $x = (t - 1)^2/4$ for $t \geq 1$.

3-13. The propagation problem

$$x = 1 \qquad\qquad t = 0$$
$$\frac{dx}{dt} = \tfrac{3}{2}x \sqrt{t} \qquad t > 0$$

has a unique solution according to theorem 1. Can this solution be expanded about the origin in powers of t?

3-14. Consider a particular solution $x = \psi(t)$ of the equation $dx/dt = f(x, t)$. Show that neighboring solutions diverge from $\psi(t)$ with increasing t when $\partial f/\partial x$ is positive and converge to $\psi(t)$ when $\partial f/\partial x$ is negative. This can be seen geometrically from a graph of x against t. For an analytical demonstration set $x = \psi + \epsilon$, and show that $d\epsilon/dt = \epsilon(\partial f/\partial x)_\xi$, where $\psi \leq \xi \leq \psi + \epsilon$. Thus ϵ grows if $\partial f/\partial x$ is positive.

Extend this analysis to the system (3-24). Show that all neighboring solutions converge if all the eigenvalues of the matrix

$$P = \begin{bmatrix} \dfrac{\partial f_1}{\partial x_1} & \dfrac{\partial f_1}{\partial x_2} & \cdots & \dfrac{\partial f_1}{\partial x_n} \\ \dfrac{\partial f_2}{\partial x_1} & \dfrac{\partial f_2}{\partial x_2} & \cdots & \dfrac{\partial f_2}{\partial x_n} \\ \cdots & \cdots & \cdots & \cdots \\ \dfrac{\partial f_n}{\partial x_1} & \dfrac{\partial f_n}{\partial x_2} & \cdots & \dfrac{\partial f_n}{\partial x_n} \end{bmatrix}$$

have negative real parts.

3-4. Iteration

Turning now to procedures for solving propagation problems, we first consider iteration, which has long[1] been used by mathematicians as a tool in proving existence theorems. Iteration can also be used to obtain numerical solutions in particular cases. The nature of the approximation is such that high accuracy in the immediate vicinity of the initial position is obtained with relatively few iterations.

Introductory Example. We consider Prob. 3-1, cooling of an object, with the formulation [see (3-4)]

$$x = 1 \qquad t = 0$$
$$\frac{dx}{dt} = -x \qquad t > 0 \qquad\qquad (3\text{-}29)$$

---

[1] For references to early work going back to 1838, see Ince, *op. cit.*, chap. 3. The method is often called Picard's method after E. Picard, who established a rigorous convergence proof. See E. Picard, "Traité d'analyse," vol. 2, Gauthiers-Villars & Cie, Paris, 1893, p. 301.

We rewrite this as an integral equation by integrating the second of (3-29) and using the first to evaluate the constant of integration.

$$x = 1 - \int_0^t x \, dt \qquad (3\text{-}30)$$

When the *true solution* is known and is placed in the right-hand side of (3-30), the operation so defined reproduces the solution. In the iteration process we place a *trial solution* in the right-hand side of (3-30) to obtain

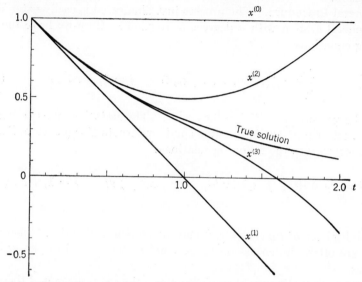

FIG. 3-7.   Comparison of the first three iterates of (3-32) with the exact solution.

an improved trial. Starting with an initial trial function $x^{(0)}$, we construct a sequence of iterates according to the following relation:

$$x^{(r+1)} = 1 - \int_0^t x^{(r)} \, dt \qquad (3\text{-}31)$$

For instance, if we take $x^{(0)} = 1$, we have

$$x^{(1)} = 1 - \int_0^t (1) \, dt \qquad = 1 - t$$

$$x^{(2)} = 1 - \int_0^t (1 - t) \, dt \qquad = 1 - t + \frac{t^2}{2} \qquad (3\text{-}32)$$

$$x^{(3)} = 1 - \int_0^t \left(1 - t + \frac{t^2}{2}\right) dt = 1 - t + \frac{t^2}{2} - \frac{t^3}{6}$$

The behavior[1] of these first three iterates is sketched in Fig. 3-7 along with the exact solution $x = e^{-t}$.

[1] For further discussion of the accuracy of approximation see Sec. 3-5 and also Exercise 3-26.

It will be noted that the iteration process here is building up the power-series expansion of the true solution. In general, when the solution to a propagation problem is expandable in power series, iteration can be used to construct the series one term at a time. In such cases the same results are obtainable by using the Taylor's-series method of the next section. The advantage of iteration is that it still can be used when the solution to a propagation problem does not permit power-series expansion.[1]

General Procedure.   The process just illustrated is easily extended to the general system with n degrees of freedom. Here we have the governing equations

$$\frac{dx_j}{dt} = f_j(x_1, x_2, \ldots, x_n, t) \qquad j = 1, \ldots, n \qquad (3\text{-}33)$$

and the prescribed initial values $x_j(0)$. The iteration procedure starts with a set of initial trial functions and constructs a sequence of iterated sets according to the following relations:

$$x_j^{(r+1)} = x_j(0) + \int_0^t f_j(x_1^{(r)}, x_2^{(r)}, \ldots, x_n^{(r)}, t) \, dt \qquad j = 1, \ldots, n$$
$$(3\text{-}34)$$

The initial set of functions $x_j^{(0)}$ may be chosen arbitrarily. In practice they are often chosen as constants equal to the prescribed set of initial values.

If the propagation problem (3-33) meets the requirements of theorem 1 for the existence of a solution in the neighborhood of the initial position, then the iterated sequence of (3-34) *will converge*[2] to that solution. More specifically, if the requirements of theorem 1 are met within the domain (3-26), the iteration is guaranteed to converge within the interval (3-27).

For example, within the domain $|x - 1| < 1$, $|t| < \infty$ the problem (3-29) meets the conditions of theorem 1, and the maximum value of $|f|$ is 2. The result just stated thus guarantees the convergence of the sequence (3-32) in the interval $|t| < \frac{1}{2}$. The sequence is actually convergent for $|t| < \infty$.

In order to perform iteration it is necessary to do integrations as in (3-32). If the functions f_j are very complicated, it may not be possible to integrate them analytically even for very simple trial solutions. In such cases the iterations may be performed approximately by dividing the t axis into finite intervals Δt and integrating the f_j numerically (e.g.,

[1] See Exercise 3-17.
[2] See, for example, Ince, *op. cit.*, chap. 3.

by Simpson's rule).   We shall encounter[1] finite difference methods which employ this device.

EXERCISES

3-15. Apply iteration to the hydraulic surge tank, Prob. 3-2, with the formulation (3-11).   Take the parameter θ to have the value unity.
 Ans. $x_1 = -1 + t - t^3/6 + t^4/12 - t^5/40 + \cdots$, $x_2 = 1 - t^2/2 + t^3/3 - t^4/8 - t^5/60 - \cdots$.

3-16. Apply iteration to the formulation of Exercise 3-3.   Take $\theta = 1$ and $\epsilon = 10$. Note the difficulties in performing the integrations in closed form.   An easy way out here is to set $(1 - t)^{-1} = 1 + t + t^2 + t^3 + \cdots$.
 Ans. $x = t^2/2 + t^3/3 + t^4/4 - 3t^5/10 - \cdots$.

3-17. The solution of Exercise 3-13 cannot be expanded in a power series about $t = 0$.   The problem does meet the requirements for a unique solution, and hence iteration will converge.   Carry out a few steps of iteration.   (The true solution is $x = e^{t^{\frac{3}{2}}}$.)

3-18. Note the formal analogy between (3-34) and the method of iteration by total steps (1-86).   This suggests an iteration procedure for propagation problems modeled after the Gauss-Seidel procedure (1-88).   Set up this process for the problem of Exercise 3-15.   Carry out a few steps, and compare the results with those of Exercise 3-15.   Note that in both procedures only one additional correct term in the solution series is provided by each iteration.   The two procedures differ only in the higher-order terms, which have not yet stabilized to their ultimate values.

3-19. If in Exercise 3-2 $T_f = 2T_a$, $C_2 = 2C_1$, $K_{f1} = K_{12} = K_{2a} = k$, and

$$K_{f2} = k/(5T_a{}^3)$$

show that the formulation can be put in the following dimensionless form:

$$\left.\begin{matrix} x_1 = 1 \\ x_2 = 1 \end{matrix}\right\} \quad t = 0$$

$$\left.\begin{matrix} \dfrac{dx_1}{dt} = -2x_1 + x_2 + 2 \\ \dfrac{dx_2}{dt} = \tfrac{1}{2}x_1 - x_2 + \dfrac{1}{2} + \dfrac{2^4 - x_2{}^4}{10} \end{matrix}\right\} \quad t > 0$$

Apply a few steps of iteration to this system.   Note the complications introduced by the nonlinear term $x_2{}^4$.   Some simplification can be obtained by observing that each iteration produces only one additional correct term and that it is therefore a waste of time to manipulate with the higher-order terms which are produced at each step.
 Ans. $x_1 = 1 + t - t^2/4 - \cdots$, $x_2 = 1 + 3t/2 - 4t^2/5 - \cdots$.

3-20. Show that the *special second-order system* of (3-23) can be iterated by the following scheme:

$$x_j{}^{(r+1)} = x_j(0) + \frac{dx_j(0)}{dt} t + \int_0^t dt \int_0^t f_j(x_1{}^{(r)}, x_2{}^{(r)}, \ldots, x_n{}^{(r)}, t)\, dt$$

Apply this method to the response of a nonlinear system, Prob. 3-3, as formulated in (3-17).   Note that each step of this iteration produces two additional correct terms

[1] The process described in Sec. 3-9 for evaluating second-category recurrence formulas is an iteration of this type.

in the series expansion of the solution (e.g., if $x^{(r)}$ is correct to terms in t^6, then $x^{(r+1)}$ will be correct to terms in t^8).

$Ans.$ $x = t^2/2! - t^4/4! + t^6/6! - (1 + 90\epsilon)t^8/8! + (1 + 1{,}350\epsilon)t^{10}/10! - (1 + 14{,}580\epsilon)t^{12}/12! + \cdots$.

3-5. Series Methods

We consider here procedures for constructing, term by term, infinite-series expansions for the solutions of propagation problems. These methods have proved to be powerful tools in mathematical analysis.[1] They may also be used to obtain numerical solutions to particular propagation problems.

We first consider *Taylor's series*, which, when they can be used, give the same results as the iteration procedure of the foregoing section. A method of extending the range of a Taylor's-series solution by *analytical continuation* is then presented, and finally we describe the expansion of a solution in terms of a *perturbation parameter*.

Taylor's Series.   The Taylor's-series expansion of a function $x_j(t)$ centered about $t = t_0$ is

$$x_j(t) = x_j(t_0) + \left(\frac{dx_j}{dt}\right)_0 \frac{t - t_0}{1!} + \left(\frac{d^2x_j}{dt^2}\right)_0 \frac{(t - t_0)^2}{2!}$$
$$+ \left(\frac{d^3x_j}{dt^3}\right)_0 \frac{(t - t_0)^3}{3!} + \cdots \quad (3\text{-}35)$$

where the subscripts on the derivatives signify that they are to be evaluated at $t = t_0$. Thus, to obtain a Taylor's-series expansion of a function, it is necessary to know only the value of the function and its derivatives at a single point.

Now consider the functions x_j which are the (unknown) solutions to the propagation problem

$$\left.\begin{array}{ll} x_j(t) = x_j(t_0) & t = t_0 \\ \dfrac{dx_j}{dt} = f_j(x_1, x_2, \ldots, x_n, t) & t > t_0 \end{array}\right\} \quad j = 1, \ldots, n \quad (3\text{-}36)$$

where the f_j meet the requirements of theorem 2, page 133. According to this theorem these solutions can be expanded as in (3-35). To obtain the coefficients, we proceed as follows: The initial conditions give us the first terms $x_j(t_0)$ directly. If these values are then inserted in the right-hand sides of the governing equations in (3-36) at $t = t_0$, we obtain the coefficients $(dx_j/dt)_0$. Next we differentiate the governing equations to get

$$\frac{d^2x_j}{dt^2} = \frac{\partial f_j}{\partial t} + \sum_{k=1}^{n} \frac{\partial f_j}{\partial x_k} \frac{dx_k}{dt} \quad j = 1, \ldots, n \quad (3\text{-}37)$$

[1] See, for example, P. Dienes, "The Taylor Series; An Introduction to the Theory of Functions of a Complex Variable," Oxford University Press, New York, 1931.

At $t = t_0$ we already have the $x_j(t_0)$ and the $(dx_j/dt)_0$ so that the right-hand sides of (3-37) can be evaluated to give the coefficients $(d^2x_j/dt^2)_0$. Continuing, we would differentiate (3-37) to obtain the third derivatives, which could be evaluated at $t = t_0$ by using the already known values of $x_j(t_0)$, $(dx_j/dt)_0$, $(d^2x_j/dt^2)_0$. In this fashion it is possible to obtain as many terms in (3-35) as is desired.

We have already noted that whenever the Taylor's-series method is applicable iteration can also be used to obtain the same result. The choice between the two procedures in a particular instance depends on the relative convenience[1] of the differentiation and evaluation process [exemplified by (3-39) and (3-40)] as compared with the integration process [exemplified by (3-32)].

Example. As an illustration we take the formulation (3-11) of the hydraulic surge tank, Prob. 3-2, for the case $\theta = 1$.

$$\left.\begin{array}{l} x_1 = -1 \\ x_2 = 1 \end{array}\right\} \quad t = 0$$

$$\left.\begin{array}{l} \dfrac{dx_1}{dt} = x_2 \\[2mm] \dfrac{dx_2}{dt} = -x_1 - x_2{}^2 \end{array}\right\} \quad t > 0 \qquad (3\text{-}38)$$

The first four derivatives of the x_j are obtained below by differentiating the governing equations.

$$
\begin{array}{ll}
x_1 = x_1 & x_2 = x_2 \\[2mm]
\dfrac{dx_1}{dt} = x_2 & \dfrac{dx_2}{dt} = -x_1 - x_2{}^2 \\[3mm]
\dfrac{d^2x_1}{dt^2} = \dfrac{dx_2}{dt} & \dfrac{d^2x_2}{dt^2} = -\dfrac{dx_1}{dt} - 2x_2\dfrac{dx_2}{dt} \\[3mm]
\dfrac{d^3x_1}{dt^3} = \dfrac{d^2x_2}{dt^2} & \dfrac{d^3x_2}{dt^3} = -\dfrac{d^2x_1}{dt^2} - 2x_2\dfrac{d^2x_2}{dt^2} - 2\left(\dfrac{dx_2}{dt}\right)^2 \\[3mm]
\dfrac{d^4x_1}{dt^4} = \dfrac{d^3x_2}{dt^3} & \dfrac{d^4x_2}{dt^4} = -\dfrac{d^3x_1}{dt^3} - 2x_2\dfrac{d^3x_2}{dt^3} - 6\dfrac{dx_2}{dt}\dfrac{d^2x_2}{dt^2}
\end{array}
\qquad (3\text{-}39)
$$

Inserting the initial values $x_1 = -1$, $x_2 = 1$ at the top of (3-39), we can work down, evaluating each derivative at $t = 0$, as follows:

$$
\begin{array}{ll}
x_1 = -1 & x_2 = 1 \\[2mm]
\left(\dfrac{dx_1}{dt}\right)_0 = 1 & \left(\dfrac{dx_2}{dt}\right)_0 = 1 - 1 = 0 \\[3mm]
\left(\dfrac{d^2x_1}{dt^2}\right)_0 = 0 & \left(\dfrac{d^2x_2}{dt^2}\right)_0 = -1 - (2)(1)(0) = -1 \\[3mm]
\left(\dfrac{d^3x_1}{dt^3}\right)_0 = -1 & \left(\dfrac{d^3x_2}{dt^3}\right)_0 = -0 - (2)(1)(-1) - (2)(0) = 2 \\[3mm]
\left(\dfrac{d^4x_1}{dt^3}\right)_0 = 2 & \left(\dfrac{d^4x_2}{dt^4}\right)_0 = -(-1) - (2)(1)(2) - (6)(0)(-1) = -3
\end{array}
\qquad (3\text{-}40)
$$

[1] See Exercises 3-24 and 3-25 for examples in which the relative convenience of the two procedures is reversed.

When these coefficients are placed in (3-35), we get the following series expansions for the solutions of (3-38):

$$x_1 = -1 + t - \frac{t^3}{6} + \frac{t^4}{12} + \cdots$$

$$x_2 = 1 - \frac{t^2}{2} + \frac{t^3}{3} - \frac{t^4}{8} + \cdots \tag{3-41}$$

Accuracy of Truncated Series. When we extract numerical values from expansions such as (3-41), we sum a limited number of terms of what are in reality infinite series. It is very desirable to have an estimate of the error involved in thus cutting, or *truncating*, infinite series. We quote[1] three theorems which are useful for this purpose.

1. If in a series the terms are alternately positive and negative and if the absolute value of the individual terms is continually decreasing, then the magnitude of the truncation error is not greater than the magnitude of the first term omitted.

2. If in a series the magnitude of every term is always less than a fixed fraction r of the magnitude of the preceding term, then the truncation error is not greater than $1/(1 - r)$ times the magnitude of the first term omitted. In particular if the magnitude of each term is less than one-tenth that of its predecessor, then the truncation error is substantially equivalent to the first term omitted.

3. If a series is a Taylor's series centered about $t = t_0$ [see (3-35)], the truncation error at $t = t_1$, when the series is truncated after the term containing $(t_1 - t_0)^m$, is not greater than

$$\left| \frac{d^{m+1}x_j}{dt^{m+1}} \right|_{max} \frac{(t_1 - t_0)^{m+1}}{(m + 1)!} \tag{3-42}$$

where the maximum magnitude of the derivative within the interval $t_0 \le t \le t_1$ is to be used.

These theorems provide upper bounds for the truncation error of a series which meets the specified requirements. A difficulty in applying the theorems to the solution of a propagation problem which has been obtained term by term is that we cannot always guarantee that all subsequent terms of the series will meet the specified requirements. For instance, the terms so far obtained in the series for x_2 in (3-41) have alternating signs. We cannot, however, say that the following terms will continue to have alternating signs. In fact, it turns out that when (3-39) and (3-40) are continued for one more step the next term in x_2 is $-t^5/60$, which has the same sign as its predecessor.

The series of (3-41) are Taylor's series, and hence the third theorem is applicable. Here the difficulty is in estimating the maximum value of a high-order derivative of an

---

[1] See, for example, H. B. Phillips, "Analytical Geometry and Calculus," Addison-Wesley Publishing Company, Cambridge, Mass., 1942, chap. XIII.

unknown solution. To illustrate this, let us estimate the truncation error of (3-41) in the interval of $0 < t < 0.2$. According to (3-42) we need the maximum magnitudes of the fifth derivatives within this interval. By continuing the scheme of (3-39) and (3-40), we can find the values of these derivatives at the beginning of the interval.

$$\left(\frac{d^5 x_1}{dt^5}\right)_0 = -3 \qquad \left(\frac{d^5 x_2}{dt^5}\right)_0 = -2 \tag{3-43}$$

To gain some idea of how these change in the interval, we can obtain their rates of change in the beginning of the interval by going one more step with (3-39) and (3-40).

$$\frac{d}{dt}\left(\frac{d^5 x_1}{dt^5}\right)_0 = \left(\frac{d^6 x_1}{dt^6}\right)_0 = -2 \qquad \left(\frac{d^6 x_2}{dt^6}\right)_0 = 47 \tag{3-44}$$

Thus $d^5 x_1/dt^5$ is getting more negative, and if there is no great change in trend, it will reach approximately -3.4 at $t = 0.2$. The change in $d^5 x_2/dt^5$ indicated by (3-43) is large. If the rate shown continues throughout the interval, $d^5 x_2/dt^5$ will go from -2 at $t = 0$ to $+7.4$ at $t = 0.2$. With these figures as a guide we might make the conservative estimate that the maximum value for both fifth derivatives would be less than 15. Then according to (3-42) the truncation error will be less than

$$15 \frac{(0.2)^5}{5!} = 0.00004 \tag{3-45}$$

i.e., we are probably safe in expecting the expansions (3-41) to be correct to four decimal places within the interval $0 < t < 0.2$.

The accuracy of a result obtained from a series such as (3-41) depends, in addition to the above considerations, on the policy adopted toward *rounding off* during the computation. For example, we might evaluate each individual term to six decimal places, sum the terms, and then round off the final result to four places, or we might round off each term separately to four places before summing. These two results may differ by as much as m units in the fourth decimal place when the number of terms summed is $2m + 1$.

Analytical Continuation. A few terms of a Taylor's series provide high accuracy close to the expansion center, but as the distance from the center increases, this accuracy rapidly deteriorates. In order to maintain a specified accuracy, more and more terms of the series must be used, the greater the distance from the expansion center.

An alternate procedure is to use several expansion centers spread out through an interval, as shown in Fig. 3-8. The solution will then consist of a sequence of series expansions each of which gives high accuracy near its own expansion center. Series such as these having different expansion centers but which represent the same function in overlapping intervals are said to provide *analytical continuation*[1] of the function.

[1] See, for example, E. T. Whittaker and G. N. Watson, "A Course of Modern Analysis," 4th ed., Cambridge University Press, New York, 1927, p. 96.

We illustrate this process for the hydraulic surge tank, Prob. 3-2. We already have the expansions (3-41) centered about $t = 0$, and we have estimated that these will be correct to the fourth decimal place up to $t = 0.2$. Let us now obtain expansions centered about $t = 0.2$. We use the basic approach described above. For the coefficients of the Taylor's series, we need the values of x_1 and x_2 and their successive derivatives evaluated at $t = 0.2$. The results of (3-39) still apply. These give the successive derivatives in chain fashion in terms of the values of x_1 and x_2. Thus all

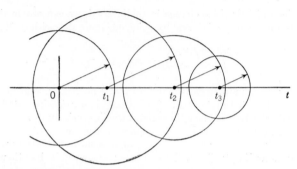

FIG. 3-8.   An extended interval covered by several overlapping series representations.

that is required is to sum (3-41) for $t = 0.2$ (we get $x_1 = -0.8012$, $x_2 = 0.9825$) and then to repeat (3-40), starting from these values.

$$x_1(0.2) = -0.8012 \qquad x_2(0.2) = 0.9825$$

$$\left(\frac{dx_1}{dt}\right)_{0.2} = 0.9825 \qquad \left(\frac{dx_2}{dt}\right)_{0.2} = -(-0.8012) - (0.9825)^2 = -0.1641$$

$$\left(\frac{d^2x_1}{dt^2}\right)_{0.2} = -0.1641 \qquad \left(\frac{d^2x_2}{dt^2}\right)_{0.2} = -(0.9825) - (2)(0.9825)(-0.1641)$$
$$= -0.6600$$

$$\left(\frac{d^3x_1}{dt^3}\right)_{0.2} = -0.6600 \qquad \left(\frac{d^3x_2}{dt^3}\right)_{0.2} = -(-0.1641) - (2)(0.9825)(-0.6600)$$
$$- (2)(-0.1641)^2 = 1.4071$$

$$\left(\frac{d^4x_1}{dt^4}\right)_{0.2} = 1.4071 \qquad \left(\frac{d^4x_2}{dt^4}\right)_{0.2} = -(-0.6600) - (2)(0.9825)(1.4071)$$
$$- (6)(-0.1641)(-0.6600) = -2.7548$$

$$(3\text{-}46)$$

These coefficients are then inserted in (3-35) to give the following analytical continuations of (3-41):

$$x_1 = -0.8012 + 0.9825(t - 0.2) - 0.0820(t - 0.2)^2 - 0.1100(t - 0.2)^3$$
$$+ 0.0586(t - 0.2)^4 + \cdots$$
$$x_2 = 0.9825 - 0.1641(t - 0.2) - 0.3300(t - 0.2)^2 + 0.2345(t - 0.2)^3$$
$$- 0.1148(t - 0.2)^4 + \cdots$$

$$(3\text{-}47)$$

A useful check on the working is obtained[1] if $t = 0$ is substituted in (3-47) to see whether the initial values $x_1 = -1$, $x_2 = 1$ are reproduced.

When an extended comparison between (3-41) and (3-47) is made, it is found that the two representations give either identical results or results which differ by one unit in the fourth decimal place (this can be ascribed to round-off) within the inter-

[1] See Exercise 3-28.

val[1] $-0.1 < t < 0.3$. Outside of this interval the two representations diverge rapidly: e.g., there is disagreement in the third decimal place outside of the interval $-0.3 < t < 0.5$.

The truncation error of (3-47) can be estimated by repeating the argument that led to (3-45). We find that four-decimal-place accuracy can be expected at $t = 0.4$. The solution could then be continued further by repeating the above process to obtain expansions centered at $t = 0.4$.

Perturbation Series.   A perturbed system is one which differs slightly from a known standard system.   Expansion in terms of a perturbation parameter provides a means for obtaining solutions to the perturbed system by utilizing properties of the standard system.   The method has often been used to indicate the behavior of *slightly* nonlinear systems and systems with *almost* constant coefficients.

The method can be illustrated[2] by considering Prob. 3-3, the response of a nonlinear spring-mass system.   If in the formulation [see (3-17)]

$$x = \frac{dx}{dt} = 0 \qquad t = 0$$
$$\frac{d^2x}{dt^2} = 1 - x - \epsilon x^3 \qquad t > 0 \tag{3-48}$$

the parameter ϵ were to vanish, the system would be linear with constant coefficients and solutions could be readily obtained.   The parameter ϵ is thus a perturbation parameter which measures how much the linear system has been perturbed.   The basic idea of the perturbation method is to seek a solution in the form of a series in powers of ϵ.

$$x = \varphi_0(t) + \epsilon\varphi_1(t) + \epsilon^2\varphi_2(t) + \cdots \tag{3-49}$$

where the $\varphi_j(t)$ are to be determined.   We evaluate them by requiring that x of (3-49) must satisfy the formulation (3-48).   The initial condition will be satisfied if at $t = 0$

$$\varphi_j = \frac{d\varphi_j}{dt} = 0 \qquad j = 0, 1, 2, \ldots \tag{3-50}$$

When (3-49) is inserted in the governing equation, we get

$$\frac{d^2\varphi_0}{dt^2} + \epsilon\frac{d^2\varphi_1}{dt^2} + \epsilon^2\frac{d^2\varphi_2}{dt^2} + \cdots = 1 - \varphi_0 - \epsilon\varphi_1 - \epsilon^2\varphi_2 - \cdots$$
$$- \epsilon(\varphi_0{}^3 + 3\epsilon\varphi_0{}^2\varphi_1 + \cdots) \tag{3-51}$$

or

$$\left(\frac{d^2\varphi_0}{dt^2} + \varphi_0 - 1\right) + \epsilon\left(\frac{d^2\varphi_1}{dt^2} + \varphi_1 + \varphi_0{}^3\right) + \epsilon^2\left(\frac{d^2\varphi_2}{dt^2} + \varphi_2 + 3\varphi_0{}^2\varphi_1\right) + \cdots = 0 \tag{3-52}$$

If this is to be satisfied identically in ϵ, then the coefficient of each power of ϵ must separately vanish.   Thus we obtain a set of governing equations for the φ_j.

---

[1] Note that this indicates that our estimate [see (3-45)] of the interval over which (3-41) maintains four-decimal-place accuracy was conservative.

[2] For a more thorough treatment see P. M. Morse and H. Feshbach, "Methods of Theoretical Physics," pt. II, McGraw-Hill Book Company, Inc., New York, 1953, p. 1001.

$$\frac{d^2\varphi_0}{dt^2} + \varphi_0 = 1$$

$$\frac{d^2\varphi_1}{dt^2} + \varphi_1 = -\varphi_0{}^3 \qquad (3\text{-}53)$$

$$\frac{d^2\varphi_2}{dt^2} + \varphi_2 = -3\varphi_0{}^2\varphi_1$$

Note that the first of these represents the unperturbed system. Each succeeding equation has the same form except for the forcing term on the right.

If the equations are solved in sequence, the right-hand side of each is made up only of functions that have already been determined. Thus, solving (e.g., by Laplace transforms) the first of (3-53) subject to the initial conditions (3-50), we find

$$\varphi_0 = 1 - \cos t \qquad (3\text{-}54)$$

Substituting this into the second of (3-53), we have

$$\frac{d^2\varphi_1}{dt^2} + \varphi_1 = -(1 - \cos t)^3 = -\frac{5}{2} + \frac{15}{4}\cos t - \frac{3}{2}\cos 2t + \frac{1}{4}\cos 3t \qquad (3\text{-}55)$$

together with the initial conditions (3-50) as the conditions for determining φ_1. The solution is

$$\varphi_1 = \tfrac{1}{32}(-80 + 65\cos t + 60t\sin t + 16\cos 2t - \cos 3t) \qquad (3\text{-}56)$$

Continuing in this manner, it is possible to obtain φ_2, φ_3, etc., and to insert these into (3-49) to obtain x.[†] When ϵ is small compared with unity, a useful approximation[1] is obtained by using only the first perturbation, i.e.,

$$x = \varphi_0 + \epsilon\varphi_1 \qquad (3\text{-}57)$$

EXERCISES

3-21. Apply the Taylor's-series method to obtain an expansion in powers of t for the solution of Prob. 3-1, cooling of an object, with the formulation (3-4) [see (3-32) for answer].

3-22. Verify the results of Exercise 3-19 by using Taylor's series.

3-23. Obtain the Taylor's-series expansion in powers of t for the solution of Prob. 3-3, response of a nonlinear spring-mass system, with the formulation (3-17) (see Exercise 3-20 for answer).

3-24. Consider the propagation problem

$$x = 0 \qquad t = 0$$
$$\frac{dx}{dt} = t + x^4 \qquad t > 0$$

Compare the difficulty involved in obtaining a series expansion in powers of t for the solution by iteration and by Taylor's series.   *Ans.* $x = t^2/2 + t^9/144 + \cdots$.

3-25. Consider the propagation problem

$$x = 1 \qquad t = 0$$
$$\frac{dx}{dt} = \sqrt{3 + x^3} \qquad t > 0$$

[†] For a discussion of the convergence of perturbation series see J. J. Stoker, "Nonlinear Vibrations," Interscience Publishers, New York, 1950, p. 223.

[1] See Exercise 3-30.

Compare the difficulty involved in obtaining a series expansion in powers of t for the solution by iteration and by Taylor's series.   *Ans.* $x = 1 + 2t + \frac{3}{4}t^2 + \cdots$.

3-26. Show that the truncation error in Exercise 3-21 is smaller than the magnitude of the first term omitted if t is less than the number of terms used. For k-place accuracy the truncation error should be less than 0.5×10^{-k}. If m terms are used, show that k-place accuracy can be guaranteed in the interval

$$0 \le t \le \left[\frac{(m+1)!}{2} 10^{-k} \right]^{1/(m+1)} = T$$

By using *Stirling's approximation*[1]

$$m! \approx \sqrt{2\pi m}\ m^m e^{-m}$$

valid for large m show that when the number of terms used is large the interval T for a given accuracy is extended by about 0.37 for each additional term included.

3-27. Estimate the number of terms that must be retained in the expansion of Exercise 3-21 to ensure three-decimal-place accuracy at $t = 0.2$. Obtain an analytic continuation of the solution in the form of an expansion in powers of $t - 0.2$. Check $x(0)$, and obtain $x(0.4)$.   *Ans.* $x(0.4) = 0.670$.

3-28. Suppose that from a given expansion of $x(t)$ centered at $t = t_0$ we have obtained an analytical continuation centered at $t = t_1$. Show that by properly arranging the computation[2] of $x(t_0)$ from the second series we can almost immediately obtain $x(2t_1 - t_0)$. We can thus simultaneously check our continuation and use it to obtain the value of x at the point which is as far ahead of t_1 as t_0 is behind. Apply this to the series of (3-47).   *Ans.* $x_1(0.4) = -0.6088$, $x_2(0.4) = 0.9382$.

3-29. Consider the hydraulic surge tank, Prob. 3-2, with the formulation (3-11). Let the parameter θ be small. Apply the perturbation series method, and obtain the first perturbation.

Ans. $x_1 = -\cos t + \sin t - \theta(1 - \cos t + \frac{2}{3}\sin t - \frac{1}{3}\sin 2t)$; $x_2 = \sin t + \cos t - \theta(\sin t + \frac{2}{3}\cos t - \frac{2}{3}\cos 2t)$.

3-30. Expand the solution (3-57) in powers of t, and verify that it agrees with the solution of Exercise 3-20, up to the term in t^{12}.

3-6. Trial Solutions with Undetermined Parameters

The procedures of the two previous sections produce approximate solutions to propagation problems in the form of analytical expressions. The character of these approximations is such that high accuracy can be obtained in the immediate neighborhood of the initial state. In this section we consider a type of procedure which sacrifices high accuracy near the initial state in order to gain a more uniform degree of accuracy over an extended interval. The approximate solutions are still in the form of analytical expressions.

Preliminary Example. To introduce the underlying concepts in a simple manner, we take Prob. 3-1, cooling of an object, with the formulation [see (3-4)]

[1] See, for example, P. Franklin, "Methods of Advanced Calculus," McGraw-Hill Book Company, Inc., New York, 1944, p. 265.

[2] This scheme is illustrated in H. Jeffreys and B. Jeffreys, "Methods of Mathematical Physics," Cambridge University Press, New York, 1946, p. 266.

$$x = 1 \qquad t = 0$$
$$\frac{dx}{dt} = -x \qquad t > 0 \qquad\qquad (3\text{-}58)$$

Let us suppose that the solution is desired in the interval $0 < t < 1$.

Our first and most important step is to select a *trial family* of approximate solutions. At present we postpone a general discussion of the factors involved in selecting this solution and arbitrarily take

$$x = 1 + c_1 t + c_2 t^2 \qquad\qquad (3\text{-}59)$$

as our trial family. The parameters c_1 and c_2 are undetermined. Different combinations of values for c_1 and c_2 represent different possible approximations. We do note that the structure of (3-59) is such that the initial condition of (3-58) is satisfied independently of the values of the parameters.

Our next step is to develop criteria for picking the "best" approximation within the family (3-59). We substitute (3-59) into the governing equation of (3-58) and form the equation *residual*, $R(t)$, as follows:

$$R = \frac{dx}{dt} + x$$
$$= 1 + c_1(1 + t) + c_2(2t + t^2) \qquad\qquad (3\text{-}60)$$

For the true solution the equation residual vanishes identically. For our restricted family (3-59) the best we can do is to adjust c_1 and c_2 so that (3-60) stays "close" to zero throughout the interval $0 < t < 1$. Several different criteria have been suggested.

1. *Collocation.*[1] We select as many locations as there are undetermined parameters and then adjust the parameters until the residual vanishes at these locations. The presumption here is that the residual does not get very far away from zero in between the locations where it vanishes. In the present example we have two parameters; so we choose two locations within the desired interval, for example, $\frac{1}{3}$ and $\frac{2}{3}$, $\frac{1}{4}$ and $\frac{3}{4}$, or $\frac{1}{2}$ and 1. For instance, using the locations $\frac{1}{3}$ and $\frac{2}{3}$, we have

$$R(\tfrac{1}{3}) = 1 + \tfrac{4}{3}c_1 + \tfrac{7}{9}c_2 = 0$$
$$R(\tfrac{2}{3}) = 1 + \tfrac{5}{3}c_1 + \tfrac{16}{9}c_2 = 0 \qquad\qquad (3\text{-}61)$$

from which we can solve for c_1 and c_2. We get $c_1 = -0.9310$, $c_2 = 0.3103$ and hence

$$x = 1 - 0.9310t + 0.3103t^2 \qquad\qquad (3\text{-}62)$$

as our approximate solution.

---

[1] R. A. Frazer, W. P. Jones, and S. W. Skan, Approximations to Functions and to the Solutions of Differential Equations, *Aeronaut. Research Comm. Rept. and Mem.* **1799** (1937).

2. *Subdomain Method.*[1]   We divide the desired interval into as many subdomains as there are adjustable parameters and then adjust the parameters until the average value[2] of the residual in each subdomain is zero.   In the present case we might select the subdomains $0 < t < \frac{1}{2}$ and $\frac{1}{2} < t < 1$ or $0 < t < \frac{2}{3}$ and $\frac{1}{3} < t < 1$.   For instance, using the first pair, we have

$$\int_0^{\frac{1}{2}} R\,dt = \frac{1}{2} + \frac{5}{8}c_1 + \frac{7}{24}c_2 = 0$$

$$\int_{\frac{1}{2}}^1 R\,dt = \frac{1}{2} + \frac{7}{8}c_1 + \frac{25}{24}c_2 = 0 \tag{3-63}$$

as the conditions for fixing c_1 and c_2.   The approximate solution which results is

$$x = 1 - 0.9474t + 0.3158t^2 \tag{3-64}$$

3. *Galerkin's Method.*[3]   Here we require that weighted averages[4] of the residual over the desired interval should vanish.   The weighting functions are taken to be the same functions of t as were used in constructing the trial family.   In our case these functions are simply t and t^2, and the Galerkin criterion is

$$\int_0^1 tR\,dt = \frac{1}{2} + \frac{5}{6}c_1 + \frac{11}{12}c_2 = 0$$

$$\int_0^1 t^2R\,dt = \frac{1}{3} + \frac{7}{12}c_1 + \frac{9}{20}c_2 = 0 \tag{3-65}$$

---

[1] C. B. Biezeno and J. J. Koch, Over een nieuwe Methode ter Berekening van vlakke Platen met Toepassing op Enkele voor de techniek Belangrijke belastings Gevallen, *Ingenieur*, **38**, 25–36 (1923). See also C. B. Biezeno and R. Grammel, "Technische Dynamik," Springer-Verlag OHG, Berlin, 1939, p. 142.

[2] The average value of a function $f(t)$ in an interval $a < t < b$ is

$$\frac{\int_a^b f(t)\,dt}{\int_a^b dt}$$

and this vanishes when the numerator is zero.

[3] B. G. Galerkin, *Vestnik Inzhenerov*, **1**, 897–908 (1915).   The method was described by H. Hencky, Eine wichtige Vereinfachung der Methode von Ritz zur angenäherten Behandlung von Variationaufgaben, *Z. angew. Math. u. Mech.*, **7**, 80–81 (1927).   See also W. J. Duncan, The Principles of Galerkin's Method, *Aeronaut. Research Comm. Rept. and Mem.* 1848 (1938).

[4] If $W(t)$ is a weighting function, the weighted average of a function $f(t)$ in an interval $a < t < b$ is

$$\frac{\int_a^b Wf\,dt}{\int_a^b W\,dt}$$

and this vanishes when the numerator is zero.   An alternate way of stating that this numerator vanishes is to say that the functions f and W are *orthogonal* to each other over the interval (a, b).

Solving these for c_1 and c_2 leads to

$$x = 1 - 0.9143t + 0.2857t^2 \tag{3-66}$$

as our approximate solution.

4. *Method of Least Squares.*[1] The parameters are here adjusted in such a way as to minimize the integral of the square of the residual over the desired interval. Thus we set

$$\frac{1}{2}\frac{\partial}{\partial c_1} \int_0^1 R^2 \, dt = \int_0^1 R \frac{\partial R}{\partial c_1} \, dt = \frac{3}{2} + \tfrac{7}{3}c_1 + \tfrac{9}{4}c_2 = 0$$

$$\frac{1}{2}\frac{\partial}{\partial c_2} \int_0^1 R^2 \, dt = \int_0^1 R \frac{\partial R}{\partial c_2} \, dt = \frac{4}{3} + \tfrac{9}{4}c_1 + \tfrac{38}{15}c_2 = 0 \tag{3-67}$$

from which we obtain $c_1 = -0.9427$, $c_2 = 0.3110$ and hence

$$x = 1 - 0.9427t + 0.3110t^2 \tag{3-68}$$

Relationship between the Criteria. We have just obtained four slightly different approximate solutions to (3-58), all within the trial family (3-59). We can compare the four approaches with respect to amount of computational labor. We note that in every case we have a set of simultaneous equations to solve. The only difference is in the way the coefficients are derived from the residual. Collocation is most direct, requiring only evaluations of the residual, whereas the other three methods require integrations. These integrations are least awkward for the subdomain method and most awkward in the method of least squares.

We can compare the various solutions with respect to accuracy of approximation. In Fig. 3-9 the differences between the four approximate solutions just obtained and the true solution $x = e^{-t}$ are shown. For comparison the truncation error of the first three terms of the Taylor's series [see (3-32)]

$$x = 1 - t + \tfrac{1}{2}t^2 \tag{3-69}$$

is also displayed. Note that (3-69) is a member of the trial family (3-59).

Studying Fig. 3-9, we see that in the interval $0 < t < 0.3$ the truncated Taylor's series gives the best approximation but that in the interval $0 < t < 1$ it must certainly be considered to yield the poorest over-all approximation. The error of approximation is more uniformly distributed in the other four cases. Although these four approximations are different, their differences are not oversignificant. The maximum error

---

[1] The method was discovered in 1795 by C. F. Gauss, although he did not present it publicly until 1809. See "Carl Friedrich Gauss Werke," vol. VII, Göttingen, 1871, p. 242. The method was independently developed and first published by A. M. Legendre, "Nouvelles méthodes pour la détermination des orbites des comètes," Paris, 1806, p. 72.

of the worst case (collocation) is only twice the maximum error of the best (least squares). Note also that for each of the four solutions there is some portion of the interval $0 < t < 1$ over which it provides the best approximation.

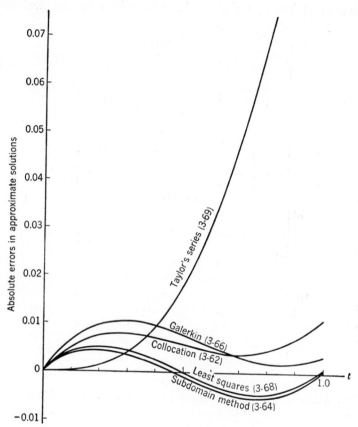

FIG. 3-9. Comparison of errors in approximate solutions to (3-58).

All four criteria can be considered[1] as special cases of the single general criterion that *weighted averages of the residual should vanish*.

The only difference between the methods lies in the choice of the weighting functions. In terms of the example just worked the generalized conditions for fixing c_1 and c_2 are

$$\int_0^1 W_1 R \, dt = 0 \qquad \int_0^1 W_2 R \, dt = 0 \qquad (3\text{-}70)$$

[1] The basis for this consideration is contained in a remark made by R. Courant in the discussion following C. B. Biezeno, Graphical and numerical methods for solving stress problems, *Proc. 1st Intern. Congr. Appl. Mech.*, Delft, 1924, pp. 3–17.

We show in Fig. 3-10 the weighting functions W_1 and W_2 required to produce the four previous results. The conditions (3-70) reduce to (3-61) when unit impulses[1] at $\frac{1}{3}$ and $\frac{2}{3}$ are used for weighting functions. Similarly (3-70) reduces to the subdomain criterion (3-63) when the weighting functions are taken as unity within the subdomains and zero outside.

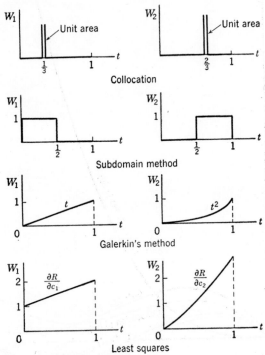

Fig. 3-10. Weighting functions which when inserted into (3-70) yield the four earlier criteria.

The Galerkin and least-squares criteria (3-65) and (3-67) are more obviously of the form (3-70).

In all approximate methods where we focus our attention on achieving a small equation residual we make the tacit assumption that a small error in satisfying the equation is reflected by only a small deviation of the approximate solution from the true solution. Systems for which this assumption is justified are often said to be *well behaved* or *well conditioned*.

[1] The unit impulse, or *Dirac delta*, $\delta(t - a)$ vanishes everywhere except at $t = a$ and satisfies

$$\int_{-\infty}^{\infty} \delta(t - a)\, dt = 1$$

General Procedure. The above procedures are easily generalized to apply to the propagation problem

$$
\left.
\begin{aligned}
x_j &= x_j(0) \qquad\qquad t = 0 \\
\frac{dx_j}{dt} &= f_j(x_1, x_2, \ldots, x_n, t) \qquad t > 0
\end{aligned}
\right\} \quad j = 1, \ldots, n \quad (3\text{-}71)
$$

for a system with n degrees of freedom. Suppose a solution is desired in the interval $0 < t < T$.

A set of n trial functions, each containing adjustable parameters, must be constructed. This is the crucial step. The trial functions should be sufficiently simple so that the operations required can be easily performed. The parameters should be inserted in such a way as to give a wide variation of possible solutions all of which satisfy the initial conditions.

For linear systems[1] it is common to take the trial solutions in the form

$$
x_j = \varphi_{j0}(t) + c_{j1}\varphi_{j1}(t) + \cdots + c_{jm}\varphi_{jm}(t) \qquad j = 1, \ldots, n \quad (3\text{-}72)
$$

where the $\varphi_{jk}(t)$ are known functions (e.g., polynomials or trigonometric or exponential functions) which satisfy the following initial conditions:

$$
\left.
\begin{aligned}
\varphi_{j0}(0) &= x_j(0) \\
\varphi_{jk}(0) &= 0 \qquad k = 1, \ldots, m
\end{aligned}
\right\} \quad j = 1, \ldots, n \quad (3\text{-}73)
$$

With this choice there are mn undetermined parameters c_{jk} which must be obtained by applying one of the weighted residual criterions.

When the trial solutions are inserted in the governing equations (3-71), n equation residuals $R_j(j = 1, \ldots, n)$ are obtained. If m adjustable parameters were used in constructing x_j, then m conditions of the weighted residual type (e.g., collocation, Galerkin, etc.) must be satisfied by R_j in $0 < t < T$. We thus obtain m algebraic equations for the parameters. When each of the n residuals is treated in this way, we finally get enough equations to solve simultaneously for all the parameters.

It is conjectured that, if the number of independent parameters in a trial family of solutions is increased without limit, then the corresponding approximate solutions obtained by these methods would converge (at least under certain conditions) to the true solution. Proof of this still remains to be given. It has been argued[2] that if any one of the weighted residual criterions were to yield convergence for a particular problem and a particular family of trial solutions then so would any other of the criterions.

In practical applications the number of adjustable parameters is usually quite limited. The accuracy attained then depends critically on the selection of the form of the trial solution.

---

[1] See Exercise 3-34 for an example of a different form of trial solution for a non-linear system.

[2] Frazer, Jones, and Skan, *op. cit.* See also L. Collatz, "Numerische Behandlung von Differentialgleichungen," Springer-Verlag OHG, Berlin, 1951, p. 184.

EXERCISES

3-31. Consider the family of trial solutions

$$x = c_1 t^2 + c_2 t^3$$

for the propagation problem

$$x = 0 \qquad t = 0$$

$$\frac{dx}{dt} = t - x \qquad t > 0$$

Evaluate the parameters by applying (a) collocation at $t = \frac{1}{2}$ and $t = 1$, (b) the subdomain method with $0 < t < \frac{1}{2}$ and $\frac{1}{2} < t < 1$ as subdomains, (c) the Galerkin method for the interval $0 < t < 1$, (d) the least-squares method for the interval $0 < t < 1$, and (e) the weighted residual criterion for the same interval, using the weighting functions $W_1 = t$ and $W_2 = t^2$.

Ans. (a) $c_1 = 0.4737$, $c_2 = -0.1053$; (b) $c_1 = 0.4853$, $c_2 = -0.1176$; (c) $c_1 = 0.4745$, $c_2 = -0.1071$; (d) $c_1 = 0.4774$, $c_2 = -0.1098$; (e) $c_1 = 0.4779$, $c_2 = -0.1103$.

3-32. Evaluate and sketch the equation residual (3-60) in the interval $0 < t < 1$ for one or more of the solutions whose errors are shown in Fig. 3-9.

3-33. Consider the trial family of solutions

$$x = 1 + c_1 t + c_2 t^2 + c_3 t^3$$

for the propagation problem

$$x = 1 \qquad t = 0$$

$$\frac{dx}{dt} = x \qquad t > 0$$

Evaluate the parameters by applying collocation at $t = \frac{1}{4}, \frac{1}{2}$, and $\frac{3}{4}$.

Ans. $c_1 = 1.0261$, $c_2 = 0.4174$, $c_3 = 0.2783$.

3-34. Apply collocation at $t = 0$ and $t = \infty$ to Exercise 3-19, with the following four-parameter trial solution:

$$x_1 = 1 + c_{11}(1 - e^{-c_{12}t})$$

$$x_2 = 1 + c_{21}(1 - e^{-c_{22}t})$$

Ans. $c_{11} = 0.931$, $c_{12} = 1.074$, $c_{21} = 0.862$, $c_{22} = 1.740$.

3-7. Finite-increment Techniques

An exact solution to a propagation problem consists of a *continuous* description of the behavior of a system. All the approximate procedures so far discussed have provided continuous (if somewhat erroneous) solutions. We shall next consider procedures in which the basic approximation involves replacing the continuous variable t by a *discrete* variable which increases stepwise in finite increments Δt. Instead of a continuous history in t we obtain only a sequence of snapshot pictures at the discrete intervals Δt.

Before turning in Sec. 3-8 to actual procedures for obtaining stepwise solutions to propagation problems, we devote this section to a consideration of some of the problems involved in handling discrete functions. *Interpolation, differentiation,* and *integration* are briefly discussed.

In Fig. 3-11a a continuous function $x(t)$ is displayed. Figure 3-11b shows a number of isolated values of this same function. The value of

$x(t)$ at the discrete station point $t = t_s$ is called x_s. The treatment here is limited to the case where the spacing between station points

$$\Delta t = t_{s+1} - t_s = h \qquad (3\text{-}74)$$

is uniform; that is, h is a constant independent of s. Our problem then is to obtain approximations for integrals, derivatives, and intermediate values of the continuous function $x(t)$ when given only a table of values x_s at discrete intervals.

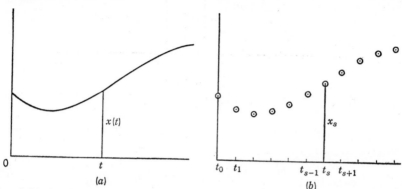

Fig. 3-11. (a) Continuous function and (b) a discrete representation of the same function.

A comprehensive treatment[1] of this problem goes far beyond the scope of this text. We consider here only the simplest nontrivial procedures. These are quite adequate for the vast majority of engineering problems.

Interpolation. The reader is undoubtedly familiar with linear interpolation[2] and interpolation by plotting the tabulated values and drawing a smooth curve through them "by eye." There also exists[3] a large family of more sophisticated methods of interpolation. In these methods an analytical expression (e.g., a polynomial or trigonometric function) which passes through several of the tabulated values is found. This expression is called an *interpolating function*. By evaluating this interpolating function in between station points we obtain approximations to the original function $x(t)$.

As a simple illustration of this we use the second-order *parabola*

$$P = At^2 + Bt + C \qquad (3\text{-}75)$$

[1] See, for example, J. J. Steffensen, "Interpolation," The Williams & Wilkins Company, Baltimore, 1927.

[2] See Exercise 3-43.

[3] For an introduction to this subject see J. B. Scarborough, "Numerical Mathematical Analysis," 2d ed., Johns Hopkins Press, Baltimore, 1950, or W. E. Milne, "Numerical Calculus," Princeton University Press, Princeton, N.J., 1949, or E. T. Whittaker and G. Robinson, "The Calculus of Observations," 2d ed., Blackie & Son, Ltd., Glasgow, 1929.

as an interpolating function. Since there are three parameters in (3-75), we can adjust them so that the parabola passes through three adjacent

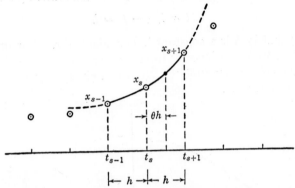

FIG. 3-12. Interpolation by means of a parabola passing through three adjacent points.

points as shown in Fig. 3-12. It is a straightforward exercise in analytical geometry[1] to obtain

$$P(t_s + \theta h) = \frac{\theta(\theta - 1)}{2} x_{s-1} + (1 - \theta^2)x_s + \frac{\theta(\theta + 1)}{2} x_{s+1} \quad (3\text{-}76)$$

as the equation of the parabola which passes through x_{s-1} when $\theta = -1$, x_s when $\theta = 0$, and x_{s+1} when $\theta = 1$. For practical applications it is useful to have a table of the coefficients in (3-76). A short table[2] which permits interpolation by tenths within the interval $t_s - 0.5h \leq t \leq t_s + 0.5h$ is given in Table 3-1.

TABLE 3-1. COEFFICIENTS FOR PARABOLIC INTERPOLATION

θ	$\frac{1}{2}\theta(\theta - 1)$	$1 - \theta^2$	$\frac{1}{2}\theta(\theta + 1)$
-0.5	0.375	0.750	-0.125
-0.4	0.280	0.840	-0.120
-0.3	0.195	0.910	-0.105
-0.2	0.120	0.960	-0.080
-0.1	0.055	0.990	-0.045
0.0	0.000	1.000	0.000
0.1	-0.045	0.990	0.055
0.2	-0.080	0.960	0.120
0.3	-0.105	0.910	0.195
0.4	-0.120	0.840	0.280
0.5	-0.125	0.750	0.375

[1] See Exercise 3-41.

[2] For more extensive tables see "Tables of Lagrangian Interpolation Coefficients," Mathematical Tables Project, National Bureau of Standards, Columbia University Press, New York, 1944.

To illustrate the use of Table 3-1, suppose that in a table of values we have the entries

t	x	
0.5	0.6065	
1.0	0.3679	(3-77)
1.5	0.2231	

and it is desired to obtain an approximation for x at $t = 1.1$. Here $h = 0.5$, and $\theta = 0.2$, and hence

$$P(t = 1.1) = (-0.080)(0.6065) + (0.960)(0.3679) + (0.120)(0.2231) = 0.3314$$
$$\text{(3-78)}$$

according to (3-76) and Table 3-1. The values (3-77) were taken from a table of the function e^{-t}. The actual value of e^{-t} at $t = 1.1$ is 0.3329.

Accuracy. Limits on the error involved in interpolation by any particular scheme can be estimated provided information is available regarding the smoothness of the original function $x(t)$. This usually consists of information about the existence and magnitude of the higher derivatives of $x(t)$.

For instance, if the first three derivatives of a function $x(t)$ are continuous, the size of the third derivative can be used as a measure of how much the function can deviate from a parabolic interpolating function. Specifically it can be shown[1] that if $P(t)$ is the parabola (3-76) passing through x_{s-1}, x_s, and x_{s+1} then

$$P(t_s + \theta h) - x(t_s + \theta h) = \frac{h^3}{6} \frac{d^3x(\tau)}{dt^3} \theta(1 + \theta)(1 - \theta) \qquad -1 \le \theta \le 1$$
$$\text{(3-79)}$$

where $d^3x(\tau)/dt^3$ is the third derivative of $x(t)$ evaluated at some (unknown) point τ in the interval $t_{s-1} \le \tau \le t_{s+1}$.

We can use (3-79) to estimate the error involved in the interpolation (3-78). The right side of (3-79) then becomes

$$\frac{(0.5)^3}{6} \frac{d^3x(\tau)}{dt^3} (0.2)(1.2)(0.8) = 0.004 \frac{d^3x(\tau)}{dt^3}$$
$$\text{(3-80)}$$

Now, if it were known that the function x was actually e^{-t}, then we would know that the third derivative was just the negative of x and hence varied from -0.6065 to -0.2231 in the interval $0.5 \le t \le 1.5$. The error predicted by (3-80) would then lie somewhere between -0.0024 and -0.0009. By direct comparison with the exact value the error in (3-78) is -0.0015.

Differentiation. We consider the problem of obtaining approximations to derivatives of the function $x(t)$ of Fig. 3-11a when we have only the values x_s at the discrete station points t_s of Fig. 3-11b. One basic

---

[1] See, for example, Milne, *op. cit.*, p. 78, or Scarborough, *op. cit.*, p. 97. See also Exercise 3-42.

approach is to pass an interpolating function through several adjacent points and then to differentiate the interpolating function.

As an example of this, we can differentiate the parabola (3-76), using $d/dt = (1/h)\, d/d\theta$ to obtain

$$\left(\frac{dP}{dt}\right)_{t_s+\theta h} = \frac{1}{h}\left(\frac{2\theta - 1}{2}\, x_{s-1} - 2\theta x_s + \frac{2\theta + 1}{2}\, x_{s+1}\right) \tag{3-81}$$

as an approximation to dx/dt at $t = t_s + \theta h$. At the station point t_s this reduces to

$$\left(\frac{dP}{dt}\right)_s = \frac{x_{s+1} - x_{s-1}}{2h} \tag{3-82}$$

which is a statement of the fact that the slope of a chord of a parabola is equal to the slope of the tangent at the mid-point of the chord. If the first three derivatives of $x(t)$ are continuous, then it can be shown[1] that

$$\left(\frac{dx}{dt}\right)_s = \frac{x_{s+1} - x_{s-1}}{2h} - \frac{h^2}{6}\frac{d^3x(\tau)}{dt^3} \tag{3-83}$$

where τ is some (unknown) value of t within the interval $t_{s-1} \leq \tau \leq t_{s+1}$. The last term on the right represents the error involved in using (3-82). It is sometimes called the *remainder term*.

The parabolic interpolating function also provides an approximation to the second derivative. Differentiating (3-81), we have

$$\left(\frac{d^2P}{dt^2}\right)_{t_s+\theta h} = \frac{1}{h^2}\,(x_{s-1} - 2x_s + x_{s+1}) \tag{3-84}$$

If $x(t)$ has continuous derivatives up to order 4, then it can be shown that the error involved in using (3-84) at $t = t_s$ is given by the remainder term on the right side of

$$\left(\frac{d^2x}{dt^2}\right)_s = \frac{x_{s-1} - 2x_s + x_{s+1}}{h^2} - \frac{h^2}{12}\frac{d^4x(\tau)}{dt^4} \tag{3-85}$$

where again τ is some value of t within the interval $t_{s-1} \leq \tau \leq t_{s+1}$.

Another approach to obtaining differentiation formulas is to express several adjacent values x_s as Taylor's series expanded about the desired value of t and then to solve for the desired derivative by algebraic elimination.

We illustrate this for the case where the derivatives are desired at a station point t_s. Assuming that $x(t)$ can be expanded in Taylor's series in the interval of interest, we have

[1] See Exercise 3-44.

$$x_{s+1} = x_s + h\left(\frac{dx}{dt}\right)_s + \frac{h^2}{2!}\left(\frac{d^2x}{dt^2}\right)_s + \frac{h^3}{3!}\left(\frac{d^3x}{dt^3}\right)_s + \frac{h^4}{4!}\left(\frac{d^4x}{dt^4}\right)_s + \cdots$$

$$x_{s-1} = x_s - h\left(\frac{dx}{dt}\right)_s + \frac{h^2}{2!}\left(\frac{d^2x}{dt^2}\right)_s - \frac{h^3}{3!}\left(\frac{d^3x}{dt^3}\right)_s + \frac{h^4}{4!}\left(\frac{d^4x}{dt^4}\right)_s + \cdots$$

$$(3\text{-}86)$$

By subtracting these we obtain

$$\left(\frac{dx}{dt}\right)_s = \frac{x_{s+1} - x_{s-1}}{2h} - \frac{h^2}{6}\left(\frac{d^3x}{dt^3}\right)_s - \frac{h^4}{120}\left(\frac{d^5x}{dt^5}\right)_s - \cdots \quad (3\text{-}87)$$

If this is cut off, or *truncated,* after the first term on the right, we get the approximation (3-82). Note the differences (and similarities) in the form of the error terms in (3-83) and (3-87). It is due to this latter derivation that the remainder term in (3-83) is often called the *truncation error.*

Returning to (3-86), we can obtain the second derivative by adding the two equations.

$$\left(\frac{d^2x}{dt^2}\right)_s = \frac{x_{s-1} - 2x_s + x_{s+1}}{h^2} - \frac{h^2}{12}\left(\frac{d^4x}{dt^4}\right)_s - \frac{h^4}{360}\left(\frac{d^6x}{dt^6}\right)_s - \cdots \quad (3\text{-}88)$$

The truncation error involved in using only the first term on the right of (3-88) should be compared with (3-85).

In many applications the important property of the truncation error is simply its dependence on the interval spacing h. Under these circumstances we write[1] in place of (3-83), for instance,

$$\left(\frac{dx}{dt}\right)_s = \frac{x_{s+1} - x_{s-1}}{2h} + 0(h^2) \quad (3\text{-}89)$$

meaning that when h is small enough the remainder term behaves essentially like a constant multiplied by h^2.

To illustrate these remarks, we consider the problem of obtaining an approximation to the first derivative at $t = 1.0$ of the function $x = e^{-t}$ tabulated below.

t	x
0.8	0.44933
0.9	0.40657
1.0	0.36788
1.1	0.33287
1.2	0.30119

Using $h = 0.2$ and (3-82), we have

$$\frac{0.30119 - 0.44933}{(2)(0.2)} = -0.37035 \quad (3\text{-}90)$$

[1] The notation $0(h^2)$ is read *(term of) order* h^2.

as an approximation to the true value of -0.36788. The error here is -0.00257. If we use the smaller interval $h = 0.1$, we get

$$\frac{0.33287 - 0.40657}{(2)(0.1)} = -0.36850 \tag{3-91}$$

with an error of -0.00062. Note that the errors in (3-91) and (3-90) are in approximately the same ratio as are the *squares* of the sizes of the intervals, that is, $1:4$. This is in accordance with (3-89). We shall see later[1] how the knowledge of the *order* of a truncation error permits us to *extrapolate* to an improved approximation.

Integration. Formulas which give approximations to the integral of a function $x(t)$ which is tabulated at intervals may be derived by passing an interpolating function through several adjacent points and then integrating the interpolating function. The simplest formula of this type is the *trapezoidal rule*, which uses a straight line as the interpolating function between x_s and x_{s+1}. The integral of $x(t)$ from t_s to t_{s+1} is then approximated by a trapezoidal area.

$$\int_{t_s}^{t_{s+1}} x(t)\, dt \approx \frac{h}{2}(x_s + x_{s+1}) \tag{3-92}$$

The error in this approximation depends on how much $x(t)$ deviates from a straight line in the interval, i.e., on the magnitude of the curvature or the size of d^2x/dt^2. This error is often called the *truncation error*. It is given[2] by

$$\frac{h}{2}(x_s + x_{s+1}) - \int_{t_s}^{t_{s+1}} x(t)\, dt = \frac{h^3}{12}\frac{d^2x(\tau)}{dt^2} \tag{3-93}$$

where τ is some value of t within the interval $t_s \le \tau \le t_{s+1}$.

When an integral over an extended range T is required, we divide T into S equal intervals $\Delta t = h = T/S$ and apply (3-93) S times in succession. The resulting formula is

$$\int_{t_0}^{t_s} x(t)\, dt = h\left(\frac{x_0}{2} + x_1 + x_2 + \cdots + x_{S-1} + \frac{x_S}{2}\right) - \frac{h^2 T}{12}\frac{d^2x(\tau)}{dt^2} \tag{3-94}$$

where $t_0 \le \tau \le t_S$. We note that, while the error in one step is of order h^3, the total error over a fixed range is of order h^2 since the number of steps is $S = T/h$.

The next formula in this family is *Simpson's rule*,[3] which uses the parabola (3-76) as an interpolating function between x_{s-1}, x_s, and x_{s+1}.

---

[1] See Sec. 3-8 and also Exercise 3-39.

[2] See Exercise 3-45.

[3] T. Simpson, "Mathematical Dissertations," London, 1743, p. 109.

Integrating (3-76), we get

$$\int_{t_{s-1}}^{t_{s+1}} P(t)\, dt = \frac{h}{3}\, (x_{s-1} + 4x_s + x_{s+1})$$

When this is used to approximate the integral of a function $x(t)$ with continuous derivatives up to order 4, the truncation error can be shown[1] to be

$$\frac{h}{3}\, (x_{s-1} + 4x_s + x_{s+1}) - \int_{t_{s-1}}^{t_{s+1}} x(t)\, dt = \frac{h^5}{90} \frac{d^4 x(\tau)}{dt^4} \qquad (3\text{-}95)$$

where $t_{s-1} \leq \tau \leq t_{s+1}$.

Simpson's rule may be used over an extended range T by dividing T into $2S$ intervals $\Delta t = h = T/2S$ and applying (3-95) S times in succession. The resulting formula is

$$\int_{t_0}^{t_{2S}} x(t)\, dt = \frac{h}{3}\, (x_0 + 4x_1 + 2x_2 + 4x_3 + 2x_4 + \cdots$$

$$+ 2x_{2S-2} + 4x_{2S-1} + x_{2S}) - \frac{Th^4}{180} \frac{d^4 x(\tau)}{dt^4} \qquad (3\text{-}96)$$

where $t_0 \leq \tau \leq t_{2S}$.

EXERCISES

3-35. The following are values of the function $x = e^t$:

t	x
0.0	1.0000
0.1	1.1052
0.2	1.2214
0.3	1.3499
0.4	1.4918

Obtain approximations to $e^{0.18}$ by linear and by parabolic interpolation.

Ans. 1.1985 and 1.1972.

3-36. Obtain an approximation to the first derivative at $t = 0.2$ of the function tabulated in Exercise 3-35 by using (3-82). Estimate the truncation error by (3-83), and compare with the actual error.

3-37. Obtain an approximation to $\int_0^{0.4} x\, dt$, where x is the function tabulated in Exercise 3-35 by using the trapezoidal rule with $h = 0.1$. Estimate the truncation error to be expected, and compare with the actual error. *Ans.* 0.4922.

3-38. Repeat Exercise 3-37, using Simpson's rule.

3-39. Show that if the error in an approximation A_1 is known to be k times the error in another approximation A_2 then the true value is given by

$$\frac{kA_2 - A_1}{k - 1}$$

---

[1] See, for example, Scarborough, *op. cit.*, p. 174.

Apply this to the approximations (3-90) and (3-91), assuming that the errors are exactly in the ratio of the squares of the interval sizes. This process is called h^2 *extrapolation*.

3-40. Approximate the integral $\int_0^{0.4} x\, dt$, where x is the function tabulated in Exercise 3-35, by using the trapezoidal rule with $h = 0.2$. Then apply the result of Exercise 3-39 to this and the approximation of Exercise 3-37 to obtain an improved approximation. *Ans. h^2 extrapolation is correct to four decimal places.*

3-41. Evaluate the constants A, B, and C in (3-75) so as to make

$$P(t_{s-1}) = x_{s-1}$$
$$P(t_s) = x_s$$
$$P(t_{s+1}) = x_{s+1}$$

Set $t = t_s + \theta h$, and verify (3-76). Interpret (3-76) as a weighted average.

3-42. Expand x_{s-1}, x_{s+1} in Taylor's series centered about t_s. Combine these according to (3-76) to obtain a series expansion for $P(t_s + \theta h)$. Expand $x(t_s + \theta h)$ in a Taylor's series, and show that the leading term in the series for $P(t_s + \theta h) - x(t_s + \theta h)$ is

$$\frac{h^3}{6}\left(\frac{d^3x}{dt^3}\right)_s \theta(1 + \theta)(1 - \theta)$$

3-43. By using the same method as in the previous example show that

$$(1 - \theta)x_s + \theta x_{s+1} = x(t_s + \theta h) + \frac{h^2}{2}\left(\frac{d^2x}{dt^2}\right)_s \theta(1 - \theta) + \cdots$$

This is the formula for *linear interpolation* including the leading term of the truncation error.

3-44. Differentiate (3-79), using $d/dt = (1/h)d/d\theta$. Since τ is unknown, $d\tau/dt$ cannot be evaluated. Show that at $t = t_s$, however, this term drops out and (3-83) is obtained.

3-45. Express $x(t_s + \theta h)$ as a Taylor's series, and integrate this term by term to obtain a series expansion for

$$\int_{\theta=0}^{1} x(t_s + \theta h)h\, d\theta$$

Use this and the series expansion for x_{s+1} to show that the leading term of the truncation error in the *trapezoidal rule* (3-93) is

$$\frac{h^3}{12}\left(\frac{d^2x}{dt^2}\right)_s$$

3-46. Apply the method of the previous example to show that the leading term of the truncation error in *Simpson's rule* (3-95) is

$$\frac{h^5}{90}\left(\frac{d^4x}{dt^4}\right)_s$$

3-47. Derive the following differentiation formulas:

$$\left(\frac{d^3x}{dt^3}\right)_s = \frac{1}{2h^3}(x_{s+2} - 2x_{s+1} + 2x_{s-1} - x_{s-2}) - \frac{h^2}{4}\left(\frac{d^5x}{dt^5}\right)_s + \cdots$$
$$\left(\frac{d^4x}{dt^4}\right)_s = \frac{1}{h^4}(x_{s+2} - 4x_{s+1} + 6x_s - 4x_{s-1} + x_{s-2}) - \frac{h^2}{6}\left(\frac{d^6x}{dt^6}\right)_s + \cdots$$

3-48. Show that

$$\left(\frac{dx}{dt}\right)_s = \frac{1}{12h}\left(-x_{s+2} + 8x_{s+1} - 8x_{s-1} + x_{s-2}\right) + \frac{h^4}{30}\left(\frac{d^5x}{dt^5}\right)_s + \cdots$$

$$\left(\frac{d^2x}{dt^2}\right)_s = \frac{1}{12h^2}\left(-x_{s+2} + 16x_{s+1} - 30x_s + 16x_{s-1} - x_{s-2}\right) + \frac{h^4}{90}\left(\frac{d^6x}{dt^6}\right)_s + \cdots$$

3-8. Introduction to Step-by-step Integration Procedures

We return to propagation problems and consider procedures for obtaining approximate solutions at discrete intervals $\Delta t = h$. These procedures are often called *finite-difference methods*. We begin by studying the simplest possible method and its associated error. We shall then consider in Sec. 3-9 three distinct families of more sophisticated methods. In the interest of simplicity the treatment in these two sections is confined to propagation problems with only a single propagating variable $x(t)$. Extensions to systems with n propagating variables are given in Sec. 3-10.

In this and the following section, then, we consider the propagation problem

$$\begin{aligned} x &= x(0) & t &= 0 \\ \frac{dx}{dt} &= f(x, t) & t &> 0 \end{aligned} \tag{3-97}$$

and desire to obtain approximate solution values x_s at discrete station points t_s. We shall take the station points to be uniformly spaced.

Euler's Method. The simplest finite-difference procedure for dealing with (3-97) is obtained by replacing $(dx/dt)_{s-1}$ with the crude approximation $(x_s - x_{s-1})/h$, where $h = \Delta t = t_s - t_{s-1}$. This leads to the recurrence formula

$$\begin{aligned} x_0 &= x(0) \\ x_s &= x_{s-1} + hf(x_{s-1}, t_{s-1}) & s &= 1, 2, \ldots \end{aligned} \tag{3-98}$$

for marching out an approximate solution to (3-97). The procedure is known as *Euler's method*.[1]

To illustrate the use of (3-98), we turn to Prob. 3-1, cooling of an object, with the formulation (3-4). In this case (3-98) is simply

$$\begin{aligned} x_0 &= 1 \\ x_s &= x_{s-1} - hx_{s-1} \end{aligned} \tag{3-99}$$

Choosing $h = 0.1$, we compute the values shown in Table 3-2. These values are plotted in Fig. 3-13 and compared with the exact continuous solution of (3-4).

---

[1] It is described in his text "Institutiones calculi integralis," St. Petersburg, **1768**. See "Leonhardi Euleri opera omnia," ser. I, vol. XI, Teubner Verlagsgesellschaft, Leipzig, 1913, p. 424.

TABLE 3-2. SOLUTION OF PROB. 3-1 BY EULER'S METHOD WHEN $h = 0.1$

s	t_s	x_s
0	0.0	1.0000
1	0.1	0.9000
2	0.2	0.8100
3	0.3	0.7290
4	0.4	0.6561
5	0.5	0.5905
6	0.6	0.5314
7	0.7	0.4783
8	0.8	0.4305
9	0.9	0.3874
10	1.0	0.3487

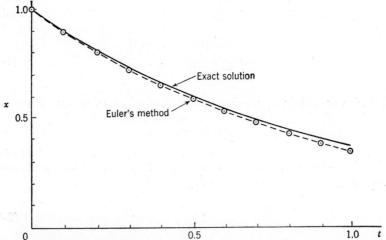

FIG. 3-13. Solution of Prob. 3-1 by Euler's method, compared with exact solution.

A Recurrence Formula for the Error.   A complete theory of error propagation in finite-difference approximations[1] is beyond the scope of this text, but some appreciation of the various factors which affect the error is almost mandatory if one is to make intelligent use of these procedures.   We give here an introduction to this topic by carefully studying the error in the simple Euler method.

We consider the finite-difference solution x_s computed from (3-98) and compare it with the exact continuous solution of (3-97).   To dis-

---

[1] For more penetrating treatments see Collatz, *op. cit.;* W. E. Milne, "Numerical Solution of Differential Equations," John Wiley & Sons, Inc., New York, 1953; M. Lotkin, The Propagation of Error in Numerical Integrations, *Proc. Am. Math. Soc.,* **5,** 869–887 (1954).

tinguish between these, we shall call the continuous solution $\psi(t)$, that is, $\psi(0) = x(0)$, and

$$\frac{d\psi}{dt} = f(\psi, t) \qquad t > 0 \tag{3-100}$$

Assuming that ψ has a continuous second derivative, the Taylor's series with remainder term for $\psi(t_s)$ is

$$\psi(t_s) = \psi(t_{s-1}) + h\left(\frac{d\psi}{dt}\right)_{s-1} + \frac{h^2}{2}\frac{d^2\psi(\tau_s)}{dt^2} \tag{3-101}$$

where $t_{s-1} \leq \tau_s \leq t_s$. Comparing this with (3-98), we see that *if* $x_{s-1} = \psi_{s-1}$ then the error in x_s would be

$$E_{t,s} = -\frac{h^2}{2}\frac{d^2\psi(\tau_s)}{dt^2} \tag{3-102}$$

This error is called the *truncation error of the recurrence formula* (3-98). It is not the total solution error because once the process is started we usually do not have $x_{s-1} = \psi_{s-1}$.

In the numerical solution according to (3-98) it is almost never possible to carry out the indicated operations exactly: we must *round off* our results to a finite number of places. Thus instead of (3-98) we actually use

$$x_s = x_{s-1} + hf(x_{s-1}, t_{s-1}) + E_{r,s} \tag{3-103}$$

where $E_{r,s}$ represents the *round-off error of the recurrence formula*. Its size depends on the manner in which the rounding off is done; e.g., if all preliminary operations such as evaluating $f(x_{s-1}, t_{s-1})$ are performed to a greater number of decimal places and the final result rounded off to k places, then $E_{r,s}$ would be a random value satisfying $|E_{r,s}| \leq 0.5 \times 10^{-k}$.

Let us define the total *solution error* ϵ_s as

$$\epsilon_s = x_s - \psi_s \tag{3-104}$$

We next obtain an expression for ϵ_s by subtracting (3-101) from (3-103). We have

$$\epsilon_s = \epsilon_{s-1} + h[f(x_{s-1}, t_{s-1}) - f(\psi_{s-1}, t_{s-1})] + E_{t,s} + E_{r,s} \tag{3-105}$$

using (3-100) and (3-102). If $f(x, t)$ has a continuous partial derivative with respect to x, the square bracket in (3-105) equals[1] $(\partial f/\partial x)_{s-1} (x_{s-1} - \psi_{s-1})$, where the partial derivative is to be evaluated at (ξ_{s-1}, t_{s-1}) and

---

[1] This is the *mean-value theorem* of the calculus. See, for example, Phillips, *op. cit.*, p. 60.

ξ_{s-1} is some value between x_{s-1} and ψ_{s-1}. We finally obtain

$$\epsilon_s = \left[1 + h\left(\frac{\partial f}{\partial x}\right)_{s-1} \right] \epsilon_{s-1} + E_{t,s} + E_{r,s} \qquad (3\text{-}106)$$

as the basic recurrence formula for the propagation of error in the Euler method. The three terms on the right of (3-106) are called, respectively, the *inherited error*, the *truncation error*, and the *round-off error*. The inherited-error term gives the contribution to the total error at t_s due to the total error at t_{s-1}. We note that for small h the inherited error is magnified[1] when $(\partial f/\partial x)_{s-1}$ is positive.

An exact solution of (3-106) is impossible because all three terms on the right involve a certain amount of uncertainty. The derivative $(\partial f/\partial x)_{s-1}$ is to be evaluated at an unknown position, the truncation error (3-102) involves a derivative of the continuous solution at an unknown location, and the round-off error is random. It is sometimes possible to make reasonable estimates of these uncertainties and then by using (3-106) to obtain an estimate of the over-all solution error.

The truncation error (3-102) involves the second derivative of ψ, which can be evaluated in terms of the function $f(\psi, t)$ in the governing equation (3-100) as follows:

$$\frac{d^2\psi}{dt^2} = \frac{d}{dt}\frac{d\psi}{dt} = \frac{d}{dt}f(\psi,t) = \frac{\partial f}{\partial \psi}\frac{d\psi}{dt} + \frac{\partial f}{\partial t} = \frac{\partial f}{\partial \psi}f + \frac{\partial f}{\partial t} \qquad (3\text{-}107)$$

This should strictly be evaluated at an unknown point $[\psi(t_s), \tau_s]$ on the exact continuous solution. If our total solution error is not too great, we can obtain a useful estimate of the truncation error by evaluating (3-107) at (x_s, t_s) on the finite-difference solution. Inserting in (3-102) gives

$$E_{t,s} \approx -\frac{h^2}{2}\left(f\frac{\partial f}{\partial x} + \frac{\partial f}{\partial t} \right)_s \qquad (3\text{-}108)$$

Similarly the derivative $(\partial f/\partial x)_{s-1}$ can be estimated by evaluating it at (x_{s-1}, t_{s-1}) instead of at the unknown point (ξ_{s-1}, t_{s-1}).

A Solution of the Recurrence Formula. Let the symbol p_s be used for the partial derivative $(\partial f/\partial x)_s$, that is,

$$\left(\frac{\partial f}{\partial x}\right)_s = p_s \qquad (3\text{-}109)$$

The recurrence formula (3-106) can then be solved by *induction*[2] as follows:

---

[1] In this connection see Exercise 3-14.

[2] A more elegant way is to use *variation of parameters*. See L. M. Milne-Thomson, "The Calculus of Finite Differences," The Macmillan Company, New York, 1933, p. 374.

$$\epsilon_0 = 0$$
$$\epsilon_1 = E_{t,1} + E_{r,1}$$
$$\epsilon_2 = (1 + hp_1)(E_{t,1} + E_{r,1}) + E_{t,2} + E_{r,2}$$
$$\epsilon_3 = (1 + hp_1)(1 + hp_2)(E_{t,1} + E_{r,1}) + (1 + hp_2)(E_{t,2} + E_{r,2})$$
$$+ E_{t,3} + E_{r,3} \tag{3-110}$$

$$\cdot \ \cdot$$

$$\epsilon_s = \sum_{j=1}^{s-1} \left\{ \prod_{k=j}^{s-1} (1 + hp_k) \right\} (E_{t,j} + E_{r,j}) + E_{t,s} + E_{r,s}$$

Note that the contributions due to truncation and round-off are independent. We shall denote them by $\epsilon_{t,s}$ and $\epsilon_{r,s}$, respectively, i.e.,

$$\epsilon_{t,s} = \sum_{j=1}^{s-1} \left\{ \prod_{k=j}^{s-1} (1 + hp_k) \right\} E_{t,j} + E_{t,s}$$

$$\epsilon_{r,s} = \sum_{j=1}^{s-1} \left\{ \prod_{k=j}^{s-1} (1 + hp_k) \right\} E_{r,j} + E_{r,s} \tag{3-111}$$

The error $\epsilon_{t,s}$ is sometimes called the truncation error of the solution We prefer to call it the *discretization error;*[1] i.e., we speak of the truncation error of a finite-difference *equation* but of the discretization error of its *solution*. The two are of course related, but even for Euler's method the relationship is not simple.

We can estimate $\epsilon_{t,s}$ fairly accurately by evaluating $E_{t,j}$ and p_k according to (3-108) and (3-109) at every step and forming the sum indicated in (3-110). This would be a laborious process. If the quantities p_k and $E_{t,j}$ do not change too much in the range of integration, we can obtain a quick crude estimate by replacing them by constant values p and E_t. We then have a simple geometric progression for $\epsilon_{t,s}$ whose sum is

$$\epsilon_{t,s} = E_t[(1 + hp)^{s-1} + (1 + hp)^{s-2} + \cdots + (1 + hp) + 1]$$
$$= E_t \frac{(1 + hp)^s - 1}{hp} \tag{3-112}$$

The contribution of the round-off errors cannot be evaluated explicitly because of the random nature of $E_{r,j}$. We can, however, do two things. We can find the *maximum possible* value of $\epsilon_{r,s}$, and by using statistical methods we can estimate the probability of $\epsilon_{r,s}$ building up to any particular smaller value.

If the greatest round-off error possible in a single step is E_r (that is, $|E_{r,j}| \leq E_r$), then the maximum possible build-up would occur when $E_{r,j} = E_r$ at every step.

---

[1] This name was suggested by W. R. Wasow, Discrete Approximations to Elliptic Differential Equations, *Z. angew. Math. u. Phys.*, **6**, 81–97 (1955).

If we again replace p_k by the constant p, we get from (3-111)

$$\text{Max }(\epsilon_{r,s}) = E_r \frac{(1 + hp)^s - 1}{hp} \tag{3-113}$$

by summing the same geometric progression as in (3-112).

If, on the other hand, we consider that $E_{r,j}$ can take any value between $-E_r$ and $+E_r$ with equal probability, we find[1] that, when the number of steps s is large, the distribution of probable values for $\epsilon_{r,s}$ is very nearly a *normal Gaussian distribution* centered about $\epsilon_{r,s} = 0$ with the *standard deviation*,

$$\sigma(\epsilon_{r,s}) = \frac{E_r}{\sqrt{3}} \sqrt{\frac{(1 + hp)^{2s} - 1}{(1 + hp)^2 - 1}} \tag{3-114}$$

This means that the probability of $|\epsilon_{r,s}|$ being smaller than (3-114) is 68.3 per cent.   The probability of $|\epsilon_{r,s}|$ being less than 2σ is 95.4 per cent, and the probability of $|\epsilon_{r,s}|$ being less than 3σ is 99.7 per cent.

Selection of the Interval Size h.   The estimates just given for $\epsilon_{t,s}$ and $\epsilon_{r,s}$ provide the basis for a study of how the error at the end of a fixed interval depends on the number of increments used to cover the interval. Suppose that a solution to (3-97) is desired over an interval $0 < t < T$ and that the Euler recurrence formula (3-98) is to be used.   The interval is divided in S equal increments so that

$$\Delta t = h = \frac{T}{S} \tag{3-115}$$

We can introduce T into the above estimates by using the identity[2]

$$(1 + hp)^S = e^{pT} \left[1 - pT \frac{hp}{2} + 0(h^2 p^2) \right] \tag{3-116}$$

Thus under the assumptions that p_s and $E_{t,s}$ can be considered as constant over the interval $0 < t < T$ the total solution error at $t = T$ is

$$\epsilon_S = \frac{E_t}{hp} \left\{ e^{pT} \left[1 - pT \frac{hp}{2} + 0(h^2 p^2) \right] - 1 \right\} + \epsilon_{r,S} \tag{3-117}$$

according to (3-112) and (3-116).   Taking an average value of (3-108) for the whole interval, we have finally

$$\epsilon_s = -\frac{h}{2p} \left(fp + \frac{\partial f}{\partial t} \right)_{\text{avg}} \left[e^{pT} - 1 - \frac{hp}{2} (pTe^{pT}) + 0(h^2 p^2) \right] + \epsilon_{r,s} \tag{3-118}$$

---

[1] The theorems of statistical theory required to derive (3-114) are given in A. Hald, "Statistical Theory with Engineering Applications," John Wiley & Sons, Inc., New York, 1952, pp. 117, 188.

[2] See Exercise 3-51.

where the maximum possible value and the standard deviation of $\epsilon_{r,s}$ in (3-118) are

$$\text{Max }(\epsilon_{r,s}) = \frac{E_r}{hp}\left\{e^{pT} - 1 - \frac{hp}{2}(pTe^{pT}) + 0(h^2p^2)\right\}$$

$$\sigma(\epsilon_{r,s}) = \frac{E_r}{\sqrt{6hp}}\{e^{2pT} - 1 - hp[(pT + \tfrac{1}{2})e^{2pT} - \tfrac{1}{2}] + 0(h^2p^2)\}^{\frac{1}{2}}$$

$$\text{(3-119)}$$

which were obtained by using (3-116) in (3-113) and (3-114). Examining (3-118) and (3-119), we see that when hp is small the discretization

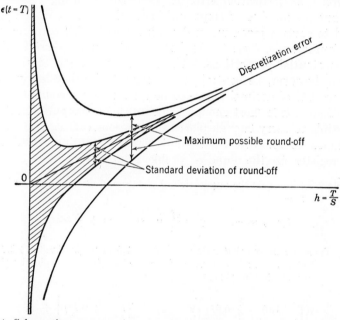

FIG. 3-14. Schematic representation of how the error at the end of a fixed interval depends on the size of the increments used to cover the interval. The probability that the error lies in the shaded region is 68.3 per cent.

error is proportional to h, while the contribution due to round-off has an upper bound inversely proportional to h and a standard deviation inversely proportional to $\sqrt{h}$. The total error then varies with h as shown in Fig. 3-14. It is a simple exercise[1] to show that the minimum of the "maximum total error" curve occurs when the round-off and truncation errors of a single step are of equal magnitude.

The following qualitative argument is sometimes given to show that the discretization error in (3-118) should be proportional to h. According to (3-102) the truncation error of a single step of the Euler recurrence

[1] See Exercise 3-52.

formula is $0(h^2)$. At the end of S steps the total error should be of the order of S times the error of a single step, and since S is inversely proportional to h, this means that total error should be $0(h)$. Arguments of this kind give no hint of the many assumptions necessary to deduce (3-118) and cannot be rigorously defended. They are useful, however, in providing tentative preliminary estimates.

Some general considerations influencing the selection of the increment size h can be drawn out of the foregoing study. These remarks apply not only to the simple Euler recurrence formula but to all finite-difference procedures. Discretization error is decreased by using smaller increments, but the number of steps and hence the amount of computation required to cover a given interval are increased. When the number of steps taken is large, there is the danger that round-off errors will build up to substantial proportions. The magnitude of the total round-off error can, however, be controlled by selecting E_r, that is, by deciding on the number of decimal places to be carried in the computation. In any problem that is short enough to permit hand computation it is usually possible to carry enough places so that round-off can be neglected. In extended computations using automatic computing machines with limited register size the round-off problem can be serious.

Example. To illustrate the application of (3-118) and (3-119), we return to Prob. 3-1. Here we have

$$f(x, t) = -x \qquad p = \frac{\partial f}{\partial x} = -1 \qquad fp + \frac{\partial f}{\partial t} = x \qquad (3\text{-}120)$$

and from Table 3-2 we see that x decreases from 1.0 to (approximately) 0.34 in the interval $0 < t < 1.0$. For the purpose of error estimation we shall take the mean value 0.67. Substituting in (3-118) yields

$$\epsilon_S(T = 1.0) = \frac{h}{2}(0.67)\left[e^{-1.0} - 1 - \frac{h}{2}e^{-1.0} + 0(h^2)\right] + \epsilon_{r,S}$$

$$\begin{cases} \text{Max}(\epsilon_{r,S}) = \frac{E_r}{h}\left[1 - e^{-1.0}\left(1 - \frac{h}{2}\right) + 0(h^2)\right] \\ \sigma(\epsilon_{r,S}) = \frac{E_r}{\sqrt{6h}}\left[1 - e^{-2.0} + \frac{h}{2}(1 + e^{-2.0}) + 0(h^2)\right]^{\frac{1}{2}} \end{cases} \qquad (3\text{-}121)$$

If $h \leq 0.1$, we make little error by neglecting h completely within the square brackets. Thus we obtain

$$\epsilon_S(T = 1.0) = -0.211h + \epsilon_{r,S}$$

$$\begin{cases} \text{Max}(\epsilon_{r,S}) = 0.632\frac{E_r}{h} \\ \sigma(\epsilon_{r,S}) = 0.380\frac{E_r}{\sqrt{h}} \end{cases} \qquad (3\text{-}122)$$

as an approximate description of the behavior of the solution error as a function of the increment size h and the round-off bound per step E_r. Applying (3-122) to the numerical solution in Table 3-2 where $h = 0.1$ and $E_r = 0.5 \times 10^{-4}$, we find

with
and
$$e_{10} = -0.0211 + \epsilon_{r,10}$$
$$\text{Max } (\epsilon_{r,10}) = 3.2 \times 10^{-4}$$
$$\sigma(\epsilon_{r,10}) = 0.60 \times 10^{-4}$$

This means that the possibility exists of round-off affecting three units in the last decimal place but that the probability is very high that no more than one unit is affected. By comparison with the exact solution of Prob. 3-1 the actual error in x_{10} in Table 3-2 is -0.0192; that is, our estimate is correct to within 10 per cent.

h^m **Extrapolation.**[1]   In the derivation of the discretization error $\epsilon_{t,s}$ [(3-112)] we have seen the important role[2] played by the factor $hp = h\ \partial f/\partial x$. In the estimate (3-118) we see that, if hp is small enough for the term $\frac{1}{2}hp(pTe^{pT})$ to be negligible in comparison with $e^{pT} - 1$, then the discretization error at $t = T$ is linearly proportional to h. If, moreover, the round-off error $\epsilon_{r,s}$ is negligible, then the entire solution error after $S = T/h$ steps should be proportional to h. To capitalize on this, we can obtain two numerical solutions using different-sized increments h_1 and h_2 and then use linear extrapolation back to $h = 0$ to obtain an improved approximation.

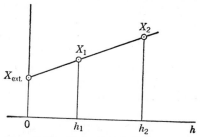

FIG. 3-15. Extrapolation from two numerical solutions whose errors are $O(h)$.

Let X_1 be the value of $x(T)$ obtained using h_1, and let X_2 be the value obtained with h_2; then

$$X_{extrap} = \frac{X_1 h_2 - X_2 h_1}{h_2 - h_1} \tag{3-123}$$

is the value obtained by linear extrapolation as shown in Fig. 3-15.

An illustration of this is furnished by the system (3-99). We already have in Table 3-2 one approximation for the value of x at $t = 1$ which was obtained by using $h = 0.1$ in (3-99). If we carry through the same computation for several other values of h, we get the approximations listed in Table 3-3. In all these approximations the round-off error is negligible in comparison with the truncation error. The values in Table 3-3 are plotted in Fig. 3-16 along with the exact solution obtained from the continuous system (3-4). By drawing a smooth curve through these points we get a graphic picture of the convergence as $h \to 0$. For small h a chord joining two points

[1] This idea was suggested by L. F. Richardson, The Approximate Arithmetical Solution by Finite Differences of Physical Problems, *Trans. Roy. Soc. London*, **A210,** 307–357 (1910). A fuller treatment was later given by L. F. Richardson and J. A. Gaunt, The Deferred Approach to the Limit, *Trans. Roy. Soc. London*, **A226,** 299–361 (1927).

[2] The factor hp is a kind of *nondimensional* measure of the increment size; e.g., if x has the dimensions of length and t of time, then $h\ \partial f/\partial x$ is dimensionless.

TABLE 3-3. APPROXIMATIONS TO $x(t = 1.0)$ OBTAINED BY THE EULER
RECURRENCE FORMULA USING VARIOUS VALUES OF h

h	$x(t = 1.0)$
1	0.0000
$\frac{1}{2}$	0.2500
$\frac{1}{3}$	0.2963
$\frac{1}{4}$	0.3164
$\frac{1}{5}$	0.3277
$\frac{1}{10}$	0.3487

on the curve can be extrapolated to $h = 0$ to provide a good approximation to the true solution. Thus, taking the values in Table 3-3 obtained for $h = \frac{1}{5}$ and $h = \frac{1}{10}$, we get

$$X_{extrap} = \frac{(0.2)(0.3487) - (0.1)(0.3277)}{0.2 - 0.1} = 0.3697 \qquad (3\text{-}124)$$

according to (3-123). Comparison with the exact solution $x = 0.3679$ reveals that the error in X_{extrap} is only 0.0018, while the errors in the values used to get X_{extrap} are -0.0402 and -0.0192.

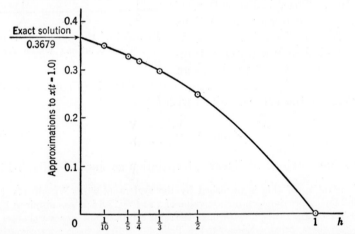

FIG. 3-16. Solution of Prob. 3-1 at $t = 1$, obtained by Euler's method, using different-sized increments h.

Even for relatively large h this extrapolation provides some improvement in the approximation. For example, if we apply (3-123) to the values in Table 3-3 obtained for $h = \frac{1}{2}$ and $h = \frac{1}{3}$ (which have errors of -0.1179 and -0.0716, respectively), we find $X_{extrap} = 0.3889$ (with error 0.0210).

This method of extrapolating the results of numerical solutions with different-sized increments can also be applied to procedures where the discretization error is $O(h^m)$ instead of $O(h)$ as it is in the simple Euler method. It is necessary only to replace h by h^m in (3-123). Thus if

X_1 is the solution at the end of an interval obtained by using $h = h_1$ and X_2 is the solution at the end of the same interval using the same recurrence formula but with $h = h_2$, the extrapolation

$$X_{extrap} = \frac{X_1 h_2{}^m - X_2 h_1{}^m}{h_2{}^m - h_1{}^m} \qquad (3\text{-}125)$$

gives an improved approximation provided that the total round-off error is negligible and that both h are small enough for the major portion of the discretization error to be proportional to h^m.

EXERCISES

3-49. Apply Euler's method to the propagation problem

$$x = 0 \qquad t = 0$$
$$\frac{dx}{dt} = 1 - x \qquad t > 0$$

Using $h = 0.1$, obtain an approximation for the value of x at $t = 0.8$.

Ans. $x_8 = 0.5695$.

3-50. Show that, if $E_{r,s}$ can take any value between $-E_r$ and $+E_r$ with equal probability, then the *standard deviation* or root-mean-square value is

$$\sigma = E_r/\sqrt{3} = 0.577 E_r$$

3-51. Verify the identity (3-116). One method is to use the binomial theorem and then to collect terms in such a way as to build up the series for $e^{pT} = e^{pSh}$. Another method is to expand the logarithm of $(1 + hp)^S$ into a series and then take the antilogarithm.

3-52. Show that the minimum value of $c_1 h + c_2/h$ occurs when $c_1 h = c_2/h$. Apply this to (3-118) and (3-119) to show that the minimum of the "maximum total error curve" in Fig. 3-14 occurs when $\epsilon_{t,s} = \max (\epsilon_{r,s})$. Then show that this occurs when $E_t = E_r$.

3-53. Apply (3-118) and (3-119) to estimate the error in Exercise 3-49. ($x = 0.5507$ at $t = 0.8$ in the exact solution.)

3-54. Show that if the error at $t = T$ by a certain procedure can be shown to be of the form

$$\epsilon = c_1 h^m + c_2 h^{2m} + \cdots$$

then an improved approximation can be obtained by extrapolating from *three* numerical solutions by the following procedure. Let X_1, X_2, and X_3 be the approximations to $x(T)$ obtained by using h_1, h_2, and h_3. Verify the extrapolation formula

$$X_{extrap} = \frac{\dfrac{h_2{}^m - h_3{}^m}{h_1{}^m} X_1 + \dfrac{h_3{}^m - h_1{}^m}{h_2{}^m} X_2 + \dfrac{h_1{}^m - h_2{}^m}{h_3{}^m} X_3}{\dfrac{h_2{}^m - h_3{}^m}{h_1{}^m} + \dfrac{h_3{}^m - h_1{}^m}{h_2{}^m} + \dfrac{h_1{}^m - h_2{}^m}{h_3{}^m}}$$

Interpret this as resulting from passing a parabola through three points where X has been plotted against h^m.

3-55. Show that for the system of Exercise 3-49 the term $\frac{1}{2}hp(pTe^{pT})$ in (3-118) is less than one-tenth of $|e^{pT} - 1|$ when $h \leq 0.3$. Obtain an Euler approximation to x

at $t = 0.8$ by using $h = 0.2$. Use this result and the answer to Exercise 3-49 to obtain an extrapolation according to (3-123). Show with, the aid of Exercise 3-53 that to obtain approximately the same accuracy by a single integration using Euler's method would require about 80 steps.

3-56. Apply Euler's method to Exercise 3-49, using $h = 0.4$. Use this result together with the results already obtained for $h = 0.2$ and $h = 0.1$ to obtain the extrapolated value for x at $t = 0.8$ according to the formula of Exercise 3-54.

Ans. $X_{extrap} = 0.5512$. This same accuracy from a single solution would require over 300 steps.

3-57. Show that if Euler's method is applied to the equation $dx/dt = px$ then an estimate of E_r (the maximum allowable round-off error per step) in order to ensure that $|\epsilon_{r,s}| \leq \frac{1}{10}|\epsilon_{t,s}|$ is given by

$$E_r \leq \frac{h^2 p^2}{20} |x_{\mathrm{avg}}|$$

where x_{avg} is an estimate of the average value of the solution in the interval of integration.

3-9. Recurrence Formulas with Higher-order Truncation Error

The Euler recurrence formula is simple and easy to apply. Its only drawback is its large truncation error. We now turn to procedures which at the price of somewhat more computation per step provide approximations with smaller discretization errors.[1] A bewildering array of such procedures have been devised. To furnish an introduction to these methods, we shall consider three major categories of recurrence formulas and shall examine a simple representative of each category.

All recurrence formulas which we shall consider for integrating the propagation problem

$$\begin{aligned} x &= x(0) & t &= 0 \\ \frac{dx}{dt} &= f(x, t) & t &> 0 \end{aligned} \tag{3-126}$$

may be considered as special cases of the formula

$$x_s = x_{s-q} + qh \text{ (weighted average of } f \text{ values)} \tag{3-127}$$

where, as before, x_s is the approximate solution at $t = t_s$ and h is the (uniform) spacing $\Delta t = t_s - t_{s-1}$. This formula can be interpreted as requiring that x_s and x_{s-q} be end points of a chord whose slope is some weighted average of slopes of tangents to curves which are solutions to

---

[1] Strictly speaking there is no guarantee that a method with higher-order error will give less error than a lower-order method when both are used with a step length of particular size. There is only the assurance that as the size of the step is diminished the error for the higher-order method decreases at a faster rate than that for the lower-order procedure. This means that for steps smaller than some critical size the higher-order methods will actually give less error.

$dx/dt = f(x, t)$. The difference between methods lies in the selection of f values to be averaged. The three categories which we consider are indicated schematically in Table 3-4.

TABLE 3-4. CATEGORIES OF RECURRENCE FORMULAS ACCORDING TO f VALUES USED IN (3-127)

Category	$f(x, t)$ evaluated at
First	. . . , $t_{s-3}, t_{s-2}, t_{s-1}$
Second	. . . , $t_{s-3}, t_{s-2}, t_{s-1}, t_s$
Third	Intermediate values between t_{s-1} and t_s

In the first two categories $f(x, t)$ is evaluated only at each station point, but in making the step to get x_s we utilize f values at previous station points. The distinction between them is based on whether or not $f(x_s, t_s)$ is required in the recurrence formula for x_s. Formulas of the first category are often called *open*, while formulas of the second category are called *closed*. In the third category $f(x, t)$ is evaluated one or more times at intermediate points within the increment $t_{s-1} < t < t_s$ in order to make the step from x_{s-1} to x_s.

Some examples of formulas of the first category are given in Table 3-5. The order of the truncation error $E_{t,s}$ is indicated. Formulas (1) to (4)

TABLE 3-5. RECURRENCE FORMULAS OF THE FIRST CATEGORY

$$x_s = x_{s-q} + qh \sum_k \beta_k f_k$$

Formula	q	β_{s-5}	β_{s-4}	β_{s-3}	β_{s-2}	β_{s-1}	Truncation error $E_{t,s}$
(1)	1				$-\dfrac{1}{2}$	$\dfrac{3}{2}$	$0(h^3)$
(2)	1			$\dfrac{5}{12}$	$-\dfrac{16}{12}$	$\dfrac{23}{12}$	$0(h^4)$
(3)	1		$-\dfrac{9}{24}$	$\dfrac{37}{24}$	$-\dfrac{59}{24}$	$\dfrac{55}{24}$	$0(h^5)$
(4)	1	$\dfrac{251}{720}$	$-\dfrac{1,274}{720}$	$\dfrac{2,616}{720}$	$-\dfrac{2,774}{720}$	$\dfrac{1901}{720}$	$0(h^6)$
(5)	2				0	1	$0(h^3)$
(6)	2			$\dfrac{1}{6}$	$-\dfrac{2}{6}$	$\dfrac{7}{6}$	$0(h^4)$
(7)	2		$-\dfrac{1}{6}$	$\dfrac{4}{6}$	$-\dfrac{5}{6}$	$\dfrac{8}{6}$	$0(h^5)$
(8)	2	$\dfrac{29}{180}$	$-\dfrac{146}{180}$	$\dfrac{294}{180}$	$-\dfrac{266}{180}$	$\dfrac{269}{180}$	$0(h^6)$
(9)	4		0	$\dfrac{2}{3}$	$-\dfrac{1}{3}$	$\dfrac{2}{3}$	$0(h^5)$

are particular cases of a family given by Adams.[1]   Formulas (5) to (8) are particular cases of a family given by Nyström.[2]   Formula (9) is due to Milne.[3]

Some examples of formulas of the second category are given in Table 3-6. The first four formulas are particular cases of a family given by

TABLE 3-6. RECURRENCE FORMULAS OF THE SECOND CATEGORY

$$x_s = x_{s-q} + qh \sum_k \beta_k f_k$$

Formula	q	β_{s-4}	β_{s-3}	β_{s-2}	β_{s-1}	β_s	Truncation error $E_{t,s}$
(1)	1				$\frac{1}{2}$	$\frac{1}{2}$	$O(h^3)$
(2)	1			$-\frac{1}{12}$	$\frac{8}{12}$	$\frac{5}{12}$	$O(h^4)$
(3)	1		$\frac{1}{24}$	$-\frac{5}{24}$	$\frac{19}{24}$	$\frac{9}{24}$	$O(h^5)$
(4)	1	$-\frac{19}{720}$	$\frac{106}{720}$	$-\frac{264}{720}$	$\frac{646}{720}$	$\frac{251}{720}$	$O(h^6)$
(5)	2			$\frac{1}{6}$	$\frac{4}{6}$	$\frac{1}{6}$	$O(h^5)$

Adams,[4] while the fifth is simply *Simpson's rule* applied to integrating differential equations. All the formulas in the first two categories are closely related to formulas for differentiation and integration of functions tabulated at equal increments, which are in turn related to interpolation formulas. For example, formulas (1) to (8) in Table 3-5 and (1) to (4) in Table 3-6 can be derived[5] in a uniform way from an interpolation formula due to Gregory[6] and to Newton.[7]

[1] F. Bashforth and J. C. Adams, "An Attempt to Test Theories of Capillary Attraction," Cambridge University Press, New York, 1883, p. 18. The formulas were originally given in terms of *differences* of f values rather than directly in terms of f values.

[2] E. J. Nyström, Über die numerischen Integration von Differentialgleichungen, *Acta Soc. Sci. Fennicae*, **L** (13) (1925).

[3] W. E. Milne, Numerical Integration of Ordinary Differential Equations, *Am. Math. Monthly*, **33**, 455–460 (1926).

[4] Bashforth and Adams, *op. cit.*

[5] See Scarborough, *op. cit.*, pp. 244–246.

[6] The formula was given in a letter to Collins in 1670. See G. A. Gibson, James Gregory's Mathematical Work; A Study Based chiefly on His Letters, *Proc. Edinburgh Math. Soc.*, **41**, 2–25 (1923).

[7] I. Newton, "Philosophiae naturalis principia mathematica," book III, lemma V, 1687. See F. Cajori's revision of the translation by A. Motte, University of California Press, Berkeley, Calif., 1934, p. 499.

Three representative formulas of the third category are shown in Table 3-7. Formula (1) was given by Runge.[1] Formula (2) is due to Heun,[2] and formula (3) is due to Kutta.[3] Formula (3) is widely known

TABLE 3-7. RECURRENCE FORMULAS OF THE THIRD CATEGORY

Formula		Truncation error $E_{t,s}$
(1)	$x_s = x_{s-1} + hf(x^*_{s-\frac{1}{2}}, t_{s-\frac{1}{2}})$ where $\quad x^*_{s-\frac{1}{2}} = x_{s-1} + \dfrac{h}{2}f(x_{s-1}, t_{s-1})$	$0(h^3)$
(2)	$x_s = x_{s-1} + h[\frac{1}{4}f(x_{s-1}, t_{s-1}) + \frac{3}{4}f(x^*_{s-\frac{2}{3}}, t_{s-\frac{2}{3}})]$ where $\quad x^*_{s-\frac{1}{3}} = x_{s-1} + \frac{2}{3}hf(x^*_{s-\frac{2}{3}}, t_{s-\frac{2}{3}})$ $\quad x^*_{s-\frac{2}{3}} = x_{s-1} + \frac{1}{3}hf(x_{s-1}, t_{s-1})$	$0(h^4)$
(3)	$x_s = x_{s-1} + h[\frac{1}{6}f(x_{s-1}, t_{s-1}) + \frac{1}{3}f(x^*_{s-\frac{1}{2}}, t_{s-\frac{1}{2}})$ $\qquad\qquad\qquad + \frac{1}{3}f(x^{**}_{s-\frac{1}{2}}, t_{s-\frac{1}{2}}) + \frac{1}{6}f(x^*_s, t_s)]$ where $\quad x^*_{s-\frac{1}{2}} = x_{s-1} + \frac{1}{2}hf(x_{s-1}, t_{s-1})$ $\quad x^{**}_{s-\frac{1}{2}} = x_{s-1} + \frac{1}{2}hf(x^*_{s-\frac{1}{2}}, t_{s-\frac{1}{2}})$ $\quad x^*_s = x_{s-1} + hf(x^{**}_{s-\frac{1}{2}}, t_{s-\frac{1}{2}})$	$0(h^5)$

as the *Runge-Kutta formula*, and methods which fall into the third category are sometimes referred to as being of Runge-Kutta type.

A Recurrence Formula of the First Category. As an example, we consider formula (5) in Table 3-5 for obtaining an approximate solution to (3-126).

$$x_0 = x(0)$$
$$x_s = x_{s-2} + 2hf(x_{s-1}, t_{s-1}) \qquad s = 2, 3, \ldots \qquad (3\text{-}128)$$

This states that a chord joining x_{s-2} and x_s has the same slope as the tangent at x_{s-1} [a relationship which would hold exactly[4] if $x(t)$ were an arc of a second-order parabola between t_{s-2} and t_s].

The recurrence formula (3-128) involves exactly the same amount of computational labor once the calculation has been started as does the Euler recurrence formula (3-98). The only difficulty with (3-128) is that it is *not self-starting*. In order to begin to use (3-128), it is necessary to obtain x_1 by some auxiliary method. The iteration and Taylor's-series

[1] C. Runge, Über die numerische Auflösung von Differentialgleichungen, *Math. Ann.*, **46**, 167–178 (1895).

[2] K. Heun, Neue Methode zur approximativen Integration der Differentialgleichungen einer unabhängigen Veränderlichen, *Z. Math. u. Physik*, **45**, 23–28 (1900).

[3] W. Kutta, Beitrag zur näherungsweisen Integration totaler Differentialgleichungen, *Z. Math. u. Physik*, **46**, 435–453 (1901).

[4] See (3-82).

methods of Secs. 3-4 and 3-5 are often employed to get the additional starting values required for recurrence formulas of the first two categories.

If $x = \psi(t)$ is the exact solution of (3-128) and ψ has a continuous third derivative, then the *truncation error of the recurrence formula* (3-128) can be shown[1] to be

$$E_{t,s} = -\frac{h^3}{3}\frac{d^3\psi(\tau)}{dt^3} \tag{3-129}$$

where $t_{s-2} \leq \tau \leq t_s$. By analogy with the argument used for the Euler recurrence formula we might expect that under certain limitations the *discretization error of the solution* at the end of a fixed interval would be $0(h^2)$ and that h^2 extrapolation could be used to advantage. This turns out to be correct, although the limitations involved require more elaborate discussion than was the case for the Euler formula. We shall consider these limitations shortly in the discussion of *stability*.

Example. To illustrate the use of (3-128), we take Prob. 3-1 with the formulation (3-4). For this case (3-128) is simply

$$\begin{aligned} x_0 &= 1.0000 \\ x_s &= x_{s-2} - 2hx_{s-1} \qquad s = 2, 3, \ldots \end{aligned} \tag{3-130}$$

We can get x_1 from the iteration solution (3-32). Thus, choosing $h = 0.1$, we find $x_1 = 0.9048$ by setting $t = 0.1$ in $x^{(3)}$ of (3-32). Then, using (3-130), we can complete Table 3-8. The actual errors in this approximate solution are shown in Fig. 3-17.

TABLE 3-8. SOLUTION OF PROB. 3-1, USING (3-130) WITH $h = 0.1$

s	t_s	x_s	$2hx_s$
0	0.0	1.0000	
1	0.1	0.9048	0.1810
2	0.2	0.8190	0.1638
3	0.3	0.7410	0.1482
4	0.4	0.6708	0.1342
5	0.5	0.6068	0.1214
6	0.6	0.5494	0.1099
7	0.7	0.4969	0.0994
8	0.8	0.4500	0.0900
9	0.9	0.4069	0.0814
10	1.0	0.3686	

A Recurrence Formula of the Second Category. The distinguishing feature of procedures of this type is the inclusion of $f(x_s, t_s)$ in the recurrence formula for x_s. Since $f(x_s, t_s)$ cannot be determined precisely until x_s is, some form of iteration is usually required to solve the recurrence

[1] See (3-83).

formula. The simplest procedure[1] within this category is formula (1), Table 3-6.

$$x_0 = x(0)$$
$$x_s = x_{s-1} + h\{\tfrac{1}{2}f(x_{s-1}, t_{s-1}) + \tfrac{1}{2}f(x_s, t_s)\} \qquad s = 1, 2, \ldots \qquad (3\text{-}131)$$

In one sense (3-131) is not representative because it does not involve f values at station points previous to t_{s-1}. It therefore requires no auxiliary starting information as do the other formulas in this category.

To solve (3-131), the following procedure is usually adopted: A preliminary estimate $x_s^{(0)}$ is obtained by guessing or by use of a recurrence formula of the first category. This is used to evaluate $f(x_s^{(0)}, t_s)$, which is then inserted in the right of (3-131) to produce $x_s^{(1)}$ on the left. If $x_s^{(1)}$ differs from $x_s^{(0)}$, the process is repeated to obtain $x_s^{(2)}$, etc. The rate of convergence depends on the magnitude of the expression $\tfrac{1}{2}h \, \partial f/\partial x$ since the correction to $x_s^{(r+1)}$ on the left of (3-131) due to a correction Δx_s in $x_s^{(r)}$ on the right is approximately $\tfrac{1}{2}h \, \Delta x_s \, \partial f/\partial x$. If $|h \, \partial f/\partial x| < 0.1$, this convergence is very rapid: each successive correction is less than one-twentieth of its predecessor. When $f(x, t)$ is a *linear* function of x, it is possible to solve (3-131) explicitly for x_s and thereby dispense with the need for iteration.

If $x = \psi(t)$ is the exact solution of (3-126) and ψ has a continuous third derivative, the *truncation error of the recurrence formula* (3-131) is[2]

$$E_{t,s} = \frac{h^3}{12} \frac{d^3\psi(\tau)}{dt^3} \qquad (3\text{-}132)$$

where $t_{s-1} \leq \tau \leq t_s$. If over a given interval the quantities $\partial f/\partial x$ and $E_{t,s}$ do not vary too much, then the *discretization error of the solution* at the end of the interval is[3] $O(h^2)$.

Example. Returning to Prob. 3-1 with the formulation (3-4), we have

$$x_0 = 1$$
$$x_s = x_{s-1} - \frac{h}{2}\{x_{s-1} + x_s\} \qquad s = 1, 2, \ldots \qquad (3\text{-}133)$$

as an illustration of (3-131). Here $f(x, t)$ is linear in x, and hence it is possible to solve the recurrence formula explicitly for x_s in terms of x_{s-1}. This would be the easiest approach in the present case. For the sake of illustration, however, we proceed to solve (3-133) by iteration. At each step we shall obtain a preliminary estimate $x_s^{(0)}$ by applying the first-category recurrence formula

$$x_s^{(0)} = x_{s-2} - 2hx_{s-1} \qquad (3\text{-}134)$$

[1] When (3-126) is written in the form $x = x_0 + \int_0^t f(x, t) \, dt$, it is clear that (3-131) is an approximation based on the *trapezoidal rule* (3-92).

[2] See (3-94).

[3] See Exercise 3-63.

which is the same as (3-130). Starting with this value, we then iterate, using

$$x_s^{(r)} = x_{s-1} - \frac{h}{2} \{x_{s-1} + x_s^{(r-1)}\} \qquad r = 1, 2, \ldots \qquad (3\text{-}135)$$

until we get convergence to x_s. The results of this computation for $h = 0.1$ are shown in Table 3-9. Except for $s = 4$ and $s = 6$ the very first application of (3-135) gives

TABLE 3-9. SOLUTION OF PROB. 3-1 WHEN (3-135) IS SOLVED BY ITERATION

s	t	$x_s^{(0)}$	$x_s^{(1)}$	$x_s^{(2)}$	$x_s^{(3)}$
0	0.0	1.0000	1.0000		
1	0.1	0.9048	0.9048		
2	0.2	0.8190	0.8186	0.8186	
3	0.3	0.7411	0.7406	0.7406	
4	0.4	0.6705	0.6700	0.6701	0.6701
5	0.5	0.6066	0.6063	0.6063	
6	0.6	0.5488	0.5485	0.5486	0.5486
7	0.7	0.4966	0.4963	0.4963	
8	0.8	0.4493	0.4490	0.4490	
9	0.9	0.4065	0.4062	0.4062	
10	1.0	0.3678	0.3675	0.3675	

the solution of (3-133) correct to four places. The errors in x_s in Table 3-9 are shown in Fig. 3-17.

A Recurrence Formula of the Third Category. Formulas of this category are distinguished by the fact that $f(x, t)$ is evaluated at intermediate positions within the interval $t_{s-1} < t < t_s$. As an example of this type we consider formula (1) of Table 3-7,

$$\left. \begin{array}{l} x_0 = x(0) \\[4pt] x_{s-\frac{1}{2}}^* = x_{s-1} + \dfrac{h}{2} f(x_{s-1}, t_{s-1}) \\[8pt] x_s = x_{s-1} + hf(x_{s-\frac{1}{2}}^*, t_{s-\frac{1}{2}}) \end{array} \right\} \qquad s = 1, 2, \ldots \qquad (3\text{-}136)$$

The first part of this recurrence formula will be recognized as the Euler method for the interval $h/2$. The value $x_{s-\frac{1}{2}}^*$ so obtained is not, however, considered to be an approximate solution value. It is used only to evaluate $f(x_{s-\frac{1}{2}}^*, t_{s-\frac{1}{2}})$, which is employed in the second part of the recurrence formula to obtain x_s. The procedure is self-starting, as is the case for all formulas of the third category.

If $x = \psi(t)$ is the exact solution of (3-126) and ψ has continuous derivatives up to the third order, then it can be shown that the *truncation error of the recurrence formula* (3-136) is

$$E_{t,s} = -\frac{h^3}{8} \left[\left(\frac{\partial f}{\partial x} \right)_{s-\frac{1}{2}} \frac{d^2\psi(\tau_1)}{dt^2} + \frac{1}{3} \frac{d^3\psi(\tau_2)}{dt^3} \right] \qquad (3\text{-}137)$$

where the partial derivative is to be evaluated at $(\xi_{s-\frac{1}{2}}, t_{s-\frac{1}{2}})$, where $\xi_{s-\frac{1}{2}}$ is some (unknown) value between $\psi(t_{s-\frac{1}{2}})$ and $\psi(t_{s-1}) + (h/2)f[\psi(t_{s-1}), t_{s-1}]$ and $t_{s-1} \le \tau_1 \le t_{s-\frac{1}{2}}$ and $t_{s-1} \le \tau_2 \le t_s$. If over a given interval the quantities $\partial f/\partial x$ and $E_{t,s}$ do not vary too much, then the *discretization error of the solution* at the end of the interval is[1] $O(h^2)$.

Example. To illustrate the use of (3-136), we again consider Prob. 3-1 with the formulation (3-4). In this case we have the following recurrence formula:

$$x_0 = 1$$

$$\left. \begin{array}{l} x^*_{s-\frac{1}{2}} = x_{s-1} - \dfrac{h}{2}x_{s-1} \\[2mm] x_s = x_{s-1} - hx^*_{s-\frac{1}{2}} \end{array} \right\} \quad s = 1, 2, \ldots \qquad (3\text{-}138)$$

Here, because $f(x, t)$ is linear in x, it is possible to eliminate $x^*_{s-\frac{1}{2}}$ and obtain a single recurrence formula giving x_s directly in terms of x_{s-1}. In general this simplification is not possible, and so for the sake of illustration we have used (3-138) as it stands in the numerical solution shown in Table 3-10. This means that each step in Table 3-10 requires twice as much computation as the corresponding step in Table 3-8.

TABLE 3-10. SOLUTION OF PROB. 3-1, USING RECURRENCE FORMULA (3-138) WITH $h = 0.1$

s	t	x_s	$x^*_{s+\frac{1}{2}}$
0	0.0	1.0000	0.9500
1	0.1	0.9050	0.8598
2	0.2	0.8190	0.7780
3	0.3	0.7412	0.7041
4	0.4	0.6708	0.6373
5	0.5	0.6071	0.5767
6	0.6	0.5494	0.5219
7	0.7	0.4972	0.4723
8	0.8	0.4500	0.4275
9	0.9	0.4072	0.3868
10	1.0	0.3685	

Stability. The approximate solutions to Prob. 3-1 obtained in Tables 3-8 to 3-10 can be compared with the known exact solution $x = e^{-t}$ and the actual solution errors ascertained. These errors have been plotted in Fig. 3-17.

We note first that all three procedures have errors of about the same order of magnitude, at least in comparison with the Euler-method solution of Table 3-2, which has a maximum error of 192 units in the fourth decimal place. The error growth for all three cases in Fig. 3-17 is somewhat erratic owing to the random round-off errors. The errors for the

[1] See Exercise 3-64.

first-category procedure, however, seem to indicate the presence of an oscillating error of increasing amplitude. We turn next to a closer examination of this phenomenon. We shall find that the oscillating error is due to an *unstable extraneous solution* of the recurrence formula. This

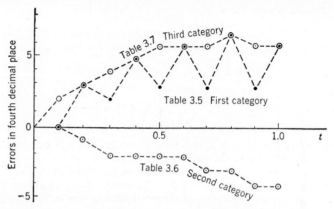

Fig. 3-17. Comparison of the actual errors in the approximate solutions of Tables 3-8 to 3-10.

extraneous solution is excited both by the initial conditions and by the round-off errors.

Consider the propagation problem

$$x = 1 \qquad t = 0$$
$$\frac{dx}{dt} = px \qquad t > 0 \tag{3-139}$$

where p is a constant. Problem 3-1 is the special case of (3-139) for which $p = -1$. The recurrence formula (3-128) when applied to (3-139) is

$$x_s = x_{s-2} + 2hpx_{s-1} \tag{3-140}$$

Let us obtain the general solution[1] to (3-140), neglecting round-off. We try a solution of the form $x_s = \lambda^s$. Substituting in (3-140) yields the equation

$$\lambda^2 - 2hp\lambda - 1 = 0 \tag{3-141}$$

whose roots are

$$\lambda_1 = hp + \sqrt{1 + h^2 p^2}$$
$$\lambda_2 = hp - \sqrt{1 + h^2 p^2} \tag{3-142}$$

The general solution to (3-140) is then given by

$$x_s = c_1 \lambda_1{}^s + c_2 \lambda_2{}^s \tag{3-143}$$

where c_1 and c_2 are constants. It thus requires two initial conditions to fix a particular solution to (3-140).

---

[1] A study of the stability of recurrence formulas using this method was given by H. Rutishauser, Über die Instabilität von Methoden zur Integration von Differentialgleichungen, *Z. angew. Math. u. Phys.*, **3**, 65–74 (1952).

Let us now evaluate the constants c_1 and c_2 for the case where $x_0 = 1$ and $x_1 = e^{hp}$. These are the values assumed by $x = e^{pt}$, the exact continuous solution of (3-139), at $t_0 = 0$ and $t_1 = h$. We have

$$1 = c_1 + c_2$$
$$e^{hp} = c_1\lambda_1 + c_2\lambda_2 \qquad\qquad (3\text{-}144)$$

which yields

$$c_1 = \frac{\sqrt{1 + h^2p^2} - hp + e^{hp}}{2\sqrt{1 + h^2p^2}} = 1 + \frac{h^3p^3}{12} + 0(h^4p^4)$$

$$c_2 = \frac{\sqrt{1 + h^2p^2} + hp - e^{hp}}{2\sqrt{1 + h^2p^2}} = -\frac{h^3p^3}{12} + 0(h^4p^4) \qquad (3\text{-}145)$$

by application of the binomial theorem and the Taylor's series for e^{hp}. The solution of the recurrence formula (3-140) which coincides with the exact continuous solution at t_0 and t_1 is then

$$x_s = \left[1 + \frac{h^3p^3}{12} + 0(h^4p^4)\right]\left[1 + hp + \frac{h^2p^2}{2} + 0(h^4p^4)\right]^s$$
$$+ (-1)^s\left[-\frac{h^3p^3}{12} + 0(h^4p^4)\right]\left[1 - hp + \frac{h^2p^2}{2} + 0(h^4p^4)\right]^s \qquad (3\text{-}146)$$

For comparison the exact solution is

$$e^{pt} = e^{shp} = \left[1 + hp + \frac{h^2p^2}{2} + \frac{h^3p^3}{6} + 0(h^4p^4)\right]^s \qquad (3\text{-}147)$$

We note that the recurrence solution (3-146) is made up of two terms. The first term by itself is a good approximation to (3-147). We call this term the *primary solution* of the recurrence formula. The second term represents an oscillation. It is called an *extraneous*, or *spurious*, solution. For small hp (3-146) may be estimated[1] as follows:

$$x_s \approx e^{shp}\left[1 - \frac{h^2p^2}{6}shp\right] + (-1)^{s+1}\frac{h^3p^3}{12}e^{-shp} \qquad (3\text{-}149)$$

When h is fixed, the amplitude of the extraneous solution *decays* or *grows* with increasing s, depending on whether p is *positive* or *negative*. Note that when p is positive the primary solution grows exponentially, while the undesired oscillation decays. The effect of the extraneous solution thus remains insignificant. When p is negative, however, the primary solution decays, while the undesired oscillation grows. In this case the extraneous solution is a potential source of gross errors. The number of independent extraneous solutions that a particular recurrence formula has is equal to the number of auxiliary initial conditions required to start using the formula. If a recurrence formula is self-starting, it has no extraneous solutions. Thus all formulas of the third category are without extraneous solutions (3-136).

When a recurrence formula applied to a particular propagation problem has one or more extraneous solutions which can grow without limit, we shall say that the recurrence formula has *unstable extraneous solutions*.

[1] See Exercise 3-62.

The analysis just given has shown that the formula (3-128) when applied to a system having

$$\frac{dx}{dt} = px \tag{3-150}$$

as its governing equation has unstable extraneous solutions when p is negative. Similar analyses can be made of the other recurrence formulas in Tables 3-5 and 3-6, with the results shown in Table 3-11.

TABLE 3-11. STABILITY OF EXTRANEOUS SOLUTIONS OF RECURRENCE FORMULAS WHEN APPLIED TO (3-150)

	Recurrence formula	$p < 0$	$p > 0$
Table 3-5	(1)	Stable	Stable
	(2)	Stable	Stable
	(3)	Stable	Stable
	(4)	Stable	Stable
	(5)	Unstable	Stable
	(6)	Unstable	Stable
	(7)	Unstable	Stable
	(8)	Unstable	Stable
	(9)	Unstable	Unstable
Table 3-6	(1)	Has no extraneous solution	
	(2)	Stable	Stable
	(3)	Stable	Stable
	(4)	Stable	Stable
	(5)	Unstable	Stable

While these results are strictly valid only for the system (3-150), where $\partial f/\partial x = p$ is constant, they may be used over a limited interval, for a system where $\partial f/\partial x$ varies to provide qualitative indication of whether or not any extraneous solution has a tendency to grow.

The fact that a recurrence formula has unstable extraneous solutions does not necessarily mean that it should not be used. Perfectly satisfactory approximations can be obtained provided proper precautions are taken. One approach is to estimate the growth of the extraneous solutions ahead of time and then to select the increment size h and the number of decimal places to be carried accordingly so that the contribution of the extraneous solutions to the over-all solution error remains small. A second approach[1] is to build a detection scheme into the computational program which signals the build-up of an oscillating pattern at an early

[1] This idea is due to L. Collatz and R. Zurmühl, Beitrage zu den Interpolations verfahren der numerischen Integration von Differentialgleichungen erster und zweiter Ordnung, Z. angew. Math. u. Mech., 22, 42–55 (1942).

stage. These oscillations are then smoothed out whenever they become noticeable.

Controlling an Unstable Extraneous Solution. To illustrate the first approach, we return to the recurrence formula (3-140) and the estimate (3-149) of the primary and extraneous solutions excited by the initial conditions. If we are interested in a solution over a fixed interval T, then with $S = T/h$ we obtain

$$x_S \approx e^{pT}\left(1 - \frac{h^2p^2}{6}\,pT\right) + (-1)^{s+1}\frac{h^3p^3}{12}\,e^{-pT} \qquad (3\text{-}151)$$

from (3-149). Thus the contribution of the extraneous solution is $O(h^3)$, while the discretization error of the primary solution is $O(h^2)$. We have

$$\left|\frac{\text{Extraneous solution}}{\text{Discretization error of primary solution}}\right| = \frac{h}{2}\frac{e^{-2pT}}{T} = \frac{e^{-2pT}}{2S} \qquad (3\text{-}152)$$

which provides an explicit measure of how small an increment must be taken to control the magnitude of the extraneous solution excited by the initial conditions.

We have so far neglected the effect of round-off errors on the extraneous solution. To investigate this, we shall reconsider the solution (3-140) when a round-off error $E_{r,s}$ is committed at each step as follows:

$$x_s = x_{s-2} + 2hpx_{s-1} + E_{r,s} \qquad (3\text{-}153)$$

The single-step round-off error $E_{r,s}$ is a random value satisfying $|E_{r,s}| \le E_r$. The contribution of round-off to the total solution error $\epsilon_{r,s}$ (see the error analysis of Euler's method, page 165) then satisfies the following relations:

$$
\begin{aligned}
\epsilon_{r,0} &= 0 \\
\epsilon_{r,1} &= E_{r,1} \\
\epsilon_{r,s} &= \epsilon_{r,s-2} + 2hp\epsilon_{r,s-1} + E_{r,s} \qquad s = 2, 3, \ldots
\end{aligned}
\qquad (3\text{-}154)
$$

It may be verified[1] that the solution of (3-154) is

$$\epsilon_{r,s} = \sum_{j=1}^{s}\frac{\lambda_1^{s+1-j} - \lambda_2^{s+1-j}}{\lambda_1 - \lambda_2}\,E_{r,j} \qquad (3\text{-}155)$$

where λ_1 and λ_2 are given by (3-142). From this we can obtain the *maximum possible value* and the *standard deviation* of $\epsilon_{r,s}$. If p is positive, the maximum round-off build-up occurs when $E_{r,s} = E_r$, while if p is negative, the maximum build-up occurs when $E_{r,s} = (-1)^s E_r$. Without going into the details[2] we obtain

$$\text{Max}\,(\epsilon_{r,S}) = \begin{cases} E_r\,\dfrac{e^{Shp} - 1}{2hp} & p > 0 \\[2mm] E_r\,\dfrac{1 - e^{-Shp}}{2hp} & p < 0 \end{cases} \qquad (3\text{-}156)$$

when $|hp|$ is small compared with unity. If $E_{r,s}$ can take any value between $-E_r$ and $+E_r$ with equal probability, then the distribution of $\epsilon_{r,s}$ values for large S approaches a *normal distribution* with average value zero and the following *standard deviation*:

$$\sigma(\epsilon_{r,S}) = E_r\,\sqrt{\frac{\sinh\,(2Shp)}{12\,hp}} \qquad (3\text{-}157)$$

[1] See Exercise 3-68.
[2] See Exercise 3-69.

The reader should compare these results with (3-119). The essential difference here is that the accumulated round-off tends to grow without limit whether p is positive or negative, i.e., whether the primary solution grows or decays.

As a means of controlling the amount of the unstable extraneous solution excited by round-off when p is negative, we can ask that max $(\epsilon_{r,s})$ should remain less than the amplitude of the extraneous solution excited by the initial conditions. Thus, equating max $(\epsilon_{r,s})$ in (3-156) with the magnitude of the extraneous solution in (3-149), we obtain

$$E_r = \frac{h^4 p^4}{6} \frac{1}{1 - e^{Shp}} \qquad p < 0 \qquad (3\text{-}158)$$

as the maximum permissible round-off per step required to ensure that the amplitude of the extraneous solution excited by round-off will be less than that excited by the initial conditions. This, together with (3-152), enables us to control the magnitude of the unstable extraneous solution of (3-140) by selecting h and E_r appropriately.

Example. Let us apply these considerations to (3-130), which is the special case of (3-140) where $p = -1$. Taking $T = 1$, we have

$$\frac{\text{Extraneous solution}}{\text{Discretization error of primary solution}} = \frac{h}{2} e^2 = 3.69h \qquad (3\text{-}159)$$

from (3-152). Therefore, if we wish the amount of the extraneous solution excited by the initial conditions to remain less than 10 per cent of the total error, we would have to take $h < 1/36.9$. In Table 3-8, where $h = 0.1$ was used, we would expect the initially excited extraneous solution to be about 37 per cent of the primary-solution discretization error. Note that this is borne out by the actual errors displayed in Fig. 3-17.

From (3-158) we get

$$E_r = \frac{h^4}{6} \frac{1}{1 - e^{-Sh}} = 0.26 \times 10^{-4} \qquad (3\text{-}160)$$

when $h = 0.1$ and $S = 10$ as the largest permissible round-off required to guarantee that the extraneous solution excited by round-off should be less than that excited by the initial conditions. In the computation of Table 3-8 we actually have

$$E_r = 0.5 \times 10^{-4}$$

Substituting in (3-156) and (3-157), we find that although max $(\epsilon_{r,10}) = 4.3 \times 10^{-4}$ (which is about twice the initially excited extraneous solution) the standard deviation is only $\delta(\epsilon_{r,10}) = 1.2 \times 10^{-4}$. This means that the probability of round-off affecting one unit in the fourth decimal place is roughly $\frac{1}{2}$ but that the probability of affecting more than two units is very small. In Fig. 3-17 the round-off contribution to ϵ_{10} is actually[1] about 1.4×10^{-4}.

Comparisons between Categories. We can draw some conclusions concerning the relative advantages of the three categories of recurrence formulas described in this section. The remarks which follow apply when the procedures being compared have the *same order of truncation error* and use the *same increment size h*.

First-category methods require less computation than methods of the other two categories. Only one evaluation of $f(x, t)$ per step is required. In the second category the number of evaluations depends on the num-

[1] See Exercise 3-70.

ber of iterations required, while in the third category it depends on the number of intermediate values such as $x^*_{s-\frac{1}{2}}$ which are necessary.

First- and third-category methods provide no means for checking. The iteration in the second category is in itself something of a check. When a first-category method is used to provide an initial trial value for a second-category method, a very valuable running check[1] is obtained.

The higher-order methods of first and second categories are not self-starting. This means that auxiliary methods must be employed at the start. It also means that the recurrence formulas have extraneous solutions, and these may possibly be unstable (see Table 3-11). Third-category methods are always self-starting. If during an integration it is desired to alter the increment size h, there is no difficulty in doing so when using a third-category method. First- and second-category methods require special tactics (such as interpolation) to get them started again.

No simple statement can be made regarding the relative magnitudes of the errors to be expected from the three categories. Even when the contributions of round-off and extraneous solutions are negligible and when the discretization error of the solution can be estimated simply in terms of the truncation error of the recurrence formula, we still find that the error depends on the particular system being integrated. Thus a second-category method may give smaller error than a third-category method for one problem and larger error for another.[2]

EXERCISES

3-58. Integrate (3-133) up to $t = 1.0$, using $h = 0.2$.   *Ans.* $x_5 = 0.3666 \cdot$ Obtain an improved approximation at $t = 1.0$ by applying h^2 *extrapolation* [(3-125)] to this and the result in Table 3-9.   *Ans.* $X_{extrap} = 0.3678$; $e^{-1.0} = 0.36788$.

3-59. Integrate (3-138) up to $t = 1.0$, using $h = 0.2$.   *Ans.* $x_5 = 0.3707$. Apply h^2 *extrapolation* to this and the result in Table 3-10 to obtain an improved approximation at $t = 1.0$.   *Ans.* $X_{extrap} = 0.3678$.

3-60. Apply formula (3) of Table 3-5 to Prob. 3-1 as formulated in (3-4). Obtain $x(t = 1.0)$, using $h = 0.2$. For starting values take $x_1 = 0.81873$, $x_2 = 0.67032$, and $x_3 = 0.54881$, which are correct to five decimal places.   *Ans.* $x_3 = 0.36800$.

3-61. Apply formula (3) of Table 3-7 to Prob. 3-1 as formulated in (3-4), using $h = 0.2$.

Ans. $x_1 = 0.81873$, $x_2 = 0.67032$, $x_3 = 0.54882$, $x_4 = 0.44933$, and $x_5 = 0.36788$, which, except for x_3, are correct to five decimal places.

3-62. If $Sh = T$, show that for small values of hp

$$(e^{hp} + ah^m p^m)^S = e^{pT}[1 + apT(hp)^{m-1} + 0(h^m p^m)]$$

[See (3-116) and the hints in Exercise 3-51.]

[1] See Exercise 3-72.

[2] See Exercises 3-73 and 3-74 for examples of systems in which the relative superiority of (3-131) and (3-136) is reversed.

3-63. Show that the propagation of error when using (3-131) is governed by

$$\epsilon_s = \epsilon_{s-1} + \tfrac{1}{2}h(p_{s-1}\epsilon_{s-1} + p_s\epsilon_s) + E_{t,s} + E_{r,s}$$

where p_s and $E_{t,s}$ are defined by (3-109) and (3-132). Verify that when p_s is constant and an average value is used for $E_{t,s}$

$$\epsilon_{r,s} = \left(1 - \frac{hp}{2}\right)^{-1} \sum_{j=1}^{s} E_{r,s}\lambda^{s-j}$$

$$\epsilon_{t,s} = \frac{h^2}{12p}\left(\frac{d^3x}{dt^3}\right)_{\text{avg}} (\lambda^s - 1)$$

where $\lambda = (1 + hp/2)/(1 - hp/2)$. If $Sh = T$, show that

$$\epsilon_{t,S} = \frac{h^2}{12p}\left(\frac{d^3x}{dt^3}\right)_{\text{avg}}\left[e^{pT}\left(1 + \frac{h^2p^2}{12}pT + \cdots\right) - 1\right]$$

3-64. Show that when (3-136) is applied to the system (3-139) the recurrence formula can be written

$$x_s = \left(1 + hp + \frac{h^2p^2}{2}\right)x_{s-1}$$

Utilizing the fact that the exact solution to this system satisfies $\psi_s = e^{hp}\psi_{s-1}$, show that

$$\epsilon_{t,s} = \epsilon_{t,s-1}\left(1 + hp + \frac{h^2p^2}{2}\right) - \frac{h^3p^3}{6}\psi(\tau_s)$$

where $t_{s-1} \leq \tau_s \leq t_s$. Verify that this is equivalent to (3-137) for this particular system. If we can use a constant average value for $E_{t,s}$, show that

$$\epsilon_{t,s} = -\frac{h^2p^2}{6}x_{\text{avg}}\frac{\lambda^s - 1}{1 + hp/2}$$

where $\lambda = 1 + hp + h^2p^2/2$. Compare this result with the previous exercise.

3-65. Write out formula (1) of Table 3-5 for the system (3-139). Try a solution of the form $x_s = \lambda^s$. Show that when hp is small $\lambda_1 = 1 + hp + h^2p^2/2 + 0(h^3p^3)$ is the root corresponding to the primary solution and $\lambda_2 = hp/2 - h^2p^2/2 + 0(h^3p^3)$ is the root corresponding to the extraneous solution and hence that the extraneous solution is stable for small hp whether p is positive or negative.

3-66. Write out formula (9) of Table 3-5 for the system (3-139), and try a solution of the form $x_s = \lambda^s$. Verify that after neglecting terms of order h^2p^2

$$\lambda_1 = 1 + hp$$
$$\lambda_2 = -(1 - \tfrac{5}{3}hp)$$
$$\lambda_3 = i(1 + \tfrac{1}{3}hp)$$
$$\lambda_4 = -i(1 + \tfrac{1}{3}hp)$$

are the roots. The root λ_1 corresponds to the primary solution. The other three correspond to extraneous solutions. Under what circumstances are the extraneous solutions unstable? What is the character of the oscillation associated with the pair (λ_3, λ_4)? Show that

$$c_3\lambda_3{}^s + c_4\lambda_4{}^s = \lambda^s\left[(c_3 + c_4)\cos\frac{\pi s}{2} + i(c_3 - c_4)\sin\frac{\pi s}{2}\right]$$

where $\lambda = 1 + \tfrac{1}{3}$ hp.

3-67. Show that in Tables 3-5 and 3-6 the sum $\Sigma\beta_k$ must equal unity by applying the general formulas to the particular system $dx/dt = 1$.

3-68. Write out $\epsilon_{r,s}$ according to (3-154) for $s = 1$, 2, and 3. Verify that these same results are obtained by evaluating the corresponding quantities from (3-155). A general proof of the correctness of (3-155) can be made on this basis by using *induction.* A more straightforward method of obtaining (3-155) is to use the *method of variation of parameters.*[1]

3-69. Verify (3-156) by substituting $E_{r,s} = E_r$ and $E_{r,s} = (-1)^s E_r$ into (3-155) and summing the geometric progressions. An alternative procedure is to insert a trial solution into (3-153), having the form

$$x_s = c_1\lambda_1^s + c_2\lambda_2^s + c_3 \qquad \text{or} \qquad x_s = c_1\lambda_1^s + c_2\lambda_2^s + c_3(-1)^s$$

3-70. Use (3-151) to estimate that for the solution of Table 3-8 at $t = 1.0$ (see Fig. 3-17) the discretization error of the primary solution is 0.00061 and that the amplitude of the extraneous solution excited by the initial conditions is 0.00023. The round-off error $\epsilon_{r,10}$ must then be -0.00014.

3-71. Use (3-146) to show that if in the system of (3-139) $\epsilon_{s-2} = \epsilon_{s-1} = 0$ then $\epsilon_{t,s}$ is correctly given by (3-129). Note that the truncation error of the recurrence formula is *not the same* as the one-step discretization error of the primary solution.

3-72. The current truncation error $E_{t,s}$ of a second-category formula can be estimated at every step[2] if the formula is solved by iteration and a first-category formula is used to obtain the initial value $x_s^{(0)}$. To illustrate this, consider (3-134) and (3-135). If ψ_s is the exact continuous solution and $\epsilon_s = x_s - \psi_s$, show that

$$x_s^{(1)} - x_s^{(0)} = \epsilon_s^{(1)} - \epsilon_s^{(0)} = \epsilon_{s-1} - \epsilon_{s-2} + \frac{h}{2}\left(3\epsilon_{s-1} - \epsilon_s^{(0)}\right)$$

$$+ \frac{5}{12}h^3\frac{d^3\psi(\tau)}{dt^3} + E_{r,s}^{(1)} - E_{r,s}^{(0)}$$

and hence if round-off is neglected and if h is small and if $|\epsilon_{s-1} - \epsilon_{s-2}|$ is small, then the truncation error (3-132) is approximately one-fifth the discrepancy between $x_s^{(1)}$ and $x_s^{(0)}$. The discrepancy $x_s^{(1)} - x_s^{(0)}$ thus serves as a running check. Any sudden change in its value signals a mistake, while a gradual increase indicates that the increment size should be decreased if accuracy is to be maintained.

3-73. Verify that $x = (1 + t)^2$ is the exact solution of the propagation problem

$$x = 1 \qquad t = 0$$
$$\frac{dx}{dt} = \frac{2x}{1+t} \qquad t > 0$$

Integrate this for one step of size h by the third-category process (3-136), and show that

$$x_1 = 1 + 2h + h^2 - \tfrac{1}{2}h^3 + 0(h^4)$$

Verify that the truncation error here is given correctly by (3-137). Show that there is no truncation error when the second-category process (3-131) is applied to this system.

3-74. Verify that $x = e^{t - \frac{1}{3}t^2}$ is the exact solution of the propagation problem

$$x = 1 \qquad t = 0$$
$$\frac{dx}{dt} = x(1 - \tfrac{2}{3}t) \qquad t > 0$$

---

[1] See Milne-Thomson, *loc. cit.*

[2] This idea is due to W. E. Milne, Numerical Integration of Ordinary Differential Equations, *Am. Math. Monthly*, **33**, 455–460 (1926). The use of formula (9), Table 3-5, and formula (5), Table 3-6, in this connection is often called *Milne's method.*

Integrate this for one step of size h by the third-category process (3-136) and for one step by the second-category process (3-131), and by comparison with the exact solution show that the truncation error in (3-136) is $0(h^4)$, while it is $0(h^3)$ in (3-131).

3-75. Integrate (3-130) up to $t = 1.0$, using $h = 0.2$. As in Exercises 3-58 and 3-59 apply h^2 *extrapolation* to this and the result for $t = 1.0$ in Table 3-8. Explain why the extrapolation produces no substantial increase in accuracy. (It is helpful to sketch the actual errors obtained with $h = 0.2$ on a graph similar to Fig. 3-17.)

3-76. Show that when any first- or second-category formula is applied to (3-139) the general solution of the recurrence formula (neglecting round-off) is

$$x_s = \sum_j c_j \lambda_j{}^s$$

where the λ_j are the roots of

$$\lambda^s - \lambda^{s-q} - qhp \sum_k \beta_k \lambda^k = 0$$

Show that when hp is small q of the λ's lie in the vicinity of the qth roots of unity and that the remainder lie in the vicinity of the origin. The root in the neighborhood of unity corresponds to the primary solution; the rest correspond to extraneous solutions. An extraneous solution is unstable if the corresponding λ lies *outside* the unit circle.

3-10. Step-by-step Integration Methods for Systems with Several Degrees of Freedom

The two preceding sections have described the essential features of finite-difference approximations for propagation problems with only one propagating variable. The extension to n propagating variables is immediate.

Let us write the formulation of the general propagation problem

$$\begin{aligned} x_j &= x_j\,(0) & t &= 0 \\ \frac{dx_j}{dt} &= f_j(x_1, x_2, \ldots, x_n, t) & t &> 0 \end{aligned} \tag{3-161}$$

in matrix notation as

$$\begin{aligned} X &= X_0 & t &= 0 \\ \frac{dX}{dt} &= F(X, t) & t &> 0 \end{aligned} \tag{3-162}$$

where

$$X = \begin{bmatrix} x_1 \\ x_2 \\ \cdots \\ x_n \end{bmatrix} \qquad F = \begin{bmatrix} f_1 \\ f_2 \\ \cdots \\ f_n \end{bmatrix} \tag{3-163}$$

The statement (3-162) is identical in form with the statement of the propagation problem for a single variable. All the recurrence formulas of Sec. 3-9 can thus be applied directly to (3-162) by simply considering them to be written in matrix notation. The discussion of errors can also be generalized.[1]

Example. The formulation (3-11) of the hydraulic surge tank, Prob. 3-2, with $\theta = 1$ is written below in matrix notation.

$$X = \begin{bmatrix} x_1 \\ x_2 \end{bmatrix} = \begin{bmatrix} -1 \\ 1 \end{bmatrix} \qquad t = 0$$

$$\frac{dX}{dt} = \begin{bmatrix} \dfrac{dx_1}{dt} \\ \dfrac{dx_2}{dt} \end{bmatrix} = \begin{bmatrix} x_2 \\ -x_1 - x_2{}^2 \end{bmatrix} = F \qquad t > 0 \tag{3-164}$$

We shall illustrate the application of the third-category formula (3-136) to this system. In matrix notation (3-136) appears as follows:

$$X_0 = X(0)$$
$$\left. \begin{aligned} X_{s-\frac{1}{2}}^* &= X_{s-1} + \frac{h}{2} F(X_{s-1}, t_{s-1}) \\ X_s &= X_{s-1} + hF(X_{s-\frac{1}{2}}^*, t_{s-\frac{1}{2}}) \end{aligned} \right\} \qquad s = 1, 2, \ldots \tag{3-165}$$

Selecting $h = 0.1$, we begin the integration of (3-164) according to (3-165).

$$X_{\frac{1}{2}}^* = \begin{bmatrix} -1 \\ 1 \end{bmatrix} + 0.05 \begin{bmatrix} 1 \\ 1 - 1 \end{bmatrix} = \begin{bmatrix} -0.95 \\ 1.00 \end{bmatrix}$$

$$X_1 = \begin{bmatrix} -1 \\ 1 \end{bmatrix} + 0.10 \begin{bmatrix} 1.00 \\ 0.95 - 1.00 \end{bmatrix} = \begin{bmatrix} -0.900 \\ 0.995 \end{bmatrix}$$

$$X_{1\frac{1}{2}}^* = \begin{bmatrix} -0.900 \\ 0.995 \end{bmatrix} + 0.05 \begin{bmatrix} 0.995 \\ 0.900 - (0.995)^2 \end{bmatrix} = \begin{bmatrix} -0.8502 \\ 0.9905 \end{bmatrix} \tag{3-166}$$

$$X_2 = \begin{bmatrix} -0.900 \\ 0.995 \end{bmatrix} + 0.10 \begin{bmatrix} 0.9905 \\ 0.8502 - (0.9905)^2 \end{bmatrix} = \begin{bmatrix} -0.8010 \\ 0.9819 \end{bmatrix}$$

These results can be compared with $X_2 = (-0.8012, 0.9825)$ obtained in (3-46) by using Taylor's series.

The amount of labor required to integrate a system with n propagating variables depends markedly on the complexity of the functions f_j in (3-161). For systems of similar complexity the total computational labor required is in direct proportion to n.

Special Second-order Systems. The formulation

$$\left. \begin{aligned} x &= x(0) \\ \frac{dx}{dt} &= x'(0) \end{aligned} \right\} \qquad t = 0$$

$$\frac{d^2x}{dt^2} = f(x, t) \qquad t > 0 \tag{3-167}$$

represents a propagation problem in a special second-order system having only one propagating variable. The discussion which follows can be

[1] See Exercises 3-85 and 3-86.

generalized to apply to systems with n propagating variables by substituting matrix notation.

There have been many recurrence formulas proposed for integrating (3-167). The simplest one (which plays the same role as Euler's method for first-order systems) is based on the parabolic differentiation formula (3-84).

$$x_{s+1} = 2x_s - x_{s-1} + h^2 f(x_s, t_s) \qquad (3\text{-}168)$$

The truncation error $E_{t,s}$ is $0(h^4)$ according to (3-85). Higher-order truncation error can be obtained from formulas fitting into the three categories discussed in Sec. 3-9.

As an example of a first-category formula we have[1]

$$x_{s+1} = 2x_s - x_{s-1} + \frac{h^2}{12}[13f(x_s, t_s) - 2f(x_{s-1}, t_{s-1}) + f(x_{s-2}, t_{s-2})] \quad (3\text{-}169)$$

The truncation error $E_{t,s}$ is $0(h^5)$. A related second-category formula is

$$x_{s+1} = 2x_s - x_{s-1} + \frac{h^2}{12}[f(x_{s+1}, t_{s+1}) + 10f(x_s, t_s) + f(x_{s-1}, t_{s-1})] \quad (3\text{-}170)$$

whose truncation error is $0(h^6)$. Third-category methods for second-order systems have also been[2] proposed.

It will be noted that (3-168) to (3-170) are *not self-starting*. Taylor's series or iteration (see Exercise 3-20) are often employed to obtain the additional starting data.

Example. Consider the response of the nonlinear spring-mass system, Prob. 3-3, with the formulation (3-17). Taking $\epsilon = 1$, we have

$$x = \frac{dx}{dt} = 0 \qquad t = 0$$
$$\frac{d^2x}{dt^2} = 1 - x - x^3 \qquad t > 0 \qquad (3\text{-}171)$$

The recurrence formula (3-168) then has the form

$$x_{s+1} = 2x_s - x_{s-1} + h^2(1 - x_s - x_s^3) \qquad s = 1, 2, \ldots \qquad (3\text{-}172)$$

We have $x_0 = 0$ but must turn elsewhere to obtain x_1. Choosing $h = 0.1$, we find from the result of Exercise 3-20 $x_1 = 0.005004$. Then, applying (3-172), we obtain, in sequence,

[1] Formulas (3-169) and (3-170) are special cases of formulas given by Störmer. They are commonly stated in terms of *differences* of f values rather than directly in terms of f values as given here. See C. Störmer, Sur les trajectoires des corpuscules électrisés dans l'espace sous l'action du magnétisme terrestre avec application aux aurores boréales, *Arch. sci. phys. et nat.*, (4)**24**, 5–18, 113–158, 221–247 (1907), and C. Störmer, Méthode d'intégration numérique des équations différentielles ordinaires, *Compt. rend. congr. intern. math.*, Strasbourg, 1920, Toulouse, privat, pp. 243–257, 1921.

[2] Nyström, *op. cit.* See also Collatz, *op. cit.*, p. 26.

$x_2 = (2)(0.005004) - 0 + (0.01)[1 - 0.005004 - (0.005004)^3]$
$\quad = 0.019958$

$x_3 = (2)(0.019958) - 0.005004 + (0.01)[1 - 0.019958 - (0.019958)^3]$
$\quad = 0.044712$

. . . . (3-173)

$x_{20} = 1.169626$
$x_{21} = 1.179604$
$x_{22} = 1.171372$

EXERCISES

3-77. Set up Euler's method with $h = 0.1$ for Prob. 3-2 as formulated in (3-164). Carry out a few steps.   *Ans.* $X_2 = (-0.80, 0.99)$.

3-78. Set up the first-category procedure (3-128) with $h = 0.1$ for Prob. 3-2 as formulated in (3-164). Use the series (3-41) to obtain X_1. Calculate X_2 from the recurrence formula.   *Ans.* $X_2 = (-0.8009, 0.9819)$.

3-79. Set up the second-category procedure (3-131) with $h = 0.1$ for the same system. Use the result of Exercise 3-78 as the initial trial $X_2^{(0)}$. Iterate to obtain X_2.
Ans. $X_2 = (-0.8013, 0.9826)$.

3-80. Use the parabolic interpolation formula (3-76) to locate and obtain the maximum value of x which is attained by the solution (3-173) in the neighborhood of $t = 2.1$.   *Ans.* $x_{max} = 1.179624$ at $t = 2.10476$.

The maximum amplitude can be checked by integrating (3-171) with respect to x. In this way show that

$$x_{max} - \frac{x_{max}^2}{2} - \frac{x_{max}^4}{4} = 0$$

and hence that $x_{max} = 1.179509$.

3-81. Use Taylor's-series expansions to show that the truncation error $E_{t,s}$ of (3-169) is $-h^5(d^5\psi/dt^5)_s/12 + O(h^6)$.

3-82. Use Taylor's-series expansions to show that the truncation error $E_{t,s}$ of (3-170) is $h^6(d^6\psi/dt^6)_s/240 + O(h^8)$.

3-83. Show that the exact solution of the propagation problem

$$x_1 = 2 \qquad x_2 = 0 \qquad t = 0$$

$$\left.\begin{array}{l} \dfrac{dx_1}{dt} = 2x_1 - x_2 \\[2mm] \dfrac{dx_2}{dt} = -x_1 + 2x_2 \end{array}\right\} \qquad t > 0$$

can be written

$$X = \begin{bmatrix} 1 \\ 1 \end{bmatrix} e^t + \begin{bmatrix} 1 \\ -1 \end{bmatrix} e^{3t}$$

Apply Euler's method, and show that (neglecting round-off) the solution is

$$X_s = \begin{bmatrix} 1 \\ 1 \end{bmatrix}(1 + h)^s + \begin{bmatrix} 1 \\ -1 \end{bmatrix}(1 + 3h)^s$$

3-84. Verify that the exact solution of the propagation problem

$$x_1 = 2 \qquad x_2 = -1 \qquad t = 0$$

$$\left.\begin{array}{l} \dfrac{dx_1}{dt} = x_2 \\[2mm] \dfrac{dx_2}{dt} = -x_1 - 2x_2 \end{array}\right\} \qquad t > 0$$

is

$$X = \begin{bmatrix} 1 \\ -1 \end{bmatrix} e^{-t} + \left\{ \begin{bmatrix} 1 \\ 0 \end{bmatrix} + t \begin{bmatrix} 1 \\ -1 \end{bmatrix} \right\} e^{-t}$$

This example is interesting because of the repeated root.   Apply Euler's method, and show that the solution (neglecting round-off) is

$$X_s = \begin{bmatrix} 1 \\ -1 \end{bmatrix} (1 - h)^s + \left\{ (1 - h) \begin{bmatrix} 1 \\ 0 \end{bmatrix} + sh \begin{bmatrix} 1 \\ -1 \end{bmatrix} \right\} (1 - h)^s$$

3-85. Show that when a first- or second-category procedure is applied to the general propagation problem (3-161) the error can be obtained from the recurrence formula

$$\epsilon_s = \epsilon_{s-q} + qh \sum_k \beta_k P_k \epsilon_k + E_{t,s} + E_{r,s}$$

where ϵ_s is a column matrix of solution errors, P_k is a square matrix[1] of elements $p_{ij} = \partial f_i / \partial x_j$ evaluated at $t = t_k$, $E_{t,s}$ and $E_{r,s}$ are column matrices of truncation and round-off errors, and q and β_k are the quantities tabulated in Tables 3-5 and 3-6.

3-86. The rate of growth of the inherited error in Exercise 3-85 depends on the properties of the *homogeneous* solution, i.e., the solution when $E_{t,s}$ and $E_{r,s}$ are set equal to zero.   When P is a constant matrix, show that homogeneous solutions can be obtained by assuming $\epsilon_s = Y_j \lambda^s$, where Y_j is an eigenvector of P.   Show that the equation for fixing λ is the same as for a single-degree-of-freedom system except that $p = \partial f / \partial x$ is replaced by p_j, which is the eigenvalue of P corresponding to the eigenvector Y_j.

3-87. Show that the truncation error of (3-168) is $h^4 [d^4 \psi(\tau)/dt^4]/12$, where $t_{s-1} \leq \tau \leq t_{s+1}$.   Note the complication of the recurrence formulas required if (3-167) is treated as a pair of first-order equations

$$\frac{dx_1}{dt} = x_2$$

$$\frac{dx_2}{dt} = f(x_1, t)$$

and the same order of truncation error is desired.

[1] See Exercise 3-14.

CHAPTER 4

EQUILIBRIUM PROBLEMS IN CONTINUOUS SYSTEMS

The first three chapters of this work treat lumped parameter systems. The remainder of the book deals with continuous systems. Both categories represent idealizations of physical reality. An important preliminary decision for the analyst is whether a particular engineering system should be represented by a discrete model or by a continuous model. We shall find in the following pages that some numerical treatments for continuous systems are actually equivalent to reconsidering the physical systems as discrete.

The present chapter contains a survey of numerical procedures used in studying equilibrium problems in continuous systems. The chapter begins with a brief formulation of several particular problems and a discussion of the general properties of equilibrium problems. The relationship to extremum problems is considered. Numerical procedures using trial solutions with adjustable parameters and methods based on finite-difference approximations are then described and illustrated for the particular problems already formulated.

4-1. Particular Examples

Four particular equilibrium problems are analyzed in this section. For each problem the governing equations and boundary conditions are established and cast into nondimensional form. The problems are:

4-1. Beam on elastic foundation.
4-2. Temperatures in a slab.
4-3. Torsion of a square section.
4-4. Steady oscillation of a nonlinear system.

The first three problems are linear, while the fourth is nonlinear. Problems 4-1 and 4-4 involve one-dimensional continua, while the other two involve two-dimensional continua.

Problem 4-1. Beam on Elastic Foundation. We consider the equilibrium of a flexible beam subjected to a distributed load while resting on a continuous elastic foundation. Figure 4-1 is a sketch of the particular physical system under consideration. A beam of length L is hinged at its ends and rests on an elastic foundation. When no load is applied, the neutral axis of the beam lies along the x axis.   The flexural stiffness is EI, the distributed load is w per unit length, and the foundation modulus[1] is k per unit length per unit deflection.   These are all taken to be constants for our particular problem.   The equilibrium problem consists in determining the deflections and bending moments throughout the length of the beam.   The governing relations are those of equilibrium, compatibility, and elasticity.   The equilibrium requirements are

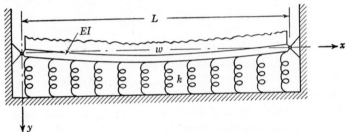

FIG. 4-1. Beam on elastic foundation subjected to distributed loading.

that the bending moment at the hinged ends should vanish and that throughout the length of the beam

$$-\frac{d^2M}{dx^2} = w - s \tag{4-1}$$

where M is the bending moment and s is the foundation reaction per unit length.   The compatibility requirements are that both the beam and the foundation should have the same deflection, $y(x)$, and that this deflection should vanish at the ends.   The elastic requirements are that the foundation reaction should be proportional to the deflection

$$s = ky \tag{4-2}$$

and that the bending moment in the beam should be proportional to the curvature.

$$M = -EI\frac{d^2y}{dx^2} \tag{4-3}$$

There are two standard ways of combining the above requirements into a single differential equation for a single variable.

In the first method the deflection $y(x)$ is considered to define the state of the system.   The elastic requirements (4-2) and (4-3) then provide

---

[1] See, for example, S. Timoshenko, "Strength of Materials," 2d ed., pt. II, D. Van Nostrand Company, Inc., New York, 1941, p. 1, for background and development of the fundamental theory.

s and M in terms of y. When these are substituted into the equilibrium relation (4-1), we obtain the following equation for $y(x)$:

$$EI \frac{d^4y}{dx^4} = w - ky \qquad (4\text{-}4)$$

With this governing equation go the conditions for zero deflection and zero bending moment at the ends.

$$\left. \begin{array}{c} y = 0 \\ \dfrac{d^2y}{dx^2} = 0 \end{array} \right\} \qquad \text{at } x = 0 \text{ and at } x = L \qquad (4\text{-}5)$$

In the second method the bending moment $M(x)$ is considered to define the state of the system. The equilibrium relation (4-1) then provides s in terms of M. A compatibility equation for determining M is obtained by eliminating y from (4-2) and (4-3) as follows:

$$M = -\frac{EI}{k} \frac{d^2s}{dx^2} \qquad (4\text{-}6)$$

Introducing s from (4-1) yields the following governing equation for $M(x)$:

$$M = -\frac{EI}{k} \frac{d^4M}{dx^4} \qquad (4\text{-}7)$$

The terminal requirements impose the following boundary conditions:

$$\left. \begin{array}{c} M = 0 \\ \dfrac{d^2M}{dx^2} = -w \end{array} \right\} \qquad \text{at } x = 0 \text{ and } x = L \qquad (4\text{-}8)$$

The solution of either of the above complementary formulations will provide a complete solution to the equilibrium problem.

Equilibrium problems in linear continuous systems can often be formulated in an alternative manner in terms of *integral equations*. The fundamental idea in this approach is that the equilibrium state resulting from a complex loading may be obtained by superposing known results for simple unit loadings. We proceed to a brief development of an integral equation for the system of Fig. 4-1.

We take as known the solution to the unit problem indicated in Fig. 4-2. The beam and end supports are the same as in Fig. 4-1, but the elastic foundation and distributed load are removed, and a single concentrated unit load is applied at $x = \xi$. The deflection at x due to a unit load at ξ is given by the function $G(x, \xi)$. This function is called the *Green's function*, or *influence function*. It is obtained in this case from elementary beam theory. We then build up the deflection of Fig. 4-1 by superposition.   At ξ the load per unit length is $w - s(\xi)$; hence, on summing over ξ we have the resultant deflection at x.

$$y(x) = \int_0^L (w - s)G(x, \xi)\,d\xi \qquad (4\text{-}9)$$

Eliminating s with the aid of (4-2) and making use of the definition of G result in the following integral equation for y:

$$y(x) = \frac{wL^4}{24EI}\left[\frac{x}{L} - 2\left(\frac{x}{L}\right)^3 + \left(\frac{x}{L}\right)^4\right] - k \int_0^L y(\xi)G(x, \xi)\,d\xi \qquad (4\text{-}10)$$

The unknown function y appears both on the left and under the integral on the right.   It is interesting to compare the differential-equation formulation, (4-4) and (4-5), with the integral equation, (4-10).   Note that

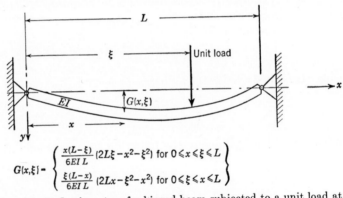

$$G(x,\xi) = \begin{cases} \dfrac{x(L-\xi)}{6EI\,L}\,(2L\xi - x^2 - \xi^2) & \text{for } 0 \leqslant x \leqslant \xi \leqslant L \\[2mm] \dfrac{\xi(L-x)}{6EI\,L}\,(2Lx - \xi^2 - x^2) & \text{for } 0 \leqslant \xi \leqslant x \leqslant L \end{cases}$$

Fig. 4-2. Deflection at x of a hinged beam subjected to a unit load at ξ.

separate boundary conditions are required for the differential equation, whereas they are implicitly included in the integral equation with the selection of $G(x, \xi)$.

For numerical treatment we consider the special case for which

$$EI = kL^4 \qquad (4\text{-}11)$$

Introducing the dimensionless variables

$$x' = \frac{x}{L} \qquad \psi = \frac{y}{wL^4/EI}$$
$$\xi' = \frac{\xi}{L} \qquad G' = \frac{G}{L^3/EI} \qquad (4\text{-}12)$$

we obtain the following nondimensional differential-equation and integral-equation formulations of the problem.   Note that we have already dropped the primes from the dimensionless quantities.   This should not cause any ambiguity. From (4-4) and (4-5) we obtain

$$\frac{d^4\psi}{dx^4} + \psi = 1$$

with $\qquad \psi = \frac{d^2\psi}{dx^2} = 0 \qquad$ at $x = 0$ and $x = 1$

(4-13)

as a differential-equation formulation, and from (4-10) we obtain

$$\psi = \frac{x - 2x^3 + x^4}{24} - \int_0^1 \psi(\xi)G(x, \xi)\, d\xi \qquad (4\text{-}14)$$

as an integral-equation formulation.

Problem 4-2. Temperatures in a Slab. We consider the steady-state temperature distribution in a homogeneous isotropic conducting slab with insulated faces when the temperatures along the edges are known. A semi-infinite slab whose long edges are maintained at temperature T_0 is shown in Fig. 4-3. Along the short edge the temperature is given by

$$T = T_0 + T_1 x(L - x) \qquad (4\text{-}15)$$

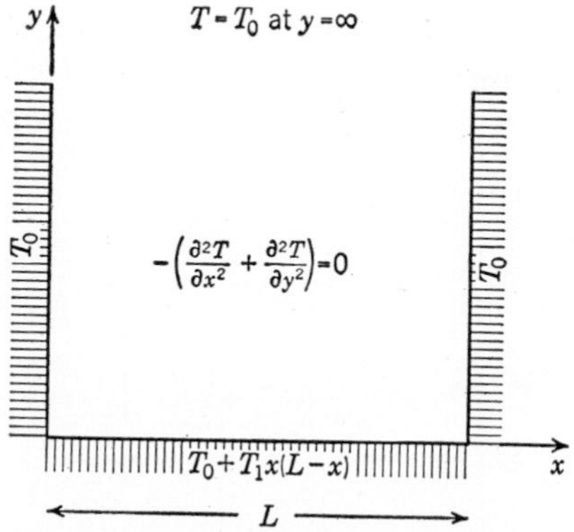

The governing equation which states the requirement for zero net heat flux into a differential region is *Laplace's equation.*[1]

FIG. 4-3. Formulation of the problem of determining the steady-state temperature distribution in a semi-infinite conducting slab.

$$-\left(\frac{\partial^2 T}{\partial x^2} + \frac{\partial^2 T}{\partial y^2}\right) = 0 \qquad (4\text{-}16)$$

The equilibrium problem consists in obtaining a temperature distribution $T(x, y)$ which satisfies (4-16) throughout the interior of the slab and which assumes the indicated values at the edges of the slab.

A nondimensional formulation can be obtained by introducing the following dimensionless variables:

$$x' = \frac{x}{L} \qquad y' = \frac{y}{L} \qquad \psi = \frac{T - T_0}{T_1 L^2} \qquad (4\text{-}17)$$

---

[1] See, for example, M. Jakob, "Heat Transfer," John Wiley & Sons, Inc., New York, 1949, p. 141. The negative sign is inserted in order to make the system *positive definite.* See p. 210.

In terms of these variables we have to determine $\psi(x, y)$ which satisfies (note that the primes have already been dropped from the dimensionless quantities)

$$-\left(\frac{\partial^2\psi}{\partial x^2} + \frac{\partial^2\psi}{\partial y^2}\right) = 0 \quad \text{in} \begin{cases} 0 < x < 1 \\ 0 < y < \infty \end{cases} \tag{4-18}$$

and which satisfies the following boundary conditions.

$$\begin{aligned} \psi(0, y) &= \psi(1, y) = 0 & \text{for } y > 0 \\ \psi(x, 0) &= x(1 - x) \\ \psi(x, \infty) &= 0 \end{aligned} \left. \begin{aligned} \\ \\ \end{aligned} \right\} \quad \text{for } 0 \le x \le 1 \tag{4-19}$$

Problem 4-3. Torsion of a Square Section. When a uniform elastic cylindrical prism is subjected to a twisting moment, there develops in

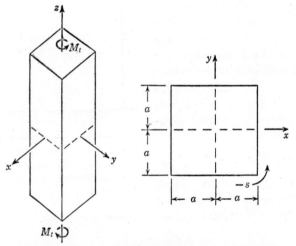

FIG. 4-4. Torsion of a prism with square cross section.

each cross section a distribution of shear stress and the prism as a whole suffers a twisting deformation. We consider the problem of determining the equilibrium state of the square prism shown in Fig. 4-4. The twisting moment is M_t, the twist angle per unit length is θ, and the shear modulus of the material is G.

In the approach of *Saint-Venant*[1] the state of the twisted prism is characterized by the warping function, $w(x, y)$, which gives the distribution of z displacements in each cross section. The equilibrium requirements impose the following conditions on w: Throughout the cross section

[1] See, for example, S. Timoshenko and J. N. Goodier, "Theory of Elasticity," 2d ed., McGraw-Hill Book Company, Inc., New York, 1951, p. 259.

w must satisfy Laplace's equation

$$-\left(\frac{\partial^2 w}{\partial x^2} + \frac{\partial^2 w}{\partial y^2}\right) = 0 \qquad (4\text{-}20)$$

while on the boundary the derivatives of w must satisfy

$$\left(\frac{\partial w}{\partial x} - \theta y\right)\frac{dy}{ds} - \left(\frac{\partial w}{\partial y} + \theta x\right)\frac{dx}{ds} = 0 \qquad (4\text{-}21)$$

where s denotes arc length along the boundary. These conditions constitute a complete formulation of the problem, for when w has been determined, the components of shear stress are given directly as follows:

$$\tau_{xz} = G\left(\frac{\partial w}{\partial x} - y\theta\right)$$
$$\tau_{yz} = G\left(\frac{\partial w}{\partial y} + x\theta\right) \qquad (4\text{-}22)$$

In the complementary approach of *Prandtl*[1] the state of the twisted prism is characterized by a *stress function,* $\psi(x, y)$, which through its definition

$$\frac{\partial \psi}{\partial x} = -\tau_{yz} \qquad \frac{\partial \psi}{\partial y} = \tau_{xz} \qquad (4\text{-}23)$$

represents a self-equilibrating stress system. Geometrical compatibility requires that ψ should satisfy

$$-\left(\frac{\partial^2 \psi}{\partial x^2} + \frac{\partial^2 \psi}{\partial y^2}\right) = 2G\theta \qquad (4\text{-}24)$$

throughout the section, and the condition of no stresses on the periphery becomes simply[2] $\psi = 0$ on the boundary. This constitutes the formulation of the problem by the complementary approach. When the stress function and hence the stresses have been determined, strains and displacements may be obtained via Hooke's law.

In practical investigations the ratio of twisting moment to twisting angle per unit length, or *torsional rigidity,* is often of more interest than the detailed distribution of stress and strain. In terms of the warping function the torsional rigidity, C, is given by

$$C = \frac{M_t}{\theta} = G \iint \left[x^2 + y^2 + \frac{1}{\theta}\left(x\frac{\partial w}{\partial y} - y\frac{\partial w}{\partial x}\right)\right] dx\, dy \qquad (4\text{-}25)$$

---

[1] L. Prandtl, Zur Torsion von prismatischen Stäben, *Physik. Z.,* **4,** 758–759 (1903).
[2] The requirement is actually only that ψ be constant. If this constant is different from zero, (4-26) becomes slightly more awkward.

while in terms of the stress function we have

$$C = \frac{M_t}{\theta} = \frac{2}{\theta} \iint \psi \, dx \, dy \qquad (4\text{-}26)$$

where in both cases the double integral extends over the cross section of the prism.

In preparation for numerical treatments we render these formulations dimensionless for the particular cross section of Fig. 4-4. Defining the nondimensional variables

$$x' = \frac{x}{a} \qquad y' = \frac{y}{a} \qquad \psi' = \frac{\psi}{G\theta a^2} \qquad C' = \frac{C}{Ga^4} \qquad (4\text{-}27)$$

the Prandtl problem becomes (after dropping the primes on the dimensionless quantities) that of evaluating the integral

$$C = 2 \int_{-1}^{1} \int_{-1}^{1} \psi \, dx \, dy \qquad (4\text{-}28)$$

for that function $\psi(x, y)$ which vanishes on the boundaries $x = \pm 1$ and $y = \pm 1$ and satisfies

$$-\left(\frac{\partial^2 \psi}{\partial x^2} + \frac{\partial^2 \psi}{\partial y^2}\right) = 2 \qquad (4\text{-}29)$$

within the square enclosed by these boundaries.

For the Saint-Venant formulation we introduce the following additional nondimensional variables

$$s' = \frac{s}{a} \qquad n' = \frac{n}{a} \qquad \Omega = \frac{w}{\theta a^2} \qquad (4\text{-}30)$$

where n measures distance along the outward directed normal to the boundary. The problem is then to evaluate the integral

$$C = \int_{-1}^{1} \int_{-1}^{1} \left\{ x^2 + y^2 + x\frac{\partial \Omega}{\partial y} - y\frac{\partial \Omega}{\partial x} \right\} dx \, dy \qquad (4\text{-}31)$$

for the function Ω which satisfies

$$-\left(\frac{\partial^2 \Omega}{\partial x^2} + \frac{\partial^2 \Omega}{\partial y^2}\right) = 0 \qquad (4\text{-}32)$$

within the same square and satisfies the boundary condition

$$\frac{\partial \Omega}{\partial n} = \frac{1}{2}\frac{\partial(x^2 + y^2)}{\partial s} \qquad (4\text{-}33)$$

on the periphery of the square.

Problem 4-4. Steady Oscillations of a Nonlinear System. In Prob. 3-3 we considered the response of a certain nonlinear mass-spring system to a suddenly applied force. This was a propagation problem for a lumped parameter system. We now return to this same system to consider a problem of steady motion. We seek to find the possible periodic motions of the system when excited by a given periodic force. This is a steady-state forced-vibration problem. Our problem consists in determining a continuous function: the displacement-time curve for a complete cycle.

If in Fig. 3-3 we take the exciting force to be

$$F(t) = 2T_0 \sin \sqrt{\frac{2T_0}{mL}} \omega t \qquad (4\text{-}34)$$

where ω is a dimensionless parameter, the equation of motion becomes

$$\frac{d^2 x}{dt^2} + x + \epsilon x^3 = \sin \omega t \qquad (4\text{-}35)$$

in terms of the nondimensional variables of (3-16). Our problem is to find functions $x(t)$ which have the period $2\pi/\omega$ and which satisfy (4-35). In the present investigation we further limit ourselves to the consideration of motions which are either in phase[1] or 180° out of phase with the excitation. This is equivalent to the assumption that the boundary value for x at $t = 0$ and $t = 2\pi/\omega$ is zero. Thus our problem is to find a solution to (4-35) in the range $0 \leq t \leq 2\pi/\omega$ and which satisfies the following boundary condition:

$$x = 0 \qquad \text{at} \begin{cases} t = 0 \\ t = \dfrac{2\pi}{\omega} \end{cases} \qquad (4\text{-}36)$$

It is of practical interest to obtain solutions for a whole range of values of the parameter ω and thus to obtain a *response curve* showing the amplitude of steady motion as a function of ω.

EXERCISES

4-1. A uniform cantilever beam of length L is clamped at the left end and subjected to a concentrated load P at the (free) right end. The beam rests on an elastic foundation of modulus k. Obtain complementary formulations of the equilibrium problem in terms of the deflection $y(x)$ and the bending moment $M(x)$. Reduce these to dimensionless form in the special case where $4EI = kL^4$.

---

[1] See, for example, S. Timoshenko, "Vibration Problems in Engineering," 2d ed., D. Van Nostrand Company, Inc., New York, 1937, p. 137.

Ans. In terms of a dimensionless bending moment M

$$\frac{d^4M}{dx^4} + 4M = 0 \quad \text{in } 0 < x < 1 \quad \text{and} \quad \left. \begin{array}{l} \dfrac{d^2M}{dx^2} = 0 \\[2mm] \dfrac{d^3M}{dx^3} = 0 \end{array} \right\} \quad \text{at } x = 0$$

$$\left. \begin{array}{l} M = 0 \\[2mm] \dfrac{dM}{dx} = 1 \end{array} \right\} \quad \text{at } x = 1$$

For the deflection formulation see Exercise 4-66.

4-2. Show that the Green's function for a uniform cantilever beam clamped at the left end is

$$G(x, \xi) = \begin{cases} \dfrac{x^2(3\xi - x)}{6EI} & \text{for } 0 \le x \le \xi \le L \\[3mm] \dfrac{\xi^2(3x - \xi)}{6EI} & \text{for } 0 \le \xi \le x \le L \end{cases}$$

By means of this show that a dimensionless integral-equation formulation of Exercise 4-1 is

$$\psi(x) = \frac{x^2(3 - x)}{6} - \frac{2}{3} \int_0^x \psi(\xi)[\xi^2(3x - \xi)]\, d\xi - \frac{2}{3} \int_x^1 \psi(\xi)[x^2(3\xi - x)]\, d\xi$$

4-3. A long electric transmission cable consists of concentric square cylinders separated by a vacuum. In a steady state the inner conductor carries a positive

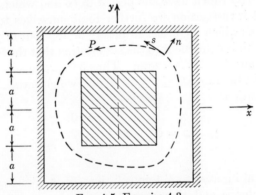

FIG. 4-5. Exercise 4-3.

charge q per unit length (normal to the sketch), and the outer conductor carries an equal negative charge. The potential φ of the electrostatic field between the conductors satisfies Laplace's equation, and the charge per unit length is given by the line integral

$$q = -\frac{1}{4\pi} \oint_P \frac{\partial \varphi}{\partial n}\, ds$$

where P is a closed path as shown in Fig. 4-5. The *capacitance* per unit length is defined as the ratio of the charge to the potential difference between conductors. Formulate the problem of determining the capacitance of the cable shown in Fig. 4-5.

Ans. When $\varphi = 1$ on inner conductor and $\varphi = 0$ on outer conductor and $\partial^2\varphi/\partial x^2 + \partial^2\varphi/\partial y^2 = 0$ in between, then $C = -\dfrac{1}{4\pi} \oint (\partial\varphi/\partial n)\, ds$.

4-4. An insulating jacket for a long oven of square cross section has the same shape as the evacuated space in the transmission cable of the previous problem. In the steady state the oven wall has a uniform high temperature, and the outside of the jacket has a uniform low temperature. The temperature T within the jacket satisfies Laplace's equation, and the total heat flux per unit length is

$$q = -k \oint_P \frac{\partial T}{\partial n}\, ds$$

where k is the conductivity of the jacket material. Study the analogy between this and the preceding exercise. What is the analog of capacitance?

4-5. A thin elastic membrane is stretched taut over a horizontal square frame of side $2a$. The uniform surface tension is T. Pressure, p, underneath the membrane causes it to assume a slightly bulged equilibrium position. Let $z(x, y)$ be the (small) displacement, and let m_x and m_y be the vertical components of the surface tension across cuts perpendicular to x and y, respectively. Show that the *equilibrium* requirement is

$$-\frac{\partial m_x}{\partial x} - \frac{\partial m_y}{\partial y} = p$$

and that for small displacements

$$m_x = T \frac{\partial z}{\partial x} \qquad m_y = T \frac{\partial z}{\partial y}$$

are the *force-deflection* relationships. The *compatibility* requirements are simply that z should be continuous and vanish at the edges. Obtain a formulation of the problem in terms of z, and study the analogy with (4-24). A complementary formulation can be obtained by introducing a *stress function* $\varphi(x, y)$ defined by

$$m_x = -\frac{\partial \varphi}{\partial y} - p\frac{x}{2} \qquad m_y = \frac{\partial \varphi}{\partial x} - p\frac{y}{2}$$

Show that the equilibrium requirement is automatically satisfied by arbitrary φ. Obtain a formulation of the problem in terms of φ, and study the analogy with (4-20) and (4-21).

Ans. $\partial^2\varphi/\partial x^2 + \partial^2\varphi/\partial y^2 = 0$ inside square, and $(\partial\varphi/\partial x - py/2)\, dy/ds - (\partial\varphi/\partial y + px/2)\, dx/ds = 0$ on the edges where s is arc length along the edge.

4-6. Deflection of an Elastic Plate. Formulate the equilibrium problem for the small deflection of a uniform thin elastic plate hinged at the periphery and subjected to a uniform transverse load p per unit area. The elastic requirements are[1]

$$M_x = -D\left(\frac{\partial^2 w}{\partial x^2} + \nu\frac{\partial^2 w}{\partial y^2}\right)$$

$$M_y = -D\left(\frac{\partial^2 w}{\partial y^2} + \nu\frac{\partial^2 w}{\partial x^2}\right)$$

$$M_{xy} = D(1 - \nu)\frac{\partial^2 w}{\partial x\, \partial y}$$

where M_x and M_y are the bending moments per unit length across cuts normal to x and y, respectively, M_{xy} is the twisting moment per unit length acting in a cut normal to x, D is the flexural rigidity, ν is Poisson's ratio, and w is the deflection of the middle surface of the plate perpendicular to the (x, y) plane. Geometric compatibility is automatically ensured if all the derivatives of w are obtained from a single-valued

[1] See S. Timoshenko, "Theory of Plates and Shells," McGraw-Hill Book Company, Inc., New York, 1940, p. 88.

function $w(x, y)$. The force-balance requirement is

$$\frac{\partial^2 M_x}{\partial x^2} + \frac{\partial^2 M_y}{\partial y^2} - 2\frac{\partial^2 M_{xy}}{\partial x\, \partial y} = -p$$

Ans. $\partial^4 w/\partial x^4 + 2\, \partial^4 w/\partial x^2\, \partial y^2 + \partial^4 w/\partial y^4 = p/D$ in the interior, and $w = \partial^2 w/\partial n^2$ $= 0$ at the periphery.

4-7. Compressible Flow around a Cylinder. In polar coordinates the *momentum* equations for an ideal (no viscosity) fluid in *steady flow* are

$$u_r \frac{\partial u_r}{\partial r} + u_\theta \frac{\partial u_r}{r\, \partial \theta} - \frac{u_\theta^2}{r} + \frac{1}{\rho}\frac{\partial p}{\partial r} = 0$$

$$u_r \frac{\partial u_\theta}{\partial r} + u_\theta \frac{\partial u_\theta}{r\, \partial \theta} + \frac{u_r u_\theta}{r} + \frac{1}{\rho}\frac{\partial p}{r\, \partial \theta} = 0$$

where u_r and u_θ are velocity components, ρ is density, and p is pressure. *Continuity* requires

$$\frac{\partial}{\partial r}\,(r\rho u_r) + \frac{\partial}{\partial \theta}\,(\rho u_\theta) = 0$$

and finally if the flow is to be *irrotational*

$$\frac{\partial}{\partial r}\,(r u_\theta) - \frac{\partial u_r}{\partial \theta} = 0$$

We are to formulate the equilibrium problem for the steady two-dimensional irrotational flow around a circular cylinder of radius a when at large distances the flow has

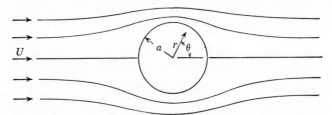

FIG. 4-6. Exercise 4-7.

uniform velocity U. This is an equilibrium problem so long as the flow remains subsonic. We assume that the fluid satisfies the relationship

$$\frac{p}{\rho^\gamma} = \text{constant}$$

and therefore p and ρ can be eliminated from the equations by introducing the velocity of sound $c = \sqrt{dp/d\rho}$. Show that

$$u_r \frac{\partial u_r}{\partial r} + u_\theta \frac{\partial u_\theta}{\partial r} + \frac{c^2}{\rho}\frac{\partial \rho}{\partial r} = 0$$

$$u_r \frac{\partial u_r}{r\, \partial \theta} + u_\theta \frac{\partial u_\theta}{r\, \partial \theta} + \frac{c^2}{\rho}\frac{\partial \rho}{r\, \partial \theta} = 0$$

and that these can be integrated to obtain

$$c^2 = c_\infty^2 + \frac{\gamma - 1}{2}\,[U^2 - u_r^2 - u_\theta^2]$$

The irrotationality condition is automatically satisfied if the velocity components are obtained from a *velocity potential* $\varphi(r, \theta)$ as follows:

$$\frac{\partial \varphi}{\partial r} = u_r \qquad \frac{\partial \varphi}{r \, \partial \theta} = u_\theta$$

Complete the formulation of the problem in terms of φ.

Ans. $(c^2 - u_r{}^2) \, \partial^2\varphi/\partial r^2 - 2u_r u_\theta \, \partial^2\varphi/r \, \partial\theta \, \partial r + (c^2 - u_\theta{}^2) \, \partial^2\varphi/r^2 \, \partial\theta^2 + (c^2 + u_\theta{}^2) \, \partial\varphi/r \, \partial r = 0$ outside the cylinder, and $\varphi = Ur \cos \theta$ as $r \to \infty$ and $\partial\varphi/\partial r = 0$ at $r = a$.

4-8. In a color kinescope a thin metal mask is interposed between the electron guns and the phosphor screen. Electron bombardment of the mask generates heat Q per unit area. This heat is conducted away through the mask to its rim and is also radiated to the walls of the vacuum envelope. Assuming that the rim of the mask and the envelope walls are at temperature T_0 and that the radiant heat transfer can be linearized, formulate the equilibrium problem for the steady-state temperature distribution in the mask.

Ans. With x and y as rectangular coordinates on the mask the temperature $T(x, y)$ must satisfy $-k(\partial^2 T/\partial x^2 + \partial^2 T/\partial y^2) + hT = Q + hT_0$ in the interior and $T = T_0$ on the periphery, where k is a conductivity and h is a (linearized) radiation coefficient.

4-2. Formulation of the General Problem

The preceding problems have been particular examples of equilibrium problems in one- and two-dimensional continua. In every case the mathematical formulation was in terms of a function, ψ, of one or two independent variables which represented the state of a physical system. The conditions for determining the equilibrium state consisted of a differential equation, ordinary or partial, together with boundary conditions. In mathematical parlance the equilibrium problem is known as a *boundary-value problem*. A generalized formulation follows. This formulation may be considered to provide a precise *definition* of what we mean by an *equilibrium problem*.

The problem is to determine the function ψ which satisfies the differential equation

$$L_{2m}[\psi] = f \tag{4-37}$$

within a domain D and meets the boundary conditions

$$B_i[\psi] = g_i \qquad i = 1, \ldots, m \tag{4-38}$$

on the boundary of D.

Domain. In the above formulation the domain D is either a one-dimensional continuum (e.g., a portion of the x axis) or a two-dimensional continuum [e.g., a portion of the (x, y) plane] as shown in Fig. 4-7. The boundary of D consists of the two end points when D is a one-dimensional continuum and consists of one or more closed curves when D is a two-dimensional continuum.

Differential Equation. The symbol $L_{2m}[\psi]$ stands for an expression containing ψ and its derivatives (ordinary or partial, as the case may be) up to order $2m$. Known functions of position in D may appear in $L_{2m}[\psi]$, but ψ and its derivatives must enter in such a manner that $L_{2m}[\psi]$ vanishes when ψ and all its derivatives are set equal to zero. If ψ and all its derivatives appear linearly, L_{2m} is said to be a *linear operator*. If some terms are nonlinear but the derivatives of order $2m$ appear linearly, L_{2m} is said to be *quasi-linear*. The symbol f stands for a prescribed function known throughout the domain D. If L_{2m} is linear, the differential equation (4-37) is said to be *homogeneous* when $f \equiv 0$.

Boundary Conditions. At each boundary point there are m independent boundary conditions. The symbols $B_i[\psi]$ stand for expressions containing ψ and its derivatives *normal to the boundary* up to order $2m - 1$

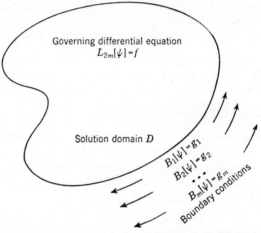

Fig. 4-7. Schematic representation of the general equilibrium problem.

to be evaluated at each boundary point. These expressions may be nonlinear, but ψ and its derivatives must enter in such a manner that $B_i[\psi]$ vanishes when ψ and its derivatives vanish. The symbols g_i stand for prescribed values known at every boundary point. If all the B_i are linear, the boundary conditions are said to be *homogeneous* when all the $g_i \equiv 0$. It is not necessary for the $B_i[\psi]$ and the g_i to be continuous along the boundary (for example, $B_2[\psi] = g_2$ might stand for $\psi = 1$ along part of the boundary and for $\partial\psi/\partial n + 2\psi = 3$ along another part of the boundary). In a one-dimensional continuum the derivatives normal to the boundary are simply the ordinary derivatives along the axis of the continuum at the end points.

Physically the operators L_{2m} and B_i describe the intrinsic properties of a passive system. They are analogous to the functions a_j on the left of (1-35) with the one difference that in a lumped-parameter system there is no mathematical distinction between an internal point and a boundary point. The equilibrium condition is simply an algebraic equation in both cases.

When f and all the g_i are zero, a solution of (4-37) and (4-38) is $\psi \equiv 0$. This represents the "no-load" equilibrium state of the system. The functions f and g_i describe the external loading to which the system is subjected. The fundamental problem of this chapter is to obtain equilibrium states corresponding to prescribed loadings.

The reader should verify that in the formulations of Sec. 4-1 the operator L_{2m} is linear in Probs. 4-1 to 4-3 and quasi-linear in Prob. 4-4. The order, $2m$, is 4 in Prob. 4-1. Note that there are two boundary conditions at each boundary point. The remaining problems have second-order equations and a single boundary condition at every boundary point. The majority of boundary conditions are homogeneous. Nonhomogeneous boundary conditions appear over parts of the boundary in (4-8), (4-19), and (4-33).

If the formulations of equilibrium problems in one-dimensional continua be compared with the formulations of propagation problems in lumped-parameter systems, it will be noted that the governing equations in both cases are ordinary differential equations.[1] The distinguishing characteristic is the type of boundary condition. In propagation problems all conditions are prescribed at one point, i.e., at the beginning of an interval. In equilibrium problems half the conditions are prescribed at one end and half the conditions are prescribed at the other end of an interval. This seemingly small difference completely alters the nature of the problem.

When we turn to two-dimensional continua and partial differential equations, we shall again note[2] a difference in type of boundary condition between equilibrium and propagation problems. In addition, we find a difference in the type of differential operator in the governing equations. A quasi-linear partial differential operator L_{2m} is said to be *elliptic, parabolic,* or *hyperbolic* according to the value of a certain relation involving the coefficients of the derivatives of order $2m$. This relation for second-order systems is given explicitly in Sec. 6-3. Here we merely state that in equilibrium problems the operators L_{2m} are of *elliptic* type. The two elliptic operators which are of greatest importance in engineering analysis are the *Laplacian,*

$$\nabla^2 = \frac{\partial^2}{\partial x^2} + \frac{\partial^2}{\partial y^2} \tag{4-39}$$

where ∇^2 is pronounced "del squared," and the *biharmonic* operator,

$$\nabla^2(\nabla^2) = \nabla^4 = \frac{\partial^4}{\partial x^4} + 2\frac{\partial^4}{\partial x^2\,\partial y^2} + \frac{\partial^4}{\partial y^4} \tag{4-40}$$

where ∇^4 is pronounced "del fourth."

[1] The equations may in fact be identical (see Exercise 4-9).
[2] See Fig. 6-13.

In Sec. 1-2 it was noted that *complementary* formulations of lumped-parameter equilibrium problems are possible. One formulation might have an advantage over the other in that a smaller number of degrees of freedom are required. Equilibrium problems in continua also permit complementary formulations. Illustrations are provided in Probs. 4-1 and 4-3. In continuous systems there is no longer any difference regarding the number of degrees of freedom since all formulations involve a continuous function, i.e., an infinite number of degrees of freedom. Possible advantage of one formulation over another now lies in the relative convenience of the particular boundary conditions or differential equations. Thus in Prob. 4-3 the formulation of Prandtl is often preferred to that of Saint-Venant because of its simpler boundary condition.

Description of Linear Systems. An equilibrium problem is said to be linear when both the governing equation (4-37) and the boundary conditions (4-38) are linear. In Chap. 1 we saw that linear lumped-parameter equilibrium problems arising from physical systems were very often *symmetric* and *positive definite* and that certain computational simplifications arose out of these properties. We now consider the analogous properties for continuous equilibrium problems. These properties are properties of the passive system; i.e., they depend on L_{2m} and B_i but not on the loading functions f and g_i.

The linear system defined by L_{2m} and $B_i (i = 1, \ldots, m)$ is said to be *symmetric*, or, more commonly, *self-adjoint*, if for any two functions u and v (sufficiently differentiable in D) which satisfy the homogeneous boundary conditions

$$B_i(u) = B_i(v) = 0 \qquad i = 1, \ldots, m \qquad (4\text{-}41)$$

we always have

$$\int_D u L_{2m}[v]\, dD = \int_D v L_{2m}[u]\, dD \qquad (4\text{-}42)$$

where the integrals are taken over the domain D and dD stands for the infinitesimal element, one- or two-dimensional, as the case may be. The same system is said to be *positive* if for any function u satisfying (4-41) we always have

$$\int_D u L_{2m}[u]\, dD \geq 0 \qquad (4\text{-}43)$$

The system is said to be *positive definite* if the equality in (4-43) holds only for $u \equiv 0$. Negative and negative definite systems can also be defined, but since negative can be turned into positive by a trivial change of sign in the original formulation,[1] it is customary to consider only the positive case.

---

[1] See footnote, page 199.

The demonstration that a particular system does or does not possess the above properties is accomplished by *integration by parts.*

To illustrate,[1] we consider Prob. 4-1 with the formulation (4-13). Here the domain D is the segment of the x axis between $x = 0$ and $x = 1$. The operator L_{2m} and the (already homogeneous) boundary conditions are as follows:

$$L_4[\psi] = \frac{d^4\psi}{dx^4} + \psi \qquad \begin{matrix} B_1[\psi] = \psi = 0 \\ \\ B_2[\psi] = \dfrac{d^2\psi}{dx^2} = 0 \end{matrix} \Bigg\} \quad \text{at} \quad \begin{cases} x = 0 \\ x = 1 \end{cases} \qquad (4\text{-}44)$$

For this case the left side of (4-42) can be transformed as follows by two integrations by parts:

$$\int_0^1 u \left(\frac{d^4v}{dx^4} + v \right) dx = \left| u \frac{d^3v}{dx^3} - \frac{du}{dx}\frac{d^2v}{dx^2} \right|_0^1 + \int_0^1 \left\{ \frac{d^2u}{dx^2}\frac{d^2v}{dx^2} + uv \right\} dx \qquad (4\text{-}45)$$

Now if u and v must satisfy the boundary conditions of (4-44), the integrated terms on the right of (4-45) vanish, leaving only the integral which is symmetrical in u and v. It is clear that the right side of (4-42) can be reduced to this same symmetrical integral by a similar transformation. Hence the system is self-adjoint. To show that the problem is positive definite, we have only to put u in place of v in (4-45) to see that (4-43) takes the following form:

$$\int_0^1 u \left\{ \frac{d^4u}{dx^4} + u \right\} dx = \int_0^1 \left\{ \left(\frac{d^2u}{dx^2} \right)^2 + u^2 \right\} dx \qquad (4\text{-}46)$$

The integral on the right, being the sum of squares, can vanish only when $u \equiv 0$.

The three linear problems formulated in Sec. 4-1 are all self-adjoint. They are all positive definite except for the Saint-Venant formulation, (4-32) and (4-33), of Prob. 4-3 which is only positive since the function, $u = $ constant, satisfies (4-43) with the equality sign.

EXERCISES

4-9. (*a*) Consider the elementary projectile-trajectory problem in which all forces except a constant gravitational force are neglected. Show that the *propagation problem* is defined by the equation

$$-\frac{d^2y}{dx^2} = f$$

where $f = g/V_0^2 \cos^2 \alpha$ and the *initial* conditions $y = 0$ and $dy/dx = \tan \alpha$ at $x = 0$.

(*b*) Consider the small deflections of a taut elastic cable stretched to a tension T and loaded by a uniform transverse load w per unit length. Show that the *equilibrium problem* is defined by the equation

$$-\frac{d^2y}{dx^2} = f$$

where $f = w/T$ and the *boundary* conditions $y = 0$ at $x = 0$ and $x = L$.

---

[1] For further illustrations see Exercises 4-11 and 4-13 to 4-15.

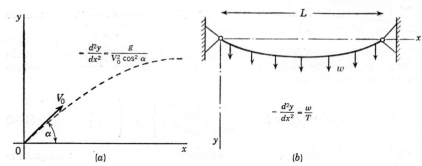

FIG. 4-8. Exercise 4-9.

4-10. Show that a nonlinear *eigenvalue* problem is formulated if we ask for those values of α for which the equation of Exercise 4-9a satisfies the boundary values of Exercise 4-9b. This is the projectile-aiming problem. How many eigenvalues are there? What are the corresponding modes?

4-11. Show that the cable-equilibrium problem of Exercise 4-9b is self-adjoint and positive definite.

4-12. The linear equilibrium problem in systems with a finite number of degrees of freedom can be written as $AX = C$ in matrix notation. Show that if X and Y are arbitrary vectors the matrix A is *symmetric* if we always have

$$X'AY = Y'AX$$

This is the converse of Exercise 2-12. It is the exact analog of the definition of *self-adjoint* for continuous systems. [*Hint.* Consider $X = (1, 0, 0, \ldots, 0)$, $Y = (0, 1, 0, \ldots, 0)$.]

4-13. Show that the system of (4-18) and (4-19) is self-adjoint and positive definite.

4-14. Show that the system of (4-29) is self-adjoint and positive definite. The integrations by parts can be performed in one direction at a time or simultaneously by using the formula of (4-70).

4-15. Show that the system of (4-32) and (4-33) is self-adjoint and positive, but not positive definite.

4-3. Mathematical Properties

The properties of well-behaved continuous systems are straightforward generalizations of the corresponding properties of systems with only a finite number of degrees of freedom. We next describe these properties and give a listing of some of the systems that can be proved to possess them. A few counterexamples are included as a reminder that a comprehensive mathematical theory is extremely difficult and is still not complete.

The well-behaved equilibrium problem has a solution, and this solution is unique; i.e., there exists one, and only one, function ψ which satisfies (4-37) throughout the domain D and which satisfies all the boundary conditions (4-38). The solution ψ is moreover a well-behaved function. It is smoother (possesses continuous derivatives of higher order) than the prescribed coefficients and functions which appear in the differential equation.

Well-behaved Systems.   We have called an equilibrium problem well behaved if it possesses a unique solution.   In general, positive definite systems are well behaved.   Proofs of existence and uniqueness have been given[1] which apply to most practical linear self-adjoint positive definite systems.   In all such proofs restrictions on the continuity of the terms appearing in the differential equation and restrictions on the continuity of the boundary conditions and restrictions on the boundary geometry itself are necessary.   These restrictions are automatically satisfied by the vast majority of equilibrium problems arising from physical problems.

For linear self-adjoint systems which are not definite, the situation is more complicated.   In general there is a unique solution, but, corresponding to the case where the determinant (1-38) of the coefficients in the lumped-parameter case vanishes, there is an exceptional case in which the problem has either no solution or an infinity of solutions.

Nonlinear systems have been less thoroughly explored.   The theorems available are usually considerably weaker than in the linear case.   For instance, for Exercise 4-7, it has been proved[2] that, if a solution ψ exists for which $L_{2m}[\psi]$ is everywhere elliptic, then that solution is unique; however, there is no way of telling in advance whether such a solution exists. Speaking physically, there is a unique solution if the flow remains subsonic, but we cannot say whether it does or not.

Counterexamples.   The general boundary-value problem can exhibit characteristics very different from the well-behaved problems described above.   The following illustrations give some indication of the possibilities.

1. *No Solution at All.*   There is no solution ψ which satisfies the following linear self-adjoint boundary-value problem.

$$\frac{d^2\psi}{dx^2} + \psi = 1 \qquad \begin{cases} \psi = 0 \text{ at } x = 0 \\ \psi = 1 \text{ at } x = \pi \end{cases} \qquad (4\text{-}47)$$

The problem is not positive definite.

2. *More than One Solution.*   The quasi-linear boundary-value problem

$$2\psi \frac{d^2\psi}{dx^2} - \left(\frac{d\psi}{dx}\right)^2 = -4 \qquad \psi = 0 \qquad \text{at } \begin{cases} x = 1 \\ x = -1 \end{cases} \qquad (4\text{-}48)$$

is satisfied by both $\psi = 1 - x^2$ and $\psi = -(1 - x^2)$.

[1] See R. Courant and D. Hilbert, "Methoden der mathematischen Physik," vol. 2, Springer-Verlag OHG, Berlin, 1937, chap. 7, for a good discussion of the mathematical restrictions.   See also D. L. Bernstein, "Existence Theorems in Partial Differential Equations," pt. H, Princeton University Press, Princeton, N.J., 1950.
[2] Courant and Hilbert, *op. cit.*, p. 276.

The linear system,

$$\frac{d^2\psi}{dx^2} + \psi = 1 \qquad \psi = 0 \qquad \text{at } \begin{cases} x = 0 \\ x = 2\pi \end{cases} \tag{4-49}$$

which differs only slightly from (4-47) has an infinity of solutions. For any value of the parameter C the function $\psi = 1 - \cos x + C \sin x$ is a solution.

These two cases have their parallel in lumped-parameter systems (see Sec. 1-3). To give only a brief hint of the additional mathematical pitfalls that are introduced by the extension to continua, consider the following two-dimensional problem: Let the domain D be the interior of the unit circle in the (x, y) plane, and let the differential equation be Laplace's equation, $-\nabla^2\psi = 0$, with the boundary condition $\psi = 0$ on the circumference. This problem is linear, self-adjoint, and positive definite. It is clear that $\psi_1 \equiv 0$ is a solution, but consider the function

$$\psi_2 = \frac{1 - (x^2 + y^2)}{(1 - x)^2 + y^2} \tag{4-50}$$

It satisfies Laplace's equation within the unit circle and vanishes on the circumference *except at the single point* $(1, 0)$, where it is indeterminate. If this point is approached from within the circle, we find that $\psi_2 \rightarrow \infty$. Thus two distinct functions satisfy the same differential equation and the same boundary condition almost everywhere. This illustrates why careful restrictions on continuity are required in mathematical proofs of uniqueness.

3. *Complex Solutions for Real Problems.*   The real quasi-linear problem

$$2\psi \frac{d^2\psi}{dx^2} - \left(\frac{d\psi}{dx}\right)^2 = 4 \qquad \psi = 0 \text{ at } \begin{cases} x = 1 \\ x = -1 \end{cases} \tag{4-51}$$

is satisfied by two distinct imaginary solutions: $\psi = i(1 - x^2)$, and $\psi = -i(1 - x^2)$.

Green's Functions.   The solution of a linear lumped-parameter equilibrium problem could be obtained directly provided the *inverse* of the coefficient matrix was known. A linear boundary-value problem with homogeneous[1] boundary conditions can similarly be treated provided the Green's function is known. Although of considerable theoretical interest, Green's functions are not often employed in the practical solution of boundary-value problems because the determination of a Green's function for a particular system generally poses a more difficult problem than the original boundary-value problem. This is analogous to the lumped-parameter case, where it is easier to solve a set of simultaneous equations for a particular right-hand side than it is to obtain the inverse matrix.

For simplicity of notation we consider the one-dimensional boundary-value problem with homogeneous boundary conditions,

$$L_{2m}(\psi) = f \qquad B_i(\psi) = 0 \qquad i = 1, \ldots, m \tag{4-52}$$

---

[1] Nonhomogeneous boundary conditions can also be handled by an extension of the method. See E. Kamke, "Differentialgleichungen, Lösungsmethoden und Lösungen," 2d ed., vol. I, Akademische Verlagsgesellschaft m.b.H, Leipzig, 1943, p. 190.

where the region R is $\alpha < x < \beta$. The Green's function $G(x, \xi)$ is defined as follows: When considered as a function of x,

1. $G(x, \xi)$ satisfies the boundary conditions of (4-52).
2. $G(x, \xi)$ satisfies the differential equation

$$L_{2m}(G) = \delta(x - \xi) \tag{4-53}$$

where $\delta(x - \xi)$ represents the *unit impulse*.[1] Thus $G(x, \xi)$ is the equilibrium configuration due to a unit load at ξ. Once $G(x, \xi)$ is known, the solution to (4-52) is given by a simple integration

$$\psi(x) = \int_\alpha^\beta f(\xi)G(x, \xi)\,d\xi \tag{4-54}$$

which is a statement of the principle of superposition.

Integral Equations. When the operator $L_{2m}(\psi)$ can be split into two parts and the Green's function corresponding to one of the parts is known, the boundary-value problem can be transformed into an integral equation. This was illustrated for Prob. 4-1, the beam on elastic foundation. There the complete operator $L_{2m}(\psi)$ was $d^4\psi/dx^4 + \psi$, and the Green's function corresponding to the operator $d^4\psi/dx^4$ was known. The differential equation (4-13) could thus be written as

$$\frac{d^4\psi}{dx^4} = 1 - \psi \tag{4-55}$$

If this is compared with (4-52), we have $1 - \psi$ playing the role of f. The solution according to (4-54) is therefore

$$\psi(x) = \int_0^1 [1 - \psi(\xi)]G(x, \xi)\,d\xi \tag{4-56}$$

where $G(x, \xi)$ is the Green's function corresponding to the operator $d^4\psi/dx^4$ and the homogeneous boundary conditions of (4-13). When the integration of G with respect to ξ is performed, the final formulation (4-14) is obtained.

The standard form of this class of integral equations is written

$$\psi(x) = p(x) + \int_\alpha^\beta \psi(\xi)G(x, \xi)\,d\xi \tag{4-57}$$

where $p(x)$ is a known function, $G(x, \xi)$ is a known Green's function, and ψ is to be determined. This formulation leads to an iteration procedure[2] which is sometimes useful in practical boundary-value problems. Start-

[1] See page 152.

[2] See, for example, F. B. Hildebrand, "Methods of Applied Mathematics," Prentice-Hall, Inc., New York, 1952, p. 421; C. Wagner, On the Solution of Fredholm Integral Equations of the Second Kind by Iteration, *J. Math. and Phys.*, **30**, 23–30 (1951).

ing with $\psi_0(x) = p(x)$, a sequence of functions, $\psi_r(x)$, is constructed according to the following law:

$$\psi_{r+1} = p(x) + \int_\alpha^\beta \psi_r(\xi)G(x, \xi)\, d\xi \tag{4-58}$$

The sequence converges to the solution of (4-57) if $p(x)$ and $G(x, \xi)$ are continuous functions and if

$$\int_\alpha^\beta \int_\alpha^\beta G^2(x, \xi)\, dx\, d\xi < 1 \tag{4-59}$$

The method is theoretically useful for both ordinary and partial boundary-value problems. In practice the successive integrations (4-58) can be carried out in closed form for only the simplest, one-dimensional cases. In more complicated cases the integrations can be performed approximately by using numerical integration.[1] This iteration process may be viewed as the analog of iteration by total steps used for lumped-parameter equilibrium problems.

EXERCISES

4-16. A taut elastic cable of length L is stretched with tension T along the x axis. Find the Green's function for small transverse deflections.

Ans. $G(x, \xi) = x(L - \xi)/LT$ for $0 \leq x \leq \xi \leq L$, and $G(x, \xi) = \xi(L - x)/LT$ for $0 \leq \xi \leq x \leq L$.

4-17. Use (4-54) and the above Green's function to obtain the deflection $y(x)$ of the cable when subjected to a uniform distributed load w per unit length.

Ans. $y = wx(L - x)/2T$.

4-18. Formulate the equilibrium problem when the cable of the preceding exercises rests on an elastic foundation of stiffness k and is subjected to a uniform distributed load w. *Ans.* $-T\, d^2y/dx^2 + ky = w$ for $0 < x < L$ and $y = 0$ at $x = 0$ and $x = L$.

4-19. Reformulate the previous exercise as an integral equation using the Green's function of Exercise 4-16.   *Ans.* $y(x) = wx(L - x)/2T - k \int_0^L y(\xi)G(x, \xi)\, d\xi$.

4-20. Show that for the particular case in which $k = T/L^2$ the formulation of Exercise 4-19 can be reduced to the following *dimensionless* statement:

$$y(x) = \frac{x(1 - x)}{2} - \int_0^x y(\xi)[\xi(1 - x)]\, d\xi - \int_x^1 y(\xi)[x(1 - \xi)]\, d\xi$$

Obtain a dimensionless formulation of Exercise 4-18 in terms of these same variables. What is the exact solution? *Ans.* $y = [\cosh(1) - 1] \sinh x/\sinh(1) - [\cosh x - 1]$.

4-21. Apply one step of the iteration (4-58) to the integral equation of Exercise 4-20 starting with $y_0 = x(1 - x)/2$.   *Ans.* $y_1 = \frac{11}{24}x(1 - x) - \frac{1}{24}x^2(1 - x)^2$.

4-22. The iteration of (4-58) can also be applied to the differential-equation formulation corresponding to a given integral equation. Verify that the same result as in the previous exercise is obtained from the following iteration scheme:

$$-\frac{d^2y_r}{dx^2} = 1 - y_{r-1} \quad \text{in } 0 < x < 1 \quad y_r = 0 \quad \text{at } \begin{cases} x = 0 \\ x = 1 \end{cases}$$

[1] See Exercise 4-23.

4-23. The integral equation (4-57) can be approximated by a set of simultaneous algebraic equations by replacing the continuum $\alpha \leq x \leq \beta$ by a discrete set of points $x_0, x_1, \ldots, x_n$ with spacing $h = (\beta - \alpha)/n$. Let ψ_j stand for $\psi(x_j)$. Show that by using the *trapezoidal rule* an approximation to (4-57) is

$$\psi_j = p(x_j) + h(\tfrac{1}{2}G_{j0}\psi_0 + G_{j,1}\psi_1 + \cdots + G_{jn-1}\psi_{n-1} + \tfrac{1}{2}G_{jn}\psi_n) \qquad j = 0, 1, \ldots, n$$

where G_{jk} stands for $G(x_j, \xi_k)$.

4-24. Apply Exercise 4-23 to the integral equation in Exercise 4-20, using $h = \tfrac{1}{2}$. Obtain an approximation to $y(\tfrac{1}{2})$.

Ans. 0.1111.

4-25. Obtain an approximation similar to that in Exercise 4-23 by using *Simpson's rule*, and apply it to the equation in Exercise 4-20, using $h = \tfrac{1}{2}$.

Ans. $y(\tfrac{1}{2}) = 0.1071$.

4-26. Apply one step of iteration to the integral equation (4-14).

Ans. $\psi_1 = \tfrac{1}{24}\{(x - 2x^3 + x^4) - (1/1{,}680)(17x - 28x^3 + 14x^5 - 4x^7 + x^8)\}$.

4-27. Apply one step of iteration to the integral equation of Exercise 4-2.

Ans. $\psi_1 = \tfrac{1}{6}\{(3x^2 - x^3) - \tfrac{1}{210}(231x^2 - 105x^3 + 7x^6 - x^7)\}$.

4-4. Extremum Problems

In Chap. 1 the fundamental equilibrium problem was formulated as a set of simultaneous equations. For many systems, however, we saw that the equilibrium problem was equivalent to an extremum problem. In the present chapter the fundamental equilibrium problem is formulated as a differential equation with boundary conditions. Here again there is an equivalent extremum problem for many systems. The practical value of this alternative formulation lies in the fact that certain important approximate procedures for solving an equilibrium problem are actually methods for "solving" the corresponding extremum problems.

The extremum problem is usually formulated as follows: A certain class of allowable functions ψ is fixed, and a means for associating a value Φ with each function is defined. To a particular function ψ there is a single value for Φ, but as ψ ranges through the class of allowable functions, the corresponding Φ values will vary, in general. The term *functional* is applied to a function, such as Φ, whose argument is a class of functions. The extremum problem is then to locate those functions ψ for which Φ remains stationary for small alterations of ψ (within the allowable class).

In studying the relationship between equilibrium and extremum problems there are two major questions:

1. For a given equilibrium problem what is the allowable class of functions and what is the structure of Φ for an equivalent extremum problem?

2. For a given functional Φ and given class of functions what is the differential equation and what are the boundary conditions for the equivalent equilibrium problem?

A complete study of these questions is not attempted here.[1] We shall

[1] For a more extended treatment, see Courant and Hilbert, *op. cit.*, vol. 1, p. 210, vol. 2, p. 473.

consider the first question to be answered by the known[1] extremum principles of physics, e.g., the principle of *minimum potential energy* and its complement, *Hamilton's principle* and its complement,[2] etc. The second question can be answered by direct manipulation, using the algorithm of the *calculus of variations*. We shall shortly present a brief account of the formal technique. This technique is employed in certain[3] approximate procedures and hence is of importance to us as a tool.

Admissible Functions. The physical extremum principles we are considering can be put in the following generalized form: Out of all states satisfying A conditions those which also satisfy B conditions give stationary values to Φ. For example, in the principle of minimum potential energy the A conditions are the requirements of *geometric compatibility*, and the B conditions are the requirements of *force balance*.

In a continuous system the state is represented by a function ψ. The requirement that the state satisfy A conditions is equivalent to certain restrictions on ψ in the domain D and on the boundary. A function ψ which meets these restrictions is said to be an *admissible* function for the extremum principle under discussion.

The restrictions on an admissible function within the domain D are usually only those of continuity or smoothness. Thus in applying the principle of minimum potential energy to the beam on elastic foundation of Prob. 4-1 geometrical compatibility within the domain is automatically ensured by taking ψ to represent the deflection of *both* beam and foundation. An admissible deflection must be sufficiently smooth to permit evaluation of the potential energy due to bending.

The boundary conditions of an equilibrium problem can be divided into two categories, depending on whether they represent A conditions or B conditions for a given extremum principle. We call the former *essential* boundary conditions and the latter *additional*[4] boundary conditions. It is usually easy to distinguish between these on the basis of physical considerations. Collatz[5] gives the following mathematical criterion: If the differential-equation formulation is of order $2m$, then the

---

[1] See, for example, P. M. Morse and H. Feshbach, "Methods of Theoretical Physics," pt. I, McGraw-Hill Book Company, Inc., New York, 1953, p. 275. An alternative to the physical approach is to apply the algorithm of the calculus of variations in reverse. A good discussion of this is given by Hildebrand, *op. cit.*, p. 177.

[2] R. A. Toupin, A Variational Principle for the Mesh-type Analysis of a Mechanical System, *J. Appl. Mech.*, **19**, 151–152 (1952).

[3] See, for example, (4-127).

[4] The *additional* boundary conditions are sometimes called *natural* boundary conditions. See R. Courant and D. Hilbert, "Methoden der mathematischen Physik," 2d ed., vol. I, Springer-Verlag OHG, Berlin, 1931, p. 179.

[5] L. Collatz, "Numerische Behandlung von Differentialgleichungen," Springer-Verlag OHG, Berlin, 1951, p. 91.

essential boundary conditions are those which can be expressed in terms of ψ and its first $m - 1$ derivatives.

An admissible function is required to satisfy the A conditions on the boundary; i.e., it must satisfy the *essential* boundary conditions. For example, when the principle of minimum potential energy is applied to beam-deflection problems, the admissible functions must satisfy the boundary conditions having to do with deflection and slope but do not necessarily have to satisfy the boundary conditions which represent prescribed end moments and end loads.

Functionals. The statement of an extremum principle implicitly contains the rules for evaluating the functional Φ for an admissible function ψ. For the principle of minimum potential energy the functional Φ *is* the total potential energy associated with a displacement state ψ. For Hamilton's principle Φ is a time integral of the difference between the kinetic and potential energies evaluated for a geometrically compatible motion.

In many cases Φ is an integral over the *domain D*, but in some cases Φ is an integral around the *boundary* of D, and in other cases Φ consists of *both* integrals over the domain and around the boundary. If ψ and its derivatives appear as squares and products (as well as linearly), Φ is said to be a *quadratic* functional.

Examples. For the beam on elastic foundation of Prob. 4-1 the total potential energy is the elastic-strain energy stored in the foundation and in the beam plus the potential energy of the loading. In terms of the original physical variables we have

$$PE = \int_0^L \frac{1}{2} ky^2 \, dx + \int_0^L \frac{EI}{2} \left(\frac{d^2y}{dx^2}\right)^2 dx - \int_0^L wy \, dx \tag{4-60}$$

For $y(x)$ to be an admissible function, it must satisfy the *essential* boundary conditions $y = 0$ at $x = 0$ and $x = L$. The principle of minimum potential energy states that (4-60) is stationary for a deflection $y(x)$ which satisfies the force-balance requirements. By introducing (4-11) and the dimensionless variables (4-12) and by setting $\Phi = EI(PE)/w^2L^5$ we obtain from (4-60) the following nondimensional functional:

$$\Phi = \int_0^1 \left\{\frac{1}{2} \psi^2 + \frac{1}{2} \left(\frac{d^2\psi}{dx^2}\right)^2 - \psi\right\} dx \tag{4-61}$$

We shall shortly verify that the extremum problem for this functional is equivalent to the equilibrium problem (4-13).

The complementary energy for the same beam is the complementary strain energy in the foundation and in the beam.

$$CE = \int_0^L \frac{1}{2k} \left(w + \frac{d^2M}{dx^2}\right)^2 dx + \int_0^L \frac{M^2}{2EI} \, dx \tag{4-62}$$

By using (4-1) to provide the foundation force we have automatically ensured the satisfaction of the force-balance requirements along the beam. In order to be admissible, $M(x)$ need satisfy only the force-balance boundary conditions $M = 0$ at $x = 0$ and $x = L$. (Note the exchange of essential boundary conditions in the complementary extremum principles.) The principle of minimum complementary energy states that (4-62) is stationary for an $M(x)$ which satisfies the conditions of geometric compatibility. It can be verified[1] that the extremum problem for (4-62) is equivalent to the equilibrium problem of (4-7) and (4-8).

For the steady-state temperature-distribution problem there is as yet no generally accepted statement of the equivalent extremum principle in the terminology of heat transfer. There is, however, the well-known *Dirichlet principle*,[2] which states that the extremum problem for

$$\Phi = \int_D \left\{ \left(\frac{\partial \psi}{\partial x}\right)^2 + \left(\frac{\partial \psi}{\partial y}\right)^2 \right\} dD \qquad (4\text{-}63)$$

for which admissible functions must assume prescribed values on the closed boundary of D is equivalent to the equilibrium problem in which ψ must satisfy Laplace's equation in D and assume the same prescribed values on the boundary of D. It is easily verified[3] that when (4-63) is applied to Prob. 4-2 the extremum problem is equivalent to the equilibrium problem of (4-18) and (4-19).

In the torsion of a square section, Prob. 4-3, let us consider the twist θ to be prescribed. Then a geometrically compatible state is represented by any (suitably continuous) warping function $w(x, y)$, and the potential energy per unit length of prism is simply the elastic-strain energy.

$$\text{PE} = \frac{G}{2} \int_{-a}^{a} \int_{-a}^{a} \left\{ \left(\frac{\partial w}{\partial x} - y\theta\right)^2 + \left(\frac{\partial w}{\partial y} + x\theta\right)^2 \right\} dx\, dy \qquad (4\text{-}64)$$

There are no essential boundary conditions. [The boundary condition (4-21) states that the periphery of the prism is stress-free; i.e., it is a force-balance condition.] Introducing the dimensionless variables of (4-27) and (4-30) and setting $\Phi = (\text{PE})/(G\theta^2 a^4)$, we obtain from (4-64) the following nondimensional functional:

$$\Phi = \frac{1}{2} \int_{-1}^{1} \int_{-1}^{1} \left\{ \left(\frac{\partial \Omega}{\partial x} - y\right)^2 + \left(\frac{\partial \Omega}{\partial y} + x\right)^2 \right\} dx\, dy \qquad (4\text{-}65)$$

We shall shortly verify that the extremum problem for (4-65) is equivalent to the Saint-Venant formulation, (4-32) and (4-33), of the equilibrium problem.

The complementary energy for the same problem consists of the complementary elastic-strain energy plus the complementary energy of the prescribed displacement θ.

$$\text{CE} = \frac{1}{2G} \int_{-a}^{a} \int_{-a}^{a} \left\{ \left(\frac{\partial \psi}{\partial x}\right)^2 + \left(\frac{\partial \psi}{\partial y}\right)^2 \right\} dx\, dy - 2\theta \int_{-a}^{a} \int_{-a}^{a} \psi\, dx\, dy \qquad (4\text{-}66)$$

By expressing the shear stress (4-23) and the twisting moment (4-26) in terms of the stress function ψ we have automatically satisfied the force-balance requirements in the interior. For ψ to be admissible, it must satisfy the *essential* boundary condition $\psi = 0$ on the periphery. Introducing the dimensionless variables (4-27) and setting $\bar{\Phi} = (\text{CE})/(G\theta^2 a^4)$, we obtain

$$\bar{\Phi} = \int_{-1}^{1} \int_{-1}^{1} \left\{ \frac{1}{2}\left(\frac{\partial \psi}{\partial x}\right)^2 + \frac{1}{2}\left(\frac{\partial \psi}{\partial y}\right)^2 - 2\psi \right\} dx\, dy \qquad (4\text{-}67)$$

---

[1] See Exercise 4-34.

[2] See, for example, R. Courant and M. Schiffer, "Dirichlet's Principle, Conformal Mapping and Minimal Surfaces," Interscience Publishers, New York, 1950.

[3] See Exercise 4-34.

as a nondimensional form of (4-66). The extremum problem for (4-67) is equivalent[1] to the Prandtl formulation (4-29) of the equilibrium problem.

An extremum problem equivalent to Prob. 4-4, the steady oscillations of a non-linear system, can be obtained by using an extension[2] of Hamilton's principle. The functional Φ is a time integral of the kinetic energy minus the potential energy of the conservative forces plus the product of the time-dependent force and its displacement. We have

$$\Phi = \int_{t_1}^{t_2} \left\{ \frac{1}{2} \left(\frac{dx}{dt} \right)^2 - \left(\frac{x^2}{2} + \epsilon \frac{x^4}{4} \right) + x \sin \omega t \right\} dt \qquad (4\text{-}68)$$

directly in terms of the nondimensional variables. According to Hamilton's principle this is stationary for a motion which satisfies Newton's second law. The admissible motions $x(t)$ must be compatible with the constraints and must coincide with the actual motion at $t = t_1$ and $t = t_2$. For the problem of finding periodic motions in phase with the excitation we set $t_1 = 0$ and $t_2 = 2\pi/\omega$ and require that admissible functions vanish at these values of t. The extremum problem so formulated is equivalent[3] to the equilibrium problem of (4-35) and (4-36).

Integral Transformations. The formal technique of the calculus of variations makes extensive use of integration by parts. For convenience we list here some of the more common formulas for integration by parts in one- and two-dimensional domains. In these formulas u and v stand for arbitrary functions, a stands for a prescribed function, α and β are the end points of the one-dimensional domain, s and n stand for distance measured along and normal to the boundary, and the closed-contour integral is always understood to be taken around the boundary of the two-dimensional domain D.

$$\int_{\alpha}^{\beta} a \frac{du}{dx} \frac{dv}{dx} \, dx = au \frac{dv}{dx} \Big|_{\alpha}^{\beta} - \int_{\alpha}^{\beta} u \frac{d}{dx} \left(a \frac{dv}{dx} \right) dx \qquad (4\text{-}69)$$

$$\int_{D} a \left\{ \frac{\partial u}{\partial x} \frac{\partial v}{\partial x} + \frac{\partial u}{\partial y} \frac{\partial v}{\partial y} \right\} dD$$
$$= \oint au \frac{\partial v}{\partial n} \, ds - \int_{D} u \left\{ \frac{\partial}{\partial x} \left(a \frac{\partial v}{\partial x} \right) + \frac{\partial}{\partial y} \left(a \frac{\partial v}{\partial y} \right) \right\} dD \qquad (4\text{-}70)$$

$$\int_{\alpha}^{\beta} a \frac{d^2u}{dx^2} \frac{d^2v}{dx^2} \, dx = a \frac{du}{dx} \frac{d^2v}{dx^2} \Big|_{\alpha}^{\beta} - u \frac{d}{dx} \left(a \frac{d^2v}{dx^2} \right) \Big|_{\alpha}^{\beta} + \int_{\alpha}^{\beta} u \frac{d^2}{dx^2} \left(a \frac{d^2v}{dx^2} \right) dx$$
$$(4\text{-}71)$$

$$\int_{D} a\nabla^2 u \nabla^2 v \, dD = \oint a \frac{\partial u}{\partial n} \nabla^2 v \, ds - \oint u \frac{\partial}{\partial n} (a\nabla^2 v) \, ds + \int_{D} u \nabla^2 (a\nabla^2 v) \, dD$$
$$(4\text{-}72)$$

---

[1] See Exercise 4-35.

[2] See, for example, S. Timoshenko and D. H. Young, "Advanced Dynamics," McGraw-Hill Book Company, Inc., New York, 1948, p. 233.

[3] See Exercise 4-36.

In addition, the following formula is sometimes helpful in two-dimensional problems:

$$\int_D \left\{ \frac{\partial u}{\partial x} - \frac{\partial v}{\partial y} \right\} dD = \oint \{v\,dx + u\,dy\} \tag{4-73}$$

This is called *Green's theorem* or *Stokes' theorem in the plane*. In all these formulas it is tacitly assumed that the functions involved possess continuous derivatives of sufficient order.

Calculus of Variations.[1] We turn now to the problem of obtaining the equilibrium problem equivalent to a given extremum problem. This will enable us to verify whether an extremum problem formulated from a physical principle is actually equivalent to the corresponding equilibrium problem. Study of this problem also provides additional insight into the structure of an equilibrium problem and clarifies the distinction between essential and additional boundary conditions.

A brief outline of the algorithm of the calculus of variations follows: Let $\Phi[\psi]$ be a functional. Let the *essential* boundary conditions which an admissible function must satisfy be

$$B_i[\psi] = g_i \qquad i = e_1, e_2, \ldots \tag{4-74}$$

We take these boundary conditions to be *linear* in ψ. Consider a *variational function*, u, which meets all the requirements of an admissible function except that it satisfies *homogeneous* essential boundary conditions.

$$B_i[u] = 0 \qquad i = e_1, e_2, \ldots \tag{4-75}$$

Now if ψ is a true extremizing function, $\psi + \epsilon u$ can be made to represent an arbitrary admissible function by suitably fixing the parameter ϵ and the variational function u. Moreover, if u is fixed and ϵ is allowed to vary, $\psi + \epsilon u$ represents a one-parameter family of admissible functions, i.e., for any value of ϵ

$$B_i[\psi + \epsilon u] = g_i \qquad i = e_1, e_2, \ldots \tag{4-76}$$

according to (4-74) and (4-75).

We next consider the values of the functional $\Phi[\psi + \epsilon u]$ for such a one-parameter family. As ϵ varies, so does Φ, but by definition Φ is to be stationary for the true extremizing function ψ. This means

$$\left(\frac{\partial \Phi[\psi + \epsilon u]}{\partial \epsilon} \right)_{\epsilon=0} = 0 \tag{4-77}$$

---

[1] For a more complete treatment see, for example, C. Fox, "Calculus of Variations," Oxford University Press, New York, 1950, and R. Courant and D. Hilbert, "Methoden der mathematischen Physik," 2d ed., vol. I, Springer-Verlag OHG, Berlin, 1931, chap. 4.

The final step in the argument is to say that (4-77) must hold for every possible family of the form $\psi + \epsilon u$, that is, for arbitrary variational functions u. The formal procedure of the calculus of variations consists in actually evaluating (4-77) for a given functional Φ and then drawing conclusions from the consideration that u can be an arbitrary variational function. In this last step it is usually necessary to rearrange (4-77), using integration by parts. The conclusions so obtained are necessary conditions for an extremum.

As a first example, we consider the functional (4-61) constructed from the potential energy of Prob. 4-1 with the essential boundary conditions $\psi = 0$ at $x = 0$ and $x = 1$. We shall show that the equilibrium problem equivalent to this extremum problem is actually (4-13), which was derived directly from the physical system. We begin by substituting $\psi + \epsilon u$ for ψ in (4-61). The variational function u must satisfy the boundary conditions $u = 0$ at $x = 0$ and $x = 1$.

$$\Phi[\psi + \epsilon u] = \int_0^1 \left\{ \frac{1}{2} (\psi + \epsilon u)^2 + \frac{1}{2} \left(\frac{d^2\psi}{dx^2} + \epsilon \frac{d^2u}{dx^2} \right)^2 - \psi - \epsilon u \right\} dx \quad (4\text{-}78)$$

Next we differentiate with respect to the parameter ϵ. Both ψ and u are treated as constants in this differentiation.

$$\frac{\partial \Phi[\psi + \epsilon u]}{\partial \epsilon} = \int_0^1 \left\{ (\psi + \epsilon u)u + \left(\frac{d^2\psi}{dx^2} + \epsilon \frac{d^2u}{dx^2} \right) \frac{d^2u}{dx^2} - u \right\} dx \quad (4\text{-}79)$$

Now, setting $\epsilon = 0$, we have

$$\left(\frac{\partial \Phi[\psi + \epsilon u]}{\partial \epsilon} \right)_{\epsilon=0} = \int_0^1 \left\{ \psi u + \frac{d^2\psi}{dx^2} \frac{d^2u}{dx^2} - u \right\} dx = 0 \quad (4\text{-}80)$$

according to (4-77). It remains to draw conclusions from the fact that (4-80) must hold for arbitrary variational functions u. Such conclusions cannot, however, be drawn directly because of the simultaneous appearance of u and d^2u/dx^2 in the integrand. We therefore integrate the second-derivative term by parts, using formula (4-71) with $a = 1$ and $v = \psi$. The statement (4-80) then reads as follows:

$$\int_0^1 u \left\{ \psi + \frac{d^4\psi}{dx^4} - 1 \right\} dx - u \frac{d^3\psi}{dx^3} \Big|_0^1 + \frac{du}{dx} \frac{d^2\psi}{dx^2} \Big|_0^1 = 0 \quad (4\text{-}81)$$

Since u must vanish at $x = 0$ and $x = 1$, the middle term on the left is zero. The requirement that the sum of the remaining terms vanish for arbitrary u leads to the conclusion that each term separately must vanish and further that this is possible only if

$$\psi + \frac{d^4\psi}{dx^4} - 1 = 0 \qquad \text{in } 0 < x < 1 \quad (4\text{-}82)$$

and if

$$\frac{d^2\psi}{dx^2} = 0 \quad \text{at} \quad \begin{cases} x = 0 \\ x = 1 \end{cases} \tag{4-83}$$

These statements are proved by using the reductio ad absurdum approach. Suppose that $(d^2\psi/dx^2)_{x=1}$, for instance, were not zero. Then it would be possible to find a u which contradicted (4-81). Such a u would be zero almost everywhere except for a sharp hook at $x = 1$. This would make the integral very small, but $(du/dx)_{x=1}$ would be large. Similarly, if it were supposed that (4-82) did not hold at some interior point, (4-81) would be contradicted by a u which is zero almost everywhere except for a bump at the suspected point. Note that (4-82) and (4-83) are the differential equation and the additional boundary conditions of (4-13). They represent the force-balance requirements for a displacement state.

For a two-dimensional example we turn to the functional (4-65) corresponding to the Saint-Venant formulation of Prob. 4-3. Here we have no essential boundary conditions so that the variational functions u are arbitrary, suitably continuous functions. The steps in the deduction are exactly the same as in the previous example.

$$\Phi[\Omega + \epsilon u] = \frac{1}{2} \int_D \left\{ \left(\frac{\partial \Omega}{\partial x} + \epsilon \frac{\partial u}{\partial x} - y \right)^2 + \left(\frac{\partial \Omega}{\partial y} + \epsilon \frac{\partial u}{\partial y} + x \right)^2 \right\} dD \tag{4-84}$$

$$\left(\frac{\partial \Phi[\Omega + \epsilon u]}{\partial \epsilon} \right)_{\epsilon=0} = \int_D \left\{ \left(\frac{\partial \Omega}{\partial x} - y \right) \frac{\partial u}{\partial x} + \left(\frac{\partial \Omega}{\partial y} + x \right) \frac{\partial u}{\partial y} \right\} dD = 0$$

$$= \int_D \left\{ \frac{\partial \Omega}{\partial x} \frac{\partial u}{\partial x} + \frac{\partial \Omega}{\partial y} \frac{\partial u}{\partial y} \right\} dD - \int_D \left\{ \frac{\partial(yu)}{\partial x} - \frac{\partial(xu)}{\partial y} \right\} dD = 0 \tag{4-85}$$

Integrating the first integral on the right with (4-70) and the second with (4-73), we obtain

$$-\int_D u \left\{ \frac{\partial^2 \Omega}{\partial x^2} + \frac{\partial^2 \Omega}{\partial y^2} \right\} dD + \oint u \left\{ \frac{\partial \Omega}{\partial n} - \frac{1}{2} \frac{\partial(x^2 + y^2)}{\partial s} \right\} ds = 0 \tag{4-86}$$

Since (4-86) must hold for arbitrary u, we must have

$$-\left(\frac{\partial^2 \Omega}{\partial x^2} + \frac{\partial^2 \Omega}{\partial y^2} \right) = 0 \tag{4-87}$$

in D and

$$\frac{\partial \Omega}{\partial n} = \frac{1}{2} \frac{\partial(x^2 + y^2)}{\partial s} \tag{4-88}$$

on the boundary. These are the same as (4-32) and (4-33) obtained in our previous formulation of the equilibrium problem.

It can be shown[1] that the extremum problem for a *quadratic* functional is equivalent to a *linear self-adjoint* equilibrium problem. Furthermore, the extremum is a *minimum*[2] when the equilibrium problem is *positive*.

---

[1] See R. Courant and D. Hilbert, "Methoden der mathematischen Physik," vol. 2, Springer-Verlag OHG, Berlin, 1937, p. 476.

[2] For verification in a particular case see Exercise 4-48.

If the equilibrium problem is positive *definite*, then the extremum problem has a *unique* minimum.

These statements are illustrated by the two preceding examples. In both cases we have linear self-adjoint equilibrium problems equivalent to extremum problems for quadratic functionals. For the beam on elastic foundation the equilibrium problem is positive definite, and Φ has a unique minimum. The Saint-Venant formulation of the torsion problem is only positive (see page 211), and we note that the minimum for Φ of (4-65) is not unique since the addition of an arbitrary constant to Ω does not alter the value of Φ.

Upper and Lower Bounds for System Properties. The equivalence between extremum problems and equilibrium problems depends only on the *stationary* property of the extremum. Although the character of the extremum (maximum, minimum, or saddle point) plays no role in the equilibrium problem per se, it is of importance for stability investigations. Knowledge of the character of an extremum is also valuable in dealing with approximate solutions. For example, if it is known that the exact solution minimizes a functional Φ, then the values of Φ corresponding to different approximate solutions can be used as a basis for comparison: the lower the value of Φ, the better the approximation.

The character of an extremum is sometimes given in the statement of a physical extremum principle. For example, the principle of minimum potential energy states that an extremum corresponding to a *stable* equilibrium configuration is a *minimum*. For linear systems we have seen that a *minimum* for the extremum problem is associated with a *positive* equilibrium problem.

In Chap. 1 it was pointed out that certain over-all system properties were closely related to the functionals of linear systems. By using complementary extremum problems it was possible to obtain upper and lower bounds for these system properties. This idea[1] is readily extended to continuous systems.

As an example we take the *torsional rigidity* in Prob. 4-3. By definition the torsional rigidity or over-all stiffness is the ratio of twisting moment to twist angle per unit length.

$$C = \frac{M_t}{\theta} \tag{4-89}$$

---

[1] Historically upper and lower bounds of this type were given first for continuous systems. See E. Trefftz, Ein Gegenstück zum Ritzchen Verfahren, *Proc. 2d Intern. Congr. Appl. Mech.*, Zurich, 1927, pp. 131–137. The procedure of Trefftz is outlined in Exercises 4-40 and 4-41. The use of complementary extremum principles is due to C. Weber, Bestimmung des Steifigkeitswertes von Körper durch zwei Näherungsverfahren, *Z. angew. Math. u. Mech.*, **11**, 244–247 (1931). For modern treatment see W. Prager and J. L. Synge, Approximations in Elasticity Based on the Concept of Function Space, *Quart. Appl. Math.*, **5**, 241–269 (1947).

In terms of these quantities the potential energy of the equilibrium configuration is simply

$$PE = \tfrac{1}{2}C\theta^2 \tag{4-90}$$

and the complementary energy of the equilibrium stress state is

$$CE = \frac{M_t^2}{2C} - M_t\theta \tag{4-91}$$

if we consider the twist angle to be prescribed. Since both the potential and complementary energies are minimized by the true equilibrium state, we have

$$C = \frac{2(PE)_{min}}{\theta^2} \tag{4-92}$$

directly from (4-90) and

$$C = -\frac{2(CE)_{min}}{\theta^2} \tag{4-93}$$

from (4-91) after eliminating M_t by (4-89). These results can also be derived[1] by analytically transforming the integrals representing the potential and complementary energy. If now we compute C from (4-92), using an approximate displacement state whose potential energy is greater than the minimum, we shall obtain too *large* a value for C. Conversely, if we compute C from (4-93), using an approximate stress state whose complementary energy is (algebraically) greater than the minimum, we shall obtain too *small* a value for C. We thus obtain upper and lower bounds for the true torsional rigidity. In terms of the dimensionless functionals Φ and $\bar{\Phi}$ of (4-65) and (4-67) and the dimensionless stiffness of (4-27), this result may be restated as follows:

$$-2\bar{\Phi} \leq C \leq 2\Phi \tag{4-94}$$

Since the functionals are *stationary* for the true state, the errors in these bounds are of second order in comparison with the errors in ψ and Ω which describe the state. This means that if, for instance, an approximate solution for ψ is available, the approximation to C furnished by (4-28) will usually[2] have first-order error, whereas $-2\bar{\Phi}$ will have second-order error.

EXERCISES

4-28. Construct the potential energy for a small displacement, $y(x)$, of a taut elastic cable of length L and tension T subjected to a uniform distributed load w per unit length. What are the essential boundary conditions?

Ans. $PE = \int_0^L \{\tfrac{1}{2}T(dy/dx)^2 - wy\}\, dx$, $y = 0$ at $x = 0$ and $x = L$.

[1] See Exercises 4-38 and 4-39.

[2] An exception in which (4-28) provides the *same* approximation as $-2\bar{\Phi}$ occurs when ψ has been obtained by the *Ritz* method (see Exercise 4-53).

4-29. Repeat Exercise 4-28 for the case where the cable is supported by an elastic foundation of stiffness k.   *Ans.* PE $= \int_0^L \{\frac{1}{2}T(dy/dx)^2 + \frac{1}{2}ky^2 - wy\}\, dx.$

4-30. Apply the algorithm of the calculus of variations to the result of Exercise 4-29 to show that the extremum problem for the potential energy is equivalent to the equilibrium problem of Exercise 4-18.

4-31. A complementary extremum problem for the cable on elastic foundation of Exercise 4-29 can be formulated as follows: Let the vertical component of the cable tension be $v = T\, dy/dx$. Show that to satisfy the force-balance requirements the foundation force must be $w + dv/dx$. Show that the complementary energy is given by

$$\text{CE} = \int_0^L \left\{ \frac{v^2}{2T} + \frac{1}{2k}\left(w + \frac{dv}{dx}\right)^2 \right\} dx$$

What are the essential boundary conditions?   *Ans.* There are none.

4-32. Apply the algorithm of the calculus of variations to the extremum problem for the complementary energy of Exercise 4-31. What is the physical significance of the additional boundary conditions?

     Ans. $-d^2v/dx^2 + kv/T = 0$, $dv/dx = -w$ at $x = 0$ and $x = L$.

4-33. Apply the algorithm of the calculus of variations to (4-62), and show that the differential equation (4-7) and the additional boundary conditions $d^2M/dx^2 = -w$ at $x = 0$ and $x = L$ are obtained as necessary conditions for stationary complementary energy. The required integration by parts is a special case of (4-71).

4-34. Show that the extremum problem with the functional (4-63) and the essential boundary conditions (4-19) is equivalent to Prob. 4-2, temperatures in a slab. The required integration by parts is a special case of (4-70).

4-35. Apply the algorithm of the calculus of variations to $\bar{\Phi}$ of (4-67). The integration by parts can be performed by using (4-70). Verify that the extremum problem for $\bar{\Phi}$ is equivalent to the equilibrium problem of (4-29).

4-36. Show that the extremum problem for (4-68) is equivalent to the equilibrium problem of (4-35) and (4-36).

4-37. Consider the linearized form of Prob. 4-4 in which $\epsilon = 0$ in (4-35). Show that this is self-adjoint but not positive or positive definite. Construct the functional for an equivalent extremum problem [see (4-68)]. Let $x(t)$ have the Fourier expansion $x = \Sigma a_n \sin n\omega t$. Show that

$$\Phi = \frac{\pi}{2\omega}\left\{ 2a_1 + \sum_1^\infty a_n{}^2(n^2\omega^2 - 1) \right\}$$

and hence that Φ is actually minimized by the solution to the equilibrium problem when $\omega > 1$.

4-38. By using Green's theorem (4-73) with $u = y\Omega$ and $v = x\Omega$ verify that

$$\int_D \left\{ x\frac{\partial\Omega}{\partial y} - y\frac{\partial\Omega}{\partial x} \right\} dD + \frac{1}{2}\oint \Omega\, \frac{\partial(x^2 + y^2)}{\partial s}\, ds = 0$$

Then show that the functional (4-65) can be transformed as follows:

$$\Phi = \frac{1}{2}\int_D \left\{ x\frac{\partial\Omega}{\partial y} - y\frac{\partial\Omega}{\partial x} + x^2 + y^2 \right\} dD - \frac{1}{2}\int_D \Omega\nabla^2\Omega\, dD + \frac{1}{2}\oint \Omega\left\{ \frac{\partial\Omega}{\partial n} - \frac{1}{2}\frac{\partial(x^2 + y^2)}{\partial x} \right\} ds$$

Use the fact that the minimizing Ω must satisfy (4-87) and (4-88) to verify the right side of (4-94).

4-39. Verify the left side of (4-94) by transforming (4-67) into

$$\bar{\Phi} = \int_D \left\{ -\frac{1}{2} \psi \nabla^2 \psi - 2\psi \right\} dD$$

and then using the fact that the minimizing ψ satisfies (4-29).

4-40. The stress function ψ in Prob. 4-3 represents an internally self-equilibrating force state. If $-\nabla^2\psi = 2G\theta$, it also represents a geometrically compatible state even though it is not the true state until the stress boundary condition $\psi = 0$ is satisfied. Show that the potential energy of a state for which $-\nabla^2\psi = 2G\theta$ is

$$\text{PE} = \frac{1}{2G} \iint \left\{ \frac{\partial^2 \psi}{\partial x^2} + \frac{\partial^2 \psi}{\partial y^2} \right\} dx\, dy$$

Finally show that the stiffness (4-25) and (4-26) is bounded by

$$\frac{1}{G} \iint \left\{ \left(\frac{\partial \psi_1}{\partial x} \right)^2 + \left(\frac{\partial \psi_1}{\partial y} \right)^2 - 2G\theta\psi_1 \right\} dx\, dy \leq C \leq \frac{1}{G} \iint \left\{ \left(\frac{\partial \psi_2}{\partial x} \right)^2 + \left(\frac{\partial \psi_2}{\partial y} \right)^2 \right\} dx\, dy$$

where $\psi_1 = 0$ on the boundary but does not necessarily satisfy $-\nabla^2\psi = 2G\theta$ and where just the opposite is true for ψ_2; that is, $-\nabla^2\psi_2 = 2G\theta$, but ψ_2 does not necessarily vanish on the boundary.

4-41. Consider the equilibrium problem of finding ψ which satisfies $\nabla^2\psi = 0$ in D and which assumes prescribed values $\bar{\psi}$ on the boundary of D. Prob. 4-2 is of this type. The extremum problem for

$$\Phi(\psi) = \int_D \left\{ \left(\frac{\partial \psi}{\partial x} \right)^2 + \left(\frac{\partial \psi}{\partial y} \right)^2 \right\} dD$$

with the essential boundary condition $\psi = \bar{\psi}$ is equivalent to this (see Exercise 4-34). If v is an approximate solution for which $v = \bar{\psi}$ on the boundary and ψ is the true solution, show that

$$\Phi(v) = \Phi(\psi) + \Phi(\psi - v)$$

and hence that $\Phi(v)$ is *stationary* and a *minimum* for the true solution. [*Hint.* Expand the last term on the right, and use (4-70) to integrate the cross-product term.]

If w is an approximate solution which satisfies $\nabla^2 w = 0$ in D, show that

$$\Phi(w) = \Phi(\psi) - \Phi(\psi - w)$$

provided that the amplitude of w is adjusted so as to make

$$\oint (w - \bar{\psi}) \frac{\partial w}{\partial n}\, ds = 0$$

Note that $\Phi(w)$ is *stationary* and a *maximum* for the true solution.

The method of *Trefftz* consists in choosing $w = \Sigma c_j \varphi_j$, where $\nabla^2 \varphi_j = 0$, and determining the c_j from the conditions

$$\oint (w - \bar{\psi}) \frac{\partial \varphi_j}{\partial n}\, ds = 0$$

Demonstrate that these conditions are equivalent to the requirement that $\Phi(\psi - w)$ should be minimized. When this method is utilized, w is determined only to within an additive constant. Verify that, if one of the φ_i is taken to be a constant, then the corresponding c_i becomes indeterminate.

4-42. Show that for the cantilever beam on elastic foundation of Exercise 4-1 dimensionless functionals representing the potential energy and complementary energy are

$$\Phi = \int_0^1 \left\{ \frac{1}{2} \left(\frac{d^2y}{dx^2} \right)^2 + 2y^2 \right\} dx - y(1)$$

$$\bar{\Phi} = \int_0^1 \left\{ \frac{1}{2} M^2 + \frac{1}{8} \left(\frac{d^2M}{dx^2} \right)^2 \right\} dx$$

where $y(x)$ is a dimensionless deflection and $M(x)$ is a dimensionless bending moment. The essential boundary conditions for Φ are $y = dy/dx = 0$ at $x = 0$, and the essential boundary conditions for $\bar{\Phi}$ are $M = 0$, $dM/dx = 1$ at $x = 1$. Apply the algorithm of the calculus of variations to show that the extremum problems for Φ and $\bar{\Phi}$ are equivalent to the equilibrium problems formulated in Exercise 4-1.

4-43. A nondimensional *stiffness* for the system of Exercise 4-42 may be defined as

$$C = \frac{L^3}{EI} \frac{P}{\delta}$$

where δ is the deflection at the free end. Show that upper and lower bounds for C are given by

$$\frac{1}{2\bar{\Phi}} \leq C \leq \frac{1}{-2\Phi}$$

4-44. Show that the *capacitance* of the cable of Exercise 4-3 is given by

$$C = \frac{1}{4\pi} \int_D \left\{ \left(\frac{\partial \varphi}{\partial x} \right)^2 + \left(\frac{\partial \varphi}{\partial y} \right)^2 \right\} dD$$

where the domain D is the space between conductors and φ is the solution of the equilibrium problem formulated in Exercise 4-3. Verify that this equilibrium problem is equivalent to the extremum problem for C when the essential boundary conditions are $\varphi = 1$ on the inner conductor and $\varphi = 0$ on the outer conductor.

4-45. Show that the potential and complementary energies for the membrane of Exercise 4-5 are

$$\text{PE} = \int_D \left\{ \frac{T}{2} \left[\left(\frac{\partial z}{\partial x} \right)^2 + \left(\frac{\partial z}{\partial y} \right)^2 \right] - pz \right\} dD$$

$$\text{CE} = \int_D \frac{1}{2T} \left[\left(\frac{\partial \varphi}{\partial x} + \frac{pv}{2} \right)^2 + \left(\frac{\partial \varphi}{\partial y} - \frac{px}{2} \right)^2 \right] dD$$

The essential boundary condition for the PE is $z = 0$. There is no essential boundary condition for the CE. Verify that the extremum problems for these are equivalent to the equilibrium problems formulated in Exercise 4-5.

4-46. The *stiffness* of the membrane of Exercise 4-5 may be defined as the ratio of the pressure against the membrane to the volume subtended by the deflected membrane.

$$C = \frac{p}{\displaystyle\int_D z \, dD}$$

Show that in terms of the stress function φ the stiffness is given by

$$C = \frac{1}{T} \int_D \left\{ \frac{p}{2} \left(x^2 + y^2 \right) + x \frac{\partial \varphi}{\partial y} - y \frac{\partial \varphi}{\partial x} \right\} dD$$

Relate the stiffness to the energies of the true equilibrium state, and show that upper and lower bounds for C are given by

$$\frac{p^2}{2(\text{CE})} \leq C \leq \frac{p^2}{-2(\text{PE})}$$

4-47. Let the cross section of a prism in torsion be geometrically similar to the area of a membrane subjected to normal pressure. Show that the membrane stiffness C_m and the torsional rigidity C_t satisfy

$$\frac{C_m L_m{}^4}{T} \frac{C_t}{G L_t{}^4} = 4$$

where L_m and L_t are corresponding linear dimensions, T is the membrane surface tension, and G is the shear modulus of the prism.

4-48. By direct expansion and integration by parts of (4-78) show that

$$\Phi(\psi + \epsilon u) = \Phi(\psi) + \frac{\epsilon^2}{2} \int_0^1 \left\{ u^2 + \left(\frac{d^2u}{dx^2}\right)^2 \right\} dx$$

and hence that Φ is *stationary* and a *minimum* for the true solution of the equilibrium problem.

4-5. Trial Solutions with Undetermined Parameters

We now begin our survey of approximate procedures for "solving" equilibrium problems in continuous systems. The basic step involved in the method of this section is the choice of a trial solution which, because of undetermined parameters or undetermined functions, actually represents a whole family of possible approximations. Once the family is fixed, there are several criterions for picking out the "best" approximation within the family. These criteria are very similar to those described in Chap. 3. There are, however, some differences when we have an equilibrium problem that is equivalent to an extremum problem.

The discussion here is limited to the system

$$L_{2m}[\psi] = f \qquad B_i[\psi] = g_i \qquad i = 1, \ldots, m \qquad (4\text{-}95)$$

in which the boundary conditions are *linear* in ψ. Furthermore we consider the trial solution to have the *linear* form

$$\psi = \varphi_0 + \sum_{j=1}^r c_j \varphi_j \qquad (4\text{-}96)$$

where the φ_j are linearly independent known functions in the domain D and the c_j are undetermined parameters. There are two basic types of criteria for fixing the c_j. In one the c_j are chosen so as to make weighted averages of the equation residual vanish, and in the other the c_j are chosen so as to give a stationary value to a functional Φ which is related to the system (4-95). In either case the result of applying the criterion is a set of r simultaneous equations for the c_j. These methods may therefore be considered as means for reducing a continuous equilibrium problem to an approximately equivalent equilibrium problem with r degrees of freedom.

Boundary Conditions. For the weighted-residual methods we select the trial solution so that it satisfies *all* the boundary conditions of (4-95). This is accomplished by choosing the functions φ_j in (4-96) so as to satisfy the following boundary conditions:

$$B_i[\varphi_0] = g_i \qquad i = 1, \ldots, m$$
$$B_i[\varphi_j] = 0 \qquad i = 1, \ldots, m, j = 1, \ldots, r \qquad (4\text{-}97)$$

Then ψ of (4-96) satisfies all the boundary conditions for arbitrary values of the c_j.

For the stationary functional method it is necessary only that ψ satisfy the *essential* boundary conditions. This is accomplished by taking

$$B_i[\varphi_0] = g_i \qquad i = e_1, e_2, \ldots$$
$$B_i[\varphi_j] = 0 \qquad i = e_1, e_2, \ldots ; j = 1, \ldots, r \qquad (4\text{-}98)$$

Weighted-residual Methods. When the trial solution (4-96) which satisfies (4-97) is inserted in (4-95), the *equation residual R* is

$$R = f - L_{2m}[\psi] = f - L_{2m}\left[\varphi_0 + \sum_{j=1}^{r} c_j\varphi_j\right] \qquad (4\text{-}99)$$

For the exact solution the residual is, of course, identically zero. Within a restricted trial family a "good" approximation may be described as one which maintains R small in some sense. The following criteria can be considered[1] as requirements that various weighted averages of R should vanish:

Collocation. The residual (4-99) is set equal to zero at r points in the domain D. This provides r simultaneous algebraic equations for the c_j. The location of the points is arbitrary but is usually such that D is covered more or less uniformly by a simple pattern.

Subdomain. The domain D is subdivided into r subdomains, usually according to a simple pattern. The integral of the residual (4-99) over each subdomain is then set equal to zero to provide r simultaneous equations for the c_j.

Galerkin. The r integrals

$$\int_D \varphi_k R \, dD \qquad k = 1, \ldots, r \qquad (4\text{-}100)$$

are set equal to zero to provide r equations for the c. The weighting functions here are the same functions used in constructing ψ.

[1] See Sec. 3-6.

Least Squares. The integral of the square of the residual is minimized with respect to the undetermined parameters to provide r simultaneous equations. If L_{2m} is a linear operator we have

$$\frac{\partial}{\partial c_k} \int_D R^2 \, dD = -2 \int_D R L_{2m}[\varphi_k] \, dD = 0 \qquad k = 1, \ldots, r \quad (4\text{-}101)$$

Stationary Functional Method. Let Φ be a functional such that the extremum problem for Φ is equivalent to the equilibrium problem (4-95). *The Ritz method*[1] consists in treating the extremum problem directly by inserting the trial family (4-96) into Φ and setting

$$\frac{\partial \Phi}{\partial c_j} = 0 \qquad j = 1, \ldots, r \qquad (4\text{-}102)$$

These r equations are solved for the c_j, and the corresponding function ψ then represents an approximate solution to the extremum problem. (It is an approximate solution because it gives Φ a stationary value only for these variations of ψ which are contained within the trial family. This solution would not in general still be an extremum if more general variations were permitted.)

An advantage of the stationary functional method is that admissible functions for the extremum problem need satisfy only essential boundary conditions. This will simplify the selection of the trial family (4-96) whenever the conditions (4-98) are less restrictive than (4-97).

Remarks. It should be emphasized that the most important (and most difficult) step in all these methods is the selection of the trial family (4-96). The purpose of the above criterions is merely to pick the "best" approximation out of a given family. Good results cannot be obtained if good approximations are not included within the trial family. Theoretically, if enough[2] independent φ_j are included in (4-96), good approximations must be contained within the family; however, the principal attraction of these methods lies in the possibility of obtaining good approximations with a limited number of adjustable parameters. In selecting the φ_j the

[1] W. Ritz, Über eine neue methode zur Lösung gewisser Variations-probleme der mathematischen Physik, *J. reine u. angew. Math.*, **135**, 1–61 (1909). The same idea had earlier been applied to eigenvalue problems by Lord Rayleigh. See J. W. Strutt, On the Theory of Resonance, *Trans. Roy. Soc. (London)*, **A161**, 77–118 (1870); and J. W. Strutt, 3d Baron Rayleigh, "The Theory of Sound," 2d ed., vol. 2, Dover Publications, New York, 1945, Appendix A. We call it the Ritz method when it is applied to equilibrium problems and the Rayleigh-Ritz method when applied to eigenvalue problems.

[2] For discussion of convergence when the number of independent φ_j approaches infinity, see Ritz, *op. cit.*, and R. A. Frazer, W. P. Jones, and S. W. Skan, Approximations to Functions and to the Solutions of Differential Equations, *Aeronaut. Research Comm. Rept. and Mem.* 1799 (1937).

analyst should give consideration to symmetry or any other special characteristics of the solution which may be known.

When the system (4-95) is *linear*, the equations for the c_j obtained by any of the weighted residual methods will be linear. The matrix of the coefficients of the c_j will always be *symmetric* in the least-squares method but will generally be unsymmetric for the other three criteria. If (4-95) is *self-adjoint*, then Galerkin's method will also lead[1] to *symmetric* equations.

When the functional Φ is *quadratic*, the Ritz method leads to *symmetric linear* equations for the c_j. When it is known that the true solution minimizes Φ, then the value of Φ may be used as a measure of the relative goodness of approximations: the lower the value of Φ, the better the approximation.

For a particular trial family (4-96) which satisfies *all* the boundary conditions of a linear-equilibrium problem the four weighted-residual criteria will in general produce slightly different approximations. If there is an equivalent extremum problem and the Ritz method is applied to the same trial family, the approximation obtained turns out to be *identical*[2] with that obtained by the Galerkin criterion. Thus Galerkin's method provides the *optimum* weighted-residual criterion in the sense that the approximation so obtained is one which renders Φ stationary for all variations within the given trial family.

There are differences in the amounts of labor required to apply these criteria. Collocation, which involves only evaluation of functions, is usually substantially shorter than the other methods which involve integration. Whether or not this is an important consideration depends on how the resulting simultaneous equations are to be solved. Thus if a standardized program for their solution on an automatic computer is available, the time spent in setting up the equations may represent the major outlay for the problem. The opposite can be true, e.g., if the system is nonlinear and the equations must be solved by hand.

Examples. We turn to Prob. 4-1, the beam on elastic foundation, and consider first a trial family with two undetermined parameters which satisfies all the boundary conditions of (4-13).

$$\psi = c_1 \sin \pi x + c_2 \sin 3\pi x \tag{4-103}$$

Since the boundary conditions are homogeneous, there is no φ_0. The equation residual for this trial is

$$R = 1 - \frac{d^4\psi}{dx^4} - \psi = 1 - c_1(\pi^4 + 1)\sin \pi x - c_2(81\pi^4 + 1)\sin 3\pi x \tag{4-104}$$

[1] See Exercise 4-49.

[2] This was shown by B. G. Galerkin, Series Solutions of Some Problems of Elastic Equilibrium of Rods and Plates (in Russian), *Vestnik Inzhenerov*, **1**, 879–908 (1915). See Exercise 4-50.

Treating this residual by the weighted-residual methods leads to simultaneous equations for c_1 and c_2. Thus if, using *collocation*, we ask that R be zero at $x = \frac{1}{4}$ and $x = \frac{1}{2}$, we obtain the equations

$$\frac{\pi^4 + 1}{\sqrt{2}} c_1 + \frac{81\pi^4 + 1}{\sqrt{2}} c_2 = 1$$

$$(\pi^4 + 1)c_1 - (81\pi^4 + 1)c_2 = 1 \tag{4-105}$$

and their solution: $c_1 = 0.012267$, $c_2 = 0.000026$. If instead, using the *subdomain method*, we ask that the integral of R over the ranges $0 < x < \frac{1}{4}$ and $\frac{1}{4} < x < \frac{1}{2}$ be zero, we obtain the equations

$$\frac{\pi^4 + 1}{\pi} \frac{\sqrt{2} - 1}{\sqrt{2}} c_1 + \frac{81\pi^4 + 1}{\pi} \frac{\sqrt{2} + 1}{3\sqrt{2}} c_2 = \frac{1}{4}$$

$$\frac{\pi^4 + 1}{\sqrt{2}\,\pi} c_1 - \frac{81\pi^4 + 1}{3\sqrt{2}\,\pi} c_2 = \frac{1}{4} \tag{4-106}$$

and their solution: $c_1 = 0.013624$, $c_2 = 0.000029$. If we use the *Galerkin method*, we find

$$\int_0^1 \sin \pi x R\, dx = \frac{2}{\pi} - \frac{\pi^4 + 1}{2} c_1 = 0$$

$$\int_0^1 \sin 3\pi x R\, dx = \frac{2}{3\pi} - \frac{81\pi^4 + 1}{2} c_2 = 0 \tag{4-107}$$

which gives $c_1 = 0.012938$, $c_2 = 0.000054$ directly. Finally, if we apply the *method of least squares*, we obtain the same[1] equations as in the Galerkin method.

As this problem is positive definite, we can obtain some measure of the relative goodness of these approximations by comparing the corresponding values of Φ given by (4-61). Substituting our trial family (4-103) into (4-61), we find

$$\Phi = \frac{\pi^4 + 1}{4} c_1{}^2 + \frac{81\pi^4 + 1}{4} c_2{}^2 - \frac{2c_1}{\pi} - \frac{2c_2}{3\pi} \tag{4-108}$$

and, using the values of c_1 and c_2 just obtained, we get the following results:

$$\begin{aligned}
\Phi(\text{collocation}) &= -0.0041115 \\
\Phi(\text{subdomain}) &= -0.0041113 \\
\Phi(\text{Galerkin}) &= -0.0041241
\end{aligned} \tag{4-109}$$

According to this the collocation and subdomain solutions are of nearly equal merit, while the Galerkin solution is superior. To obtain a more comprehensive view of the accuracy of approximation, the deflections (4-103) for the three solutions above were compared, point by point throughout the domain, with the exact solution.[2] It was found that the deflection as given by the collocation solution was consistently too small, with an error ranging from 5 to 6 per cent of the true local deflection. The subdomain solution was consistently too large, with errors ranging from $4\frac{1}{2}$ to $5\frac{1}{2}$ per

---

[1] In general, the Galerkin and least-squares methods give different results in equilibrium problems. The reason for the equivalence in this case and also the reason for the lack of coupling terms in (4-107) is that the φ_i in (4-103) happen to be eigenfunctions for $L_{2m}[\psi] = \lambda \psi$. See Exercise 5-59.

[2] M. Hetenyi, "Beams on Elastic Foundation," University of Michigan Press, Ann Arbor, Mich., 1946, p. 60.

cent. The errors of the Galerkin solution were of fluctuating sign, with the largest error of the order of 0.1 per cent.

If the Ritz method were applied to the above trial family, we would minimize (4-108) with respect to c_1 and c_2. As the reader may show, this leads to the same equations as obtained by the Galerkin method.

We next consider a trial family in which the φ_j ($j = 1, 2$) satisfy only the *essential* boundary conditions; that is, $\varphi_j = 0$ at $x = 0$ and $x = 1$.

$$\psi = c_1 x(1 - x) + c_2 x^2(1 - x)^2 \qquad (4\text{-}110)$$

The additional boundary conditions, $d^2\psi/dx^2 = 0$ at $x = 0$ and $x = 1$, are satisfied by (4-110) only in the special case where $c_1 = c_2$. The Ritz method may be used to obtain a good approximation out of this family. When (4-110) is substituted into the functional (4-61), we obtain

$$\Phi = \frac{1}{2} \left\{ \frac{121}{30} c_1{}^2 + \frac{1}{70} c_1 c_2 + \frac{101}{126} c_2{}^2 \right\} - \frac{c_1}{6} - \frac{c_2}{30} \qquad (4\text{-}111)$$

and the following equations for determining c_1 and c_2:

$$\begin{aligned} \frac{\partial \Phi}{\partial c_1} &= \frac{121}{30} c_1 + \frac{1}{140} c_2 - \frac{1}{6} = 0 \\ \frac{\partial \Phi}{\partial c_2} &= \frac{1}{140} c_1 + \frac{101}{126} c_2 - \frac{1}{30} = 0 \end{aligned} \qquad (4\text{-}112)$$

The solution of (4-112) is $c_1 = 0.041249$, $c_2 = 0.041217$. Note that, since these are not exactly equal, the additional boundary conditions are violated. Nevertheless the approximation to the deflection provided by this solution is quite good. A point-by-point comparison against the exact deflection shows an error distribution of fluctuating sign with the maximum error of the same order of magnitude as in the foregoing Galerkin solution. Another measure of the goodness of approximation is furnished by the value of Φ itself. Substituting our values of c_1 and c_2 back into (4-111), we find $\Phi = -0.0041244$, which is actually lower than any of the values (4-109). Finally it should be remarked that, since the trial family (4-110) does not satisfy *all* the boundary conditions, the weighted-residual methods *cannot* be applied. If they are tried, meaningless results may be obtained.[1]

Upper and Lower Bounds for a Functional. We take Prob. (4-3), torsion of a square section, as a two-dimensional illustration in which the value of the functional in the associated extremum problem is itself of practical importance. We have already seen in (4-94) that the *torsional rigidity* can be bounded above and below in terms of the complementary functionals Φ of (4-65) and $\bar{\Phi}$ of (4-67).

Turning first to Φ, we note that there are no essential boundary conditions for the warping function Ω. Symmetry considerations, however, lead to the conclusion that $\Omega = 0$ on the axes $x = 0$, $y = 0$ and on $x = \pm y$. We therefore select the following one-parameter family which has these symmetry properties:

$$\Omega = c_1 xy(x^2 - y^2) \qquad (4\text{-}113)$$

[1] See Exercise 4-60.

In evaluating the functional (4-65) for this trial we can make use of the symmetry to reduce the integral over the square to four times the integral over the first quadrant. We thus obtain

$$\Phi = 2 \int_0^1 \int_0^1 \{[c_1 y(3x^2 - y^2) - y]^2 + [c_1 x(x^2 - 3y^2) + x]^2\} \, dx \, dy$$
$$= \tfrac{48}{35}c_1{}^2 - \tfrac{16}{15}c_1 + \tfrac{4}{3} \tag{4-114}$$

which attains a minimum value of $\tfrac{152}{135}$ when $c_1 = \tfrac{7}{18}$. Therefore we have

$$C \leq (2)(\tfrac{152}{135}) = 2.2519 \tag{4-115}$$

as the right side of (4-94).

For the complementary functional (4-67) we note the essential boundary condition $\psi = 0$ on the periphery of the squares. A one-parameter family which meets this requirement and which has the appropriate symmetry is

$$\psi = c_1(1 - x^2)(1 - y^2) \tag{4-116}$$

Substituting in (4-67) yields

$$\Phi = 4 \int_0^1 \int_0^1 \{2c_1{}^2 x^2(1 - y^2)^2 + 2c_1{}^2 y^2(1 - x^2)^2$$
$$- 2c_1(1 - x^2)(1 - y^2)\} \, dx \, dy \tag{4-117}$$
$$= \tfrac{128}{45}c_1{}^2 - \tfrac{32}{9}c_1$$

which attains a minimum value of $-\tfrac{10}{9}$ when $c_1 = \tfrac{5}{8}$. Inserting this result in the left side of (4-94), we obtain

$$C \geq -(2)(-\tfrac{10}{9}) = 2.2222 \tag{4-118}$$

Combining this result with (4-115) leads to the following upper and lower bounds for the torsional rigidity:

$$2.2222 \leq C \leq 2.2519 \tag{4-119}$$

The true value is known[1] to be $C = 2.2495$.

Trial Solutions with Undetermined Functions. The foregoing procedures transformed a given continuous equilibrium problem into an approximately equivalent lumped-parameter system. We now describe analogous procedures which may be used to replace a two-dimensional continuous equilibrium problem with an approximately equivalent system consisting of a finite number of one-dimensional problems. The basic step involves selecting a trial family of the form

$$\psi = \sum_{j=1}^r c_j \varphi_j \tag{4-120}$$

---

[1] Timoshenko and Goodier, *op. cit.*, p. 278.

where the φ_j are linearly independent known functions in the two-dimensional domain and the c_j are undetermined functions of a single variable; e.g., in a domain in the (x, y) plane the c_j might be undetermined functions of y alone, while the φ_j are known functions, perhaps of x alone or, more generally, of both x and y. If the φ_j are suitably chosen and suitable boundary conditions are imposed on the c_j, then (4-120) will satisfy the boundary conditions of a given two-dimensional equilibrium problem. A set of r simultaneous ordinary differential equations for determining the c_j may then be obtained by applying the weighted-residual techniques or by applying the algorithm of the calculus of variations to a related functional. A general treatment[1] is not attempted here. Instead the methods are briefly illustrated by application to Prob. 4-2 as formulated in (4-18) and (4-19).

We begin with a trial family with only one undetermined function.

$$\psi = x(1 - x)c_1(y) \tag{4-121}$$

All the boundary conditions of (4-19) will be satisfied if we require that $c_1(y)$ satisfy the following:

$$\begin{aligned} c_1(0) &= 1 \\ c_1(\infty) &= 0 \end{aligned} \tag{4-122}$$

The equation residual, $R = \nabla^2\psi$, is

$$R = -2c_1 + x(1 - x)\frac{d^2c_1}{dy^2} \tag{4-123}$$

from which we can obtain second-order ordinary differential equations for $c_1(y)$. For instance, using collocation, we might ask that R should vanish along $x = \frac{1}{3}$ (which, because of the symmetry, means that R also vanishes on $x = \frac{2}{3}$). The equation for $c_1(y)$ becomes

$$\frac{d^2c_1}{dy^2} - 9c_1 = 0 \tag{4-124}$$

and the solution satisfying (4-122) is

$$c_1 = e^{-3y} \tag{4-125}$$

The other weighted-residual methods may be applied to obtain similar, although slightly different, numerical results.

An alternative approach is available if the functional for an equivalent extremum problem is known. For Prob. 4-2 we have the functional (4-63). Substituting the trial family (4-121) and carrying out the x integration, we find

$$\Phi = \int_0^\infty \left\{ \frac{1}{3}c_1^2 + \frac{1}{30}\left(\frac{dc_1}{dy}\right)^2 \right\} dy \tag{4-126}$$

---

[1] The first method of this type made use of the related extremum problem. It was given by L. V. Kantorovich, Sur une méthode directe de la solution approximative de problème du minimum d'une intégrale double (in Russian), *Bull. acad. sci. U.R.S.S.*, [7], 647–652 (1933). The equivalence to Galerkin's method was shown by L. V. Kantorovich, Application of Galerkin's Method to the So-called Procedure of Reduction to Ordinary Differential Equations (in Russian with English summary), *Prikl. Math. Mech.*, **6**, 31–40 (1942).

as a functional depending on $c_1(y)$. Since the true ψ minimizes (4-63), we seek, as the best approximation within our family, that function c_1 which minimizes (4-126) subject to the boundary conditions (4-122). Applying the algorithm of the calculus of variations, we have

$$\left(\frac{\partial \Phi[c_1 + \epsilon u]}{\partial \epsilon}\right)_{\epsilon=0} = \int_0^\infty \left\{\frac{2}{3} c_1 u + \frac{2}{30} \frac{dc_1}{dy} \frac{du}{dy}\right\} dy = 0 \tag{4-127}$$

according to (4-77). Here u is a variational function for the problem of determining c_1; that is, it is an arbitrary function vanishing at $y = 0$ and $y = \infty$. The first derivative term is integrated by parts, yielding

$$\int_0^\infty \frac{2}{3} u \left\{c_1 - \frac{1}{10} \frac{d^2c_1}{dy^2}\right\} dy = 0 \tag{4-128}$$

which, because of the arbitrariness of u, means that c_1 satisfies the following equation:

$$\frac{d^2c_1}{dy^2} - 10c_1 = 0 \tag{4-129}$$

The solution satisfying (4-122) is

$$c_1 = e^{-3.1623y} \tag{4-130}$$

It is interesting to compare the values of Φ corresponding with the solutions (4-125) and (4-130). Substituting these c_1 into (4-126) yields

$$\begin{aligned} \Phi \text{ (collocation)} &= 0.10556 \\ \Phi \text{ (minimal)} &= 0.10541 \end{aligned} \tag{4-131}$$

Finally we outline the solution of the same problem, using the Galerkin treatment with a trial family having two undetermined functions.

$$\psi = x(1 - x)c_1(y) + x^2(1 - x)^2 c_2(y) \tag{4-132}$$

This will satisfy all the boundary conditions of (4-19) if we impose the following conditions on c_1 and c_2:

$$\begin{aligned} c_1(0) &= 1 & c_2(0) &= 0 \\ c_1(\infty) &= 0 & c_2(\infty) &= 0 \end{aligned} \tag{4-133}$$

The equation residual, $R = \nabla^2\psi$, is

$$R = -2c_1 + (2 - 12x + 12x^2)c_2 + x(1 - x)\frac{d^2c_1}{dy^2} + x^2(1 - x)^2 \frac{d^2c_2}{dy^2} \tag{4-134}$$

and by setting

$$\begin{aligned} \int_0^1 x(1 - x)R \, dx &= 0 \\ \int_0^1 x^2(1 - x)^2 R \, dx &= 0 \end{aligned} \tag{4-135}$$

according to Galerkin's criterion, we obtain the following simultaneous equations for c_1 and c_2:

$$\begin{aligned} \frac{1}{30}\frac{d^2c_1}{dy^2} - \tfrac{1}{3}c_1 + \frac{1}{140}\frac{d^2c_2}{dy^2} - \tfrac{1}{15}c_2 &= 0 \\ \frac{1}{40}\frac{d^2c_1}{dy^2} - \tfrac{1}{15}c_1 + \frac{1}{630}\frac{d^2c_2}{dy^2} - \tfrac{2}{105}c_2 &= 0 \end{aligned} \tag{4-136}$$

The solution satisfying (4-133) is

$$c_1 = 0.8035e^{-3.1416y} + 0.1965e^{-10.1059y}$$
$$c_2 = 0.9105(e^{-3.1416y} - e^{-10.1059y}) \tag{4-137}$$

When these values are put in (4-132) and the functional (4-63) is evaluated, we find $\Phi = 0.10508$, which should be compared with (4-131).

Application to Nonlinear System. When trial solutions with undetermined parameters are set up for nonlinear systems, the preceding techniques yield nonlinear simultaneous algebraic equations for the parameters. Although no general theory exists, good results have been obtained in individual cases. We consider here only Prob. 4-4.

Forced vibrations of nonlinear single-degree-of-freedom systems similar to Prob. 4-4 have received considerable attention from various authors. The first treatment[1] used the *collocation* method applied to a one-parameter trial family. Other procedures of the weighted-residual type, including the *subdomain method*[2] and the *method of least squares*,[3] have been applied to one-parameter trial solutions. Collocation with a two-parameter trial family was suggested by Den Hartog.[2] The method of Ritz, which uses a related extremum problem, has been applied by Duffing[4] and Silverman.[5] We give a brief outline of this latter procedure.

We have the functional (4-68), which is equivalent to the equilibrium problem of (4-35) and (4-36). The essential boundary condition is $x = 0$ at $t = 0$ and $t = 2\pi/\omega$. A one-parameter trial family which meets this requirement is

$$x = c_1 \sin \omega t \tag{4-138}$$

Substituting in (4-68) yields

$$\Phi = \int_0^{2\pi/\omega} \left\{ \frac{c_1^2 \omega^2}{2} \cos^2 \omega t - \frac{c_1^2}{2} \sin^2 \omega t - \epsilon \frac{c_1^4}{4} \sin^4 \omega t + c_1 \sin^2 \omega t \right\} dt$$
$$= \frac{\pi}{\omega} \left(\frac{c_1^2 \omega^2}{2} - \frac{c_1^2}{2} - \frac{3c_1^4}{16} \epsilon + c_1 \right) \tag{4-139}$$

---

[1] O. Martienssen, Über neue Resonanzerscheinungen in Wechselstromkreisen, *Physik. Z.*, 448–460 (1910). The method was more explicitly described by R. Rüdenberg, Einige unharmonische Schwingungsformen mit grosser Amplitude, *Z. angew. Math. u. Mech.*, **3**, 454–467 (1923).

[2] J. P. Den Hartog, The Amplitudes of Non-harmonic Vibrations, *J. Franklin Inst.*, **216**, 459–473 (1933).

[3] G. Schwesinger, On One Term Approximations of Forced Non-harmonic Vibrations, *J. Appl. Mech.*, **17**, 202–208 (1950).

[4] G. Duffing, "Erzwungene Schwingungen bei veränderlicher Eigenfrequenz," Brunswick, 1918, p. 130.

[5] I. K. Silverman, On Forced Pseudo-harmonic Vibrations, *J. Franklin Inst.*, **217**, 743–745 (1934).

To make Φ stationary, we set $\partial\Phi/\partial c_1 = 0$, thus obtaining the following nonlinear relation:

$$\tfrac{3}{4}\epsilon c_1{}^3 + (1 - \omega^2)c_1 - 1 = 0 \qquad (4\text{-}140)$$

For a fixed value of the nonlinearity parameter ϵ it is possible to obtain from (4-140) a curve of amplitude vs. frequency as sketched in Fig. 4-9. Although this represents only an approximate solution, it appears[1] that there are true solutions with very nearly the same response. Note that for large ω there are three possible amplitudes; i.e., there is not a unique

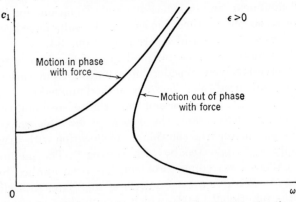

Fig. 4-9. Approximate response curve for steady-state forced vibrations of nonlinear spring-mass system.

solution.   Further study of this system indicates that there exist other steady-state periodic solutions quite different in form from (4-138).   For certain ranges of ϵ and ω there are *subharmonics* (i.e., periodic motions with periods which are integral multiples of $2\pi/\omega$) and there are *superharmonics*[2] (i.e., motions of period $2\pi/\omega$ in which the amplitude of some harmonic is large compared with the amplitude of the fundamental).

EXERCISES

4-49. When the equilibrium problem (4-95) is linear, the weighted-residual methods applied to the trial family (4-96) all lead to equations for the c_j having the following form:

$$\begin{bmatrix} a_{11} & a_{12} & \cdots & a_{1r} \\ a_{21} & a_{22} & \cdots & a_{2r} \\ \cdots & \cdots & \cdots & \cdots \\ a_{r1} & a_{r2} & \cdots & a_{rr} \end{bmatrix} \begin{bmatrix} c_1 \\ c_2 \\ \cdots \\ c_r \end{bmatrix} = \begin{bmatrix} b_1 \\ b_2 \\ \cdots \\ b_r \end{bmatrix}$$

[1] For further discussion see S. Timoshenko, "Vibration Problems in Engineering," 2d ed., D. Van Nostrand Company, Inc., New York, 1937, pp. 137–150, and K. O. Friedrichs and J. J. Stoker, Forced Vibrations of Systems with Non-linear Restoring Force, *Quart. Appl. Math.*, **1**, 97–115 (1943).

[2] See Friedrichs and Stoker, *op. cit.*, and C. R. Wylie, Jr., On the Forced Vibrations of Non-linear Springs, *J. Franklin Inst.*, **236**, 273–284 (1943).

Show that for *collocation*

$$a_{kj} = L_{2m}[\varphi_j(P_k)] \qquad b_k = f(P_k) - L_{2m}[\varphi_0(P_k)]$$

where the P_k are the r locations selected. Show that for the *subdomain* method

$$a_{kj} = \int_{D_k} L_{2m}[\varphi_j]\, dD \qquad b_k = \int_{D_k} \{f - L_{2m}[\varphi_0]\}\, dD$$

where the D_k are the r subdomains selected. Show that for the *Galerkin* method

$$a_{kj} = \int_D \varphi_k L_{2m}[\varphi_j]\, dD \qquad b_k = \int_D \varphi_k(f - L_{2m}[\varphi_0])\, dD$$

and that for the *least-squares* method

$$a_{kj} = \int_D L_{2m}[\varphi_k]L_{2m}[\varphi_j]\, dD \qquad b_k = \int_D L_{2m}[\varphi_k](f - L_{2m}[\varphi_0])\, dD$$

Note that in every case the matrix A has to do with the passive characteristics of the system and that the matrix B has to do with the "loading" in the domain and at the boundary. Show that A is symmetric for the Galerkin method if the system (4-95) is self-adjoint.

4-50. If the extremum problem for a quadratic functional Φ is equivalent to a linear equilibrium problem, then

$$\frac{\partial \Phi(\psi + \epsilon u)}{\partial \epsilon} = \int_D u\{L_{2m}[\psi + \epsilon u] - f\}\, dD + \text{boundary terms}$$

The boundary terms will vanish if $\psi + \epsilon u$ is required to satisfy *all* the boundary conditions of the equilibrium problem. Furthermore, these statements are valid whether or not ψ is the true solution just so long as the boundary conditions are satisfied. Verify this for (4-79).

Show that when the *Ritz* and *Galerkin* methods are applied to such systems with a trial family (4-96) satisfying (4-97) they lead to identical equations for determining the c_j.

4-51. Consider the taut cable on elastic foundation of Exercise 4-18. Set up a two-parameter trial family.

(a) Obtain the equations for finding the parameters by all four weighted-residual methods.

(b) Apply the Ritz method, using the functional of Exercise 4-29. Verify the equivalence with Galerkin's method.

4-52. Consider Prob. 4-1 formulated in (4-7) and (4-8) in terms of the bending moment.

(a) Construct a trial family for the weighted-residual methods. Note that a φ_0 is necessary because of the nonhomogeneous boundary condition.

(b) Show that the trial family

$$M = c_1 x(L - x) + c_2 x^2(L - x)^2$$

is admissible for the Ritz method applied to the complementary energy (4-62). Evaluate c_1 and c_2 by the Ritz method for the case where $EI = kL^4$.

Ans. $c_1 = 0.495884w$, $c_2 = -0.004419w/L^2$. The maximum M is $0.123695wL^2$ as compared with $0.123690wL^2$ in the exact solution. This solution is geometrically incompatible; that is, $y_{\text{beam}} \neq y_{\text{foundation}}$. The maximum error in $y_{\text{foundation}} =$

242 EQUILIBRIUM PROBLEMS IN CONTINUOUS SYSTEMS

$(w + d^2M/dx^2)/k$ is 4.7 per cent and in y_{beam} obtained by integrating (4-3) is 0.36 per cent when compared with the exact solution and the error expressed as a percentage of the central deflection. Note increase in accuracy obtained when approximate solution is *integrated*.

4-53. Let the general trial family (4-96) be applied to the Prandtl formulation of Prob. 4-3. Show that if the c_j are obtained by applying the Ritz method to the functional $\bar{\Phi}$ of (4-67) then the torsional rigidity computed from (4-28) will be identical with $-2\bar{\Phi}$.

4-54. Repeat Exercise 4-53 for the Saint-Venant formulation of Prob. 4-3; that is, show that if the c_j are obtained by applying the Ritz method to $\bar{\Phi}$ of (4-65) then (4-31) will be identical with $2\bar{\Phi}$.

4-55. Show that the trial family

$$\psi = c_1(1 - x^2)(1 - y^2) + c_2 x^2 y^2 (1 - x^2)(1 - y^2)$$

is suitable for the Prandtl formulation of Prob. 4-3, using either weighted-residual methods or the Ritz method applied to (4-67). Evaluate the c_j by any method, and obtain an approximation to C of (4-28).

Ans. $c_1 = 0.61058$, $c_2 = 0.50481$ by Ritz method and $C \geq 2.2466$.

4-56. Can the weighted-residual methods be applied to the trial family

$$\Omega = c_1 xy(x^2 - y^2) + c_2 x^3 y^3 (x^2 - y^2)$$

for the Saint-Venant formulation of Prob. 4-3? Evaluate the c_j, and obtain an approximation to the torsional rigidity.

Ans. $c_1 = 0.346667$, $c_2 = 0.103125$ by Ritz method and $C \leq 2.24952$.

4-57. Use the results of the two previous exercises and the analogy between the torsion problem and the *membrane* problem (see Exercise 4-47) to show that the stiffness of the membrane of Exercise 4-5 satisfies

$$1.7782 \leq C \leq 1.7805$$

4-58. Construct suitable trial families to be used with the Ritz method for the cantilever beam on elastic foundation of Exercise 4-42. Using only one-parameter families, obtain crude upper and lower bounds for the *stiffness* of Exercise 4-43.

Ans. Using $y = c_1 x^2$ and $M = x - 1 + c_1(x - 1)^2$, we find $3\frac{5}{9} \leq C \leq 4\frac{4}{5}$.

4-59. Recast the upper and lower bounds of Trefftz in Exercise 4-40 in terms of the nondimensional variables (4-27). Show that (4-116) is a suitable one-parameter trial family for ψ_1. Show that

$$\psi_2 = -\tfrac{1}{2}(x^2 + y^2) + c_1(x^4 - 6x^2 y^2 + y^4)$$

is a suitable trial family for ψ_2. Minimize the upper bound with respect to c_1, and verify that the result so obtained is identical with (4-115).

4-60. The trial family (4-110) satisfies only the essential boundary conditions for Prob. (4-1). It is not suitable for weighted residual methods.

(a) Verify that if collocation at $x = \frac{1}{2}$ is tried with the single parameter c_1 the deflection obtained is about 100 times too large.

(b) Verify that if collocation at $x = \frac{1}{4}$ and $x = \frac{1}{2}$ is tried with the parameters c_1 and c_2 the deflection obtained is *negative*.

4-61. Show that when $a = 1$ in Fig. 4-5 a suitable trial family for the *capacitance* problem is

$$\varphi = 2 - x + (2 - x)(x - 1)\Sigma c_j f_j(x, y) \qquad \begin{cases} 0 \leq y \leq x \\ 1 \leq x \leq 2 \end{cases}$$

with symmetrical values in the other seven-eighths of the domain. The functions f_j can be arbitrary. Select a particular trial, and obtain an approximation to the capacitance.

Ans. $C \leq 0.8627$ by using the result of Exercise 4-44 on the one-parameter family with $f_1 = y^2$.

4-62. Lower bounds for the capacitance in the previous exercise can be obtained by applying the method of *Trefftz* (described in Exercise 4-41) to the trial family

$$\varphi = c_1 \ln (x^2 + y^2) + \Sigma c_i f_i(x, y)$$

where the f_j must be solutions of Laplace's equation. Symmetry considerations indicate that the real parts of the complex functions $(x + iy)^m$ with $m = 4, 8, 12, \ldots$ would be likely trials. The $\ln (x^2 + y^2)$ term is necessary here. Show that if φ were made up only of f_j having no singularities inside the inner conductor then Trefftz's procedure would lead to the result $\varphi \equiv 0$. Physically the nonsingular potentials can represent only charge-free regions, but the inner conductor must carry a charge in the capacitance problem.

4-6. Finite-difference Methods

The ultimate purpose of most of the procedures described in the previous section is the reduction of a continuous-equilibrium problem into a system with a finite number of degrees of freedom. A good approximation can be obtained from a limited number of parameters if the analyst exercises enough ingenuity. In the present section we consider a class of procedures which again have as their purpose the reduction of continuous systems to "equivalent" lumped-parameter systems. These procedures are on the whole much more straightforward and may be successfully applied to a far wider class of problems. The price that is paid for this simplicity is the much larger number of degrees of freedom that usually results.

The basic approximation involves[1] the replacement of a continuous domain D by a pattern of discrete points within D as shown in Fig. 4-10. Instead of obtaining a continuous solution for ψ defined throughout D we find approximations to ψ only at these isolated points. When necessary, intermediate values, derivatives, or integrals may be obtained from our discrete solution by interpolation techniques.

The reduction of the governing equation and boundary conditions for the continuous domain to the governing equations for the discrete replacement may be accomplished *physically* or *mathematically*. In the *physical*

[1] For two-dimensional systems the first application of finite-difference methods was given by C. Runge, Über eine Methode die partielle Differentialgleichung $\Delta u =$ constans numerish zu integrieren, *Z. Math. u. Phys.*, **56**, 225–232 (1908). For one-dimensional systems the finite-difference approach has been deeply ingrained in engineering calculations for a much longer time. For instance, the graphical string polygon method for finding deflections of beams [O. Mohr, Beitrag zur Theorie der Holz und Eisenkonstruktion, *Z. Arch.-u. Ing.-Ver. Hannover* (1868)] makes essential use of the finite-difference approximation.

approach the discrete model is invested with lumped physical characteristics of the continuous system. The governing equations are then
obtained from the physical laws applied directly to the lumped-parameter
model. In the *mathematical* approach the continuous formulation is
reduced to a discrete formulation by simply replacing derivatives with
finite-difference approximations.

As examples of the former treatment a continuous elastic foundation
under a beam could be replaced by a number of discrete springs of such
stiffness and with such spacing that the resulting displacement of the

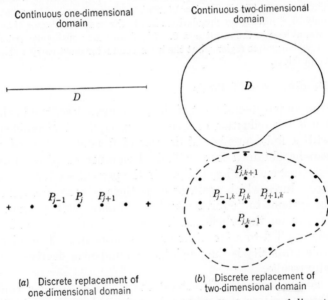

Continuous one-dimensional
domain

Continuous two-dimensional
domain

D

D

P_{j-1} P_j P_{j+1}

$P_{j,k+1}$

$P_{j-1,k}$ $P_{j,k}$ $P_{j+1,k}$

$P_{j,k-1}$

(a) Discrete replacement of
one-dimensional domain

(b) Discrete replacement of
two-dimensional domain

Fig. 4-10. Approximation of a continuous domain by an array of discrete points.

beam is substantially unaltered; or a continuous membrane subjected to
pressure could be replaced by a network of strings subjected to point
loads at the intersections; or a heat-conducting slab could be replaced
by a network of heat-conducting rods. The governing equations are then
obtained by writing the equilibrium equations for the beam at each spring,
or by balancing forces at each intersection of the string network, or by
balancing the heat flow in and out of each junction of the rod network.
The physical procedure may be used successfully by a specialist in one
particular field[1] and is often qualitatively useful in providing motivation;
however, when the formulation of the continuous problem is already
available, the mathematical procedure is simpler and more flexible. The
treatment which follows is restricted to the mathematical approach.

[1] See, for example, G. M. Dusinberre, "Numerical Analysis of Heat Flow,"
McGraw-Hill Book Company, Inc., New York, 1949.

Either procedure leads to a system of n simultaneous algebraic equations if n discrete points are involved. These equations will be linear or nonlinear according to whether the original continuous system was linear or nonlinear. The methods of Chap. 1 are then applied to complete the solution.

Computational Molecules. In preparation for replacing differential equations with finite-difference relations, we list[1] several basic finite-difference expressions in one and two dimensions. In one dimension we consider only the case of equally spaced points, and in two dimensions we consider only the square[2] network, or lattice. In both cases the spacing dimension will be called h.

In Fig. 4-10a let the points . . . , P_{j-1}, P_j, P_{j+1}, . . . be separated by a distance h, and let the value of $\psi(x)$ at P_j be denoted by ψ_j. Then, for instance, the first derivative at P_j may be approximated by the following finite-difference expression,

$$\left(\frac{d\psi}{dx}\right)_j = \frac{\psi_{j+1} - \psi_{j-1}}{2h} + 0(h^2) \qquad (4\text{-}141)$$

as shown in (3-83) and (3-89). This relation may be represented pictorially by a computational molecule[3] as shown in Fig. 4-11. The num-

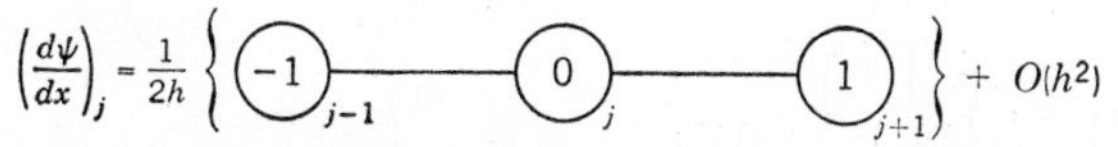

$$\left(\frac{d\psi}{dx}\right)_j = \frac{1}{2h}\left\{ \boxed{-1}_{j-1} \quad\text{———}\quad \boxed{0}_j \quad\text{———}\quad \boxed{1}_{j+1} \right\} + O(h^2)$$

FIG. 4-11. Computational molecule representation of formula (4-141).

bers in the positions, P_j, represent the multipliers that are to be applied to the values of ψ at these stations. The corresponding formulas for the first four[4] derivatives and the integral by Simpson's rule[5] are given in Fig. 4-12 in the form of computational molecules.

Similar formulas for approximating partial derivatives can easily be obtained.[6] The most commonly used molecules for square networks of spacing h are shown in Fig. 4-13. The integration formula given is the two-dimensional form of Simpson's rule.

[1] The method of derivation of these formulas and their truncation errors is the same as that used in Sec. 3-7.

[2] Triangular and hexagonal lattices have been studied by R. V. Southwell, "Relaxation Methods in Theoretical Physics," Oxford University Press, New York, 1946.

[3] This terminology is due to W. G. Bickley, Finite Difference Formulae for the Square Lattice, *Quart. J. Mech. Appl. Math.*, **1**, 35–42 (1948). Alternate designations are *computational stencil* or *computational lozenge*. See W. E. Milne, "Numerical Solution of Differential Equations," John Wiley & Sons, Inc., New York, 1953, p. 131.

[4] See (3-85) and Exercise 3-47.

[5] See (3-96).

[6] See Exercise 4-71.

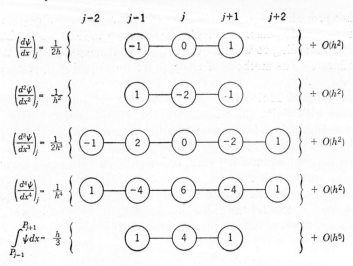

FIG. 4-12. Computational molecules for common one-dimensional operators.

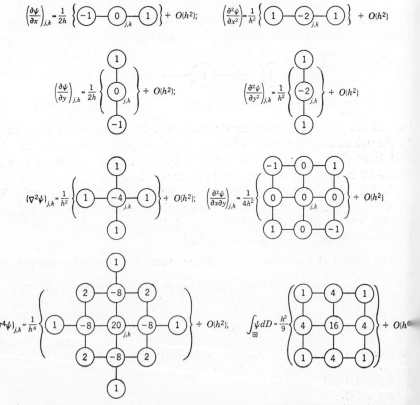

FIG. 4-13. Computational molecules for common two-dimensional operators.

Application to Equilibrium Problems. We now turn to the problem of replacing the differential equation and boundary conditions of a given continuous-equilibrium problem with algebraic equations for a discrete approximation. As soon as the appropriate basic computational molecules are available, the process is exceedingly simple and direct. At each internal point the finite-difference approximation to the governing differential equation provides an algebraic equation connecting the values of ψ at the several neighboring points. Exceptional situations can arise only near the boundaries, where it is possible that not all the neighboring points of a computational molecule will lie within D. It is then necessary to introduce finite-difference approximations to the given boundary conditions and thereby to eliminate the need for a value that lies outside D. The procedure is illustrated in the following examples.

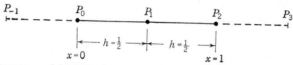

FIG. 4-14. Discrete model for the beam on elastic foundation with one internal point and two auxiliary external points.

One-dimensional Example. Consider Prob. 4-1, the beam on elastic foundation, with the formulation (4-13). We begin with the crudest possible finite-difference approximation; i.e., we replace the continuous domain $0 < x < 1$ by the single point P_1 at $x = \frac{1}{2}$ as shown in Fig. 4-14. To accommodate the span of the computational molecule to be used, two additional points P_{-1} and P_3 are located outside the domain at $x = -\frac{1}{2}$ and $x = \frac{3}{2}$. Our problem now consists in solving a one-degree-of-freedom system for ψ_1. An algebraic equation for ψ_1 can be obtained by combining finite-difference approximations to the differential equation and boundary conditions of (4-13). This may be accomplished in a pictorial manner as shown in Fig. 4-15, where computational molecules constructed with the aid of Fig. 4-12 are placed directly across from the continuous relations which they approximate.

The finite-difference boundary conditions provide us with enough information to fix the boundary values $\psi_0 = \psi_2 = 0$ and to eliminate the additional values ψ_{-1} and ψ_3 since according to the third and fourth lines of Fig. 4-15, $\psi_{-1} = \psi_3 = -\psi_1$. Inserting these values into the fifth line of the figure and setting $h = \frac{1}{2}$ gives

$$-16\psi_1 + 0 + [(6)(16) + 1]\psi_1 + 0 - 16\psi_1 = 1 \qquad (4\text{-}142)$$

from which $\psi_1 = 0.01538$. This constitutes the complete solution. Instead of a continuous deflection curve we get only an approximation to the deflection at the center. The remarkable thing is that, in spite of the crudeness of approximation, the single value that we do get is only 20 per cent larger than the exact solution $\psi(\frac{1}{2}) = 0.01288$. We thus have a simple approximation which possesses no fine-grain detail but does give an order of magnitude result with a minimum of calculation.

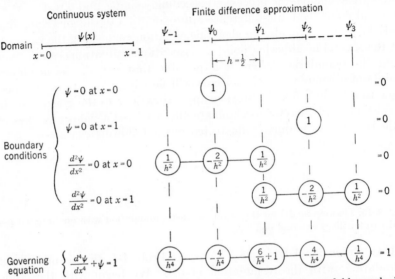

FIG. 4-15. Construction of the governing equations for the discrete model by replacing derivatives in the continuous formulation by finite-difference approximations.

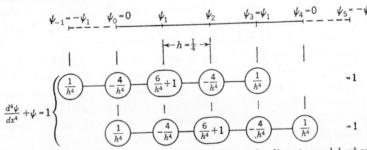

FIG. 4-16. Construction of the governing equations for the discrete model when there are three internal points.

Next consider briefly the same system but with three internal points with spacing $h = \frac{1}{4}$ as shown in Fig. 4-16. Because of symmetry we have $\psi_3 = \psi_1$. The treatment of the boundary points and the additional points outside the boundary is exactly the same as before. The results of applying the boundary conditions are indicated directly on the figure. The molecule corresponding to the differential equation is written twice, centered on P_1 and P_2. These lead to the simultaneous equations

$$-256\psi_1 + 0 + 1{,}537\psi_1 - 1{,}024\psi_2 + 256\psi_1 = 1 \qquad (4\text{-}143)$$
$$0 - 1{,}024\psi_1 + 1{,}537\psi_2 - 1{,}024\psi_1 + 0 = 1$$

which have the following solution:

$$\psi_1 = 0.00965$$
$$\psi_2 = 0.01352 \tag{4-144}$$

This solution provides approximations to ψ at the center and quarter points. Although giving more detail than the previous approximation, this is still a far cry from a continuous solution. The error at the center is now only 5 per cent.

It should now be clear how this procedure can be extended to finer and finer subdivisions. The treatment at the boundaries is always the same, independent of the number of internal points. If there are n independent ψ_j values in the domain, the molecule corresponding to the differential equation is written n times centered on n points P_j corresponding to the independent ψ_j. This provides n simultaneous equations from which the ψ_j can be obtained.

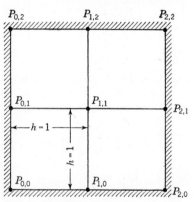

FIG. 4-17. Discrete model for torsion of a square section with only one internal point.

Two-dimensional Example. We next obtain crude finite-difference approximations for Prob. 4-3, torsion of a square section. Consider first the network shown in Fig. 4-17, which has only one internal point, $P_{1,1}$. The boundary condition $\psi = 0$ fixes the value of $\psi_{j,k}$ at all points except at the center. An equation for $\psi_{1,1}$ may be obtained by writing a finite-difference approximation to the governing differential equation (4-29).

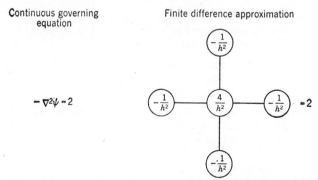

Continuous governing equation

Finite difference approximation

$$-\nabla^2\psi = 2$$

FIG. 4-18. Governing equation for discrete model obtained by replacing Laplacian operator by finite-difference approximation.

Making use of Fig. 4-13, we can represent this pictorially as shown in Fig. 4-18. Now, if we imagine the molecule of Fig. 4-18, with $h = 1$, centered on $P_{1,1}$ of Fig. 4-17, we have, since the boundary values are all zero, simply

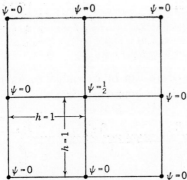

$$4\psi_{1,1} = 2 \qquad (4\text{-}145)$$

FIG. 4-19. Distribution of ψ values according to the finite-difference solution for the model of Fig. 4-17.

as the finite-difference approximation to $-\nabla^2\psi = 2$. The complete finite-difference solution for ψ then appears as in Fig. 4-19. In this problem it is of interest to obtain the torsional rigidity, C, which according to (4-28) is given by twice the integral of ψ over the square. An approximation to C can be obtained from Fig. 4-19 with the aid of the two-dimensional Simpson's rule of Fig. 4-13.

$$C = 2\,\frac{(1)^2}{9}\left\{\begin{array}{l}(1)(0) + \ (4)(0) + (1)(0)\\(4)(0) + (16)(\tfrac{1}{2}) + (4)(0)\\(1)(0) + \ (4)(0) + (1)(0)\end{array}\right\} = \frac{16}{9} = 1.7778 \quad (4\text{-}146)$$

This result is 21 per cent lower than the true value 2.2495.

Consider next a network with $h = \tfrac{1}{2}$. This introduces nine internal points as shown in Fig. 4-20. The boundary condition $\psi = 0$ again fixes the value of $\psi_{j,k}$ at all the boundary points. Nine simultaneous equations for the internal $\psi_{j,k}$ may be obtained by centering the molecule of Fig. 4-18 on each of the internal points. A great saving can be had in this case by noting that because of symmetry we have

$$\begin{array}{l}\psi_{1,1} = \psi_{3,1} = \psi_{3,3} = \psi_{1,3}\\\psi_{2,1} = \psi_{3,2} = \psi_{2,3} = \psi_{1,2}\end{array} \qquad (4\text{-}147)$$

Hence there are actually only three independent $\psi_{j,k}$, which we may take as $\psi_{1,1}$, $\psi_{2,1}$, and $\psi_{2,2}$. Centering the molecule of Fig. 4-18 with $h = \tfrac{1}{2}$ on the points $P_{1,1}$, $P_{2,1}$, and $P_{2,2}$ provides the following three equations:

$$\begin{array}{rcl}16\psi_{1,1} - \ 8\psi_{2,1} & = 2\\-8\psi_{1,1} + 16\psi_{2,1} - \ 4\psi_{2,2} & = 2\\- 16\psi_{2,1} + 16\psi_{2,2} & = 2\end{array} \qquad (4\text{-}148)$$

The reader should take the time to verify these. As an aid the position of the molecule corresponding to the second of (4-148) is indicated in Fig. 4-20. The solution of (4-148) has been obtained and the resulting values of $\psi_{j,k}$ for one-quarter of the total domain are shown in Fig. 4-21. For the torsional rigidity we need the integral of ψ over the domain. Applying the two-dimensional Simpson's rule to Fig. 4-21 (which gives one-quarter of the total integral), we get

$$C = (2)(4)\,\frac{(\tfrac{1}{2})^2}{9}\left\{\begin{array}{l}(1)(0) + (4)(0.43750) \ + (1)(0.56250)\\(4)(0) + (16)(0.34375) + (4)(0.43750)\\(1)(0) + (4)(0) \qquad\quad + (1)(0)\end{array}\right\} = 2.1250 \quad (4\text{-}149)$$

for the torsional rigidity. This is less than 6 per cent too low.

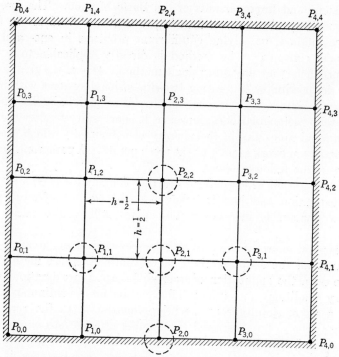

FIG. 4-20. Discrete model for torsion of a square section with nine internal points.

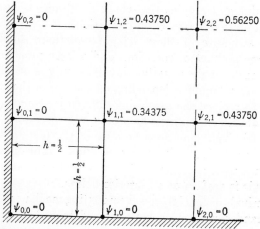

FIG. 4-21. Distribution of ψ values according to the finite-difference solution for the model of Fig. 4-20.

Extensions and Improvements of the Basic Method. The foregoing illustrations have displayed the basic principles involved in the finite-difference method for solving equilibrium problems in one- and two-dimensional continua. The method is directly applicable to a wider class of problems than any other single method. One of the great advantages is the possibility of treating two-dimensional systems with irregularly shaped boundaries. In principle irregular boundaries are no more difficult than rectangular ones; however, it is necessary to alter the basic computational molecules in the vicinity of a boundary which does not cut the square network exactly at the mesh points. A paragraph devoted to this question follows shortly.

Another question which deserves further comment concerns the error of approximation and how it behaves as the number of mesh points within a domain is increased. This point will also be dealt with shortly.

At this point, however, we turn to a question which is of great practical importance for the actual working of the finite-difference method. As we have seen, the application of finite-difference approximations to the boundary conditions and differential equation for a continuous system lead to a set of simultaneous algebraic equations. In the very crude approximations so far illustrated the number of equations obtained was small, but when greater detail and accuracy are required, the number of equations obtained rapidly becomes very large. For instance, it is not unusual to require a square network with 10 points on each side; i.e., one which leads to 100 simultaneous equations. If efficient means for solving such systems of equations are not available, the utility of finite-difference methods remains severely restricted. We therefore consider practical procedures for solving the sets of simultaneous equations which are obtained in finite-difference approximations.

The special structure of these sets of equations should be noted. Although a large number of equations and a large number of variables are involved, each single equation involves only a small number of unknowns, the exact number depending on the size of the computational molecule used to replace the governing differential equation of the continuous system. Furthermore there is a strong family resemblance among all of the single equations because of the fact that they are all approximations to the same governing equation. This is especially true in linear systems with constant coefficients where the identical computational molecule is applied at all interior points.

Iteration methods using high-speed computing machines and relaxation methods for hand computation have seemed to be the most attractive procedures for treating both linear and nonlinear systems.

Iteration. The methods used here are basically the same as those described in Chap. 1. Both iteration by *total steps*[1] and iteration by *single steps*[2] have been employed and their speeds of convergence studied[3] in particular cases. It has been shown[4] that if the original continuous system is positive definite, then the set of finite-difference equations is also positive definite and that therefore[5] iteration by single steps must always converge; however, the convergence is generally so slow that hand computation is out of the question in systems of appreciable size.

We shall here illustrate the method of iteration by single steps as applied to Prob. 4-3, torsion of a square section. The finite-difference approximation to the governing differential equation is given diagrammatically in Fig. 4-18. Written out, this takes the following form:

$$\frac{1}{h^2} \{4\psi_{j,k} - \psi_{j+1,k} - \psi_{j,k+1} - \psi_{j-1,k} - \psi_{j,k-1}\} = 2 \qquad (4\text{-}150)$$

In the iteration process (4-150) is used to define an improved value of $\psi_{j,k}$. The basic single step of iteration consists in replacing the current value of $\psi_{j,k}$ by the improved value

$$\psi'_{j,k} = \frac{h^2}{2} + \frac{1}{4} \{\psi_{j+1,k} + \psi_{j,k+1} + \psi_{j-1,k} + \psi_{j,k-1}\} \qquad (4\text{-}151)$$

The latest improvements for the ψ values on the right-hand side are always employed. To illustrate, we use the network of Fig. 4-20 with $h = \frac{1}{2}$ and nine interior points. For simplicity we shall make no use of symmetry but shall just apply (4-151) systematically at each of the nine internal points, $P_{j,k}$ in sequence, starting in the top row and working across each row from left to right. Beginning with all initial values zero, the iteration was carried out on a machine which rounded off all numbers to four decimal

[1] L. F. Richardson, The Approximate Arithmetical Solution by Finite Differences of Physical Problems Involving Differential Equations with an Application to the Stresses in a Masonry Dam, *Trans. Roy. Soc. (London)*, **A210**, 307–357 (1910). This is the earliest work on the application of iteration to the solution of continuous-equilibrium problems by finite differences. In it a fairly sophisticated technique of iteration by whole steps is presented which has recently received considerable study. See, for example, S. P. Frankel, Convergence Rates of Iterative Treatments of Partial Differential Equations, *Math. Tables and Other Aids to Computation*, **4**, 65–75 (1950).

[2] H. Liebmann, Die angenäherte Ermittelung harmonischer Funktionen und Konformer Abbildung, *Sitzber. math.-physik. Kl. bayer. Akad. Wiss. München*, **3**, 385–416 (1918). Considering the finite-difference approximation to Laplace's equation, Liebmann suggested a mixed method in which the outer ring of points were approximated by a total step of iteration but these new values were used in approximating the next ring of points, etc. In current literature Liebmann's name is now associated with any method of iteration by single steps in which a fixed sequence is followed.

[3] Frankel, *op. cit.* D. Young, Iterative Methods for Solving Partial Difference Equations of Elliptic Type, *Trans. Amer. Math. Soc.*, **76**, 92–111 (1954).

[4] R. Courant, Über Randwertaufgaben bei partieller Differenzengleichungen, *Z. angew. Math. u. Mech.*, **6**, 322-325 (1926) (see also Exercise 4-74).

[5] See Sec. 1-6.

places.   The results are indicated in Fig. 4-22.   The $\psi_{j,k}$ after one cycle of nine single steps is shown together with the distributions after 9 and 13 cycles.   The fourteenth cycle produced no change in any of the $\psi_{j,k}$.

This example illustrates the following result given by Frankel.[1]   When the iteration scheme just described is applied to a reotangular domain, ph by qh, the number of cycles n_c required to obtain convergence to k

0	0	0				0	0	0	
0	0	0	0	0	0	.1250	.1562	.1640	0
0	0	0	0	0	0	.1562	.2031	.2168	0
0	0	0	0	0	0	.1640	.2168	.2334	0
0	0	0				0	0	0	

| Initial distribution | | | | | | After 1 cycle | | | |

0	0	0				0	0	0	
0	.3422	.4360	.3430	0	0	.3437	.4374	.3437	0
0	.4360	.5610	.4368	0	0	.4374	.5624	.4374	0
0	.3430	.4368	.3434	0	0	.3437	.4374	.3437	0
	0	0	0			0	0	0	

| After 9 cycles | | | | | | After 13 cycles | | | |

FIG. 4-22. Results of iteration.   Each cycle consists of nine single steps carried out in sequence, starting at the upper left.

decimal places is approximately

$$n_c = \frac{k/2}{-\log\left[\frac{1}{2}(\cos \pi/p + \cos \pi/q)\right]} \tag{4-152}$$

In Fig. 4-22 we have $p = q = 4$ and $k = 4$, which according to (4-152) would require 13.3 cycles.

For a given rectangle the numbers p and q are inversely proportional to h.   Thus, as h is diminished, the number of internal points increases in proportion to $1/h^2$.   Furthermore it is not difficult to show[2] from (4-152) that the number of cycles of iteration required also increases in proportion to $1/h^2$ so that the total time required increases in proportion to $1/h^4$.

A similar analysis of systems involving the *biharmonic operator* ∇^4 shows that the number of cycles of iteration required increases in proportion to $1/h^4$ so that the total time increases in proportion to $1/h^6$.

The slow convergence of iteration makes the use of high-speed automatic computing machines almost mandatory.   There are two properties of the method which should be balanced against the slow rate of convergence in considering the possibility of machine computation.   These

[1] Frankel, *op. cit.*

[2] See Exercise 4-75.   A procedure in which the number of cycles required increases only as fast as log $(1/h)$ is given by D. W. Peaceman and H. H. Rachford, Jr., *J. Soc. Indust. Appl. Math.*, **3**, 28–41 (1955).

properties are modest storage requirements and simplicity of programming. If there are n internal points, only n values need be retained at any instant during the calculation and the systematic alteration of these values by formulas such as (4-151) is relatively easy to program.

Relaxation. Various schemes have been put forward for accelerating the convergence of iteration. Some of these techniques may be adapted for automatic programming,[1] while others[2] require more in the way of judgment from the computer and are therefore better suited to hand computation. In the latter category the method which has received the most thorough development is *relaxation*.[3] As we have seen in Chap. 1, the principal features of a relaxation method include the introduction of *residuals* and the successive correction of the unknowns which produce changes in the residuals according to an *operations table*. A maximum of judgment is required from the computer since no fixed order of operations is specified. In a sense the computer is left free to devise his own scheme for accelerating the convergence. The relaxation method has proved to be quite powerful for solving equilibrium problems involving the second-order operators d^2/dx^2 and ∇^2. Here most of the known convergence-accelerating devices have been translated into relaxation terminology, and with a little practice a computer soon learns to take advantage of them to such an extent that rates of convergence are obtained which are considerably faster than those possible from fixed routines. More complicated problems such as those involving the operators d^4/dx^4 and ∇^4 may also be treated effectively by a skilled computer, but it must be admitted that the necessary skill is not so easily achieved.

As in other finite-difference methods the relaxation procedure begins with the selection of a finite-difference network. Initial trial values of ψ are assigned to the internal mesh points. The residual at a point is then defined as the amount by which the governing finite-difference molecule centered on that point is not satisfied. Thus, to take Prob. 4-3, torsion of a square section, as an illustration, the governing finite-difference equation (4-150) is used to define the residual $R_{j,k}$ as follows:

$$R_{j,k} = 2h^2 + \{\psi_{j+1,k} + \psi_{j,k+1} + \psi_{j-1,k} + \psi_{j,k-1} - 4\psi_{j,k}\} \quad (4\text{-}153)$$

The ψ values and R values are recorded on a diagram of the network as shown in Fig. 4-23. At each mesh point the corresponding ψ value

---

[1] See Frankel, *op. cit.*, and Peaceman and Rachford, *op. cit.*

[2] For an extended bibliography see T. J. Higgins, A Survey of the Approximate Solution of Two-dimensional Physical Problems by Variational Methods and Finite Difference Procedures, in L. E. Grinter (ed.), "Numerical Methods of Analysis in Engineering," The Macmillan Company, New York, 1949, pp. 169–198.

[3] Relaxation was first applied to the problems of this chapter in 1938. See R. V. Southwell, "Relaxation Methods in Engineering Science," Oxford University Press, New York, 1940, chap. 10, and "Relaxation Methods in Theoretical Physics," Oxford University Press, New York, 1946.

is placed on the left, the R value on the right. In the ensuing computation *increments* to ψ will be written above ψ, while new *values* of R

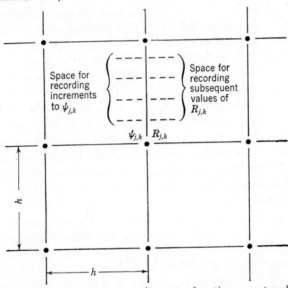

FIG. 4-23. Recording scheme for carrying out relaxation on network diagram.

will be placed above the original value. Next an operations table is required to facilitate the computation of the changes in R values caused by increments to the ψ values. The operations are most conveniently represented in the form of operational molecules. Thus the basic unit operation of adding one unit to $\psi_{j,k}$ causes changes in the residuals at $P_{j,k}$ and the four surrounding points. By centering (4-153) on these five points in turn we find the increments in residuals indicated in Fig. 4-24.

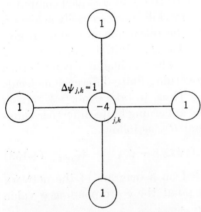

FIG. 4-24. Basic unit operation for the residual (4-153). Unit addition to ψ at the central point results in the circled residual increments.

The simplest relaxation technique would consist in applying suitable multiples of this operation, point by point, throughout the domain in an effort to reduce all residuals to zero. If this single point operation were used to reduce the residual at the point, $P_{j,k}$, exactly to zero, then the operation would be identical with that involved in iteration by single steps. Perhaps the first convergence-accelerating technique one learns is that it does not pay to bring this residual exactly to zero because subsequent operations on neighboring points soon cause a partial return of the residual. For this reason the skilled computer uses a judicious amount of *overrelaxation* or *underrelaxation*. For example, if the residual at a certain point were 75, adding 18.75 to ψ at this point would reduce the residual exactly to zero according to Fig. 4-24; adding 20 to ψ (which would leave a residual at the central point of -5) would be overrelaxing.

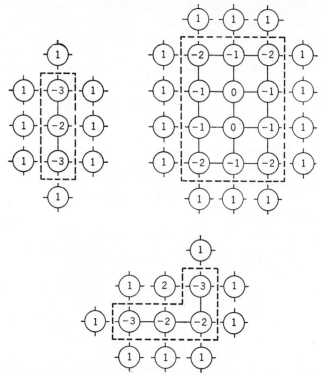

FIG. 4-25. Block relaxation operators. Unit addition to ψ within the dotted subdomains results in the circled residual increments.

Whether overrelaxation or underrelaxation is more appropriate in a particular instance depends on the distribution of residuals throughout the domain and also on the special properties of the system[1] being treated.

In addition to the basic unit operation, *extended operations* which involve simultaneously changing several ψ values are employed. The most common of these is the *block operation*, which gives the same increment to ψ at all the points in a certain subdomain. Several typical examples are shown in Fig. 4-25. In each case all the points within the

[1] Overrelaxation is generally beneficial for systems involving d^2/dx^2 or ∇^2; in fact, it has been shown that, if iteration by single steps is modified so that each step corresponds to a certain fixed amount of overrelaxation, the number of cycles of iteration required for a given accuracy is proportional to $1/h$ as against $1/h^2$ for the ordinary procedure (see Frankel, *op. cit.*).

dotted block have had their ψ values increased by unity. The circled numbers are the increments to the residuals at the corresponding points. These block operations are derived by superposing the basic unit operations of Fig. 4-24. The reader should try constructing a few of his own. Note that a point outside the block has a residual increment of $+1$ for every one of its neighbors which is inside the block and that a point inside the block has a residual increment of -1 for every one of its neighbors which is outside the block. In each operation the total of positive increments has the same magnitude as the total of negative increments. These operations applied in the interior of a network can thus effect only a redistribution of individual residuals but cannot change the algebraic sum of all the residuals. The residual total *is altered*, however, when one of the circled points is a boundary point.

The actual relaxation should be performed on a sheet of paper which will stand considerable erasure because the spaces left for recording the work in Fig. 4-23 are seldom adequate for the entire computation. A good grade of tracing paper is recommended. The network can then be drawn in once and for all on the reverse side of the paper and will not be obliterated by the erasures. Other suggestions are included in the following illustrative example.

Example. We return to Prob. 4-3, torsion of a square section, and consider a network with $h = \frac{1}{4}$ as shown in Fig. 4-26. There are 49 internal points in the whole domain, but because of the eightfold symmetry of the square we can restrict our attention to the 10 points in the shaded triangle. We have only to imagine that any operation applied at a point such as $P_{3,2}$ is simultaneously accompanied by the same operation at the 7 image points shown in the figure. Our next step is to select an initial trial set of ψ values. It is advisable to obtain a preliminary solution on a coarser grid to provide the order of magnitude and general configuration of the expected solution. Suppose the solution of Fig. 4-21 for $h = \frac{1}{2}$ is available. We enter these ψ values (rounded off to two places) on the network of Fig. 4-27 as indicated. The remaining ψ values are written in by roughly interpolating between these

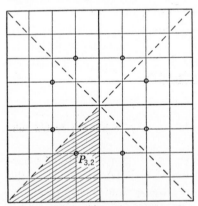

FIG. 4-26. Network for relaxation solution of torsion of a square section.

values and the given boundary values. It is understood that the values recorded here apply to all symmetrical image points in the whole domain of Fig. 4-26. Next we calculate the residuals at the 10 points of Fig. 4-27, using (4-153), which for $h = \frac{1}{4}$ becomes

$$R_{j,k} = 0.125 + \psi_{j+1,k} + \psi_{j,k+1} + \psi_{j-1,k} + \psi_{j,k-1} - 4\psi_{j,k} \qquad (4\text{-}154)$$

Care must be taken to use the correct image-point values in computing the residuals for points on the border of the triangle. The R values are recorded to the right of each point. Note that initially the largest residual is 0.025 and that positive residuals predominate. A final preliminary step, taken to avoid decimals, is to scale up both ψ and R values by a factor of 1,000. This has been done in Fig. 4-28.

The actual relaxation might begin with the block of 20 units shown in Fig. 4-28. The increments (+20) to the ψ values are recorded above the initial ψ values. The increments to the residuals are mentally added to the initial residuals, and the actual new residuals are recorded above the old ones. In such a block operation the only points whose residuals are affected are those just inside and just outside of the block. Thus the outside points $P_{2,1}$, $P_{3,1}$, and $P_{4,1}$ all receive an increment $+20$, while the

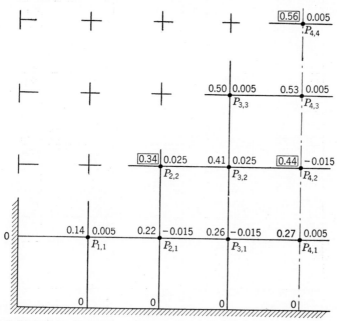

FIG. 4-27. Initial trial values of ψ. Values from coarse-grid solution are boxed. Other values are interpolated.

inside points $P_{3,2}$ and $P_{4,2}$ receive an increment -20. The inside corner point $P_{2,2}$ gets the double increment -40. The next operation might be an addition of five to all interior ψ values, i.e., the second block shown in Fig. 4-28. The serious reader would do well to take up a pencil and carry out this operation. Note that, since no residuals are carried at the boundary points, the only residuals affected by this operation are those in the outer ring of points, that is, $R_{1,1}$, $R_{2,1}$, $R_{3,1}$, and $R_{4,1}$. The third step might well consist of a single point operation at $P_{4,2}$. The reader is encouraged to carry on from here.

The actual relaxation is not self-checking. Therefore it is suggested that when the working has progressed to the point where the largest residual is about 10 per cent of what was initially the largest residual the relaxation should be halted, the increments to ψ should be totaled to find the present ψ values, and the residuals should be recomputed from these with the aid of (4-154). If no mistakes have been made, these new residuals will check with those obtained in the working. If there is a slight discrepancy, nothing is lost and the work proceeds anew, treating the new set of ψ values as a new initial trial.   To avoid fractions, it is convenient to scale up both ψ and R values by a factor of 10 each time the residuals are decimated.

Continuing with the example, it was possible in 30 operations[1] to obtain the distribution of ψ and R values shown in Fig. 4-29.   The largest residual here represents less than one-thousandth of the largest initial residual.   Positive and negative residuals are essentially counterbalanced.   At three stages during the computation the residuals were recomputed with a desk calculator.   Only mental arithmetic was used in the

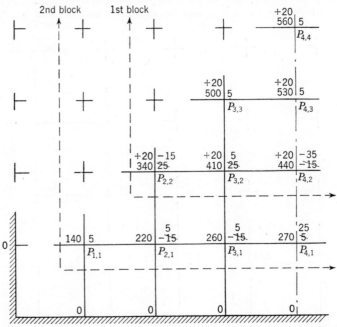

FIG. 4-28. The first step of relaxation consists of a block operation of 20 units.

actual relaxation.   The total computing time was nearly equally divided between checking the residuals and performing the relaxation.

By applying Simpson's rule to this distribution we find[2] $C = 2.2182$ for the torsional rigidity, which is 1.4 per cent too low.

Irregular Boundaries.   In the cases previously considered it has been possible to fill the continuous domain D with a uniformly spaced network in such a way that the boundary of D intersected the network only at mesh points.   For each independent ψ value in the network a finite-difference molecule approximating the continuous governing differential equation was applied.   For points near the boundary it was sometimes

---

[1] If iteration by single steps had been employed without making allowances for the symmetry, 57 cycles (each of 49 single steps) would have been required to obtain convergence to four decimal places.

[2] See Exercise 4-76.

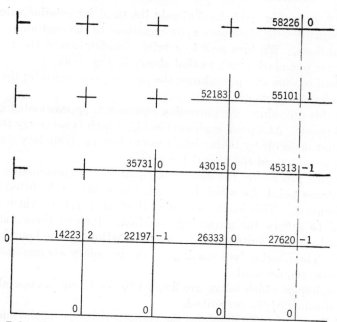

FIG. 4-29. Relaxation solution of torsion problem.   To avoid decimals, a scale factor of 10^5 has been introduced.

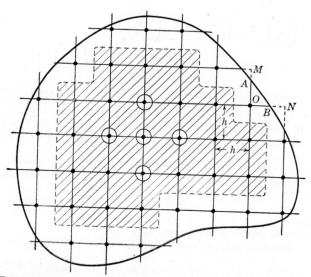

FIG. 4-30. Uniform grid applied to irregular domain.   Standard molecule can be centered at each point of shaded interior region.

necessary, as in Fig. 4-15, to eliminate the need for exterior values of ψ by introducing finite-difference approximations to the continuous boundary conditions. We turn now to a brief consideration of the treatment of irregular boundaries such as that shown in Fig. 4-30.

We shall outline two procedures: the first is more accurate; the second is somewhat simpler.

1. In this procedure[1] the governing equation is approximated at every internal point. At a point such as O in Fig. 4-30 it is necessary to modify the regular molecule by taking into account *both* the boundary conditions and the nonstandard spacings OA and OB.

2. In this procedure[2] the governing equation is approximated only at those interior points for which the regular molecule can be fitted without any alteration. This leaves an outer ring[3] of ψ values which are not required to satisfy the governing equation. Instead these values are required to satisfy finite-difference approximations to the boundary conditions. The nonstandard spacings of the boundary are incorporated in these latter requirements.

The examples which follow are limited to the Laplacian operator with ψ prescribed or $\partial\psi/\partial n$ prescribed.

First Procedure. To illustrate the construction of irregular molecules, we shall obtain the molecule corresponding to $\nabla^2\psi$ (see Fig. 4-18) which is to be applied at the point O in Fig. 4-31. In the first case suppose that the boundary values of ψ are prescribed. Then ψ_A and ψ_B are known. We assume that ψ can be expanded in the two-dimensional Taylor's series

$$\psi = \psi_0 + x\left(\frac{\partial\psi}{\partial x}\right)_0 + y\left(\frac{\partial\psi}{\partial y}\right)_0 + \frac{x^2}{2}\left(\frac{\partial^2\psi}{\partial x^2}\right)_0 + xy\left(\frac{\partial^2\psi}{\partial x\partial y}\right)_0 + \frac{y^2}{2}\left(\frac{\partial^2\psi}{\partial y^2}\right)_0 + \cdots \quad (4\text{-}155)$$

where the origin of coordinates has been temporarily shifted to point O. The points A, B, C, and D with coordinates $(0, \theta_A h)$, $(\theta_B h, 0)$, $(0, -h)$, and $(-h, 0)$, respectively, when substituted into (4-155) provide four equations for the first and second derivatives of ψ at O. When these

[1] The method was given by Sh. Mikeladze, Numerische Integration der Gleichungen vom elliptischen und parabolischen Typus (in Russian with German summary), *Izvest. Akad. Nauk S.S.S.R.*, Ser. Matem., **5**, 57–74 (1941).

[2] See S. Gerschgorin, Fehlerabschätzung für das Differenzenverfahren zur Lösung partiellen Differentialgleichungen, *Z. angew. Math. u. Mech.*, **10**, 373–382 (1930), and the extension by L. Collatz, Bemerkungen zur Fehlerabschätzung für das Differenzenverfahren bei partiellen Differentialgleichungen, *Z. angew. Math. u. Mech.*, **13**, 55–57 (1933).

[3] In Fig. 4-30 this ring is made up of points such as O which lie *inside* the boundary. Alternatively auxiliary points such as M and N can be established *outside* the boundary.

are solved simultaneously for the second derivatives and their values inserted in $\nabla^2\psi$, we find

$$(\nabla^2\psi)_O = \frac{2}{h^2}\left[\frac{\psi_A}{\theta_A(1+\theta_A)} + \frac{\psi_B}{\theta_B(1+\theta_B)}\right.$$
$$\left. + \frac{\psi_C}{1+\theta_A} + \frac{\psi_D}{1+\theta_B} - \frac{(\theta_A+\theta_B)\psi_O}{\theta_A\theta_B}\right] + 0(h) \quad (4\text{-}156)$$

which may be represented pictorially as shown in Fig. 4-32.

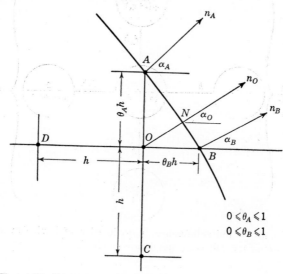

FIG. 4-31. Geometry of a typical point near the boundary.

As a second case we briefly indicate how this same point would be treated if the normal derivative $\partial\psi/\partial n$ were prescribed on the boundary. The derivatives of ψ would then be known along the three normals n_A, n_B, and n_O which make the angles α_A, α_B, and α_O as shown in Fig. 4-31. First we note from the geometry of the normal that

$$\frac{\partial\psi}{\partial n} = \frac{\partial\psi}{\partial x}\cos\alpha + \frac{\partial\psi}{\partial y}\sin\alpha \quad (4\text{-}157)$$

Next by differentiation of (4-155) we have

$$\frac{\partial\psi}{\partial x} = \left(\frac{\partial\psi}{\partial x}\right)_0 + x\left(\frac{\partial^2\psi}{\partial x^2}\right)_0 + y\left(\frac{\partial^2\psi}{\partial x\,\partial y}\right)_0 + \cdots$$
$$\frac{\partial\psi}{\partial y} = \left(\frac{\partial\psi}{\partial y}\right)_0 + x\left(\frac{\partial^2\psi}{\partial x\,\partial y}\right)_0 + y\left(\frac{\partial^2\psi}{\partial y^2}\right)_0 + \cdots \quad (4\text{-}158)$$

Using these, we can express the normal derivatives at A, B, and N in terms of the first and second derivatives of ψ at O and geometrical quantities available from Fig. 4-31. Thus for point A we have

$$\left(\frac{\partial\psi}{\partial n}\right)_A = \left[\left(\frac{\partial\psi}{\partial x}\right)_0 + \theta_A h\left(\frac{\partial^2\psi}{\partial x\,\partial y}\right)_0\right]\cos\alpha_A$$
$$+\left[\left(\frac{\partial\psi}{\partial y}\right)_0 + \theta_A h\left(\frac{\partial^2\psi}{\partial y^2}\right)_0\right]\sin\alpha_A + \cdots \quad (4\text{-}159)$$

If we join to this the two similar expressions for $(\partial\psi/\partial n)_B$ and $(\partial\psi/\partial n)_N$ plus the Taylor's series for ψ_C and ψ_D, we have a system of five simultaneous equations for the

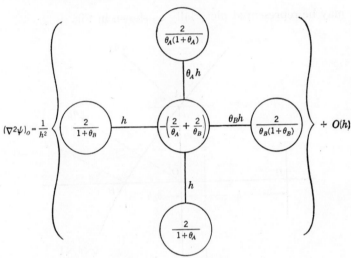

FIG. 4-32. Molecule for finite-difference approximation of Laplacian at point O of Fig. 4-31 when boundary values ψ_A and ψ_B are specified.

first and second derivatives of ψ at O. These can then be solved for the second derivatives and the Laplacian formed. While this is simple in principle, it will be recognized that a considerable amount of work is involved in repeating this process at every point which is within distance h of the boundary.

Second Procedure. We consider the same point O shown again in Fig. 4-33. In this procedure the Laplacian molecule is centered on points such as D and E but not at O. At O we use instead a finite-difference approximation to the boundary condition. If ψ is specified on the boundary, we know ψ_B. Then by using linear interpolation from B to D we obtain

$$\psi_O = \frac{\theta_B}{1+\theta_B}\psi_D + \frac{1}{1+\theta_B}\psi_B \quad (4\text{-}160)$$

as the relation to be satisfied at point O. We could equally well obtain ψ_O by linear interpolation from A to C; or we could use the average of the two interpolations.

If $\partial\psi/\partial n$ is specified on the boundary, we know $(\partial\psi/\partial n)_N$. This can be approximated by the finite difference

$$\frac{\psi_O - \psi_F}{OF} \tag{4-161}$$

The value of ψ_F can be approximated by linear interpolation between D and E, and the various distances can be expressed in terms of h and the angle α. In this way we find

$$\psi_O = \psi_D(1 - \tan \alpha) + \psi_E \tan \alpha + h \left(\frac{\partial\psi}{\partial n}\right)_N \sec \alpha \tag{4-162}$$

as the relation to be satisfied at point O.

Introduction to Error Analysis. The analysis of the errors in finite-difference approximations for equilibrium problems is similar to the

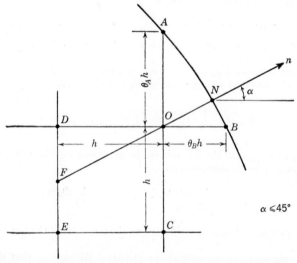

Fig. 4-33. Geometry of a typical point near the boundary.

analysis for propagation problems outlined in Secs. 3-8 and 3-9. As before, there are two distinct kinds of error. There is the *discretization error* due to the basic approximation of replacing the continuous problem by the discrete model, and there is an additional error whenever the discrete equations are not solved exactly. This latter error is indicated by the presence of *residuals* when the relaxation method is used. It is also present in machine iteration solutions since the iteration is only continued until there is no change out to a certain number of decimal places. The subsequent changes which would occur if the iteration were continued without any limitation on the number of decimal places can be considered as *round-off errors*. The size of the finite-difference increment h affects these two kinds of errors in opposite sense. The discretization error decreases as h is made smaller, but the round-off error generally increases.

As an introduction to the type of analysis required, we shall consider in detail Prob. 4-1, the beam on elastic foundation, for which we have already obtained numerical solutions with $h = \frac{1}{2}$ and $h = \frac{1}{4}$ in Figs. 4-15 and 4-16. The formulation of the continuous problem for $\psi(x)$ is

$$\frac{d^4\psi}{dx^4} + \psi = 1 \qquad 0 < x < 1 \qquad \begin{aligned} \psi &= 0 & \begin{cases} x = 0 \\ x = 1 \end{cases} \\ \frac{d^2\psi}{dx^2} &= 0 & \begin{cases} x = 0 \\ x = 1 \end{cases} \end{aligned} \qquad (4\text{-}163)$$

Let the unit interval be divided into S subdivisions so that $h = 1/S$. Then the formulation for the discrete function ψ_j is

$$\psi_{j-2} - 4\psi_{j-1} + 6\psi_j - 4\psi_{j+1} + \psi_{j+2} + h^4\psi_j = h^4 \qquad j = 1, \ldots, S-1 \qquad (4\text{-}164)$$

$$\psi_0 = \psi_S = 0 \qquad \psi_{-1} - 2\psi_0 + \psi_1 = \psi_{S-1} - 2\psi_S + \psi_{S+1} = 0$$

The continuous solution $\psi(x)$ does not satisfy (4-164). If we substitute $\psi(jh)$ for ψ_j in (4-164) we find[1] that we must add the *equation truncation error*

$$E_{t,j} = -\frac{h^6}{6}\frac{d^6\psi(\xi_j)}{dx^6} \qquad j = 1, \ldots, S-1 \qquad (4\text{-}165)$$

where $x_{j-2} \le \xi_j \le x_{j+2}$, to the left side of the governing equation and we must add the *boundary-condition truncation errors*

$$E_{t,0} = -\frac{h^4}{12}\frac{d^4\psi(\xi_0)}{dx^4} \qquad E_{t,S} = -\frac{h^4}{12}\frac{d^4\psi(\xi_S)}{dx^4} \qquad (4\text{-}166)$$

where $x_{-1} \le \xi_0 \le x_1$ and $x_{S-1} \le \xi_S \le x_{S+1}$, to the left sides of the last two boundary conditions.

Let V_j be an approximate solution to (4-164). We assume that the boundary conditions have been met exactly but that there are residuals $E_{r,j}$ in the governing equations; that is, V_j satisfies

$$V_{j-2} - 4V_{j-1} + 6V_j - 4V_{j+1} + V_{j+2} + h^4V_j = h^4 + E_{r,j}$$
$$j = 1, \ldots, S-1 \qquad (4\text{-}167)$$
$$V_0 = V_S = 0 \qquad V_{-1} - 2V_0 + V_1 = V_{S-1} - 2V_S + V_{S+1} = 0$$

We are interested in the behavior of the solution error

$$\epsilon_j = V_j - \psi(jh) \qquad (4\text{-}168)$$

To obtain the equations governing ϵ_j, we subtract from (4-167) the corresponding relations for the continuous solution (which contain the truncation terms given above) and introduce (4-168). In this way we get

[1] See Exercise 3-46 and Eq. (3-85).

$$\epsilon_{j-2} - 4\epsilon_{j-1} + 6\epsilon_j - 4\epsilon_{j+1} + \epsilon_{j+2} + h^4\epsilon_j = E_{t,j} + E_{r,j}$$

$$j = 1, \ldots, S - 1 \qquad (4\text{-}169)$$

$$\epsilon_0 = \epsilon_S = 0 \qquad \begin{cases} \epsilon_{-1} - 2\epsilon_0 + \epsilon_1 = E_{t,0} \\ \epsilon_{S-1} - 2\epsilon_S + \epsilon_{S+1} = E_{t,S} \end{cases}$$

Notice that this formulation defines a discrete *equilibrium problem* for the error ϵ_j. The system is the same discrete model of the beam on elastic foundation as in (4-164). The only difference is in the loading. Instead of the uniform loading [represented by the term h^4 on the right of the governing equation in (4-164)] we have a loading along the beam proportional to the sum of the truncation and residual errors, and in addition there are end loadings corresponding to applied moments proportional to the boundary truncation errors.

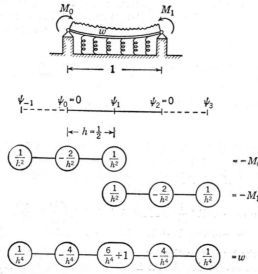

Fig. 4-34. Crude model of a beam on elastic foundation subjected to distributed load and to end bending moments. This is used to estimate the errors in Prob. 4-1.

In principle (4-169) could be solved explicitly to give ϵ_j in terms of $E_{t,j}$ and $E_{r,j}$, which would provide a complete picture of the behavior of the error. In practice only rough estimates of the error are usually required, and these can be obtained quite simply from (4-169). What is needed is some quantitative estimate of the deflection resulting from a loading. This can be got from either the continuous system or the discrete model, whichever is more convenient. We shall use the crude discrete model of Fig. 4-34 with only one interior point to obtain these data. The subsequent behavior of the error as h is diminished can be argued from (4-169) by observing the way in which h enters into the various terms.

Therefore we consider the system shown in Fig. 4-34. It is a model of a beam subjected to a dimensionless distributed load w and dimensionless terminal bending moments M_0 and M_1. The solution for ψ_1 is easily found to be

$$\psi_1 = 0.0154w + 0.0615(M_0 + M_1) \qquad (4\text{-}170)$$

Returning to (4-169), we now estimate the truncation errors $E_{t,j}$. According to (4-165) and (4-166) we need the fourth and sixth derivatives of the (unknown) continuous solution. However, from (4-163) we have

$$\frac{d^4\psi}{dx^4} = 1 - \psi$$
$$\frac{d^6\psi}{dx^6} = -\frac{d^2\psi}{dx^2}$$

(4-171)

so that

$$E_{t,0} = E_{t,s} \approx -\frac{h^4}{12}$$
$$E_{t,j} = \frac{h^6}{6}\frac{d^2\psi(\xi_j)}{dx^2}$$

(4-172)

The order of magnitude of the second derivative can be estimated from any of our earlier continuous approximations [e.g., (4-103) or (4-110)], or, even simpler, we can apply the finite-difference approximation for a second derivative to our discrete solutions of Figs. 4-15 and 4-16. Thus from Fig. 4-15 we find

$$\left(\frac{d^2\psi}{dx^2}\right)_1 \approx \frac{\psi_0 - 2\psi_1 + \psi_2}{h^2} = \frac{0 - (2)(0.01538) + 0}{(\frac{1}{2})^2} = -0.123 \qquad (4\text{-}173)$$

If we use this estimate at every station, we have

$$E_{t,j} \approx -0.0205h^6 \qquad (4\text{-}174)$$

Next we compare (4-169) with Fig. 4-34. We note that $(1/h^4)(E_{t,j} + E_{r,j})$ plays the role of w and that $(1/h^2)E_{t,0}$ and $(1/h^2)E_{t,s}$ play the roles of M_0 and M_1. If we consider the discretization error only (i.e., set $E_{r,j} = 0$ temporarily) and put the estimates (4-172) and (4-174) into (4-170), we find

$$\epsilon_t \approx -0.0154\frac{1}{h^4}0.0205h^6 + 0.0615\frac{1}{h^2}\left(\frac{h^4}{12} + \frac{h^4}{12}\right)$$
$$= -0.00032h^2 + 0.01025h^2 = 0.00993h^2$$

(4-175)

as an estimate of the total discretization error. Note that the contribution from the boundary condition overshadows[1] that from the governing equation. If we apply (4-175) to the cases $h = \frac{1}{2}$ and $h = \frac{1}{4}$, we get $\epsilon_t = 0.00244$ and 0.00062. Now these cases have already been worked in Figs. 4-15 and 4-16 and the actual errors at the mid-point of the beam can be obtained by comparison with the exact solution. We find the values $\epsilon_t = 0.00250$ and 0.00064, which are in surprisingly good agreement with our rough estimates.

When we consider the effect of the residuals $E_{r,j}$, we cannot reach such definite conclusions. We can, however, estimate the *maximum possible* round-off error when $|E_{r,j}| \leq E_r$ and also the *standard deviation* of the round-off error when the $E_{r,j}$ take any value between $-E_r$ and $+E_r$ with equal probability. Thus if, for instance,[2]

$$|E_{r,j}| \leq 0.5 \times 10^{-k} \qquad (4\text{-}176)$$

then the maximum round-off error would occur when all the $E_{r,j}$ had the same sign and each was exactly equal to the limiting value. Using the model of Fig. 4-34 and

---

[1] See Exercise 4-77 for a procedure which takes advantage of this fact.

[2] This would mean no residual to k decimal places. If the equations were solved using iteration by single steps until there was no further change out to k decimal places, then the residuals might be six times as large as (4-176) (see Exercise 4-78).

letting $(1/h^4)E_{r,j}$ play the role of w (with $M_0 = M_1 = 0$), we obtain

$$\text{Max } |\epsilon_r| = 0.0154 \frac{1}{h^4} 0.5 \times 10^{-k}$$
$$= \frac{0.0077}{h^4} \times 10^{-k} \tag{4-177}$$

from (4-170) as the maximum contribution to the error possible from residuals satisfying (4-176). Note how this grows with an increasing number of subdivisions. In fact, if the contribution due to round-off is not to increase as we go to a larger number of subdivisions, S, we must increase the number of decimal places, k, to which we work so as to satisfy

$$k = 4 \log_{10} S + \log_{10} \frac{0.0077}{\text{max } |\epsilon_r|} \tag{4-178}$$

When the above analysis is repeated for the probable round-off error when $E_{r,j}$ is a random value within the limits of (4-176), it can be shown[1] that the standard deviation of ϵ_r satisfies a relation having the form of (4-177) except that h^4 in the denominator is replaced by $h^{\frac{5}{2}}$; i.e., the probable value of ϵ_r grows somewhat more slowly than the maximum possible value as the number of subdivisions is increased.

The foregoing analysis applies only to a particular problem. The procedures followed and the conclusions obtained can, however, be readily extended to any equilibrium problem in one or two dimensions. For such extensions the reader is referred to the texts of Collatz[2] and Milne.[3]

h^2 **Extrapolation.** The preceding analysis has shown that, when the molecules of Fig. 4-12 which have $O(h^2)$ error are used to approximate the governing equation and boundary conditions, the total solution error is also $O(h^2)$ provided that the round-off or residual contribution is negligible and that certain higher derivatives of the solution are well behaved. If all these assumptions are justified, we can obtain an improved approximation from two or more finite-difference solutions by h^2 extrapolation [see (3-125)].

For example, the central deflection of the beam on elastic foundation just discussed was found to be 0.01538 for $h = \frac{1}{2}$ and 0.01352 for $h = \frac{1}{4}$. Applying h^2 extrapolation to these values, we get

$$\frac{(\frac{1}{2})^2(0.01352) - (\frac{1}{4})^2(0.01538)}{(\frac{1}{2})^2 - (\frac{1}{4})^2} = 0.01290 \tag{4-179}$$

which differs by only 0.00002 from the central deflection of the exact continuous solution.

Another illustration is provided by the torsional stiffness C in Prob. 4-3. From networks with $h = 1, \frac{1}{2},$ and $\frac{1}{4}$ (see Figs. 4-19, 4-21, and 4-29)

[1] See Exercises 4-79 and 4-80.
[2] L. Collatz, "Numerische Behandlung von Differentialgleichungen," Springer-Verlag OHG, Berlin, 1951, pp. 120, 284.
[3] Milne, *op. cit.*, p. 216.

we have previously obtained the stiffness values 1.7778, 2.1250, and 2.2182. These values have been plotted against h^2 in Fig. 4-35. Note that the curve joining these values with the stiffness of the exact continuous solution is very nearly a straight line. This constitutes a graphic[1] proof of the validity of h^2 extrapolation in this case.

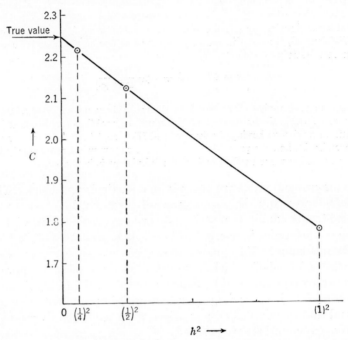

FIG. 4-35. Illustrating h^2 extrapolation for the torsional stiffness C of a square section.

Further Discussion of Errors. The study of errors in finite-difference approximations is still in a state of active development. The early mathematical convergence proofs[2] are being reexamined[3] from the practical viewpoint of numerical analysis.

Some interesting results concerning h^2 extrapolation have recently been obtained. For Laplace's equation in a rectangle with boundary values

[1] For quantitative results see Exercise 4-81.

[2] J. LeRoux, Sur le problème de Dirichlet, *J. math. pures et appl.*, [6]**10**, 189–230 (1914); H. B. Phillips and N. Wiener, Nets and the Dirichlet Problem, *J. Math. and Phys.*, **2**, 105–124 (1923); R. Courant, K. Friedrichs, and H. Lewy, Über die partieller Differenzengleichungen der mathematischen Physik, *Math. Ann.*, **100**, 32–74 (1928).

[3] See, for example, J. L. Walsh and D. Young, On the Degree of Convergence of Solutions of Difference Equations to the Solution of the Dirichlet Problem, *J. Math. and Phys.*, **33**, 80–93 (1954); W. R. Wasow, Discrete Approximations to Elliptic Differential Equations, *Z. angew. Math. u. Phys.*, **6**, 81–97 (1955).

specified on the periphery Walsh and Young[3] have shown that the discretization error is not necessarily $O(h^2)$. They show that the behavior of the discretization error is intimately bound up with the smoothness of the prescribed boundary values. If the boundary values are reasonably well behaved (e.g., there may even be finite discontinuities provided the jumps occur at mesh points), then the discretization error is actually $O(h^2)$. However, they exhibit a case where the boundary values are continuous (but otherwise quite pathological) and for which the discretization error approaches zero *arbitrarily slowly* with h.

There are some recent results concerning irregular two-dimensional boundaries. Wasow[1] has shown that, for well-behaved cases in which the boundary values are specified, h^2 extrapolation will be successful when the first procedure of page 262 is employed but that h^2 extrapolation will not in general be successful when the second procedure is used, For the case where the normal derivative is specified at the boundary. Batschelet[2] has shown that in the second procedure of page 262 part of the discretization error is $O(h)$. It might be expected that in the first procedure the entire discretization error would be $O(h^2)$ and that h^2 extrapolation would be successful, but this is still an open question.

Another question concerns the relative economy of computation and accuracy of $O(h^2)$ procedures using small h as compared with higher-order procedures[3] utilizing larger values of h. This question is not so clear-cut for equilibrium problems as it is for propagation problems. Arguments for both approaches can be given,[4] but a comprehensive judgment has not yet been given.

EXERCISES

4-63. Obtain a finite-difference solution to the problem of the taut elastic cable formulated in Exercise 4-9b by subdividing the length of the cable into four intervals. Verify that the finite-difference solution gives the same deflections as the continuous solution. Why is there no discretization error?

---

[1] *Op. cit.*

[2] E. Batschelet, Über die numerische Auflösung von Randwert problemen bei elliptischen partiellen Differentialgleichungen, *Z. angew. Math. u. Phys.*, **3**, 165–193 (1952).

[3] These have been proposed by L. Collatz, Das Differenzenverfahren mit höherer Approximation für lineare Differentialgleichungen, *Schr. Math. Sem. Inst. Angew. Math. Univ. Berlin*, **3**, 1–34 (1935), and L. Fox, Some Improvements in the Use of Relaxation Methods for the Solution of Ordinary and Partial Differential Equations, *Proc. Roy. Soc. (London)*, **A190**, 31–59 (1947).

[4] See R. V. Southwell, The Quest for Accuracy in Computations Using Finite Differences, in L. E. Grinter (ed.), "Numerical Methods of Analysis in Engineering," The Macmillan Company, New York, 1949, pp. 66–74, and L. Fox, The Use of Large Intervals in Finite-difference Equations, *Math. Tables and Other Aids to Computation*, **7**, 14–18 (1953).

4-64. Carry the solutions of Figs. 4-15 and 4-16 to Prob. 4-1 one step further by obtaining the solution for $h = \frac{1}{8}$.

$Ans.$ $\psi_1 = 0.005074$, $\psi_2 = 0.009302$, $\psi_3 = 0.012080$, $\psi_4 = 0.013046$.

4-65. Apply h^2 extrapolation to (4-144) and the results of Exercise 4-64 to obtain an improved value for the central deflection. $Ans.$ 0.01288.

4-66. The dimensionless formulation for the deflection of the cantilever beam on elastic foundation of Exercise 4-1 is

$$\frac{d^4y}{dx^4} + 4y = 0 \qquad 0 < x < 1$$

$$\left.\begin{aligned} y &= 0 \\ \frac{dy}{dx} &= 0 \end{aligned}\right\} \quad \text{at } x = 0$$

$$\left.\begin{aligned} \frac{d^2y}{dx^2} &= 0 \\ \frac{d^3y}{dx^3} &= -1 \end{aligned}\right\} \quad \text{at } x = 1$$

Set up finite-difference approximations to the governing equation and boundary conditions for a general increment size h. Actually solve the equations for the end deflection in the cases $h = 1$ and $h = \frac{1}{2}$.

$Ans.$ 0.2500 and 0.2577.

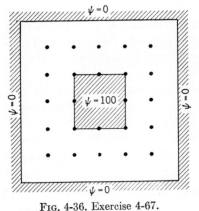

4-67. Obtain approximate solutions to Laplace's equation at the mesh points shown in Fig. 4-36. Make use of symmetry.

$Ans.$ 20.83, 41.67, and 45.83.

4-68. Resolve Exercise 4-67 with a network twice as fine. Use relaxation to solve the finite-difference equations.

$Ans.$ 20.21, 39.30, and 45.16 at the locations for which the answers are given in Exercise 4-67.

4-69. Set up a finite-difference procedure for solving Exercise 4-3 for a general network spacing h. How is the line integral

Fig. 4-36. Exercise 4-67.

to be approximated? Obtain approximations for the capacitance from networks with $h = a$, $a/2$, and $a/4$. $Ans.$ $C = 0.954$, 0.862, and 0.838.

4-70. The formulation of the equilibrium problem of a hinged square plate subjected to a uniform transverse load (see Exercise 4-6) is shown in Fig. 4-37 in terms of dimensionless variables. Set up finite-difference approximations to the equation and boundary conditions for a general network spacing h. Make use of symmetry. Actually solve for the central deflection for the cases $h = 1$, $\frac{1}{2}$, and $\frac{1}{4}$.

$Ans.$ 0.0625, 0.0645, and 0.0647.

4-71. Expand $\psi(x, y)$ in a two-dimensional Taylor's series. (a) Verify that the truncation error for the $(\nabla^2\psi)_{j,k}$ molecule of Fig. 4-13 is

$$-\frac{h^2}{12}\left(\frac{\partial^4\psi}{\partial x^4} + \frac{\partial^4\psi}{\partial y^4}\right)_{j,k} - \frac{h^4}{360}\left(\frac{\partial^6\psi}{\partial x^6} + \frac{\partial^6\psi}{\partial y^6}\right)_{j,k} - \cdots$$

(b) Verify that the truncation error for the $(\nabla^4\psi)_{j,k}$ molecule of Fig. 4-13 is

$$-\frac{h^2}{6}\left(\frac{\partial^6\psi}{\partial x^6} + \frac{\partial^6\psi}{\partial x^4\,\partial y^2} + \frac{\partial^6\psi}{\partial x^2\,\partial y^4} + \frac{\partial^6\psi}{\partial y^6}\right) - \cdots$$

4-72. Verify the computational molecule shown in Fig. 4-38.

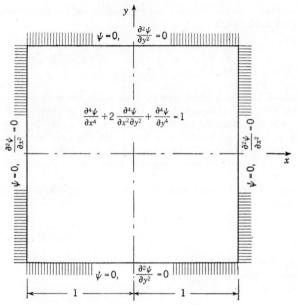

Fig. 4-37. Exercise 4-70.

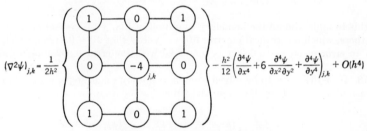

Fig. 4-38. Exercise 4-72.

4-73. Verify the computational molecule shown in Fig. 4-39. Note that this can be obtained from a superposition of the molecules of Exercises 4-71a and 4-72. Show that, if this is applied to the equation $\nabla^2\psi = $ constant, then the truncation error is $O(h^4)$.

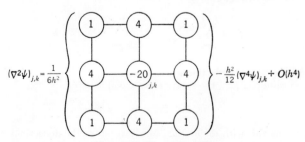

Fig. 4-39. Exercise 4-73.

4-74. Show that a finite-difference approximation for the dimensionless potential energy (4-61) of Prob. 4-1 is

$$\Phi = h \sum_{j=1}^{S-1} \left\{ \frac{1}{2} \psi_j{}^2 + \frac{1}{2h^4} (\psi_{j-1} - 2\psi_j + \psi_{j+1})^2 - \psi_j \right\}$$

with the *essential* boundary conditions $\psi_0 = \psi_S = 0$. Note that the quadratic part of Φ is positive definite. Minimize Φ with respect to ψ_j, and verify that the governing equation of (4-164) is obtained for $2 \leq j \leq S - 2$. Consider $j = 1$ separately, and note that the minimization condition includes both the governing equation and the *additional* boundary condition. The system (4-164) (when the auxiliary values ψ_{-1} and ψ_{S+1} are eliminated) is then a positive definite set of algebraic equations for ψ_i ($j = 1, \ldots, S - 1$).

4-75. By expanding the cosine and logarithm in (4-152) into series show that when p and q are large

$$n_c \approx \frac{p^2 q^2 k \ln 10}{\pi^2 (p^2 + q^2)}$$

Show that this is $O(1/h^2)$.

4-76. Apply the two-dimensional Simpson's rule to Fig. 4-29. Verify that the approximation to (4-28) is $C = 2.21817$.

4-77. Show that

$$\left(\frac{d^2\psi}{dx^2} \right)_0 = \frac{1}{12h^2} (11\psi_{-1} - 20\psi_0 + 6\psi_1 + 4\psi_2 - \psi_3) + O(h^3)$$

Use this to approximate the boundary condition in Fig. 4-15, instead of the formula used there, which has truncation error $O(h^2)$. Verify that $\psi_1 = 0.01285$, which represents an error of only -0.00003 as compared with the error 0.00250 obtained from Fig. 4-15.

4-78. If (4-164) were to be solved using iteration by single steps, we would replace the current value of ψ_j by the improved value

$$\psi_j{}^1 = \frac{1}{6 + h^4} (h^4 - \psi_{j-2} + 4\psi_{j-1} + 4\psi_{j+1} - \psi_{j+2})$$

When we reach the stage where there is no change in any ψ_j out to k decimal places, show that there might still be a residual $E_{r,j}$ [as defined in (4-167)] as large as 3.0×10^{-k}.

4-79. The solution of the taut elastic cable of Exercise 4-63 can be given by using a finite form of the Green's function of Exercise 4-16.

$$\psi_j = \sum_{k=1}^{j-1} \frac{k(S-j)}{S^2} h + \sum_{k=j}^{S-1} \frac{j(S-k)}{S^2} h$$

Verify this for the central deflection when $S = 1/h = 4$. If in the solution of Exercise 4-63 there are residuals $E_{r,j}$, the solution error will satisfy

$$-\epsilon_{j-1} + 2\epsilon_j - \epsilon_{j+1} = E_{r,j}$$

and by comparing this with the original formulation for the deflections we see that ϵ_j plays the role of the deflection caused by a loading of intensity $E_{r,j}/h^2$. The Green's function then yields

$$\epsilon_j = \sum_{k=1}^{j-1} \frac{k(S-j)}{S^2} \frac{E_{r,k}}{h} + \sum_{k=j}^{S-1} \frac{j(S-k)}{S^2} \frac{E_{r,k}}{h}$$

Show that, if $E_{r,j} = E_r$ for all j, then

$$\text{Max } \epsilon_j = \frac{E_r}{8h^2}$$

4-80. If the x_j, $j = 1, \ldots, n$, are a set of independent random variables each having zero average value and standard deviation $\sigma(x)$, then[1] the standard deviation of the linear function

$$F = \Sigma a_j x_j$$

is

$$\sigma(F) = \sigma(x)(\Sigma a_j{}^2)^{\frac{1}{2}}$$

Apply this theorem to the summation of the previous problem when the $E_{r,j}$ have zero average and standard deviation[2] $\sigma(E_{r,j}) = E_r/\sqrt{3}$. Show that

$$\sigma(\epsilon_{S/2}) = \frac{E_r}{12}\left(\frac{1}{h^3} + \frac{2}{h}\right)^{\frac{1}{2}} \approx \frac{E_r}{12h^{\frac{3}{2}}}$$

[*Hint.* $\Sigma 1^2 + 2^2 + \cdots + n^2 = n(n+1)(2n+1)/6$.]

4-81. Apply h^2 extrapolation to the values of C in Fig. 4-35, using (a) the values for $h = 1$ and $h = \frac{1}{2}$; (b) the values for $h = \frac{1}{2}$ and $h = \frac{1}{4}$; (c) all three values and the formula of Exercise 3-54.   *Ans.* (a) 2.2407; (b) 2.2493; (c) 2.2498.

[1] See, for example, A. Hald, "Statistical Theory with Engineering Applications," John Wiley & Sons, Inc., New York, 1952, pp. 117, 188.

[2] See Exercise 3-50.

CHAPTER 5

EIGENVALUE PROBLEMS IN CONTINUOUS SYSTEMS

Eigenvalue problems are closely related to equilibrium problems. In both cases the configuration of a system is to be determined. In an equilibrium problem this is all that is required. In an eigenvalue problem it is necessary to determine a scalar parameter or eigenvalue in addition to the configuration or mode. In an equilibrium problem the configuration is completely determined, whereas in an eigenvalue problem only the relative amplitudes of a mode can be obtained. Eigenvalue problems are further complicated by the existence of a whole family of eigenvalues, each having a different mode. In many practical investigations, however, attention is centered on one particular mode, usually that associated with the smallest eigenvalue.

Linear eigenvalue problems in one- and two-dimensional continua are treated in the present chapter. The chapter begins with the analysis and formulation of four particular problems. We then review the general properties of these problems and give a brief outline of the classical theory. This introduction is followed by a survey of numerical procedures available for approximate solution. Iteration, trial solutions with adjustable parameters, and finite-difference methods are described and illustrated, using the particular problems formulated at the beginning of the chapter.

5-1. Particular Examples

The mathematical formulations of four particular eigenvalue problems are presented in this section. The problems are:

5-1. Vibrating string.
5-2. Buckling of a column.
5-3. Rectangular waveguide.
5-4. Buckling of a plate.

All the problems are linear. The first two are one-dimensional; the other two are two-dimensional.

Problem 5-1. Vibrating String. This is *the* classical eigenvalue problem in a continuous system. The system shown in Fig. 5-1 consists of a uniform, taut string of length L which is fixed at the ends. The string has mass per unit length ρ and is stretched by a large tension T. The

276

eigenvalue problem consists in determining the natural frequencies and modal configurations of small transverse vibrations. The general motion of a taut string is governed by the *wave equation*,[1]

$$\frac{\partial^2 y}{\partial x^2} = \frac{\rho}{T} \frac{\partial^2 y}{\partial t^2} \tag{5-1}$$

where $y(x, t)$ is the transverse displacement of the string, x is distance along the string, and t is time. By definition a *natural* mode of vibration

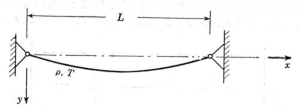

FIG. 5-1. Taut string executes natural vibration when transverse motion of all points is synchronized.

is one in which all points execute simple harmonic motion at the same frequency and in the same phase. Thus in natural vibration we have

$$y(x, t) = \psi(x) \sin (\omega t + \varphi) \tag{5-2}$$

where $\psi(x)$ is an amplitude function, ω is a *natural* frequency, and φ is a phase angle. By substituting (5-2) into (5-1) we find the following governing differential equation for the amplitude function of a natural vibration:

$$-\frac{d^2 \psi}{dx^2} = \frac{\rho \omega^2}{T} \psi \tag{5-3}$$

In addition, the amplitude of any vibration must vanish at the ends of the string.

$$\psi = 0 \qquad \text{at } x = 0 \text{ and } x = L \tag{5-4}$$

Equation (5-3) and the boundary conditions (5-4) constitute a mathematical formulation of the eigenvalue problem. The problem is to find the natural-mode amplitude functions, ψ, and the corresponding natural frequencies, ω, which satisfy these requirements.

An alternative formulation analogous with the inverse method for lumped-parameter systems leads to an integral equation. The elastic properties of the system are studied separately to obtain an influence function or Green's function, $G(x, \xi)$. This function gives the deflection

[1] See, for example, J. P. Den Hartog, "Mechanical Vibrations," 4th ed., McGraw-Hill Book Company, Inc., New York, 1956, p. 136. Equation (5-1) may be derived from that of Exercise 4-9b by considering $w = -\rho \partial^2 y / \partial t^2$ as an "inertia loading."

at x due to a unit transverse load at ξ as shown in Fig. 5-2.   A governing
equation for natural vibrations is now obtained by assuming that the
string is vibrating according to (5-2) which sets up an oscillating inertia
loading per unit length with amplitude, $\rho\omega^2\psi(\xi)$, at ξ.   By superposition
the total deflection amplitude at x is given by the following integral:

$$\psi(x) = \int_0^L \rho\omega^2\psi(\xi)G(x, \xi)\, d\xi \qquad (5\text{-}5)$$

This formulation should be compared with that of the differential equation (5-3).   In both cases the amplitude ψ and frequency ω are to be

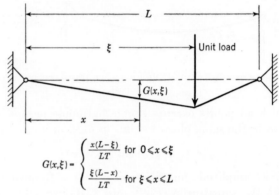

$$G(x,\xi) = \begin{cases} \dfrac{x(L-\xi)}{LT} & \text{for } 0 \leqslant x \leqslant \xi \\[2mm] \dfrac{\xi(L-x)}{LT} & \text{for } \xi \leqslant x \leqslant L \end{cases}$$

Fig. 5-2. Green's function for an elastic string under tension T.

determined.   The boundary conditions (5-4) are a necessary supple-
ment to the differential equation, whereas the boundary conditions are
implicitly included in the integral equation when the influence function
is constructed.

The above formulations can be cast into nondimensional form in terms
of the following dimensionless variables:

$$x' = \frac{x}{L} \qquad \psi' = \frac{\psi}{L} \qquad \lambda = \frac{\rho\omega^2L^2}{T}$$
$$\xi' = \frac{\xi}{L} \qquad G' = \frac{G}{L/T} \qquad\qquad (5\text{-}6)$$

The differential equation (5-3) becomes (after dropping the primes on the
dimensionless quantities)

$$-\frac{d^2\psi}{dx^2} = \lambda\psi \qquad 0 < x < 1 \qquad (5\text{-}7)$$

with the boundary conditions

$$\psi = 0 \qquad \text{at } x = 0 \text{ and } x = 1 \qquad (5\text{-}8)$$

while the integral equation (5-5) takes the following form:

$$\psi(x) = \lambda \int_0^1 \psi(\xi)G(x, \xi)\, d\xi \tag{5-9}$$

Problem 5-2. Buckling of a Column. A flexible beam of length L and uniform flexural rigidity EI is clamped at one end and hinged at the other as shown in Fig. 5-3. An elastic-stability problem is posed[1] when a longitudinal force, P, is applied. The problem is to determine the critical value of P for which the unbent position of the beam first ceases to be stable. Under the critical load neutral equilibrium is possible with a nonvanishing deflection $y(x)$. The eigenvalue problem consists in determining the deflection mode shape and the critical load.

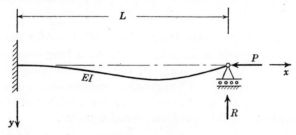

FIG. 5-3. Flexible beam held in equilibrium in slightly buckled position by the critical compressive load P.

If we assume that the beam is in equilibrium in the deflected state of Fig. 5-3, the elastic relation between bending moment and curvature is

$$-EI\frac{d^2y}{dx^2} = Py + R(L - x) \tag{5-10}$$

where R is the transverse reaction. Differentiating twice eliminates R as follows:

$$\frac{d^4y}{dx^4} = -\frac{P}{EI}\frac{d^2y}{dx^2} \tag{5-11}$$

The deflection must also satisfy the following boundary conditions:

$$\left.\begin{array}{l} y = 0 \\ \dfrac{dy}{dx} = 0 \end{array}\right\} \quad \text{at } x = 0 \qquad \left.\begin{array}{l} y = 0 \\ \dfrac{d^2y}{dx^2} = 0 \end{array}\right\} \quad \text{at } x = L \tag{5-12}$$

The differential equation (5-11) and the boundary conditions (5-12) constitute a formulation of the eigenvalue problem.

This formulation can be made dimensionless in terms of the following nondimensional quantities:

---

[1] See, for example, S. Timoshenko, "Theory of Elastic Stability," McGraw-Hill Book Company, Inc., New York, 1936, p. 88.

$$x' = \frac{x}{L} \qquad \psi = \frac{y}{L} \qquad \lambda = \frac{PL^2}{EI} \tag{5-13}$$

The problem consists in finding the smallest eigenvalue λ and the corresponding function ψ which satisfy (the prime has been dropped from the dimensionless length)

$$\frac{d^4\psi}{dx^4} = -\lambda \frac{d^2\psi}{dx^2} \qquad 0 < x < 1 \tag{5-14}$$

and the following boundary conditions:

$$\left.\begin{array}{c} \psi = 0 \\ \dfrac{d\psi}{dx} = 0 \end{array}\right\} \quad \text{at } x = 0 \qquad \left.\begin{array}{c} \psi = 0 \\ \dfrac{d^2\psi}{dx^2} = 0 \end{array}\right\} \quad \text{at } x = 1 \tag{5-15}$$

Problem 5-3. Rectangular Waveguide. We consider two particular types of electromagnetic wave propagation[1] in a long conducting cylinder of rectangular cross section. In the *transverse magnetic wave* the magnetic-field vector has no longitudinal component, while the longitudinal component of the electric-field vector vanishes on the walls of the guide and satisfies the two-dimensional equation

$$-\nabla^2\psi = \lambda\psi \qquad \nabla^2 \equiv \frac{\partial^2}{\partial x^2} + \frac{\partial^2}{\partial y^2} \tag{5-16}$$

throughout the cross section where λ is a frequency parameter. In the *transverse electric wave* the longitudinal component of the electric-field vector vanishes, while the longitudinal component of the magnetic-field vector satisfies (5-16). At the walls, however, it is the normal derivative of the longitudinal component of the magnetic field which must be zero. The problem of ascertaining these modes of propagation thus consists in obtaining functions $\psi(x, y)$ and corresponding eigenvalues λ which satisfy (5-16) and the appropriate boundary condition. The particular problem to be treated here involves a rectangular section with sides in the ratio of $2:1$. Equation (5-16) may be considered as already in nondimensional form if we take the waveguide section to have sides of one and two units of length as shown in Fig. 5-4. The boundary conditions indicated in Fig. 5-4 are those appropriate for a transverse electric wave. For a transverse magnetic wave we must solve the same equation but with the following boundary condition:

$$\psi = 0 \qquad \text{on} \quad \begin{cases} x = \pm 1 \\ y = \pm \tfrac{1}{2} \end{cases} \tag{5-17}$$

[1] See, for example, J. C. Slater and N. H. Frank, "Electromagnetism," McGraw-Hill Book Company, Inc., New York, 1947, p. 135.

Problem 5-4. Buckling of a Plate. We consider a thin elastic rectangular plate hinged along all four edges and subjected to a uniform longitudinal compression as shown in Fig. 5-5. We seek the buckling

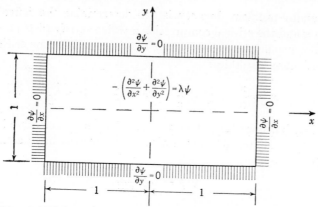

$$\frac{\partial \psi}{\partial y} = 0$$

$$-\left(\frac{\partial^2 \psi}{\partial x^2} + \frac{\partial^2 \psi}{\partial y^2}\right) = \lambda \psi$$

$$\frac{\partial \psi}{\partial x} = 0$$

$$\frac{\partial \psi}{\partial x} = 0$$

$$\frac{\partial \psi}{\partial y} = 0$$

Fig. 5-4. Dimensionless formulation of the problem of determining the transverse electric modes in a rectangular waveguide.

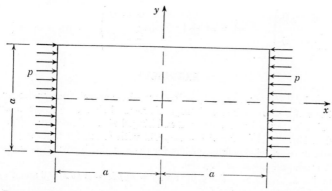

Fig. 5-5. Rectangular plate held in equilibrium in slightly bent position by the critical compressive load p.

load, i.e., the smallest compression for which the unbent configuration ceases to be stable. If we assume that the plate is in equilibrium in a slightly bent state under the critical compression, p per unit length, the transverse deflection, $w(x, y)$, must satisfy[1]

$$\nabla^4 w = -\frac{p}{D} \frac{\partial^2 w}{\partial x^2} \qquad \nabla^4 \equiv \frac{\partial^4}{\partial x^4} + 2\frac{\partial^4}{\partial x^2 \, \partial y^2} + \frac{\partial^4}{\partial y^4} \qquad (5\text{-}18)$$

where D is the flexural rigidity of the plate. The conditions of hinged support also require that

[1] See, for example, Timoshenko, *op. cit.*, p. 324.

$$\left.\begin{array}{l} w = 0 \\ \dfrac{\partial^2 w}{\partial x^2} = 0 \end{array}\right\} \quad \text{at } x = \pm a \qquad \left.\begin{array}{l} w = 0 \\ \dfrac{\partial^2 w}{\partial y^2} = 0 \end{array}\right\} \quad \text{at } y = \pm \dfrac{a}{2} \quad (5\text{-}19)$$

The eigenvalue problem thus consists in determining the deflection configuration w and the critical compression p which satisfy (5-18) and (5-19).

A nondimensional formulation may be achieved in terms of the following dimensionless variables:

$$x' = \frac{x}{a} \qquad y' = \frac{y}{a} \qquad \psi = \frac{w}{a} \qquad \lambda = \frac{pa^2}{D} \qquad (5\text{-}20)$$

The problem becomes that of finding the smallest eigenvalue λ and the corresponding function ψ which satisfies (the primes have already been dropped)

$$\nabla^4 \psi = -\lambda \frac{\partial^2 \psi}{\partial x^2} \qquad (5\text{-}21)$$

within the rectangle $-1 < x < 1$, $-\frac{1}{2} < y < \frac{1}{2}$ and satisfies the following boundary conditions on the edges:

$$\left.\begin{array}{l} \psi = 0 \\ \dfrac{\partial^2 \psi}{\partial x^2} = 0 \end{array}\right\} \quad \text{at } x = \pm 1 \qquad \left.\begin{array}{l} \psi = 0 \\ \dfrac{\partial^2 \psi}{\partial y^2} = 0 \end{array}\right\} \quad \text{at } y = \pm \tfrac{1}{2} \quad (5\text{-}22)$$

EXERCISES

5-1. Formulate the eigenvalue problem for the natural frequencies and modal configurations of a uniform beam of length L, flexural rigidity EI, and mass per unit length ρ which is clamped at the left end and hinged at the right end. Compare with (5-11) and (5-12).   *Ans.* $d^4\psi/dx^4 = \rho\omega^2\psi/EI$ with (5-12) for boundary conditions.

5-2. Formulate the eigenvalue problem for the buckling load of the cantilever beam on elastic foundation shown in Fig. 5-6.

    Ans. $EI\, d^4y/dx^4 + ky = -P\, d^2y/dx^2$ and $y = dy/dx = 0$ at $x = 0$ and $d^2y/dx^2 = 0$, $-EI\, d^3y/dx^3 = P\, dy/dx$ at $x = L$.

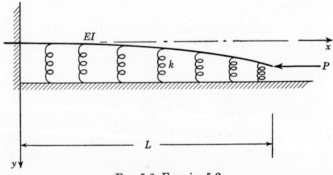

Fɪɢ. 5-6. Exercise 5-2.

5-3. Reduce the formulation of Exercise 5-2 to dimensionless form for the case where $4EI = kL^4$.

Ans. $d^4\psi/dx^4 + 4\psi = -\lambda\, d^2\psi/dx^2$, $0 < x < 1$; $\psi = d\psi/dx = 0$ at $x = 0$; $d^2\psi/dx^2 = 0$, $-d^3\psi/dx^3 = \lambda\, d\psi/dx$ at $x = 1$.

5-4. Formulate the eigenvalue problem for the natural torsional vibrations of the system shown in Fig. 5-7. The elastic relation between twisting moment M_t and angle of twist φ is

$$M_t = GI_p \frac{\partial\varphi}{\partial x}$$

and the mass moment of inertia per unit length of shaft is ρI_p, where G is the shear modulus, ρ is the mass per unit volume, and I_p is polar area moment of inertia of the shaft section. The mass moment of inertia of the disk is J.

Ans. $-G\, d^2\psi/dx^2 = \rho\omega^2\psi$ and $\psi = 0$ at $x = 0$ and $GI_p\, d\psi/dx = J\omega^2\psi$ at $x = L$.

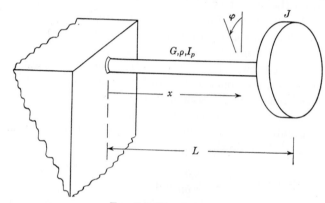

FIG. 5-7. Exercise 5-4.

5-5. Reduce the formulation of Exercise 5-4 to dimensionless form in the case where $J = \rho I_p L$. *Ans.* $-d^2\psi/dx^2 = \lambda\psi$, $0 < x < 1$; $\psi = 0$ at $x = 0$; $d\psi/dx = \lambda\psi$ at $x = 1$.

5-6. A uniform elastic membrane has mass per unit area ρ and is stretched with surface tension T over a rigid frame. The edge of the frame is a plane curve C enclosing the domain D. Formulate the eigenvalue problem for the natural transverse vibrations of the membrane. *Ans.* $-\nabla^2\psi = \rho\omega^2\psi/T$ in D and $\psi = 0$ on C.

5-7. Study the analogy between Exercise 5-6 and the problem of determining the transverse magnetic modes of a waveguide. How would a membrane have to be supported if the eigenvalue problem for its natural vibrations is to be analogous with the problem of determining the transverse electric modes of a waveguide?

5-8. Consider the problem of acoustic resonance in a two-dimensional room, i.e., a large room with low ceiling. Acoustic variations in pressure satisfy[1] the wave equation

$$\nabla^2 p + \frac{1}{c^2}\frac{\partial^2 p}{\partial t^2} = 0$$

where c is the velocity of sound. If the walls are perfectly reflecting, verify that the eigenvalue problem for standing waves is completely analogous with that for the transverse electric modes of a waveguide (see Fig. 5-4).

[1] See, for example, P. M. Morse, "Vibration and Sound," 2d ed., McGraw-Hill Book Company, Inc., New York, 1948, p. 389.

5-2. Formulation of the General Problem

The mathematical formulations of the problems of the previous section may all be considered as special cases of the following general formulation:

The problem is to find one or more constants, λ, and corresponding functions, ψ, such that a differential equation

$$M_{2m}[\psi] = \lambda N_{2n}[\psi] \tag{5-23}$$

is satisfied throughout a domain D and boundary conditions

$$B_i[\psi] = \lambda C_i[\psi] \qquad i = 1, \ldots, m \tag{5-24}$$

are satisfied on the boundary of D.

Domain.   In the above formulation the domain D may be either a one- or a two-dimensional continuum similar to the domains considered in Sec. 4-2 for equilibrium problems.

Differential Equation.   The symbols M_{2m} and N_{2n} stand for linear homogeneous differential operators of order $2m$ and $2n$, respectively, with $m > n$. These operators are similar to the operators L_{2m} encountered in the linear-equilibrium problems of Chap. 4.

Boundary Conditions.   At each boundary point there are m independent conditions of the following form,

$$b_{i0}\psi + b_{i1}\frac{\partial\psi}{\partial n} + \cdots + b_{i(2m-1)}\frac{\partial^{2m-1}\psi}{\partial n^{2m-1}}$$
$$= \lambda\left(c_{i0}\psi + c_{i1}\frac{\partial\psi}{\partial n} + \cdots + c_{i(2m-1)}\frac{\partial^{2m-1}\psi}{\partial n^{2m-1}}\right) \tag{5-25}$$

where the b_{ij} and c_{ij} are known. The constant λ did not appear in the boundary conditions (that is, $C_i[\psi] \equiv 0$) in the illustrative problems of Sec. 5-1. The text discussion is also limited to this case. The treatment of the case $C_i[\psi] \not\equiv 0$ is indicated, however, in a series[1] of exercises for the reader.

When the constant λ does not appear in the boundary conditions, the eigenvalue problem, (5-23) and (5-24), is said to be *self-adjoint* if for any two functions, u and v, which do satisfy the boundary conditions (but which are otherwise arbitrary) both the following statements are true:

$$\int_D uM_{2m}[v]\,dD = \int_D vM_{2m}[u]\,dD$$
$$\int_D uN_{2n}[v]\,dD = \int_D vN_{2n}[u]\,dD \tag{5-26}$$

---

[1] For examples of systems in which the constant λ does enter into the boundary conditions, see Exercises 5-3 and 5-5. Further discussion of these systems is given in Exercises 5-12 to 5-16, 5-19, 5-20, 5-31 to 5-33, 5-48, 5-49, 5-60, and 5-71.

The operator M_{2m} is said to be *positive* if, for any such function u, we have

$$\int_D uM_{2m}[u]\,dD \geq 0 \qquad (5\text{-}27)$$

If the equality holds only for $u \equiv 0$, the operator is said to be *positive definite*. Exactly similar remarks apply to the operator N_{2n}. The eigenvalue problem itself is said to be *positive definite* if *both* M_{2m} and N_{2n} are positive definite. As in Sec. 4-2 the demonstration of whether or not (5-26) or (5-27) is satisfied in a particular case is accomplished by using integration by parts. Throughout this chapter we consider only problems which are self-adjoint and for which the operator N_{2n} is positive definite.

All the problems of Sec. 5-1 are self-adjoint. Problems 5-1, 5-2, and 5-4 are also positive definite. In the transverse electric wave of Prob. 5-3 the equality in (5-27) holds for the function $u \equiv$ constant, and therefore the problem formulated in Fig. 5-4 is only positive.

Special Eigenvalue Problem. In Chap. 2 the eigenvalue problem was said to be special when the matrix B had the special form of a diagonal matrix with positive elements. There is a class of continuous eigenvalue problems which have analogous properties and to which the same label is applied. When the constant λ does not appear in the boundary conditions, a continuous eigenvalue problem is said to be *special* if the operator N_{2n} has the special form

$$N_{2n}[\psi] = g \cdot \psi \qquad (5\text{-}28)$$

where g is a prescribed continuous function which is positive throughout the domain D. Problems 5-1 and 5-3 are examples of special eigenvalue problems.

EXERCISES

5-9. Verify that the formulation (5-7) and (5-8) for Prob. 5-1 is self-adjoint and positive definite.

5-10. Verify that the formulation (5-14) and (5-15) for Prob. 5-2 is self-adjoint and positive definite.

5-11. Verify that the formulation (5-21) and (5-22) for Prob. 5-4 is self-adjoint and positive definite. The integration by parts can be performed with the aid of (4-70).

5-12. Consider the general eigenvalue problem (5-23) and (5-24). At each boundary point there are m boundary conditions. Separate these into the conditions, $\bar{B}_i[\psi] = \lambda \bar{C}_i[\psi]$, which contain λ and those, $B_i[\psi] = 0$, which do not. Such a system is *self-adjoint* if, for any two functions u and v which satisfy all the conditions of the form $B_i[\psi] = 0$ but which are otherwise arbitrary, we always have

$$\int_D \{uM_{2m}[v] - vM_{2m}[u]\}\,dD + \sum \oint \{u\bar{B}_i[v] - v\bar{B}_i[u]\}\,ds = 0$$

$$\int_D \{uN_{2n}[v] - vN_{2n}[u]\}\,dD + \sum \oint \{u\bar{C}_i[v] - v\bar{C}_i[u]\}\,ds = 0$$

where the line integral and summation indicate that at every boundary point every condition which contains λ is included. Show that for the system of Exercise 5-3 these statements take the following form:

$$\int_0^1 \left\{ u\left(\frac{d^4v}{dx^4} + 4v\right) - v\left(\frac{d^4u}{dx^4} + 4u\right) \right\} dx - \left[u\frac{d^3v}{dx^3} - v\frac{d^3u}{dx^3} \right]_{x=1} = 0$$

$$\int_0^1 \left\{ u\left(-\frac{d^2v}{dx^2}\right) - v\left(-\frac{d^2u}{dx^2}\right) \right\} dx + \left[u\frac{dv}{dx} - v\frac{du}{dx} \right]_{x=1} = 0$$

Use integration by parts to verify that these are actually true for arbitrary u and v which satisfy

$$u = \frac{du}{dx} = v = \frac{dv}{dx} = 0 \qquad \text{at } x = 0$$

$$\frac{d^2u}{dx^2} = \frac{d^2v}{dx^2} = 0 \qquad \text{at } x = 1$$

5-13. Use the previous exercise to verify that the system of Exercise 5-5 is self-adjoint.

5-14. In the general system of Exercise 5-12 the operator M_{2m} is said to be *positive* if, for any function u which satisfies all the conditions $B_i[u] = 0$, the expression

$$\int_D u M_{2m}[u]\, dD + \sum \oint u\bar{B}_i[u]\, ds$$

is never negative. If this expression vanishes only when u is identically zero, then M_{2m} is said to be *positive definite*. A corresponding statement applies to N_{2n}. The eigenvalue problem itself is said to be *positive definite* if both M_{2m} and N_{2n} are positive definite. Verify that the system of Exercise 5-3 is positive definite.

5-15. Show that the system of Exercise 5-5 is positive definite.

5-16. The designation *special* can be applied to an eigenvalue problem in which the constant λ appears in the boundary conditions provided that $N_{2n}[\psi]$ has the special form of (5-28) and provided that in all boundary conditions $\bar{B}_i[\psi] = \lambda \bar{C}_i[\psi]$ which contain λ the form of $\bar{C}_i[\psi]$ is simply $c_{i0}\psi$, where c_{i0} is positive. Show that the eigenvalue problem of Exercise 5-3 is not a special eigenvalue problem but that the problem of Exercise 5-5 is.

5-3. Mathematical Properties

The eigenvalue problem

$$M_{2m}[\psi] = \lambda N_{2n}[\psi] \qquad \text{in } D$$
$$B_i[\psi] = 0 \qquad i = 1, \ldots, m \text{ on boundary}$$

$$(5\text{-}29)$$

where D is a finite bounded domain, where M_{2m} and N_{2n} are self-adjoint differential operators of order $2m$ and $2n$, respectively, with $m > n$, where N_{2n} is positive definite, and where the constant λ does not[1] enter into the boundary conditions, has properties which correspond in almost every detail with the discrete eigenvalue problem considered in Chap. 2. Heuristically these properties can be obtained by simply letting the num-

---

[1] The case where λ does enter into the boundary conditions is treated in Exercises 5-19 and 5-20.

ber of degrees of freedom in the discrete system increase without limit. Rigorous establishment[1] of these properties is more difficult since pathological behavior is possible if continuity and smoothness restrictions are inadequate.

The conditions enumerated above are sufficient to ensure that real solutions to (5-29) do exist. There are in fact an infinity of solutions. Each solution consists of a scalar, λ_j, called an *eigenvalue* and a function, ψ_j, called an *eigenfunction*, or *mode*, which satisfy

$$M_{2m}[\psi_j] = \lambda_j N_{2n}[\psi_j]$$
$$B_i[\psi_j] = 0 \qquad i = 1, \ldots, m \tag{5-30}$$

Since both the differential equation and the boundary conditions are homogeneous, the function $\psi = 0$ satisfies (5-30) for any value of λ. This is referred to as the *trivial solution* and is explicitly ruled out as a proper solution to the eigenvalue problem. Furthermore the homogeneous nature of (5-30) implies that, if ψ_j satisfies (5-30), then so does any scalar multiple of ψ_j; that is, the eigenvalue problem does not fix the absolute magnitude of the mode. We shall consider any function proportional to ψ_j to be essentially the same mode.

An eigenvalue λ_j is said to be *distinct*, or *nonrepeated*, if ψ_j is uniquely determined to within a multiplicative factor. If there are r linearly independent functions which satisfy (5-30) with a particular λ_j, then λ_j is said to be an eigenvalue of *multiplicity* r. The eigenvalue problem (5-29) cannot have eigenvalues of multiplicity greater than $2m$. The distribution of the eigenvalues λ_j for a particular system is often referred to as the *spectrum* of the system. For the class of systems we are considering the spectrum consists of an infinite number of *discrete* real values. If M_{2m} is positive definite, then all the eigenvalues are greater than zero. If M_{2m} is only positive, then $\lambda = 0$ is also an eigenvalue.

Orthogonality of the Modes. The eigenfunctions corresponding to different eigenvalues satisfy[2] the following relations:

$$\int_D \psi_j M_{2m}[\psi_k] \, dD = 0$$
$$\int_D \psi_j N_{2n}[\psi_k] \, dD = 0 \tag{5-31}$$

These relations express the *orthogonality* of the modes with respect to M_{2m} and N_{2n}. In the case of an eigenvalue of multiplicity greater than 1 (that is, $\lambda_j = \lambda_k$) linearly independent eigenfunctions are not necessarily

[1] See R. Courant and D. Hilbert, "Methoden der mathematischen Physik," 2d ed., vol. 1, Springer-Verlag OHG, Berlin, 1931, and L. Collatz, "Eigenvertaufgaben mit technischen Anwendung," Akademische Verlagsgesellschaft M.G.H., Leipzig, 1949, chaps. 2, 3.

[2] For proof see Exercise 5-17.

orthogonal; however, it is always possible[1] to select a set of linearly independent eigenfunctions for a repeated eigenvalue which will be orthogonal.

It is sometimes convenient to have the amplitudes of the modes scaled up to some reference standard. For theoretical investigations the standard usually chosen is normality with respect to N_{2n}. A mode, ψ_j, is said to be *normalized* with respect to N_{2n} when

$$\int_D \psi_j N_{2n}[\psi_j]\, dD = 1 \tag{5-32}$$

A family of modes, ψ_j, is said to be *orthonormal* when all modes satisfy both (5-31) and (5-32).

As an illustration, consider Prob. 5-1 with the formulation (5-7) and (5-8). Here we have $0 < x < 1$ as the domain D with the following operators:

$$M_2[\psi] = -\frac{d^2\psi}{dx^2}$$
$$N_0[\psi] = \psi \tag{5-33}$$
$$B_1[\psi] = \psi = 0 \qquad \text{at} \begin{cases} x = 0 \\ x = 1 \end{cases}$$

The exact solution may be displayed as follows:

$$\begin{array}{cccc} \lambda_1 = \pi^2 & \lambda_2 = 4\pi^2 & \lambda_3 = 9\pi^2 & \cdots \\ \psi_1 = \sin \pi x & \psi_2 = \sin 2\pi x & \psi_3 = \sin 3\pi x & \cdots \end{array} \tag{5-34}$$

All the eigenvalues are distinct. The spectrum consists of the discrete set of values, π^2, $(2\pi)^2$, $(3\pi)^2$, ... The reader can readily verify that the modes satisfy (5-31) but do not satisfy (5-32) until each ψ_j has been multiplied by the factor $\sqrt{2}$.

Expansion Theorem. The totality of modes ψ_j for a given system is an infinite set of functions. When a function u defined in D can be represented by a convergent series of the form

$$u = \sum_j c_j \psi_j \tag{5-35}$$

where the c_j are constant coefficients, it is said to be *expanded in terms of the eigenfunctions.* Speaking roughly, we can say that any well-behaved function can be expanded in terms of the mode system of an eigenvalue problem of the type we are considering. A more precise specification of what is meant by a well-behaved function turns out to be a very delicate question. For instance, when the modes of (5-34) are placed in (5-35) we have the *Fourier sine series*, which has been studied extensively. For

---

[1] See the discussion of repeated eigenvalues in Sec. 2-4.

this system it is known[1] that functions which are piecewise smooth but which may have a finite number of simple discontinuities in du/dx and in u itself can be expanded in convergent sine series. There are, however, functions continuous everywhere which cannot[2] be so expanded.

When a function u does permit expansion in terms of eigenfunctions, then the coefficients c_j in (5-35) can be computed from the following formula

$$c_j = \int_D u N_{2n}[\psi_j]\, dD \qquad (5\text{-}36)$$

if ψ_j is normalized with respect to N_{2n}.

Counterexamples. When eigenvalue problems do not meet all the conditions listed under (5-29), then the properties just described cannot be guaranteed. Some examples illustrating different behavior are given below:

No Solution. There is no function, except $\psi \equiv 0$, which satisfies[3]

$$\frac{d^4\psi}{dx^4} = -\lambda \frac{d^2\psi}{dx^2} \qquad \text{in } 0 < x < 1$$

$$\psi = \frac{d\psi}{dx} = 0 \qquad \text{at } x = 0 \qquad (5\text{-}37)$$

$$\frac{d^2\psi}{dx^2} = \frac{d^3\psi}{dx^3} = 0 \qquad \text{at } x = 1$$

no matter what the value of λ. It is interesting to note how closely this resembles the formulation (5-14) and (5-15) of Prob. (5.2). The only difference is that the condition $\psi = 0$ at $x = 1$ is replaced by $d^3\psi/dx^3 = 0$. This may be physically interpreted[4] as requiring that transverse displacements be allowed under the point of application of the buckling load but that during such displacements the buckling load must ride with the beam always remaining tangent.[5] The operator M_4 in (5-37) is both self-adjoint and positive definite, but N_2 is neither.

Complex Eigenfunctions. The eigenvalue problem

$$\frac{d^2\psi}{dx^2} + \frac{\cos^2 x}{1 + \cos^2 x}\,\psi = \lambda\,\frac{\cos x}{1 + \cos^2 x}\,\psi \qquad \text{in } 0 < x < \pi \qquad \frac{d\psi}{dx} = 0 \qquad \text{at} \begin{cases} x = 0 \\ x = \pi \end{cases}$$

$$(5\text{-}38)$$

has the eigenvalue $\lambda = i = \sqrt{-1}$ and the corresponding eigenfunction $\psi = i + \cos x$. Here both M_2 and N_0 are self-adjoint, but neither is positive definite.

[1] See, for example, R. V. Churchill, "Fourier Series and Boundary Value Problems," McGraw-Hill Book Company, Inc., New York, 1941, chap. 4.

[2] See A. Zygmund, "Trigonometric Series," Monografje Matematyczne, Warsaw, 1935, p. 167.

[3] This and the next example are given by Collatz, *op. cit.*, pp. 41, 57.

[4] A. Pflüger, "Stabilitätsprobleme der Elastokinetik," Springer-Verlag OHG, Berlin, 1950, p. 218.

[5] The system actually has a critical load, but it must be determined from kinetic rather than static considerations. See M. Beck, Die Knicklast des einseitig eingespannten, tangential gedrückten Stabes, *Z. angew. Math. u. Phys.*, **3**, 225–228 (1952).

Continuous Spectra. The following special case of *Schrödinger's equation,*[1]

$$-\frac{d^2\psi}{dx^2} - \frac{2\psi}{x} = \lambda\psi \quad \text{in } 0 < x < \infty \tag{5-39}$$

together with the boundary conditions that ψ should vanish at $x = 0$ and $x = \infty$ is an eigenvalue problem which fixes certain energy levels for the hydrogen atom. The spectrum can be shown to consist of the discrete values -1, $-(\frac{1}{2})^2$, $-(\frac{1}{3})^2$, . . . , plus

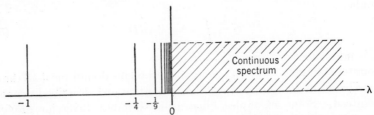

FIG. 5-8. Spectrum for Schrödinger's equation includes discrete eigenvalues plus a continuous band.

the continuous range of values $0 \leq \lambda < \infty$ as sketched in Fig. 5-8. Both operators M_2 and N_0 are self-adjoint, and N_0 is positive definite, but M_2 is not positive, and the domain D is unbounded. It is still possible[2] to expand well-behaved functions in terms of the eigenfunctions, but the expansion is made up of an infinite series corresponding to the discrete spectrum *plus an integral corresponding to the continuous spectrum.*

Another interesting case is the equation

$$-\frac{d^2\psi}{dx^2} + 4 \cos 2x\psi = \lambda\psi \quad \text{in } 0 < x < \infty \tag{5-40}$$

with the boundary conditions $\psi = 0$ at $x = 0$ and $x = \infty$. This is an eigenvalue

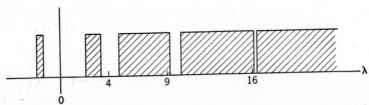

FIG. 5-9. Spectrum for Mathieu's equation is made up of continuous bands.

problem whose spectrum consists[3] of separated continuous bands as sketched in Fig. 5-9. Here again M_2 and N_0 are self-adjoint, and N_0 is positive definite, but M_2 is not positive, and the domain D is unbounded.

[1] See, for example, G. Joos, "Theoretical Physics," transl. by I. M. Freeman, G. E. Stechert & Company, New York, 1934, p. 658.

[2] See, for example, E. C. Titchmarsh, "Eigenfunction Expansions Associated with Second-order Differential Equations," Oxford University Press, New York, 1946, p. 87.

[3] Equation (5-40) is a Mathieu equation. The spectrum can be deduced from Fig. 8 of N. W. McLachlan, "Theory and Application of Mathieu Functions," Oxford University Press, New York, 1947.

Integral-equation Formulation. When λ does not enter into the boundary conditions of the eigenvalue problem,

$$M_{2m}[\psi] = \lambda N_{2n}[\psi]$$
$$B_i[\psi] = 0 \qquad i = 1, \ldots, m \tag{5-41}$$

and the Green's function corresponding to the operator M_{2m} and the boundary conditions of (5-41) is known, the problem may be cast into an integral equation. For simplicity of notation we consider here a one-dimensional domain $\alpha < x < \beta$. In this case the problem (5-41) takes the form

$$\psi(x) = \lambda \int_\alpha^\beta G(x, \xi) N_{2n}[\psi(\xi)] \, d\xi \tag{5-42}$$

which is analogous to the inverse form (2-33) for lumped-parameter eigenvalue problems. The integral formulation is useful for making theoretical deductions,[1] and also is a good starting point for numerical solutions, provided the Green's function is known.[2] In Prob. 5-1, the vibrating string, the integral equation (5-9) was formulated, and in Sec. 5-5 its application to iteration will be illustrated.

Enclosure Theorem for Special Eigenvalue Problem. Let u be a function which satisfies the boundary conditions for the following special eigenvalue problem:

$$M_{2m}[\psi] = \lambda g\psi$$
$$B_i[\psi] = 0 \qquad i = 1, \ldots, m \tag{5-43}$$

The function l defined by

$$M_{2m}[u] = lgu \tag{5-44}$$

then plays a role similar to the set of l_j defined by (2-45) for lumped-parameter problems. If u is an eigenfunction ψ_j, then l is a constant equal to λ_j. If u is not an eigenfunction, then l varies throughout the domain D, but the set of values taken by l possess the following *enclosure property:* If l_{max} and l_{min} are the upper and lower bounds for l within the domain D, then there always exists[3] at least one eigenvalue λ_j such that

$$l_{min} \leq \lambda_j \leq l_{max} \tag{5-45}$$

As an example, consider Prob. 5-1 and the formulation (5-7), (5-8). If we choose the function $u = (x - x^2) + (x - x^2)^2$ which satisfies the

[1] See, for example, W. V. Lovitt, "Linear Integral Equations," McGraw-Hill Book Company, Inc., New York, 1924, and Courant and Hilbert, vol. 1, chap. 3.
[2] See, for example, R. L. Bisplinghoff, H. Ashley, and R. L. Halfman, "Aeroelasticity," Addison-Wesley Publishing Company, Cambridge, Mass., 1955, p. 146.
[3] Collatz, *op. cit.*, p. 126.

boundary conditions, we obtain

$$l = \frac{M_2[u]}{u} = \frac{-d^2u/dx^2}{u} = \frac{12}{1 + x - x^2} \tag{5-46}$$

This function is sketched in Fig. 5-10. The values of l_{min} and l_{max} are found to be 9.6 and 12.0, respectively. These values enclose the eigenvalue $\lambda_1 = \pi^2 = 9.8696$.

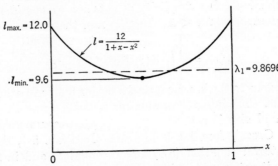

FIG. 5-10. Illustrating the enclosure theorem.

EXERCISES

5-17. To prove the *orthogonality* relations, multiply both sides of the equation in (5-30) by ψ_k, and integrate over D. Then rewrite (5-30) for ψ_k and λ_k, multiply by ψ_j, and integrate over D. Now subtract these two integrated expressions, and use the self-adjoint property to obtain

$$(\lambda_j - \lambda_k) \int_D \psi_j N_{2n}[\psi_k] \, dD = 0$$

From this deduce the relations (5-31).

5-18. Verify the formula (5-36) for the expansion coefficients by following a procedure analogous with that used for the lumped-parameter case [see (2-43) and Exercise 2-22].

5-19. Consider the general eigenvalue problem of Exercise 5-12, in which λ occurs in the boundary conditions. If ψ_j is a mode and λ_j is the corresponding eigenvalue, we have

$$\begin{aligned}
&(a) \quad M_{2m}[\psi_j] = \lambda_j N_{2n}[\psi_j] \qquad \text{in } D \\
&(b) \quad \bar{B}_i[\psi_j] = \lambda_j \bar{C}_i[\psi_j] \quad\left.\right\} \quad \text{on boundary} \\
&(c) \quad B_i[\psi_j] = 0
\end{aligned}$$

in the notation of Exercise 5-12. To prove the orthogonality relations, proceed as in Exercise 5-17. Multiply (a) by ψ_k, and integrate over D. Multiply (b) by ψ_k, and sum over all boundary points where the conditions (b) are specified. Add these two expressions. Now repeat the whole process with j and k interchanged, and use the self-adjoint relations of Exercise 5-12 to obtain

$$(\lambda_j - \lambda_k) \left\{ \int_D \psi_j N_{2n}[\psi_k] \, dD + \sum \oint \psi_j \bar{C}_i[\psi_k] \, ds \right\} = 0$$

From this deduce the orthogonality relations to be used in place of (5-31).

5-20. Carry out the procedure outlined in the preceding exercise for the particular problem of Exercise 5-3. Show that the orthogonality relations are

$$\left. \int_0^1 \psi_j \left(\frac{d^4\psi_k}{dx^4} + 4\psi_k \right) dx - \left(\psi_j \frac{d^3\psi_k}{dx^3} \right)_{x=1} = 0 \right\}$$
$$\left. - \int_0^1 \psi_j \frac{d^2\psi_k}{dx^2} dx + \left(\psi_j \frac{d\psi_k}{dx} \right)_{x=1} = 0 \right\} \qquad \lambda_j \neq \lambda_k$$

Verify that alternative orthogonality relations are

$$\left. \int_0^1 \left\{ \frac{d^2\psi_j}{dx^2} \frac{d^2\psi_k}{dx^2} + 4\psi_j\psi_k \right\} dx = 0 \right\}$$
$$\left. \int_0^1 \frac{d\psi_j}{dx} \frac{d\psi_k}{dx} dx = 0 \right\} \qquad \lambda_j \neq \lambda_k$$

5-21. Verify that $u = 3x - 5x^3 + 3x^5 - x^6$ satisfies the boundary conditions for Prob. 5-1, the vibrating string. Apply the *enclosure theorem* to this function, and show that an eigenvalue lies between 9.837 and 10.000.

5-22. Formulate the eigenvalue problem for the buckling of the hinged flexible column shown in Fig. 5-11. There is a uniformly distributed vertical loading of intensity w per unit length. Until buckling occurs, the load is divided equally between the upper and lower springs. Show that in dimensionless variables the formulation is

$$\frac{d^4\psi}{dx^4} = \lambda \left\{ \frac{d\psi}{dx} + \left(x - \frac{1}{2} \right) \frac{d^2\psi}{dx^2} \right\} \qquad 0 < x < 1$$
$$\psi = 0 \qquad \text{at } x = 0, \ x = 1$$
$$\frac{d^2\psi}{dx^2} = 0 \qquad \text{at } x = 0, \ x = 1$$

where $\lambda = wL^3/EI$.

5-23. Study the operator N_2 in the preceding exercise. Show that it is self-adjoint but not positive. Show that

$$\int_0^1 u N_2[u] \, dx$$

vanishes for any function u which is symmetric or antisymmetric about the center of the column. The eigenvalue problem has both positive and negative eigenvalues. Verify this physically, and prove it mathematically by showing that, when λ is replaced by $-\lambda$ and x replaced by $1 - x$, the same formulation is arrived at.

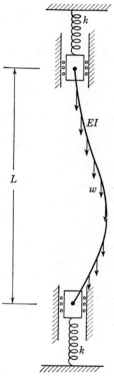

Fig. 5-11. Exercise 5-22.

5-4. Extremum Principles for Eigenvalues

The application of extremum principles to continuous eigenvalue problems is quite analogous to that for lumped-parameter problems. There is one difference in that trial modes for discrete systems are unrestricted, whereas a trial mode for a continuous system must satisfy certain boundary conditions. Furthermore different forms of an extremum principle require different boundary conditions.

A Particular Example. We begin by applying the principle of minimum potential energy to Prob. 5-2, buckling of a column. An *admissible*[1] (i.e., geometrically compatible) deflection of the beam in Fig. 5-3 must satisfy the *essential*[1] boundary conditions

$$\left. \begin{array}{l} y = 0 \\ \dfrac{dy}{dx} = 0 \end{array} \right\} \quad \text{at } x = 0 \qquad\qquad y = 0 \quad \text{at } x = L \qquad (5\text{-}47)$$

The total potential energy is the strain energy of bending plus the potential energy[2] of the load P.

$$\text{PE} = \int_0^L \frac{EI}{2} \left(\frac{d^2y}{dx^2} \right)^2 dx - P \int_0^L \frac{1}{2} \left(\frac{dy}{dx} \right)^2 dx \qquad (5\text{-}48)$$

By introducing the dimensionless variables (5-13) we obtain

$$\Phi = \frac{\text{PE}}{EI/2L} = \int_0^1 \left(\frac{d^2v}{dx^2} \right)^2 dx - \lambda \int_0^1 \left(\frac{dv}{dx} \right)^2 dx \qquad (5\text{-}49)$$

as the nondimensional potential energy associated with an *admissible* function v.

Now the principle of minimum potential energy says that those states which also meet the force-balance requirements give stationary values to (5-49). The buckling modes are such equilibrium states, and moreover they are states of neutral equilibrium. Thus, if λ is the true eigenvalue and $v = \psi$ is the corresponding buckling mode, the whole family of functions $\theta\psi$, where θ is a parameter which may be positive, negative, or zero, are also buckling modes. Since (5-49) is stationary for each member of this family, we deduce that the actual value of (5-49) must remain constant as we vary the parameter θ. Finally, from the case $\theta = 0$, it is clear that this value must be zero. Hence for buckling modes the total potential energy is zero. By setting (5-49) equal to zero we obtain the quotient

$$\frac{\displaystyle\int_0^1 (d^2\psi/dx^2)^2 \, dx}{\displaystyle\int_0^1 (d\psi/dx)^2 \, dx} = \lambda \qquad (5\text{-}50)$$

A corresponding quotient can be constructed for any admissible function v. To see the properties of such a quotient, we return to (5-49)

---

[1] See Sec. 4-4.

[2] This is $-P\delta$, where δ is the in-line displacement of P. For the computation of δ see Exercise 5-26.

and suppose that v differs from a buckling mode ψ by infinitesimals of the first order. The principle of minimum potential energy tells us that, if λ is the true eigenvalue, then Φ will differ from its stationary value (zero, in this case) by an infinitesimal of the second order. Thus, on solving (5-49) for the quotient, we get

$$\lambda_R = \frac{\int_0^1 (d^2v/dx^2)^2 \, dx}{\int_0^1 (dv/dx)^2 \, dx} = \lambda + \frac{\Phi}{\int_0^1 (dv/dx)^2 \, dx} \tag{5-51}$$

which differs from λ by an infinitesimal of second order. We call (5-51) *Rayleigh's quotient*[1] and denote it by λ_R. We have just established that λ_R is *stationary* when v is in the neighborhood of an eigenfunction and actually takes on the eigenvalue when v is the eigenfunction.

Generalization. Suppose that the extremum problem for the functional

$$\Phi[\psi, \lambda] = \Phi_m[\psi] - \lambda\Phi_n[\psi] \tag{5-52}$$

where *admissible* functions ψ must satisfy certain *essential* boundary conditions, is equivalent to the formulation

$$\begin{aligned} M_{2m}[\psi] &= \lambda N_{2n}[\psi] \\ B_i[\psi] &= 0 \qquad i = 1, \ldots, m \end{aligned} \tag{5-53}$$

By equivalent we mean that the essential boundary conditions for (5-52) are included in (5-53) and that the additional boundary conditions and differential equation of (5-53) are obtained by applying the algorithm of the calculus of variations to (5-52), treating λ as a constant. The subscripts m and n on the functionals Φ_m and Φ_n merely indicate that they are related to the operators M_{2m} and N_{2n}, respectively. [A common form for these functionals is an integral over D of a quadratic function of ψ and its derivatives up to order m or n, as the case may be; e.g., see (5-49).]

Now if ψ_p is an eigenfunction of (5-53) and λ_p the corresponding eigenvalue, we can evaluate $\Phi[\psi_p, \lambda_p]$ by repeating the argument of page 294; i.e., since Φ is stationary for any multiple of ψ_p, it must remain at some fixed value and finally this value must be zero. Thus, on setting $\Phi[\psi_p, \lambda_p] = 0$ in (5-52), we find

$$\lambda_p = \frac{\Phi_m[\psi_p]}{\Phi_n[\psi_p]} \tag{5-54}$$

We next consider $\Phi[v, \lambda_p]$, where v is an admissible function in the neighborhood of ψ_p. The stationary property of (5-52) tells us that $\Phi[v, \lambda_p]$ is of second order when v is given a first-order deviation from ψ_p. Therefore *Rayleigh's quotient*

$$\lambda_R = \frac{\Phi_m[v]}{\Phi_n[v]} \tag{5-55}$$

[1] See footnote, page 83.

is *stationary* in the neighborhood of ψ_p since from (5-52) we have

$$\lambda_R = \lambda_p + \frac{\Phi[v, \lambda_p]}{\Phi_n[v]} \tag{5-56}$$

Alternative Form of Rayleigh's Quotient. If we apply integration by parts to the numerator and denominator of λ_R as given by (5-51), we get

$$\lambda_R = \frac{\int_0^1 v(d^4v/dx^4)\, dx + (dv/dx)(d^2v/dx^2)\Big|_0^1 - v(d^3v/dx^3)\Big|_0^1}{\int_0^1 v(-d^2v/dx^2)\, dx + v(dv/dx)\Big|_0^1} \tag{5-57}$$

Now, if v is required to satisfy only the *essential* boundary conditions of (5-47), this reduces to

$$\lambda_R = \frac{\int_0^1 v\frac{d^4v}{dx^4}\, dx + \left(\frac{dv}{dx}\frac{d^2v}{dx^2}\right)_{x=1}}{\int_0^1 v(-d^2v/dx^2)\, dx} \tag{5-58}$$

but if v is required to satisfy *all* the boundary conditions of (5-15), we get the further reduction

$$\lambda_R = \frac{\int_0^1 v(d^4v/dx^4)\, dx}{\int_0^1 v(-d^2v/dx^2)\, dx} \tag{5-59}$$

It may be recognized that this form contains the operators M_4 and N_2 of the governing differential equation (5-14).

Another approach to this same quotient is to take the general differential equation satisfied by a mode ψ_p and its eigenvalue λ_p,

$$M_{2m}[\psi_p] = \lambda_p N_{2n}[\psi_p] \tag{5-60}$$

multiply both sides by ψ_p, and integrate over D to get

$$\lambda_p = \frac{\int_D \psi_p M_{2m}[\psi_p]\, dD}{\int_D \psi_p N_{2n}[\psi_p]\, dD} \tag{5-61}$$

Note that (5-61) is formally similar to the lumped-parameter quotient (2-61). If now we replace ψ_p by a noneigenfunction v, we call[1] the corresponding quotient Rayleigh's quotient.

$$\lambda_R = \frac{\int_D v M_{2m}[v]\, dD}{\int_D v N_{2n}[v]\, dD} \tag{5-62}$$

The quotient (5-59) is precisely of this form.

The preceding derivation of (5-62) from (5-60) does not bring out the stationary property of λ_R. For the particular case which led to (5-59)

---

[1] This is common usage. Rayleigh appears to have used only the form that follows from an extremum principle.

we saw that λ_R was stationary *provided* the functions v satisfied *all* the boundary conditions. A separate proof[1] of the stationary property of (5-62) along the same lines as that for (2-62) can be given provided the expansion theorem can be applied. As in (2-66) λ_R can be considered as a weighted average of the true eigenvalues. This means that λ_R must always lie between the largest and smallest of the true eigenvalues. In positive continuous systems there is no largest eigenvalue, but there is a smallest, λ_1, which equals zero if the system is only positive or is greater than zero if the system is positive definite. In either case we shall always have $\lambda_1 \leq \lambda_R$.

To recapitulate, we have seen that, when Rayleigh's quotient is obtained in the form (5-55) from an extremum principle, it has the stationary property for trial functions which are *admissible* for the extremum principle (i.e., functions which satisfy the *essential* boundary conditions), whereas if Rayleigh's quotient is obtained formally from the governing differential equation as in (5-62), the stationary property holds only for trial functions which satisfy *all* the boundary conditions. In any particular case the connection between the two forms of λ_R and the difference in boundary requirements can be established by integration-by-parts as illustrated by (5-57).

Applications. A common engineering application of Rayleigh's quotient involves simply an intuitive guess at the expected mode shape, followed by an evaluation of λ_R (analytically, graphically, or numerically) to obtain an estimate of the eigenvalue. The accuracy of this procedure clearly depends on the choice of the trial mode. The stationary property of λ_R tends to discount small deviations from the true mode so that surprisingly good estimates can be obtained when the trial mode does have the over-all shape of the desired mode.

To illustrate, we return to Prob. 5-2, buckling of a column. A plausible trial buckling mode is the deflection curve of the same beam when it is loaded by a uniformly distributed load (e.g., its own weight). Omitting a constant factor, this curve[2] is

$$v = x^2(1 - x) - \tfrac{2}{3}x^3(1 - x) \qquad (5\text{-}63)$$

which of course satisfies *all* the boundary conditions, and hence either (5-51) or (5-59) may be used to evaluate $\lambda_R = 21.00$. Since the system is positive definite, we know that $\lambda_1 \leq 21.00$. In actual fact $\lambda_1 = 20.19$.

If in (5-63) we replace the coefficient $\tfrac{2}{3}$ by an adjustable parameter c, we have

$$v = x^2(1 - x) - cx^3(1 - x) \qquad (5\text{-}64)$$

which satisfies the *essential* boundary conditions (5-47) for any value of c but satisfies *all* the boundary conditions only when $c = \tfrac{2}{3}$. Inserting this trial in (5-51) [we can

[1] A proof which makes no use of the expansion theorem is given by Collatz, *op. cit.*, p. 116.

[2] See Exercise 5-27.

no longer use (5-59)], we obtain λ_R as a function of c as shown in Fig. 5-12. By examining this function[1] we find that the best approximation to λ_1 is $\lambda_R = 20.92$, which occurs when $c = 0.635$; that is, a function which does not satisfy *all* the boundary conditions furnishes a better approximation than (5-63), which does.

Upper and Lower Bounds. In addition to Rayleigh's quotient there are upper and lower bounds for an eigenvalue which are also stationary in the neighborhood of the corresponding eigenfunction. These bounds

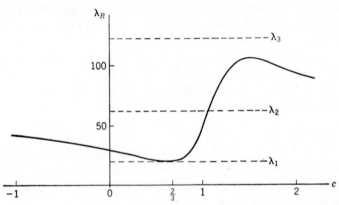

FIG. 5-12. Values of Rayleigh's quotient for the one-parameter family of admissible functions (5-64).

are useful when a high degree of precision is required and rough approximations to the adjacent eigenvalues are available. We give a formulation[2] of these bounds, which is directly analogous to the lumped-parameter formulation of Sec. 2-11. If u and v are approximations to the mode ψ_p of the eigenvalue problem (5-29) which satisfy all the boundary conditions and which satisfy

$$M_{2m}[u] = N_{2n}[v] \qquad (5\text{-}65)$$

throughout D, then λ_p is bounded above and below as follows:

$$\lambda_R - \frac{Q - \lambda_R^2}{\lambda_{p+1} - \lambda_R} \leq \lambda_p \leq \lambda_R + \frac{Q - \lambda_R^2}{\lambda_R - \lambda_{p-1}} \qquad (5\text{-}66)$$

where

$$\lambda_R = \frac{\displaystyle\int_D u M_{2m}[u]\, dD}{\displaystyle\int_D u N_{2n}[u]\, dD} \qquad (5\text{-}67)$$

$$Q = \frac{\displaystyle\int_D v N_{2n}[v]\, dD}{\displaystyle\int_D u N_{2n}[u]\, dD} \qquad (5\text{-}68)$$

---

[1] See page 318 for treatment of this point.

[2] This is essentially the formulation given independently by W. Kohn and T. Kato (see footnote, page 115). Alternative formulations are presented in Exercises 5-36 and 5-37.

and λ_{p-1} and λ_{p+1} are the algebraically nearest eigenvalues below and above λ_p. The quantity $Q - \lambda_R^2$ is an infinitesimal of second order when u and v are given first-order deviations from ψ_p.

Proof of this formulation may be given by an extension of the lumped-parameter argument of Sec. 2-11. The validity of the argument requires that the expansion theorem (5-35) holds and that the following associated series are convergent:

$$\sum_j c_j{}^2, \qquad \sum_j c_j{}^2\lambda_j, \qquad \sum_j c_j{}^2\lambda_j{}^2 \tag{5-69}$$

In applying these bounds it is not necessary to know λ_{p-1} and λ_{p+1} precisely since (5-66) is relatively insensitive to these values. It is necessary, however, for u to be a sufficiently close approximation to ψ_p so that λ_R is actually in between λ_{p-1} and λ_{p+1}; that is, there must exist positive constants δ_+ and δ_- which satisfy the following inequalities:

$$\lambda_{p+1} - \lambda_R > \delta_+$$
$$\lambda_R - \lambda_{p-1} > \delta_- \tag{5-70}$$

These constants may then be used in place of the denominators in (5-66).

Example. We consider Prob. 5-1, vibrating string, with the formulation (5-7) and (5-8). As an approximation to the second mode, $\psi_2 = \sin 2\pi x$, we select the cubic polynomial, $v = x(x - \frac{1}{2})(x - 1)$. This satisfies all the boundary conditions. The function u which satisfies (5-65) is then determined by the following equilibrium problem:

$$M_2[u] \equiv -\frac{d^2u}{dx^2} = x\left(x - \frac{1}{2}\right)(x - 1) \equiv N_0[v]$$

$$B_i[u] = u = 0 \qquad \text{at } \begin{cases} x = 0 \\ x = 1 \end{cases} \tag{5-71}$$

The exact solution is readily found to be

$$u = \frac{x}{20}\left(x - \frac{1}{2}\right)(x - 1)\left(\frac{1}{3} + x - x^2\right) \tag{5-72}$$

and the quotients (5-67) and (5-68) are evaluated as follows:

$$\lambda_R = \frac{\displaystyle\int_0^1 u(-d^2u/dx^2)\,dx}{\displaystyle\int_0^1 u^2\,dx} = \frac{1/36,600}{1/1,330,566} = 39.60 \tag{5-73}$$

$$Q = \frac{\displaystyle\int_0^1 v^2\,dx}{\displaystyle\int_0^1 u^2\,dx} = \frac{\frac{1}{840}}{1/1,330,560} = 1,584.00 \tag{5-74}$$

We then find $Q - \lambda_R^2 = 15.84$, and it remains only to estimate the constants (5-70). Now $\lambda_1 \approx 10$, and $\lambda_3 \approx 90$; so we allow some safety factor if we choose $\delta_+ = 40$ and

$\delta_- = 25$.   Substituting these results into (5-66), we obtain the following bounds on λ_2:

$$39.60 - \frac{15.84}{40} \leq \lambda_2 \leq 39.60 + \frac{15.84}{25} \tag{5-75}$$

$$39.20 \leq \lambda_2 \leq 40.23$$

The true value of λ_2 is $4\pi^2 = 39.48$.

EXERCISES

5-24. Apply Hamilton's principle [see (2-55)] to Prob. 5-1, vibrating string, to obtain

$$\lambda_R = \frac{\displaystyle\int_0^1 (dv/dx)^2\, dx}{\displaystyle\int_0^1 v^2\, dx}$$

and its stationary property.   Integrate by parts to show that this is equivalent to a quotient having the form of (5-62) derivable directly from the differential equation (5-7).   Since *all* the boundary conditions here are also *essential*, the two quotients are completely interchangeable.

5-25. Show that Rayleigh's quotient for a trial function which satisfies all the boundary conditions for Prob. 5-3, rectangular waveguide, is

$$\lambda_R = \frac{\displaystyle\int_D v(-\nabla^2 v)\, dD}{\displaystyle\int_D v^2\, dD}$$

by using (5-62).   Integrate by parts, using (4-70), to obtain

$$\lambda_R = \frac{\displaystyle\int_D \{(\partial v/\partial x)^2 + (\partial v/\partial y)^2\}\, dD}{\displaystyle\int_D v^2\, dD}$$

These two forms are equivalent for the transverse magnetic wave.   Why?   For the transverse electric wave the second form is stationary for arbitrary v, whereas the first is stationary only if v is required to satisfy the *additional* boundary condition $\partial v/\partial n = 0$.

5-26. An inextensible rod, originally straight and of length L, is bent into a shallow curve.   The ends are now a distance $L - \delta$ apart (δ is very small).   Starting with the fact that

$$L = \int_0^{L-\delta} \sqrt{1 + \left(\frac{dy}{dx}\right)^2}\, dx$$

show that for small deflections δ is given approximately by

$$\delta = \int_0^L \frac{1}{2}\left(\frac{dy}{dx}\right)^2 dx$$

5-27. A beam clamped at one end and simply supported at the other as in Fig. 5-3 carries a uniformly distributed transverse load of intensity w per unit length.   Show that the deflection is given by

$$y = \frac{w}{16EI} x^2(L - \tfrac{2}{3}x)(L - x)$$

and thus verify (5-63).

5-28. The function $v = x - x^2$ satisfies the boundary conditions (5-8) for Prob. 5-1.   Obtain λ_R for this trial, using the results of Exercise 5-24.   *Ans.* $\lambda_R = 10.00$.

5-29. Repeat the preceding exercise with $v = x - x^2 + (x - x^2)^2$.

Ans. $\lambda_R = 9.8710$.

5-30. The function $v = (1 - x^2)(1 - 4y^2)$ satisfies the boundary conditions (5-17) for the transverse magnetic wave of Prob. 5-3.   Evaluate λ_R for this trial, using the results of Exercise 5-25.

Ans. $\lambda_R = 12.500$; it is known that $\lambda_1 = 12.377$ for this system.

5-31. Consider the general eigenvalue problem of Exercise 5-12, in which λ occurs in the boundary conditions.   Show that

$$\lambda_R = \frac{\displaystyle\int_D vM_{2m}[v]\, dD + \sum \oint v\bar{B}_i[v]\, ds}{\displaystyle\int_D vN_{2n}[v]\, dD + \sum \oint v\bar{C}_i[v]\, ds}$$

in the notation of Exercise 5-12, reduces to λ_j when $v \equiv \psi_j$.   This is the natural extension of (5-62).   The stationary property can be proved as in (2-66) by using an extended form of the expansion theorem or by integration by parts from an extremum principle, as illustrated in the two following exercises.

5-32. Consider the eigenvalue problem of Exercises 5-4 and 5-5.   The elastic potential energy of a twisted shaft is

$$\frac{1}{2} \int_0^L GI_p \left(\frac{\partial \varphi}{\partial x}\right)^2 dx$$

and the total kinetic energy is

$$\frac{1}{2} \int_0^L \rho I_p \left(\frac{\partial \varphi}{\partial t}\right)^2 dx + \frac{1}{2} J \left[\frac{\partial \varphi}{\partial t}\right]_{x=L}^2$$

Apply Hamilton's principle to obtain the following dimensionless form of Rayleigh's quotient:

$$\lambda_R = \frac{\displaystyle\int_0^1 (dv/dx)^2\, dx}{\displaystyle\int_0^1 v^2\, dx + (v^2)_{x=1}} = \frac{\Phi_m}{\Phi_n}$$

Integrate by parts to get the form indicated in Exercise 5-31.   Evaluate λ_R for $v = x$.

Ans. $\lambda_R = \tfrac{3}{4}$.

5-33. Consider the eigenvalue problem of Exercises 5-2 and 5-3.   Apply the principle of minimum potential energy as in (5-51) to obtain

$$\lambda_R = \frac{\displaystyle\int_0^1 \{(d^2v/dx^2)^2 + 4v^2\}\, dx}{\displaystyle\int_0^1 (dv/dx)^2\, dx} = \frac{\Phi_m}{\Phi_n}$$

Integrate by parts to get the form indicated in Exercise 5-31.   Carefully observe the different roles played by the *essential* boundary conditions, the *additional* boundary conditions, and the boundary conditions which contain λ.

5-34. Apply the algorithm of the calculus of variations to the functional

$$\Phi_m - \lambda\Phi_n = \int_D (\nabla^2\psi)^2 \, dD - \lambda \int_D \left(\frac{\partial\psi}{\partial x}\right)^2 dD$$

where D is the domain of Prob. 5-4, buckling of a plate, and *admissible* functions must satisfy the *essential* boundary condition $\psi = 0$ on the boundary. Use (4-72) in the integration by parts, and show that this extremum problem is equivalent to the formulation (5-21) and (5-22).

5-35. Obtain the two forms of Rayleigh's quotient in the preceding exercise. Which form applies to the trial function $v = (1 - x^2)(1 - 4y^2)$? Evaluate λ_R for this v.   *Ans.* $\lambda_R = 71.00$.

5-36. An alternative formulation of the Kohn-Kato bounds (5-66) which admits some weakening of the restrictions on u and v is applicable when λ_R is available in the form (5-55). If u is an *admissible* function and v is any function which satisfies (5-65)

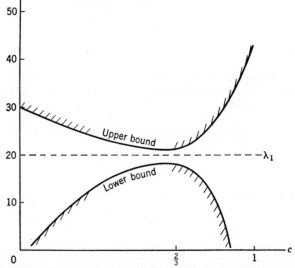

FIG. 5-13. Exercise 5-36. Kohn-Kato bounds for Prob. 5-2, buckling of a column, based on the one-parameter family (5-64).

in D and satisfies only enough of the boundary conditions to ensure that $\Phi_n[v]$ is positive definite, then the inequalities (5-66) still hold provided λ_R and $Q - \lambda_R^2$ are computed as follows:

$$\lambda_R = \frac{\Phi_m[u]}{\Phi_n[u]}$$

$$Q - \lambda_R^2 = \frac{\Phi_n[v - \lambda_R u]}{\Phi_n[u]}$$

In Prob. 5-2, buckling of a column, the required form of λ_R is (5-51). Show that if u is given by (5-64) then

$$v = 12cx(1 - x)$$

meets the above requirements. Obtain upper and lower bounds for λ_1 by using $c = \frac{2}{3}$ and assuming that $\delta_+ = 40$.

Ans. $18.025 \leq \lambda_1 \leq 21.000$. The bounds for other values of c are shown in Fig. 5.13.

5-37. An intermediate form of the Kohn-Kato bounds applies when u satisfies *all* the boundary conditions but v satisfies only the *essential* boundary conditions. This case is covered by the preceding exercise, but show that the text formulation can also be applied if one modification is made: the numerator of (5-68) is replaced by $\Phi_n[v]$. Apply this form of the bounds to Prob. 5-2 with

$$u = \tfrac{1}{60}(3x^2 - x^3 - 5x^4 + 3x^6)$$
$$v = x^2 - x^3$$

Ans. $20.007 \leq \lambda_1 \leq 20.526$.

5-5. Iteration

In Sec. 2-7 we saw how iteration could be used to approximate to either the mode corresponding to the eigenvalue of greatest absolute value or the mode corresponding to the eigenvalue of smallest absolute value. The analogous iteration for self-adjoint positive definite continuous systems is described in this section. Since such systems have a smallest eigenvalue λ_1 but no largest eigenvalue, there is only one iteration procedure and it yields convergence to the mode ψ_1 corresponding to λ_1. The principal features of the process, including the Schwarz quotients, the extrapolation technique, and the treatment of the intermediate eigenvalues, are the same as for lumped-parameter systems.

Performing the Iteration. We consider the self-adjoint positive definite eigenvalue problem with the following formulation:

$$M_{2m}[\psi] = \lambda N_{2n}[\psi]$$
$$B_i[\psi] = 0 \qquad i = 1, \ldots, m \tag{5-76}$$

The iteration procedure entails constructing a sequence of functions

$$v_0, v_1, \ldots, v_r, \ldots \tag{5-77}$$

which satisfy

$$\left. \begin{array}{l} M_{2m}[v_r] = N_{2n}[v_{r-1}] \\ B_i[v_r] = 0 \qquad i = 1, \ldots, m \end{array} \right\} \qquad r = 1, 2, \ldots \tag{5-78}$$

The initial trial, v_0, may be arbitrary, although from a practical standpoint it is useful to select a function which has the general form of the expected mode ψ_1. Note that each step of iteration according to (5-78) involves the solution of an equilibrium problem for v_r. As we have seen in Chap. 4, this is, in itself, a considerable task whenever the operator M_{2m} and the domain D are not both very simple. In any system for which the expansion theorem is valid, the lumped-parameter proof of Sec. 2-7 can be readily extended to prove[1] that in general the sequence (5-77) converges to a multiple of ψ_1. If v_0 is orthogonal to ψ_1, then theoretically the sequence (5-77) will converge to ψ_2, and if v_0 is orthogonal

---

[1] A proof that does not depend on the expansion theorem is given by Collatz, *op. cit.*, p. 163.

to both ψ_1 and ψ_2, the sequence will converge to ψ_3, etc. In practice, if the iteration is performed approximately and if special precautions to the contrary are not taken, small components of ψ_1 will inevitably be introduced and magnified so that the process eventually converges to ψ_1 regardless of the initial trial.

Example. Consider Prob. 5-1 with the formulation (5-7) and (5-8). If we select the initial trial function, $v_0 = x - x^2$, the equilibrium problem for v_1 becomes

$$-\frac{d^2 v_1}{dx^2} = x - x^2$$

$$v_1 = 0 \quad \text{at} \begin{cases} x = 0 \\ x = 1 \end{cases} \tag{5-79}$$

according to (5-78). The operator M_{2m} is sufficiently simple here so that we can solve for v_1 exactly by obtaining the general integral,

$$v_1 = \frac{x^4}{12} - \frac{x^3}{6} + C_1 x + C_2 \tag{5-80}$$

and fixing the constants of integration such that the boundary conditions are satisfied. We thus obtain

$$v_1 = \tfrac{1}{12}(x - 2x^3 + x^4) \tag{5-81}$$

and after a similar computation we get

$$v_2 = \tfrac{1}{360}(3x - 5x^3 + 3x^5 - x^6) \tag{5-82}$$

as the next iterate.

In addition to the differential-equation formulation for this example we had the following integral-equation formulation:

$$\psi(x) = \lambda \int_0^1 \psi(\xi) G(x, \xi)\, d\xi \tag{5-83}$$

This form is especially convenient for iteration. The successive iterates are defined as follows:

$$v_{r+1}(x) = \int_0^1 v_r(\xi) G(x, \xi)\, d\xi \tag{5-84}$$

This is actually the integral-equation representation of the equilibrium problem (5-78). The analogy with inverse matrix iteration [compare (5-83) with (2-87)] should also be observed. To illustrate, we recompute v_1, starting from the same initial trial $v_0 = x - x^2$.

$$v_1 = \int_0^1 (\xi - \xi^2) G(x, \xi)\, d\xi \tag{5-85}$$

Making use of the definition of $G(x, \xi)$ as given in Fig. 5-2 and Eq. (5-6), we have

$$\begin{aligned} v_1 &= \int_0^x (\xi - \xi^2)\xi(1 - x)\, d\xi + \int_x^1 (\xi - \xi^2)x(1 - \xi)\, d\xi \\ &= (1 - x)\left(\frac{x^3}{3} - \frac{x^4}{4}\right) + x\left(\frac{1}{12} - \frac{x^2}{2} + \frac{2x^3}{3} - \frac{x^4}{4}\right) \\ &= \tfrac{1}{12}(x - 2x^3 + x^4) \end{aligned} \tag{5-86}$$

as before. Thus, when the Green's function is known explicitly, it is possible to perform the iteration exactly.

Approximating the Eigenvalue. A sequence of constants which approach λ_1 in the limit may be associated with the sequence of functions (5-77) in several ways. We may use the ratio v_{r-1}/v_r evaluated at any particular point within the domain D. This is analogous to using the ratio of corresponding elements of successive vectors in lumped-parameter systems. For the special eigenvalue problem the totality of these ratios (i.e., if we imagine v_{r-1}/v_r evaluated for all points within D) possess the *enclosure property* described in Sec. 5-3. The approximation to λ_1 obtained in this manner has, in general, the same order of error as the iterated function v_r. Approximations to λ_1 which have second-order error may be obtained from the quotients of Rayleigh and Schwarz. Because of the considerable computation involved in each single step of iteration in continuous systems the increased accuracy available from these quotients is attractive.

The *Schwarz quotients*, μ_k, for the sequence (5-77) may be defined as follows:[1]

$$\mu_0 = \frac{\int_D v_0 M_{2m}[v_0]\, dD}{\int_D v_0 N_{2n}[v_0]\, dD} = \lambda_R[v_0]$$

$$\mu_1 = \frac{\int_D v_0 N_{2n}[v_0]\, dD}{\int_D v_1 N_{2n}[v_0]\, dD}$$

$$\mu_2 = \frac{\int_D v_1 N_{2n}[v_0]\, dD}{\int_D v_1 N_{2n}[v_1]\, dD} = \lambda_R[v_1] \tag{5-87}$$

$$\mu_3 = \frac{\int_D v_1 N_{2n}[v_1]\, dD}{\int_D v_2 N_{2n}[v_1]\, dD}$$

when v_0 satisfies *all* the boundary conditions. Several alternative forms are available. For instance, it can be shown that[2]

$$\mu_1 = \frac{\int_D v_0 M_{2m}[v_1]\, dD}{\int_D v_0 N_{2n}[v_1]\, dD} = \frac{\int_D v_0 N_{2n}[v_0]\, dD}{\int_D v_1 M_{2m}[v_1]\, dD} \tag{5-88}$$

[1] These quotients are analogous with the lumped-parameter quotients defined in Exercise 2-49.

[2] See Exercise 5-42.

The principal properties of the Schwarz quotients are[1] that each μ_k is stationary in the neighborhood of an eigenfunction and that, for a fixed v_0 which is not orthogonal to ψ_1, the sequence μ_k decreases monotonically, approaching λ_1 in the limit.

As an illustration the Schwarz quotient μ_1 is computed according to (5-87) for the example of page 304, as follows:

$$\mu_1 = \frac{\int_0^1 (x - x^2)(x - x^2)\,dx}{\int_0^1 \frac{1}{12}(x - 2x^3 + x^4)(x - x^2)\,dx} = \frac{\frac{1}{30}}{17/5{,}040} = 9.882393 \qquad (5\text{-}89)$$

In a similar manner the first five Schwarz quotients for the same example have been computed and are tabulated with their errors in Table 5-1. The enclosure theorem

TABLE 5-1

Rayleigh quotients	Schwarz quotients	Numerical value	Approximate error
$\lambda_R[v_0]$	μ_0	10.000000	1.0 in 10^2
	μ_1	9.882393	1.3 in 10^3
$\lambda_R[v_1]$	μ_2	9.870968	1.3 in 10^4
	μ_3	9.869754	1.5 in 10^5
$\lambda_R[v_2]$	μ_4	9.869621	1.7 in 10^6

when applied to v_1 yields[2] $9.6 \leq \lambda_1 \leq 12.0$ and when applied to v_2 yields[3] $9.837 \leq \lambda_1 \leq 10.000$.

Extrapolation. Aitken's δ^2 extrapolation formula (2-95) may be applied to the approximate eigenvalues which have been associated (by any fixed method) with three successive iterates. For instance, if we consider the Rayleigh quotients for the three iterates in the above example, we would extrapolate as follows:

$$\lambda_{extrap} = \frac{\mu_0\mu_4 - \mu_2^2}{\mu_0 - 2\mu_2 + \mu_4} \qquad (5\text{-}90)$$

In order to evaluate (5-90), it is necessary to use more precise values of μ_2 and μ_4 than are given in Table 5-1 since both numerator and denominator involve subtraction of nearly equal quantities. When this is done,

---

[1] See Collatz, *op. cit.*, p. 161. The iterates $v_1, v_2, \ldots$, necessarily satisfy *all* the boundary conditions. The quotient μ_1 of (5-87) still has the stationary property if v_0 satisfies only enough boundary conditions to ensure that

$$\int_D v_0 N_{2n}[u]\,dD = \int_D u N_{2n}[v_0]\,dD \qquad \text{and} \qquad \int_D v_0 N_{2n}[v_0]\,dD > 0$$

where u satisfies *all* the boundary conditions but is otherwise arbitrary.

[2] See Fig. 5-9.

[3] See Exercise 5-21.

we obtain $\lambda_{extrap} = 9.869607$, which is in error by 3 parts in 10^7. The success of this extrapolation depends on the geometric convergence of the iteration process and on the assumption that the error in the iterates is small and largely due to one extraneous component. The same assumptions are sufficient[1] to show that Aitken's formula may be applied to three successive Schwarz quotients. Since only two successive iterates are required, this is especially attractive in cases where the actual iteration is difficult.

As an illustration, suppose that only v_0 and v_1 are available in the above example. The quotients μ_0, μ_1, and μ_2 in Table 5-1 could be computed and an extrapolation obtained as follows:

$$\lambda_{extrap} = \frac{\mu_0\mu_2 - \mu_1^2}{\mu_0 - 2\mu_1 + \mu_2} = 9.869739 \tag{5-91}$$

Note that this provides a slightly better approximation than the next Schwarz quotient, μ_3, of Table 5-1.

Intermediate Eigenvalues. The basic iteration procedure converges to the mode corresponding to the smallest eigenvalue of a positive definite system. As we saw in Chap. 2, the basic method can be modified so as to provide other modes by (1) orthogonalizing the initial trial with respect to the known modes and, if necessary, continually purifying the successive iterates and (2) constructing a polynomial operator which has the same modes as the original system but which has altered eigenvalues. Both these methods extend to continuous systems.

A systematic procedure of the first type[2] may be described as follows: Suppose the modes $\psi_1, \ldots, \psi_{p-1}$ corresponding to the $p - 1$ smallest eigenvalues are known. The sequence of functions v_r defined in (5-92) will then converge to a multiple of ψ_p. Each step consists of an iteration followed by a purification.

$$\begin{aligned} M_{2m}[v_r^*] &= N_{2n}[v_{r-1}] \\ B_i[v_r^*] &= 0 \qquad i = 1, \ldots, m \\ v_r &= v_r^* - \sum_{j=1}^{p-1} \frac{\displaystyle\int_D v_r^* N_{2n}[\psi_j]\, dD}{\displaystyle\int_D \psi_j N_{2n}[\psi_j]\, dD}\, \psi_j \end{aligned} \tag{5-92}$$

When the preceding modes, ψ_j, have themselves been determined approximately, the process is subject to the same degradation of accuracy as was noted for matrix systems.

As an almost trivial example of this method, we shall obtain an approximation to the mode, ψ_2, for the transverse electric wave of Prob. 5-3. This system is only posi-

---

[1] See Exercise 5-46.

[2] J. J. Koch, Bestimmung höherer kritischer Drehzahlen schnell laufender Wellen, *Proc. 2d Intern. Congr. Appl. Mech.*, Zurich, 1926, pp. 213–218.

tive, which means that $\lambda_1 = 0$. The corresponding mode is $\psi_1 = 1$. Ordinary iteration fails for such systems, but if the sequence of iterates is kept orthogonal to ψ_1, they will converge to ψ_2. The iteration and purification of (5-92) take the following form for this problem:

$$-\nabla^2 v_r^* = v_{r-1}$$

$$\frac{\partial v_r^*}{\partial n} = 0 \quad \text{on boundary of Fig. 5-4} \tag{5-93}$$

$$v_r = v_r^* - \frac{1}{2}\int_{-1}^{1}\int_{-\frac{1}{2}}^{\frac{1}{2}} v_r^* \, dx \, dy$$

The purification condition here is equivalent to requiring that v_r have zero average value. Choosing[1] the initial trial $v_0 = 3x - x^3$, we find

$$v_1^* = \tfrac{1}{20}(25x - 10x^3 + x^5) + C_1$$

with C_1 arbitrary, and hence $v_1 = \tfrac{1}{20}(25x - 10x^3 + x^5)$. Rayleigh's quotient for v_1 is 2.467438, which differs from λ_2 by 1.5 parts in 10^5.

The method of *polynomial operators*[2] permits iteration to be used to find ψ_p when the $p - 1$ smallest eigenvalues are known (it is not necessary to know the corresponding modes). For simplicity of exposition we restrict the discussion to the particular case provided by Prob. 5-1. We introduce an operational symbol $G[\psi]$ defined as follows:

$$G[\psi] = \int_0^1 \psi(\xi)G(x, \xi) \, d\xi \tag{5-94}$$

The governing integral equation (5-83) and the ordinary iteration formula (5-84) have the following appearance in this symbolism:

$$\psi = \lambda G[\psi]$$
$$v_{r+1} = G[v_r] \tag{5-95}$$

Iteration with G leads to convergence, as we have seen, to the mode corresponding to the smallest λ, that is, the largest value of $1/\lambda$. If, instead of iterating with G, we iterate with a polynomial function of G, we get convergence to that mode for which the same polynomial function of $1/\lambda$ has greatest absolute value. A demonstration of this can be given along the same lines as the proof in Sec. 2-8 for matrix polynomial operators. Let us investigate the values of a few simple polynomials in $1/\lambda$ for the eigenvalues of our system. In Fig. 5-14 the spectrum of $1/\lambda$ for Prob. 5-1 is sketched. The polynomials $P_1 = 1/\lambda$, $P_2 = 1/\lambda$ $(a - 1/\lambda)$, and $P_3 = 1/\lambda$ $(a - 1/\lambda)$ $(b - 1/\lambda)$, where a and b are approximations to $1/\lambda_1$ and $1/\lambda_2$, have been superimposed on the spectrum.

[1] See Exercise 5-39.

[2] G. F. Carrier, On the Determination of the Eigenfunctions of Fredholm Equations, *J. Math. and Phys.*, **27**, 82–83 (1948); H. Bückner, Ein unbeschränkt andwendbares Iterations verfahren für Fredholmsche Integral gleichungen, *Math. Nachr.*, **2**, 304–313 (1949).

It is clear from the figure that P_1 has its greatest value for $1/\lambda_1$, that P_2 has its greatest value for $1/\lambda_2$, and that P_3 has its greatest value for $1/\lambda_3$. Hence iteration with the polynomial operators G, $G(a - G)$, and $G(G - a)(G - b)$ will yield convergence to ψ_1, ψ_2, and ψ_3, respectively.

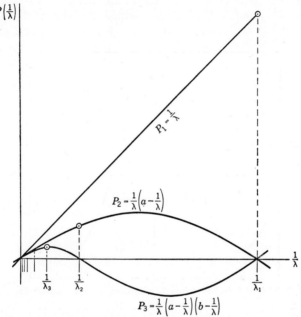

FIG. 5-14. Polynomials which produce convergence toward the first, second, and third modes, respectively.

As an illustration we shall perform an iteration with the polynomial operator $G(a - G)$. Returning to the definition, (5-94), of the symbol G, we have the following explicit formula for the iteration $v_1 = G(a - G)v_0$:

$$v_1(x) = \int_0^1 \left\{ av_0(\xi) - \int_0^1 v_0(t)G(\xi, t)\, dt \right\} G(x, \xi)\, d\xi \qquad (5\text{-}96)$$

This same iteration can be represented alternatively as the solution of successive differential equations in the following manner:

$$M_2[av_0 - v_1^*] = N_0[v_0] \qquad v_1^* = 0 \qquad \text{at} \quad \begin{cases} x = 0 \\ x = 1 \end{cases}$$

$$M_2[v_1] = N_0[v_1^*] \qquad v_1 = 0 \qquad \text{at} \quad \begin{cases} x = 0 \\ x = 1 \end{cases} \qquad (5\text{-}97)$$

According to Fig. 5-14, if v_0 is not orthogonal to ψ_2 and if a is a reasonably good approximation to $1/\lambda_1$, continued iteration with these formulas should yield convergence to a multiple of ψ_2. If v_0 is orthogonal to ψ_2, the process will converge to ψ_3 provided no component of ψ_2 is permitted to enter during the computation.   In this system the odd modes, ψ_1, ψ_3, ψ_5, . . . , are symmetric about the mid-point of the string, and the even modes, ψ_2, ψ_4, . . . , are antisymmetric, and all symmetric functions are orthogonal to all the even modes.   Thus, if we choose for v_0 a symmetric function and if we choose for a a good approximation to $1/\lambda_1$, we should get convergence to ψ_3.   To test the method, one step of iteration according to (5-96) has been performed with $v_0 = x(1 - x)$ and $a = 1/\mu_4$ (μ_4 is

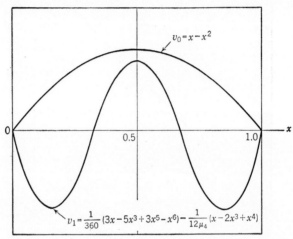

$$v_0 = x - x^2$$

$$v_1 = \frac{1}{360}(3x - 5x^3 + 3x^5 - x^6) - \frac{1}{12\mu_4}(x - 2x^3 + x^4)$$

Fig. 5-15. Initial trial and first iterate obtained by using polynomial operator P_2.

the approximation to λ_1 listed in Table 5-1).   The initial trial function v_0 and the first iterate are sketched (not to the same scale) in Fig. 5-15. Although v_0 has the general shape of the first mode, v_1 is already a fairly good approximation to ψ_3.   Rayleigh's quotient for v_1 is 90.05, which is only 1.4 per cent higher than $\lambda_3 = 9\pi^2$.

EXERCISES

5-38. Apply one step of iteration to the initial trial $v_0 = x^2 - x^3$ for Prob. 5-2 with the formulation (5-14) and (5-15).   *Ans.* $v_1 = (3x^2 - x^3 - 5x^4 + 3x^5)/60$.

5-39. Apply two steps of iteration according to (5-93) to the initial trial $v_0 = 6x$ for the transverse electric wave of Prob. 5-3.

$$Ans. \ v_1 = 3x - x^3, \ v_2 = (25x - 10x^3 + x^5)/20.$$

5-40. The initial trial in Exercise 5-38 is an *admissible* function for the extremum problem (5-49), but it does not satisfy *all* the boundary conditions (5-15).   Verify that it does meet the requirements of the footnote on page 306.   (Use integration by parts.)   Evaluate the Rayleigh's quotient (5-51) for v_0.   *Ans.* $\lambda_R = 30$. Compute the Schwarz quotient μ_1 according to (5-87).   *Ans.* $\mu_1 = 21.54$.

5-41. The initial trial in Exercise 5-39 does not satisfy all the boundary conditions of Fig. 5-4, but it is an admissible function for the second form of λ_R in Exercise 5-25. It also meets the requirements of the footnote on page 306.   Evaluate $\lambda_R[v_0]$.

$$Ans. \ \lambda_R = 3.$$

Compute some of the Schwarz quotients according to (5-87).

$$Ans.\ \mu_1 = 2\tfrac{1}{2};\ \mu_2 = 2\tfrac{8}{17};\ \mu_3 = 2\tfrac{29}{62};\ \mu_4 = 2\tfrac{323}{691}.$$

5-42. Verify the alternative forms of μ_1 given in (5-88) by using the self-adjoint property together with (5-78).

5-43. For the *special* eigenvalue problem where N_{2n} has the form (5-28) show that the Schwarz quotient μ_1, which ordinarily is expressed in terms of both v_0 and v_1, can be expressed entirely in terms of v_1 as follows:

$$\mu_1 = \frac{\displaystyle\int_D \frac{1}{g}\,(M_{2m}[v_1])^2\,dD}{\displaystyle\int_D v_1 M_{2m}[v_1]\,dD}$$

Note that v_1 must satisfy *all* the boundary conditions.

5-44. Show that a dimensionless formulation for the vibrating beam of Exercise 5-1 is

$$\frac{d^4\psi}{dx^4} = \lambda\psi \qquad 0 < x < 1 \qquad \begin{cases} \psi = \dfrac{d\psi}{dx} = 0 \text{ at } x = 0 \\[2mm] \psi = \dfrac{d^2\psi}{dx^2} = 0 \text{ at } x = 1 \end{cases}$$

This is a *special* eigenvalue problem. Apply the quotient of the preceding exercise to the trial function (5-63) to obtain an upper bound for λ_1.

$$Ans.\ \mu_1 = 320;\ \lambda_1 \text{ is known to be } 237.721.$$

5-45. Show that a dimensionless formulation of the problem of determining the natural vibrations for a uniform *free-free* beam is

$$\frac{d^4\psi}{dx^4} = \lambda\psi \qquad 0 < x < 1 \qquad \begin{cases} \dfrac{d^2\psi}{dx^2} = \dfrac{d^3\psi}{dx^3} = 0 \text{ at } x = 0 \\[2mm] \dfrac{d^2\psi}{dx^2} = \dfrac{d^3\psi}{dx^3} = 0 \text{ at } x = 1 \end{cases}$$

Verify that $\psi_1 = 1$ and $\psi_2 = x - \tfrac{1}{2}$ are eigenfunctions. What is the physical interpretation of these modes? Outline an iteration procedure which would converge to ψ_3. Verify that $v_0 = x(1-x) - \tfrac{1}{6}$ is orthogonal to ψ_1 and ψ_2. Apply one step of iteration to v_0, and obtain an approximation to λ_3.

$Ans.\ v_1 = (-3 + 14x - 70x^4 + 84x^5 - 28x^6)/10{,}080;\ \mu_1 = 504.0;\ \lambda_3$ is known to be 500.564.

5-46. Assume that an initial trial vector v_0 of (5-77) can be expanded in a convergent series of *orthonormal* eigenfunctions.

$$v_0 = \sum_{j=1}^{\infty} c_j\psi_j$$

Show that the Schwarz quotient μ_k is given by

$$\mu_k = \lambda_1 \frac{1 + \displaystyle\sum_{j=2}^{\infty} \left(\frac{c_j}{c_1}\right)^2 \left(\frac{\lambda_1}{\lambda_j}\right)^{k-1}}{1 + \displaystyle\sum_{j=2}^{\infty} \left(\frac{c_j}{c_1}\right)^2 \left(\frac{\lambda_1}{\lambda_j}\right)^{k}}$$

If the infinite sums are small in comparison with unity and if moreover the principal contribution to them occurs for a single value of j, show that the Schwarz quotients have approximately the form (2-86) and hence that δ^2 extrapolation may be applied to advantage to three successive μ_k.

5-47. Apply δ^2 *extrapolation* to μ_2, μ_3, and μ_4 of Exercise 5-41.

Ans. 2.467402; the true value is $\pi^2/4 = 2.467401$. Note that to get six decimal places in this result the individual Schwarz quotients must be evaluated to nine decimal places.

5-48. Iteration may be applied to systems in which the eigenvalue occurs in the boundary conditions by using the following procedure (the notation is that of Exercise 5-12):

$$M_{2m}[v_r] = N_{2n}[v_{r-1}] \quad \text{in } D$$

$$\left.\begin{array}{c} B_i[v_r] = 0 \\ \bar{B}_i[v_r] = \bar{C}_i[v_{r-1}] \end{array}\right\} \quad \text{on the boundary}$$

Apply this to the system of Exercise 5-5, starting with $v_0 = x$.

$$Ans. \ v_1 = \tfrac{3}{2}x - \tfrac{1}{6}x^3; \ v_2 = \tfrac{49}{24}x - \tfrac{1}{4}x^3 + \tfrac{1}{120}x^5.$$

5-49. The definition of the Schwarz quotient μ_1 of (5-87) can be extended to include systems in which the eigenvalue occurs in the boundary conditions by introducing the following modifications:

$$\mu_1 = \frac{\displaystyle\int_D v_0 N_{2n}[v_0]\,dD + \sum \oint v_0 \bar{C}_i[v_0]\,ds}{\displaystyle\int_D v_1 N_{2n}[v_0]\,dD + \sum \oint v_1 \bar{C}_i[v_0]\,ds}$$

The notation is that of Exercise 5-12. The modification of the other μ_k follows the same pattern. Obtain the Schwarz quotients for the preceding exercise.

$$Ans. \ \mu_1 = \tfrac{20}{27} = 0.7407; \ \mu_2 = \tfrac{567}{766} = 0.7402.$$

5-6. Trial Solutions with Undetermined Parameters

All the methods of this section begin with the choice of a trial family of solutions containing r undetermined parameters and end up with an r-degree-of-freedom eigenvalue problem which is an approximation to the given continuous eigenvalue problem. The actual criteria employed to effect the reduction have much in common with those described in Sec. 4-5. There are weighted-residual procedures which make use of the differential-equation formulation, and there are procedures which capitalize on the stationary properties of the Rayleigh and Schwarz quotients.

We consider the linear self-adjoint positive continuous eigenvalue problem in a domain D,

$$M_{2m}[\psi] = \lambda N_{2n}[\psi] \qquad B_i[\psi] = 0 \qquad i = 1, \ldots, m \qquad (5\text{-}98)$$

in which the eigenvalue λ does not enter[1] into the boundary conditions. A trial solution v is assumed in the linear form,

$$v = \sum_{j=1}^{r} c_j\varphi_j \tag{5-99}$$

[1] For the treatment of the case where λ occurs in the boundary conditions see Exercise 5-60.

where the φ_j are linearly independent known functions in the domain and the c_j are undetermined parameters. As in all such procedures the choice of the φ_j requires considerable imagination if a maximum of information is to be obtained with a limited number of parameters. The boundary conditions to be satisfied by the φ_j depend on the particular criterion employed. All the practically important procedures lead to a matrix eigenvalue problem

$$AC = \lambda BC \qquad (5\text{-}100)$$

where C is a column matrix of the unknown c_j and A and B are square matrices of r rows and columns. The eigenvalues of (5-100) provide approximations to r of the eigenvalues of (5-98). The eigenvectors of (5-100) provide sets of parameters which when substituted back into (5-99) yield approximations to the corresponding r modes of (5-98). The differences in the following methods lie only in the manner in which the elements of the square matrices, A and B, are obtained.

Weighted-residual Procedures. For this class of procedures the φ_j in the trial family (5-99) are taken as functions which satisfy *all* the boundary conditions. The equation residual is then defined as follows:

$$R = -M_{2m}[v] + \lambda N_{2n}[v] = -\sum_{j=1}^{r} c_j\{M_{2m}[\varphi_j] - \lambda N_{2n}[\varphi_j]\} \qquad (5\text{-}101)$$

If the family (5-99) contains any true eigenfunctions, then there exist sets of c_j and values of λ for which R vanishes identically in the domain D. If no eigenfunctions are contained in the family, approximations may be obtained by requiring that r weighted averages of R be zero.

Using *collocation*, one would select r locations in D and require that R be zero at these points. The resulting r equations are of the form (5-100) with the elements of A and B given by

$$a_{jk} = M_{2m}[\varphi_k(P_j)] \qquad b_{jk} = N_{2n}[\varphi_k(P_j)] \qquad (5\text{-}102)$$

where the $P_j(j = 1, \ldots, r)$ are the selected points. Using the *subdomain method*, one would select r subdomains D_j and set the average of R equal to zero in each of these domains. The resulting equations have the form of (5-100) with

$$a_{jk} = \int_{D_j} M_{2m}[\varphi_k]\, dD \qquad b_{jk} = \int_{D_j} N_{2n}[\varphi_k]\, dD \qquad (5\text{-}103)$$

Using the *Galerkin method*, one selects the φ_j of (5-99) as weighting functions and sets the r weighted averages of R equal to zero. The resulting equations are again of the form (5-100) with

$$a_{jk} = \int_{D} \varphi_j M_{2m}[\varphi_k]\, dD \qquad b_{jk} = \int_{D} \varphi_j N_{2n}[\varphi_k]\, dD \qquad (5\text{-}104)$$

The *least-squares* criterion may also be applied by asking that the partial derivatives with respect to the c_j of the integral of R^2 over D be zero; however, the resulting equations no longer have the form (5-100). The equations contain[1] factors of λ^2 as well as λ. Although it is still possible to obtain useful approximations[2] by this method, the labor involved is considerably greater than in the other weighted-residual procedures.

In the collocation and subdomain procedures the matrices A and B are not in general symmetric. This means that the eigenvalues and eigenvectors of (5-100) cannot be guaranteed to be real. In the Galerkin procedure the matrices A and B will be symmetric. This may be seen by applying the self-adjoint relations (5-26) to the elements (5-104). Furthermore, if the operator N_{2n} is positive definite, then it can be shown[3] that the matrix B will be positive definite. A similar result holds for the operator M_{2m} and the matrix A. Thus, of all the weighted-residual procedures, only the Galerkin method leads to a lumped-parameter eigenvalue problem of the class considered in Chap. 2. Note that in the Galerkin method, if the φ_j are orthogonal to each other with respect to M_{2m} (or N_{2n}), then A (or B) will be a diagonal matrix.

Stationary-quotient Procedures. In the Rayleigh-Ritz[4] method a trial family (5-99) is selected in which the φ_j are *admissible functions;* i.e., they need satisfy only the *essential* boundary conditions. This trial is substituted into the extremum-principle form of Rayleigh's quotient (5-55) and the sets of c_j which give stationary values are sought. We have

$$\frac{\partial \lambda_R}{\partial c_j} = \frac{\partial}{\partial c_j}\left(\frac{\Phi_m}{\Phi_n}\right) = \frac{1}{\Phi_n{}^2}\left(\Phi_n \frac{\partial \Phi_m}{\partial c_j} - \Phi_m \frac{\partial \Phi_n}{\partial c_j}\right) = 0 \qquad (5\text{-}105)$$

provided

$$\frac{\partial \Phi_m}{\partial c_j} = \lambda_R \frac{\partial \Phi_n}{\partial c_j} \qquad (5\text{-}106)$$

When r such equations are evaluated for a particular case, we obtain a system of the form (5-100). It can be shown that the matrices A and B

---

[1] See Exercise 5-51.

[2] In general the solution leads to complex values of λ. The real parts of these roots furnish the approximation. See R. A. Frazer, W. P. Jones, and S. W. Skan, Approximations to Functions and to the Solutions of Differential Equations, *Aeronaut. Research Comm. Rept. and Mem.* 1799 (1937).

[3] See Exercise 5-52.

[4] The method was employed by Rayleigh in 1870 in a computation of the natural frequency of a resonator with open end. See J. W. Strutt, 3d Baron Rayleigh, "The Theory of Sound," 2d ed., vol. II, Dover Publications, New York, 1945, Appendix A. The method itself was independently presented and studied for both equilibrium and eigenvalue problems by W. Ritz, Über eine neue Methode zur Lösung gewisser Variationsprobleme der mathematischen Physik, *J. reine u. angew. Math.*, **135**, 1–61 (1909).

will be symmetric and that B will be positive definite if the eigenvalue problem is self-adjoint and positive. If the φ_j actually satisfy all the boundary conditions, then the Rayleigh-Ritz procedure leads to precisely the same[1] equations (5-100) as the Galerkin method.

A similar criterion to be applied to a trial solution with undetermined parameters can be based on the stationary property of the *Schwarz quotient* μ_1. A method proposed by Grammel[2] may be considered in this light. The treatment here is restricted to the *special* positive definite eigenvalue problem

$$M_{2m}[\psi] = \lambda g \psi \qquad B_i[\psi] = 0 \qquad i = 1, \ldots, m \qquad (5\text{-}107)$$

where g is a positive function in the domain D.

It can be shown[3] that for this problem μ_1 can be written in the form

$$\mu_1 = \frac{\int_D (1/g)(M_{2m}[v])^2 \, dD}{\int_D v M_{2m}[v] \, dD} \qquad (5\text{-}108)$$

and that this is stationary in the neighborhood of an eigenfunction provided the functions v satisfy *all* the boundary conditions. We can therefore introduce a trial solution (5-99) for v into (5-108) and obtain equations for the parameters in the manner of Rayleigh-Ritz. The functions φ_j in (5-99) must satisfy all the boundary conditions. We have $\partial \mu_i / \partial c_j = 0$ if

$$\frac{\partial}{\partial c_j} \int_D \frac{1}{g} \left(\sum_{k=1}^{r} c_k M_{2m}[\varphi_k] \right)^2 dD = \mu_1 \frac{\partial}{\partial c_j} \int_D \left(\sum_{k=1}^{r} c_k \varphi_k \right) \left(\sum_{k=1}^{r} c_k M_{2m}[\varphi_k] \right) dD$$

$$(5\text{-}109)$$

by analogy with (5-106). Carrying out the indicated differentiations and letting j run from 1 to r lead to a set of equations of the form (5-100) with μ_1 playing the role of λ and the elements of the matrices A and B given as follows:

[1] See Exercise 5-53.

[2] R. Grammel, Ein neues Verfahren zur Lösung technischer Eigenwertprobleme, *Ing.-Arch.*, **10**, 35–46 (1939). Grammel's basic idea consisted in joining a step of iteration to the Galerkin process. The connection with the Schwarz quotient is due to L. Collatz, "Eigenwertprobleme und ihre numerische Behandlung," Akademische Verlagsgesellschaft m.b.H., Leipzig, 1945, p. 232. For a more general stationary principle which leads to the same results in special cases see E. Reissner, Complementary Energy Procedure for Flutter Calculations, *J. Aeronaut. Sci.*, **16**, 316–317 (1949), and the commentary by A. I. vandeVooren and J. H. Greidanus, Complementary Energy Method in Vibration Analysis, *J. Aeronaut. Sci.*, **17**, 454–455 (1950).

[3] See Exercise 5-43.

$$a_{jk} = \int_D \frac{1}{g} M_{2m}[\varphi_j] M_{2m}[\varphi_k] \, dD$$

$$b_{jk} = \int_D \varphi_j M_{2m}[\varphi_k] \, dD$$

(5-110)

Note that A and B will be symmetric.

Suppose that both the Rayleigh-Ritz and Grammel procedures are applied to the same problem (5-107) with the same trial family (5-99). This family necessarily satisfies *all* the boundary conditions for the Grammel method to be applicable. Let the approximate solution obtained by the Rayleigh-Ritz method be

$$\lambda_{1,R} \leq \lambda_{2,R} \leq \cdots \leq \lambda_{r,R}$$
$$v_{1,R}, \quad v_{2,R}, \quad \ldots, \quad v_{r,R}$$

(5-111)

where the $v_{p,r}$ are the approximate modes obtained by substituting the set of c_j corresponding to $\lambda_{p,R}$ back into (5-99). Similarly let the approximate solution by the Grammel procedure be

$$\lambda_{1,G} \leq \lambda_{2,G} \leq \cdots \leq \lambda_{r,G}$$
$$v_{1,G}, \quad v_{2,G}, \quad \ldots, \quad v_{r,G}$$

(5-112)

Then it may be shown[1] that the $v_{p,R}$ are orthogonal with respect to both M_{2m} and g, while the $v_{p,G}$ are orthogonal with respect to M_{2m} but not, in general, orthogonal to g; that is,

$$\left.\begin{array}{l} \int_D v_{p,R} M_{2m}[v_{q,R}] \, dD = 0 \\[2mm] \int_D g v_{p,R} v_{q,R} \, dD = 0 \end{array}\right\} \quad \text{if } \lambda_{p,R} \neq \lambda_{q,R}$$

$$\left.\begin{array}{l} \int_D v_{p,G} M_{2m}[v_{q,G}] \, dD = 0 \\[2mm] \int_D g v_{p,G} v_{q,G} \, dD \neq 0 \end{array}\right\} \quad \text{if } \lambda_{p,G} \neq \lambda_{q,G}$$

(5-113)

Furthermore, if the first r true eigenvalues of the continuous system are $\lambda_1 \leq \lambda_2 \cdots \leq \lambda_r$, then[2] the following inequalities exist:

$$\lambda_1 \leq \lambda_{1,R} \leq \lambda_{1,G}$$
$$\lambda_2 \leq \lambda_{2,R} \leq \lambda_{2,G}$$
$$\cdots \cdots \cdots \cdots$$
$$\lambda_r \leq \lambda_{r,R} \leq \lambda_{r,G}$$

(5-114)

---

[1] See Exercises 5-54 and 5-56.

[2] The inequalities involving the $\lambda_{p,R}$ are proved in L. Collatz, "Eigenwertaufgaben mit technischen Andwendung," Akademische Verlagsgesellschaft m.b.H., Leipzig, 1949, p. 217. The inequalities involving the $\lambda_{p,G}$ can be proved by an extension of the same argument, using the fact that $\lambda_R = \mu_2$ and that $\mu_2 \leq \mu_1$.

Thus the Rayleigh-Ritz method produces a more accurate approximation from the same trial solution. The Grammel process may, however, be simpler to carry out, as the coefficients (5-110) are usually easier to compute than the corresponding coefficients in the Rayleigh-Ritz method. The Grammel procedure is often begun, not by selecting the φ_j of (5-99), but by selecting a set of simpler functions and performing a single step[1] of iteration on each to obtain the φ_j.

Examples. We consider first Prob. 5-1 with the formulation (5-7) and (5-8). As a trial solution we take the following three-parameter family:

$$v = c_1 x(1 - x) + c_2 x^2(1 - x) + c_3 x^3(1 - x) \qquad (5\text{-}115)$$

Note that each φ_j satisfies the boundary conditions. The equation residual for this trial is

$$R = -M_2[v] + \lambda N_0[v] = \frac{d^2v}{dx^2} + \lambda v$$
$$= -2c_1 + (2 - 6x)c_2 + (6x - 12x^2)c_3 + \lambda\{(x - x^2)c_1 + (x^2 - x^3)c_2 \qquad (5\text{-}116)$$
$$+ (x^3 - x^4)c_3\}$$

By applying the weighted-residual procedures to this we obtain matrix eigenvalue problems for λ and the c_j. Thus, using *collocation*, we select the three locations, $x = \frac{1}{4}$, $x = \frac{1}{2}$, and $x = \frac{3}{4}$, and set $R = 0$ at these points to obtain

$$\begin{bmatrix} 2 & -\frac{1}{2} & -\frac{3}{4} \\ 2 & 1 & 0 \\ 2 & \frac{5}{2} & \frac{9}{4} \end{bmatrix} \begin{bmatrix} c_1 \\ c_2 \\ c_3 \end{bmatrix} = \lambda \begin{bmatrix} \frac{3}{16} & \frac{3}{64} & \frac{3}{256} \\ \frac{1}{4} & \frac{1}{8} & \frac{1}{16} \\ \frac{3}{16} & \frac{9}{64} & \frac{27}{256} \end{bmatrix} \begin{bmatrix} c_1 \\ c_2 \\ c_3 \end{bmatrix} \qquad (5\text{-}117)$$

where the elements of the square matrices, A and B, are given by (5-102). With considerably more computation we can apply the *Galerkin method* and obtain

$$\begin{bmatrix} \frac{1}{3} & \frac{1}{6} & \frac{1}{10} \\ \frac{1}{6} & \frac{2}{15} & \frac{1}{10} \\ \frac{1}{10} & \frac{1}{10} & \frac{3}{35} \end{bmatrix} \begin{bmatrix} c_1 \\ c_2 \\ c_3 \end{bmatrix} = \lambda \begin{bmatrix} \frac{1}{30} & \frac{1}{60} & \frac{1}{105} \\ \frac{1}{60} & \frac{1}{105} & \frac{1}{168} \\ \frac{1}{105} & \frac{1}{168} & \frac{1}{252} \end{bmatrix} \begin{bmatrix} c_1 \\ c_2 \\ c_3 \end{bmatrix} \qquad (5\text{-}118)$$

where the elements of the square matrices are given by (5-104). Since this problem has the form (5-107), we may also apply the *Grammel method* to get

$$\begin{bmatrix} 4 & 2 & 2 \\ 2 & 4 & 4 \\ 2 & 4 & \frac{24}{5} \end{bmatrix} \begin{bmatrix} c_1 \\ c_2 \\ c_3 \end{bmatrix} = \lambda \begin{bmatrix} \frac{1}{3} & \frac{1}{6} & \frac{1}{10} \\ \frac{1}{6} & \frac{2}{15} & \frac{1}{10} \\ \frac{1}{10} & \frac{1}{10} & \frac{3}{35} \end{bmatrix} \begin{bmatrix} c_1 \\ c_2 \\ c_3 \end{bmatrix} \qquad (5\text{-}119)$$

in which the elements of A and B are given by (5-110). If the *Rayleigh-Ritz method* were used, the system obtained would be identical with the Galerkin system. Note that the collocation system (5-117) is unsymmetric but that both the Galerkin and Grammel systems, (5-118) and (5-119), are symmetric. Note also that the matrix B of the Grammel system is identical with the matrix A of the Galerkin system.

---

[1] For certain vibrating beam problems Grammel, *op. cit.*, shows that only a half step of iteration is required. In this case the original set of functions represent displacements, and the iteration is carried only as far as the bending moments corresponding to the first iterates (see Exercise 5-57).

The solutions of these three systems have been obtained, and the approximate eigenvalues are compared with the true eigenvalues in Table 5-2. Note that the

TABLE 5-2

	Exact continuous solution	Collocation approximation	Galerkin approximation	Grammel approximation
λ_1	9.8696	9.9678	9.8698	9.8751
λ_2	39.4784	32.0000	42.0000	60.0000
λ_3	88.8246	51.3656	102.1302	170.1249

eigenvalues of the Galerkin (identical with Rayleigh-Ritz) and Grammel solutions satisfy the inequalities of (5-114). In this instance the collocation system, although unsymmetric, yielded three real solutions. If, instead of selecting the locations $\frac{1}{4}$, $\frac{1}{2}$, and $\frac{3}{4}$, the locations 0, $\frac{1}{2}$, 1 are chosen, the resulting three-degree-of-freedom system has[1] only a single eigenvalue and a single corresponding mode in contradistinction to the systems studied in Chap. 2.

In the above example the trial family satisfied all the boundary conditions, and hence all criteria were applicable. We next consider Prob. 5-2 with the formulation (5-14) and (5-15) and make use of a trial solution which is made up of *admissible* functions. The family

$$v = c_1 x^2 (1 - x) + c_2 x^3 (1 - x) \qquad (5\text{-}120)$$

satisfies the *essential* boundary conditions of (5-15) for all values of c_1 and c_2, but the additional boundary condition, $d^2\psi/dx^2 = 0$ at $x = 1$, is satisfied only when $c_2/c_1 = -\frac{2}{3}$. The Rayleigh-Ritz method is still applicable to such a trial. The appropriate form of Rayleigh's quotient is given by (5-51) and the Rayleigh-Ritz equations (5-106) become

$$\frac{\partial}{\partial c_j} \int_0^1 \left(\frac{d^2v}{dx^2} \right)^2 dx = \lambda_R \frac{\partial}{\partial c_j} \int_0^1 \left(\frac{dv}{dx} \right)^2 dx \qquad j = 1, 2 \qquad (5\text{-}121)$$

Substituting (5-120) into (5-121) and carrying out the indicated operations lead to the following eigenvalue problem for λ_R, c_1, and c_2:

$$\begin{bmatrix} 8 & 8 \\ 8 & \frac{48}{5} \end{bmatrix} \begin{bmatrix} c_1 \\ c_2 \end{bmatrix} = \lambda_R \begin{bmatrix} \frac{4}{15} & \frac{1}{5} \\ \frac{1}{5} & \frac{6}{35} \end{bmatrix} \begin{bmatrix} c_1 \\ c_2 \end{bmatrix} \qquad (5\text{-}122)$$

The eigenvalues of (5-122) are found to be 20.92 and 107.08, whereas the first two true eigenvalues of the continuous system are 20.19 and 59.68. It may be noted that the trial family (5-120) is essentially the same as the one-parameter family (5-64) used to construct the curve sketched in Fig. 5-12. The above computation has, in fact, been equivalent to finding the minimum and maximum values of this curve.

As a final example, we consider Prob. 5-4, buckling of a plate, with the formulation (5-21) and (5-22). An extremum-problem form of Rayleigh's quotient for this problem is[2]

$$\lambda_R = \frac{\Phi_m}{\Phi_n} = \frac{\displaystyle\int_D (\nabla^2 v)^2 \, dD}{\displaystyle\int_D \left(\frac{\partial v}{\partial x} \right)^2 dD} \qquad (5\text{-}123)$$

[1] Frazer, Jones, and Skan, *op. cit.*

[2] See Exercise 5-34. An alternative form is **given** by the principle of minimum potential energy (see Exercise 5-58).

where *admissible* functions v must satisfy the *essential* boundary condition, $v = 0$ on the boundary. We shall apply the Rayleigh-Ritz method to the following trial family:

$$v = c_1(1 - x^2)(1 - 4y^2) + c_2 x(1 - x^2)(1 - 4y^2) \tag{5-124}$$

Note that the essential boundary condition is satisfied for all values of c_1 and c_2. We next substitute the trial solution (5-124) into (5-123) and seek the stationary values according to (5-106).

$$\frac{\partial}{\partial c_j} \int_D (\nabla^2 v)^2 \, dD = \lambda_R \frac{\partial}{\partial c_j} \int_D \left(\frac{\partial v}{\partial x} \right)^2 dD \tag{5-125}$$

Carrying out the indicated operations leads to the following matrix eigenvalue problem:

$$\begin{bmatrix} \frac{9088}{45} & 0 \\ 0 & \frac{1664}{21} \end{bmatrix} \begin{bmatrix} c_1 \\ c_2 \end{bmatrix} = \lambda_R \begin{bmatrix} \frac{128}{45} & 0 \\ 0 & \frac{128}{75} \end{bmatrix} \begin{bmatrix} c_1 \\ c_2 \end{bmatrix} \tag{5-126}$$

This example is exceptional in that all the coupling terms in the square matrices are zero. This means that the functions used in constructing (5-124) are separately better approximations to eigenfunctions than any linear combination of the two. The approximate eigenvalues obtained from (5-126) are simply the Rayleigh quotients of these two functions.

$$\lambda_{1,R} = \frac{\frac{1664}{21}}{\frac{128}{75}} = 46.43 \qquad \lambda_{2,R} = \frac{\frac{9088}{45}}{\frac{128}{45}} = 71.00 \tag{5-127}$$

It is interesting to observe that the smaller quotient is associated with the function $x(1 - x^2)(1 - 4y^2)$, which has a node across the middle of the plate. According to (5-114) we can say that the two smallest true eigenvalues λ_1 and λ_2 must satisfy the following inequalities:

$$\lambda_1 \leq 46.43 \qquad \lambda_2 \leq 71.00 \tag{5-128}$$

In actual fact[1] $\lambda_1 = 39.478$, and $\lambda_2 = 61.685$.

EXERCISES

5-50. Apply the Rayleigh-Ritz or Galerkin criterion to the trial mode

$$v = c_1 x(1 - x) + c_2 x^2(1 - x)^2$$

for Problem 5-1, vibrating string. Show that the matrix eigenvalue problem is

$$\begin{bmatrix} \frac{1}{3} & \frac{1}{15} \\ \frac{1}{15} & \frac{2}{105} \end{bmatrix} \begin{bmatrix} c_1 \\ c_2 \end{bmatrix} = \lambda \begin{bmatrix} \frac{1}{30} & \frac{1}{140} \\ \frac{1}{140} & \frac{1}{630} \end{bmatrix} \begin{bmatrix} c_1 \\ c_2 \end{bmatrix}$$

Verify that the solution to this provides a good approximation to the first mode and a poor approximation to the third mode. Why is the second mode omitted?

Ans. $\lambda_1 = 9.8698$, $(c_1, c_2) = (0.8825, 1.0000)$; $\lambda_2 = 102.13$, $(c_1, c_2) = (-0.2158, 1.0000)$. The exact solution is given by (5-34).

5-51. Consider the trial mode

$$v = c_1 x(1 - x)$$

for Prob. 5-1. Obtain approximations to the corresponding eigenvalue by using (*a*) collocation at $x = \frac{1}{2}$; (*b*) the subdomain method using $0 < x < 1$; (*c*) Galerkin's criterion; (*d*) the method of least squares.

Ans. (*a*) 8; (*b*) 12; (*c*) 10; (*d*) $10 \pm i \sqrt{20}$.

[1] See Timoshenko, *op. cit.*, p. 330.

5-52. In the Galerkin method with the matrices B and C defined in (5-100) and (5-104) show that

$$C'BC = \int_D v N_{2n}[v]\, dD$$

where v is given by (5-99). Prove that, if N_{2n} is positive definite, then the matrix B is also positive definite.

5-53. Verify that when λ_R is taken in the form (5-62), which is valid when v satisfies *all* the boundary conditions, the Rayleigh-Ritz criterion (5-105) leads to the identical matrix eigenvalue problem as the Galerkin criterion (5-104).

5-54. If the matrix column $C_p(c_1, \ldots, c_r)$ is associated with the Rayleigh-Ritz approximation $c_{p,R}$ of (5-111), show that

$$C'_p A C_q = \int_D v_{p,R} M_{2m}[v_{q,R}]\, dD$$

where the elements of A are given by (5-104), and hence that the Rayleigh-Ritz modes $v_{p,R}$ are orthogonal with respect to M_{2m}. Show also that these modes are orthogonal with respect to N_{2n}.

5-55. Apply Grammel's criterion to the trial mode of Exercise 5-50. Show that the matrix eigenvalue problem is

$$\begin{bmatrix} 4 & 0 \\ 0 & \frac{4}{5} \end{bmatrix}\begin{bmatrix} c_1 \\ c_2 \end{bmatrix} = \lambda \begin{bmatrix} \frac{1}{3} & \frac{1}{15} \\ \frac{1}{15} & \frac{2}{105} \end{bmatrix}\begin{bmatrix} c_1 \\ c_2 \end{bmatrix}$$

and that its solution is $c_1 = (0.9295, 1.0000)$, $\lambda_1 = 9.8751$; $c_2 = (-0.2152, 1.0000)$, $\lambda_2 = 170.125$.

5-56. Adapt the procedure of Exercise 5-54 to show that the Grammel modes $v_{p,G}$ of (5-112) are orthogonal with respect to M_{2m}. By direct evaluation in the example of Exercise 5-55 verify that the Grammel modes are *not* orthogonal with respect to $N_{2n} \equiv g \equiv 1$.

5-57. The differential equation for uniform beam vibration is

$$\frac{d^4\psi}{dx^4} = \lambda\psi \qquad 0 < x < 1$$

in dimensionless variables (see Exercise 5-44). Let

$$v_0 = \Sigma c_j \varphi_j$$
$$v_{\frac{1}{2}} = \Sigma c_j \varphi_j^*$$
$$v_1 = \Sigma c_j \varphi_j^{**}$$

be an initial trial deflection, a *half iterate* (which may be considered as a dimensionless bending moment), and a complete iterate which satisfy

$$\frac{d^2 v_{\frac{1}{2}}}{dx^2} = v_0 \qquad \frac{d^2 v_1}{dx^2} = v_{\frac{1}{2}} \qquad 0 < x < 1$$

The functions φ_j, φ_j^*, and φ_j^{**} satisfy corresponding equations. Show that the coefficients (5-110) based on the iterate v_1 can be expressed as follows in terms of the half iterate:

$$a_{jk} = \int_0^1 \frac{d^2\varphi_j^*}{dx^2}\frac{d^2\varphi_k^*}{dx^2}\, dx \qquad b_{jk} = \int_0^1 \varphi_j^* \varphi_k^*\, dx$$

Grammel's method consists in selecting the φ_j, obtaining the φ_j^* by integration, and then evaluating the elements of the matrices A and B.

The φ_j^* must satisfy the boundary conditions for bending moment. For statically determinant beams two conditions on the bending moment (and its first derivative, shear force) are available, and thus the φ_j^* are completely determined by half an iteration. For statically indeterminant beams the φ_j^* are not completely determined, and it is necessary to continue the iteration in order to evaluate the constants of integration appearing in φ_j^*.

5-58. If the principle of minimum potential energy is used in Prob. 5-4, buckling of a plate, a Rayleigh's quotient is obtained[1] which has the form of (5-123) except that Φ_m is replaced by

$$\Phi_m = \int_D \left\{ (\nabla^2\psi)^2 - 2(1 - \nu) \left[\frac{\partial^2\psi}{\partial x^2} \frac{\partial^2\psi}{\partial y^2} - \left(\frac{\partial^2\psi}{\partial x\, \partial y} \right)^2 \right] \right\} dD$$

where ν is Poisson's ratio. Repeat Exercise 5-34 with this functional, and show that it leads to the same differential equation. The difference between the extremum problems appears only in the boundary conditions. If the essential boundary condition is prescribed, then the *additional* or *natural* boundary condition for Exercise 5-34 is $\nabla^2\psi = 0$, while here it is

$$\frac{\partial^2\psi}{\partial n^2} + \nu \frac{\partial^2\psi}{\partial s^2} = 0$$

where n and s are normal and tangent to the boundary. Show that, when $\psi = 0$ on the boundary, the two conditions become identical.

5-59. Show that if the φ_j of (4-96) are eigenfunctions of

$$L_{2m}[\psi] = \lambda\psi \qquad \text{in } D$$
$$B_i[\psi] = 0 \qquad i = 1, \ldots, m \text{ on the boundary}$$

then the Galerkin and least-squares criteria (4-100) and (4-101) are equivalent. Show furthermore that the matrix of coefficients will be diagonal.

5-60. The weighted-residual methods of this section can be extended to systems in which λ occurs in the boundary by considering a boundary residual

$$\Sigma(-B_i[v] + \lambda\tilde{C}_i[v])$$

in addition to the residual (5-101), by restricting v to functions which satisfy all the boundary conditions that do not contain λ and by making obvious alterations in the criteria, e.g., in collocation only (5-101) is used at interior locations and only the boundary residual used at boundary locations; in Galerkin's method the coefficients (5-104) become

$$a_{jk} = \int_D \varphi_j M_{2m}[\varphi_k]\, dD + \sum \oint \varphi_j \bar{B}_i[\varphi_k]\, ds$$

$$b_{jk} = \int_D \varphi_j N_{2n}[\varphi_k]\, dD + \sum \oint \varphi_j \bar{C}_i[\varphi_k]\, ds$$

The notation here is that of Exercise 5-12. Apply (a) collocation at $x = \frac{1}{2}$ and $x = 1$ and (b) Galerkin's criterion to the trial $v = c_1 x + c_2 x^2$ for Exercise 5-5.

Ans. (a) $\lambda_1 = 0.7830$, $\lambda_2 = 10.2170$; (b) $\lambda_1 = 0.7407$, $\lambda_2 = 12.0000$.

[1] See, for example, Timoshenko, *op. cit.*, p. 305.

5-7. Finite-difference Methods

In the preceding section continuous eigenvalue problems were reduced to discrete approximations by choosing trial solutions with r parameters.

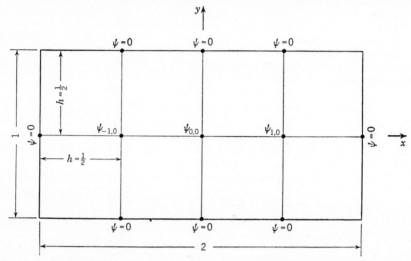

FIG. 5-16. Discrete approximation to domain of continuous-waveguide problem with $h = \frac{1}{2}$. There are only three interior values of ψ.

In this section continuous problems are more directly transformed into discrete approximations by replacing continuous domains with networks of isolated points and by replacing differential equations and boundary

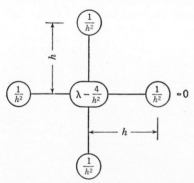

FIG. 5-17. Finite-difference approximation to $\nabla^2\psi + \lambda\psi = 0$.

conditions with finite-difference approximations. The actual reduction from a continuous system to a discrete approximation is the same here as it was for equilibrium problems in Sec. 4-6. The resulting discrete problem is, however, an eigenvalue problem of the type considered in Chap. 2.

Introductory Example. Consider the transverse magnetic wave in Prob. 5-3 with the formulation (5-16) and (5-17). Let us replace the continuous domain with the crude network shown in Fig. 5-16. The boundary condition $\psi = 0$ has already been entered on Fig. 5-16. The governing equation (5-16) is replaced by the finite-difference approximation of Fig. 5-17 which makes use of the molecule for $\nabla^2\psi$ given in Fig. 4-13. The truncation error is $0(h^2)$.

If now we center Fig. 5-17 on $\psi_{-1,0}$, $\psi_{0,0}$, and $\psi_{1,0}$ of Fig. 5-16, we get the following three equations:

$$\left(\lambda - \frac{4}{h^2}\right)\psi_{-1,0} + \frac{1}{h^2}\psi_{0,0} = 0$$

$$\frac{1}{h^2}\psi_{-1,0} + \left(\lambda - \frac{4}{h^2}\right)\psi_{0,0} + \frac{1}{h^2}\psi_{1,0} = 0 \qquad (5\text{-}129)$$

$$\frac{1}{h^2}\psi_{0,0} + \left(\lambda - \frac{4}{h^2}\right)\psi_{1,0} = 0$$

Putting $h = \frac{1}{2}$ and rewriting in matrix notation yield

$$\begin{bmatrix} 4 & -1 & 0 \\ -1 & 4 & -1 \\ 0 & -1 & 4 \end{bmatrix} \begin{bmatrix} \psi_{-1,0} \\ \psi_{0,0} \\ \psi_{1,0} \end{bmatrix} = \frac{\lambda}{4} \begin{bmatrix} \psi_{-1,0} \\ \psi_{0,0} \\ \psi_{1,0} \end{bmatrix} \qquad (5\text{-}130)$$

Thus we have reduced the continuous problem to an approximating matrix eigenvalue problem. The solution to (5-130) is

$$\begin{bmatrix} 0.7071 \\ 1.0000 \\ 0.7071 \end{bmatrix} \qquad \begin{bmatrix} 1.0 \\ 0.0 \\ -1.0 \end{bmatrix} \qquad \begin{bmatrix} -0.7071 \\ 1.0000 \\ -0.7071 \end{bmatrix} \qquad (5\text{-}131)$$

$$\lambda_1 = 10.343 \qquad \lambda_2 = 16.0 \qquad \lambda_3 = 21.657$$

These represent crude approximations to the first three modes of the continuous system. The true eigenvalues are known to be $\lambda_1 = 12.337$, $\lambda_2 = 32.076$, and $\lambda_3 = 71.555$.

General Remarks. The discussion in Sec. 4-6 concerning the applicability of finite-difference methods to irregularly shaped domains and the use of h^2 extrapolation carries over to eigenvalue problems essentially without modification. The fact that finite-difference solutions of high accuracy and great detail require huge amounts of computation is even more true for eigenvalue problems.

The discrete eigenvalue problems which result from making the finite-difference approximation can be solved by any of the methods described in Chap. 2. When only the smallest eigenvalue (and corresponding mode) of a positive definite system are required, iteration appears to be quite attractive. This procedure is ideally suited to high-speed automatic computing machines. Iteration can be adapted to hand computation by using relaxation to obtain each successive iterate. Other relaxation methods based on the stationary property of Rayleigh's quotient and on the enclosure theorem have also been employed with some success. These procedures will be illustrated in the examples which follow.

When a network with n mesh points is used to approximate a continuous domain D, the discrete system has n eigenvalues, whereas the continuous system has an infinity of eigenvalues. An interesting question is: "How many of the discrete eigenvalues can be expected to provide useful approximations to the corresponding continuous eigenvalues?" For example, in the preceding illustration, we obtained three finite-difference eigenvalues but only λ_1 was even close (16 per cent error) to the corresponding continuous eigenvalue. A precise answer here is apparently difficult to give,[1] but rough rules of thumb such as the following are current: If in a finite-difference mode there are less than two or three mesh points between nodal lines, then that mode and its eigenvalue should not be expected to provide useful approximations.

Iteration. We return to Prob. 5-3, rectangular waveguide, and consider a network with $h = \frac{1}{4}$; i.e., the mesh spacing is half that of Fig. 5-16. Instead of 3 interior points there are now 21. If we restrict our consideration to the fundamental mode, we can use symmetry and thereby limit ourselves to the 8 points, $P_{j,k}$ in the first quadrant shown in Fig. 5-18. If now the molecule of Fig. 5-17 were centered on each of these points, a matrix eigenvalue problem similar to (5-130) would be obtained. What we propose to do is to solve this problem by iteration without actually ever writing out the matrices.

We begin by selecting the elements of an initial trial vector, V_0. These are entered directly on a sketch of the domain as shown in Fig. 5-19. The values chosen were obtained by rough interpolation from the solution (5-131) for $h = \frac{1}{2}$. The iteration rule can be obtained from the continuous iteration process (5-78), which for this system is

$$-\nabla^2 v_r = v_{r-1} \qquad r = 1, 2, \ldots \qquad (5\text{-}132)$$

We have only to replace the continuous operator ∇^2 by the finite-difference molecule of Fig. 4-13 to obtain the discrete iteration process illustrated in Fig. 5-20. The elements of V_r are thus obtained from those of V_{r-1} by

---

[1] In certain special cases an exact answer can be given (see Exercises 5-67 to 5-69). The question is further complicated by the multiplicity of discrete models for any given continuous system. Here we have only considered finite-difference approximations to the differential-equation formulation which have $0(h^2)$ truncation error. Approximations with higher-order truncation error can be used (see L. Collatz, "Eigenwertaufgaben mit technischen Andwendung," Akademische Verlagsgesellschaft m.b.H., Leipzig, 1949, p. 346). In systems for which the Green's function is known, alternative discrete models can be constructed by approximating the integral on the right of (5-42) in terms of the Green's function evaluated at discrete intervals (influence coefficients). Models of great accuracy can be constructed by an extension of this idea [see M. Rauscher, Station Functions and Air Density Variations in Flutter Analysis, *J. Aeronaut. Sci.*, **16**, 345–353 (1949), in which a discrete model with only two station points gives very close approximations to the first two natural modes of a vibrating beam].

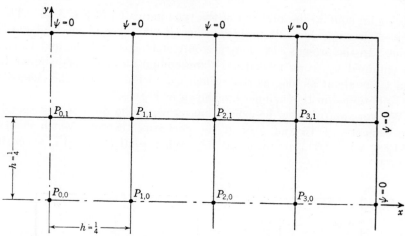

FIG. 5-18. First quadrant of waveguide approximated by network with $h = \frac{1}{4}$.

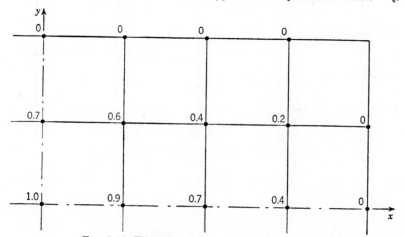

FIG. 5-19. Elements of the initial trial vector V_0.

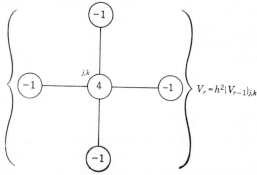

FIG. 5-20. Schematic indication of the discrete iteration process.

solving an equilibrium problem of the type indicated in Fig. 4-18. This in itself is a considerable problem when the number of mesh points is large. In the present case there are only eight mesh points, and V_1 can be obtained by using relaxation. The computation is very similar to and takes about as long as that which led to Fig. 4-29. The elements of V_1 obtained in this manner are shown in Fig. 5-21.

We can approximate the eigenvalue by using the ratio of corresponding elements of V_0 and V_1 and the enclosure theorem, which yields[1] $9.819 \leq \lambda \leq 13.092$, or by using Rayleigh's quotient[2] for V_1, which is

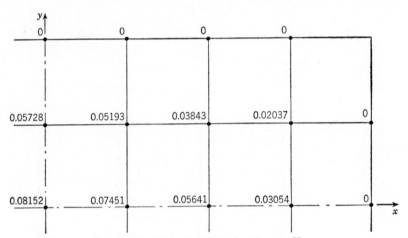

FIG. 5-21. Elements of the first iterate V_1.

11.846.   The exact eigenvalue for the network of Fig. 5-18 is $\lambda_1 = 11.808$. The true continuous eigenvalue is $\lambda_1 = 12.337$. If we apply h^2 extrapolation to $\lambda_1 = 10.343$ obtained in (5-131) for $h = \frac{1}{2}$ and to the Rayleigh quotient of V_1 just obtained for $h = \frac{1}{4}$, we get 12.347.

Relaxation Based on Rayleigh's Quotient.   We use the same example to illustrate a method proposed by Southwell[3] which is not restricted to any particular mode and which can have very rapid convergence in the hands of an experienced computer. In the notation of the continuous system we associate a *residual* function r with a trial function v and a trial eigenvalue Λ as follows:

$$r = \nabla^2 v + \Lambda v \tag{5-133}$$

[1] See Exercise 5-70.
[2] See Exercise 5-72.
[3] D. N. deG. Allen, L. Fox, H. Motz, and R. V. Southwell, Free Transverse Vibrations of Membranes with an Application (by Analogy) to Two-dimensional Oscillations in an Electro-magnetic System, *Trans. Roy. Soc. (London)*, **A239**, 488–500 (1945) (see also page 113).

If v is actually an eigenfunction ψ_p, then r will be a multiple of v when $\Lambda \neq \lambda_p$ and will be identically zero when $\Lambda = \lambda_p$. The relaxation process consists in altering v with intent to make r proportional to v. As the process proceeds, Λ is occasionally recomputed as the Rayleigh quotient of the current v. In the finite-difference approximation the relation (5-133) takes the form shown in Fig. 5-22.

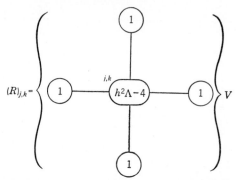

FIG. 5-22. Definition of the residual for Southwell's relaxation method.

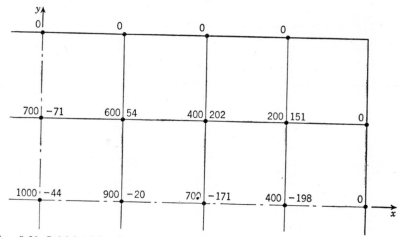

FIG. 5-23. Initial trial values entered at the left of each mesh point; corresponding residual entered at the right.

Starting with the same initial trial of Fig. 5-19 we first compute[1] its Rayleigh quotient, 12.10. Taking this for Λ, we have $h^2\Lambda - 4 = -3.244$ as the central coefficient in the operator of Fig. 5-22. Applying this operator to the trial vector, we get the residual values shown at the right of each mesh point in Fig. 5-23. A scale factor of 1,000 has been introduced to avoid decimals. Relaxation is now begun with a unit-

[1] See Exercise 5-72.

operation molecule based on Fig. 5-22.   The aim is to make the residual pattern similar to the trial pattern.   After a dozen steps the stage shown in Fig. 5-24 is reached.   The residuals, while not strictly proportional to the trial values, at least have a similar pattern: large values at the center and small values at the edges.   We then stop and compute Rayleigh's

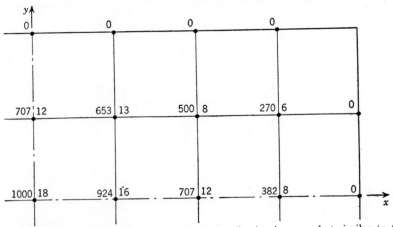

FIG. 5-24. After some relaxation the residual distribution is somewhat similar to the distribution of trial values.

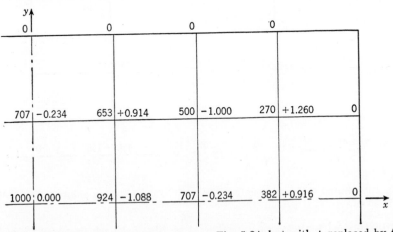

FIG. 5-25. Residuals for same trial values as Fig. 5-24, but with Λ replaced by the Rayleigh quotient of these trial values.

quotient for this trial.   We find 11.808.   When we set $\Lambda = 11.808$, the central coefficient in Fig. 5-22 becomes -3.262.   Recomputing the residuals with this new operator yields the results shown in Fig. 5-25.   Note that the magnitude of the residuals has been decreased and that their distribution no longer has the same pattern as the trial.   To continue, we would again relax aiming at a residual distribution similar to that of the trial values. With each succeeding recomputation of Λ by this process the order of magnitude of the error is *cubed*.[1] The method applies to any mode. All that is necessary when a computer first begins to get a residual pattern similar to the trial pattern is for him to decide whether or not he desires to converge on a mode of that general configuration.

Relaxation Based on the Enclosure Theorem. The procedure described at the beginning of Sec. 2-10 is immediately applicable to the finite-difference approximation of the continuous special eigenvalue problem[2] with the following governing equation:

$$M_{2m}[\psi] = \lambda g \psi \qquad (5\text{-}134)$$

At each mesh point one records the trial value $v_{j,k}$, the finite-difference approximation to $M_{2m}[v]$, the product $(gv)_{j,k}$, and the quotient

$$l_{j,k} = \frac{(M_{2m}[v])_{j,k}}{(gv)_{j,k}} \qquad (5\text{-}135)$$

The $v_{j,k}$ are then altered with intent to equalize the $l_{j,k}$. At any stage there is an eigenvalue of the discrete approximation enclosed by the maximum and minimum $l_{j,k}$.

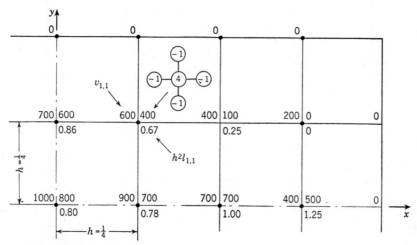

Fig. 5-26. Starting point of relaxation computation. The various entries are labeled at $P_{1,1}$. The aim is to equalize the $l_{j,k}$.

For example, in the transverse magnetic wave of Prob. 5-3 where $g = 1$ and $M_{2m} = -\nabla^2$ the computation would be laid out as shown in Fig. 5-26. At the upper left of the point $P_{j,k}$ we record the trial value $v_{j,k}$.

[1] S. H. Crandall, Iterative Procedures Related to Relaxation Methods for Eigenvalue Problems, *Proc. Roy. Soc.* (*London*), **A207**, 416–423 (1951).

[2] Vazsonyi's original presentation of the method was in terms of this application (see page 111).

At the upper right the result of applying the finite-difference approximation to $-h^2\nabla^2$ is recorded, and their quotient, obtained to two figures on a slide rule, is placed at the lower right. The set of trial values here is the same as that in Figs. 5-19 and 5-23. The enclosure theorem tells us that there is an eigenvalue for which $0 \leq h^2\lambda \leq 1.25$. Each step of the relaxation process consists in altering one or more $v_{j,k}$. The corresponding changes in the upper right entries are obtained by using a unit operation based on the molecule shown for $-h^2\nabla^2$. New quotients are then read from the slide rule. The serious student should reread the description accompanying Table 2-1 and then try a few steps for himself directly on Fig. 5-26. In eight single steps the author reached the stage

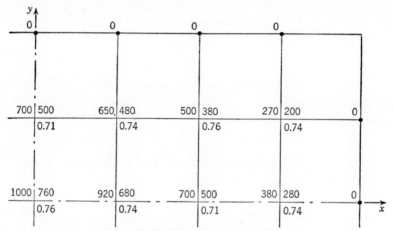

FIG. 5-27. Possible result of relaxation.

shown in Fig. 5-27. Here $h^2\lambda$ is bounded by 0.71 and 0.76; that is, $11.4 \leq \lambda \leq 12.2$. Rayleigh's quotient[1] for this set of $v_{j,k}$ is 11.809.

This process may be applied equally well to any mode. It is especially useful for preliminary studies to obtain approximate locations of the eigenvalues and rough indications of corresponding mode shapes. This same procedure may be used to explore the behavior of more general systems. For a system with the equation

$$M_{2m}[\psi] = \lambda N_{2n}[\psi] \tag{5-136}$$

it is necessary to obtain finite-difference approximations to $M_{2m}[v]$ and $N_{2n}[v]$ at each mesh point and their quotients

$$l_{j,k} = \frac{(M_{2m}[v])_{j,k}}{(N_{2n}[v])_{j,k}} \tag{5-137}$$

The relaxation consists in manipulating the $l_{j,k}$ toward equality by altering the $v_{j,k}$. This is possible, although not so intuitively simple, when

[1] See Exercise 5-72.

the enclosure property no longer exists. At present there are many eigenvalue problems for which the validity of the enclosure theorem remains an open question. Problems 5-2 and 5-4, buckling of a column and buckling of a plate, fall into this category. The procedure just described can[1] nevertheless be employed to advantage. In these two cases the numerical evidence suggests that the enclosure theorem does actually apply.

EXERCISES

5-61. Consider finite-difference approximations to Prob. 5-1, vibrating string, as formulated in (5-7) and (5-8). Show that with $h = \frac{1}{4}$ the matrix eigenvalue problem

$$\begin{bmatrix} 2 & -1 & 0 \\ -1 & 2 & -1 \\ 0 & -1 & 2 \end{bmatrix} \begin{bmatrix} \psi_1 \\ \psi_2 \\ \psi_3 \end{bmatrix} = h^2\lambda \begin{bmatrix} \psi_1 \\ \psi_2 \\ \psi_3 \end{bmatrix}$$

is an approximation with truncation error $0(h^2)$. Verify that the eigenvalues of this discrete system are $\lambda_1 = 9.3726$, $\lambda_2 = 32$, and $\lambda_3 = 54.6274$.

5-62. Consider finite-difference approximations to Prob. 5-2, buckling of a column, as formulated in (5-14) and (5-15). Note the treatment of the boundary conditions. Show that with $h = \frac{1}{3}$ the matrix eigenvalue problem

$$\begin{bmatrix} 7 & -4 \\ -4 & 5 \end{bmatrix} \begin{bmatrix} \psi_1 \\ \psi_2 \end{bmatrix} = h^2\lambda \begin{bmatrix} 2 & -1 \\ -1 & 2 \end{bmatrix} \begin{bmatrix} \psi_1 \\ \psi_2 \end{bmatrix}$$

is an approximation with truncation error $0(h^2)$. Verify that the smallest eigenvalue of the discrete system is $\lambda_1 = 16.0627$.

5-63. Consider finite-difference approximations to Prob. 5-4, buckling of a plate, as formulated in (5-21) and (5-22). Show that an $0(h^2)$ approximation is represented by

$$\begin{bmatrix} 17 & -8 & 1 \\ -8 & 18 & -8 \\ 1 & -8 & 17 \end{bmatrix} \begin{bmatrix} \psi_{-1,0} \\ \psi_{0,0} \\ \psi_{1,0} \end{bmatrix} = h^2\lambda \begin{bmatrix} 2 & -1 & 0 \\ -1 & 2 & -1 \\ 0 & -1 & 2 \end{bmatrix} \begin{bmatrix} \psi_{-1,0} \\ \psi_{0,0} \\ \psi_{1,0} \end{bmatrix}$$

when $h = \frac{1}{2}$. Obtain the eigenvalues of the discrete system. Note that the lowest eigenvalue is not associated with the simple node-free mode.
Ans. $\lambda_1 = 32$, $\lambda_2 = 34.3431$, $\lambda_3 = 45.6569$.

5-64. Obtain an approximate eigenvalue for Exercise 5-62 by using only one interior point (that is, $h = \frac{1}{2}$), and then apply h^2 extrapolation to this and the result of Exercise 5-62. *Ans.* 19.313.

5-65. Apply the relaxation method described in conjunction with (5-137) to Exercise 5-62, when $h = 0.2$. Finite-difference operators corresponding to M_{2m} and N_{2n} are

$$\text{①} + \text{⊝④} + \text{⑥} + \text{⊝④} + \text{①}$$

$$\text{⊝①} + \text{②} + \text{⊝①}$$

Their quotient is $h^2\lambda$.
Ans. After 10 steps $h^2\lambda$ values were 0.745, 0.745, 0.742, 0.746, that is, $18.55 \leq \lambda \leq$ 18.65 if the enclosure theorem is valid.

[1] See Exercises 5-65 and 5-66.

5-66. Apply the relaxation method described in conjunction with (5-137) to Exercise 5-63 when $h = \frac{1}{4}$. Use symmetry to simplify problem for lowest eigenvalue. *Ans.* $\lambda_1 = 37.488$; h^2 extrapolation with the result of Exercise 5-63 is within $\frac{1}{2}\%$ of the true continuous value $\lambda_1 = 39.478$.

5-67. Consider Prob. 5-1, vibrating string, with the formulation (5-7) and (5-8). If $h = 1/S$, show that

$$-\psi_{j-1} + 2\psi_j - \psi_{j+1} = h^2\lambda\psi_j \qquad j = 1, \ldots, S-1 \qquad \begin{cases} \psi_0 = 0 \\ \psi_S = 0 \end{cases}$$

is a finite-difference approximation with truncation error $0(h^2)$. A solution to this in general terms can be obtained[1] by trying $\psi_j = e^{j\theta}$. Verify that there are $S - 1$ solutions of the following form:

$$\psi_j = \sin j\left(\frac{p\pi}{S}\right) \qquad \lambda = 4S^2 \sin^2\left(\frac{p\pi}{2S}\right) \qquad p = 1, \ldots, S-1$$

Compare these with the solution (5-34) of the continuous system. Show that

$$\frac{(\lambda_p)_{\text{discrete}}}{(\lambda_p)_{\text{continuous}}} = 1 - h^2\frac{p^2\pi^2}{12} + \cdots$$

Verify that when $p/S = \frac{1}{3}$ there are exactly 3 intervals between nodes of the eigenfunction and that the discrete eigenvalue is in error by about 10% and that when $p/S = \frac{1}{2}$ this error is about 20%.

5-68. Repeat Exercise 5-67 for a vibrating uniform beam with hinged ends. The continuous problem in dimensionless variables is

$$\frac{d^4\psi}{dx^4} = \lambda\psi \qquad 0 < x < 1 \qquad \begin{cases} \psi = 0, \dfrac{d^2\psi}{dx^2} = 0 \text{ at } x = 0 \\ \psi = 0, \dfrac{d^2\psi}{dx^2} = 0 \text{ at } x = 1 \end{cases}$$

and the finite-difference approximation is

$$\psi_{j-2} - 4\psi_{j-1} + 6\psi_j - 4\psi_{j+1} + \psi_{j+2} = h^4\lambda\psi_j \qquad j = 1, \ldots, S-1$$
$$\psi_0 = \psi_S = 0 \qquad \psi_{-1} = -\psi_1 \qquad \psi_{S+1} = -\psi_{S-1}$$

Verify that the discrete solutions have the following form:

$$\psi_j = \sin j\frac{p\pi}{S} \qquad \lambda = 16S^4 \sin^4\frac{p\pi}{2S} \qquad p = 1, \ldots, S-1$$

Show that

$$\frac{(\lambda_p)_{\text{discrete}}}{(\lambda_p)_{\text{continuous}}} = 1 - h^2\frac{p^2\pi^2}{6} + \cdots$$

and hence that the discretization error in an eigenvalue for the beam is about twice that for the corresponding eigenvalue of the string.

5-69. Verify that $\psi = \cos \pi x/2 \cos \pi y$ is the exact fundamental mode for the transverse magnetic wave of Prob. 5-3. Show that $\psi_{j,k} = \cos j\pi/2S \cos k\pi/S$ is the corresponding discrete solution when $h = 1/S$ and that

$$\frac{(\lambda_1)_{\text{discrete}}}{(\lambda_1)_{\text{continuous}}} = 1 - h^2\frac{17\pi^2}{240} + \cdots$$

[1] See, for example, F. B. Hildebrand, "Methods of Applied Mathematics," Prentice-Hall, Inc., New York, 1952, p. 237.

5-70. Apply the enclosure theorem to Figs. 5-19 and 5-21. Verify that $l_{\min}$ and $l_{\max}$ are obtained at $P_{3,1}$ and $P_{3,0}$, respectively.

5-71. Finite-difference Approximation to Rayleigh's Quotient. Consider the network of Fig. 5-16 for Prob. 5-3, rectangular waveguide. The extremum-principle form of Rayleigh's quotient for the continuous system is (see Exercise 5-25)

$$\lambda_R = \frac{\int_D \left\{ \left(\frac{\partial v}{\partial x}\right)^2 + \left(\frac{\partial v}{\partial y}\right)^2 \right\} dD}{\int_D v^2 \, dD}$$

Show that by using an $0(h^2)$ approximation to $\partial v/\partial x$ at the center of each shaded square in Fig. 5-28a, we get

$$\tfrac{1}{2}(v_{-1,1} - v_{-2,1})^2 + \tfrac{1}{2}(v_{0,1} - v_{-1,1})^2 + \tfrac{1}{2}(v_{1,1} - v_{0,1})^2 + \tfrac{1}{2}(v_{2,1} - v_{1,1})^2$$
$$+ (v_{-1,0} - v_{-2,0})^2 + (v_{0,0} - v_{-1,0})^2 + (v_{1,0} - v_{0,0})^2 + (v_{2,0} - v_{1,0})^2$$
$$+ \tfrac{1}{2}(v_{-1,-1} - v_{-2,-1})^2 + \tfrac{1}{2}(v_{0,-1} - v_{-1,-1})^2 + \tfrac{1}{2}(v_{1,-1} - v_{0,-1})^2 + \tfrac{1}{2}(v_{2,-1} - v_{1,-1})^2$$

as an approximation to $\int (\partial v/\partial x)^2 \, dD$. Obtain a similar approximation to $\int (\partial v/\partial y)^2 \, dD$ by using approximations to $\partial v/\partial y$ at the centers of the shaded squares in Fig. 5-28b. Using the subdivision of Fig. 5-28c, show that

$$h^2\{\tfrac{1}{4}v_{-2,1}{}^2 + \tfrac{1}{2}v_{-1,1}{}^2 + \tfrac{1}{2}v_{0,1}{}^2 + \tfrac{1}{2}v_{1,1}{}^2 + \tfrac{1}{4}v_{2,1}{}^2 + \tfrac{1}{2}v_{-2,0}{}^2 + v_{-1,0}{}^2 + v_{0,0}{}^2 + v_{1,0}{}^2$$
$$+ \tfrac{1}{2}v_{2,0}{}^2 + \tfrac{1}{4}v_{-2,-1}{}^2 + \tfrac{1}{2}v_{-1,-1}{}^2 + \tfrac{1}{2}v_{0,-1}{}^2 + \tfrac{1}{2}v_{1,-1}{}^2 + \tfrac{1}{4}v_{2,-1}{}^2\} \quad \text{(i)}$$

is an approximation to $\int v^2 \, dD$.

(a) If all the boundary values are zero as in the transverse magnetic wave, write out the finite-difference approximation to λ_R, and show that the conditions for it to be stationary with respect to $v_{-1,0}$, $v_{0,0}$, and $v_{1,0}$ are just Eqs. (5-130). Show also that the same numerator in λ_R is obtained from

$$\sum_{j,k} v_{j,k}(-h^2 \nabla^2 v)_{j,k} \qquad \text{(ii)}$$

where $(-h^2\nabla^2 v)_{j,k}$ stands for the molecule $4v_{j,k} - v_{j+1,k} - v_{j,k+1} - v_{j-1,k} - v_{j,k-1}$ and the summation is over all interior points.

(b) If there are no essential boundary conditions as in the transverse electric wave, write out the finite-difference approximation to λ_R, and show that the conditions for it to be stationary with respect to all 15 mesh values provide the governing equations with the additional boundary conditions incorporated. Show also that the same numerator in λ_R is obtained from (ii) provided the additional boundary conditions are introduced by means of auxiliary boundary points (for example, $v_{1,2} = v_{1,0}$) and that the summation is over all 15 mesh points as indicated in Fig. 5-28c, the contribution of each point being in proportion to the area associated with it; i.e., the coefficients of $\tfrac{1}{4}$, $\tfrac{1}{2}$, 1 follow the same pattern as in the summation (i).

5-72. The formula

$$\lambda_R = \frac{\Sigma v_{j,k}(-h^2 \nabla^2 v)_{j,k}}{h^2 \Sigma v_{j,k}}$$

derived in the preceding exercise for $h = \tfrac{1}{2}$ may be extended to other values of h. It gives the finite-difference Rayleigh quotient for the transverse magnetic wave when the summation is over all interior mesh points. Apply this to the sets of values shown in Figs. 5-19, 5-21, 5-24, and 5-27. Note the effect of the symmetry on the summations.

Ans. 12.101, 11.846, 11.808, and 11.809.

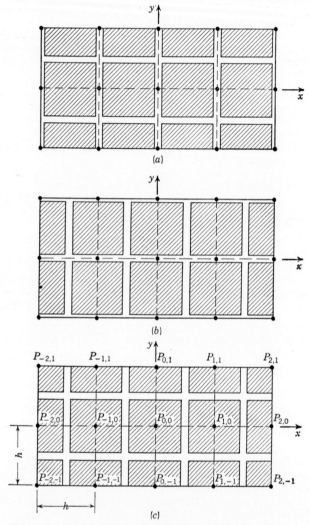

FIG. 5-28. Exercise 5-71.   Subdivisions used for evaluating

$$(a) \quad \int_D \left(\frac{\partial v}{\partial x}\right)^2 dD$$

$$(b) \quad \int_D \left(\frac{\partial v}{\partial y}\right)^2 dD$$

$$(c) \quad \int_D v^2 \, dD$$

5-73. For the problem of Exercise 5-5, which contains λ in the boundary conditions, show that a finite-difference approximation is

$$\psi_0 = 0$$
$$-\psi_{j-1} + 2\psi_j - \psi_{j+1} = h^2\lambda\psi_j \qquad j = 1, \ldots, S - 1$$
$$-2\psi_{S-1} + 2\psi_S = (h^2 + 2h)\lambda\psi_S$$

where $h = 1/S$. Obtain approximations to the smallest eigenvalue by using $S = 1$ and $S = 2$. *Ans.* 0.6667 and 0.7208.

5-74. Trial Eigenvalue Method for Beam Vibration. The governing equation for natural modes $y(x)$ and natural frequencies ω of a beam of varying stiffness EI and varying mass per unit length ρ is

$$\frac{d^2}{dx^2}\left(EI\frac{d^2y}{dx^2}\right) = \rho\omega^2 y \qquad 0 < x < L$$

There are two homogeneous boundary conditions (depending on the nature of the support) at each end. Let

$$y_1 = y$$
$$y_2 = \frac{dy}{dx}$$
$$y_3 = EI\frac{d^2y}{dx^2}$$
$$y_4 = \frac{d}{dx}\left(EI\frac{d^2y}{dx^2}\right)$$

be four variables proportional to the deflection, slope bending moment, and shear force. Show that the values of these four variables at $x = L$ are determined from their values at $x = 0$ by solving the following propagation problem of the type considered in Chap. 3:

$$
\left.\begin{matrix} y_1(0) \\ y_2(0) \\ y_3(0) \\ y_4(0) \end{matrix}\right\}
\rightarrow
\left\{\begin{matrix} \dfrac{dy_1}{dx} = y_2 \\[2mm] \dfrac{dy_2}{dx} = \dfrac{y_3}{EI} \\[2mm] \dfrac{dy_3}{dx} = y_4 \\[2mm] \dfrac{dy_4}{dx} = \rho\omega^2 y_1 \end{matrix}\right\}
\rightarrow
\left\{\begin{matrix} y_1(L) \\ y_2(L) \\ y_3(L) \\ y_4(L) \end{matrix}\right.
$$

If ω is assumed to be a fixed value, this is a linear system; i.e., the values at $x = L$ depend linearly on the initial values at $x = 0$. In the vibration problem let two of the values at $x = 0$ be prescribed to be zero. Let the two which are unprescribed be $y_\alpha(0)$ and $y_\beta(0)$. At $x = L$ let two values, $y_\xi(L)$ and $y_\eta(L)$ say, be required to vanish. Now, if we solve the above propagation problem for $y_\xi(L)$ and $y_\eta(L)$ starting with arbitrary values for $y_\alpha(0)$ and $y_\beta(0)$, we obtain, because of the linearity,

$$y_\xi(L) = Ay_\alpha(0) + By_\beta(0)$$
$$y_\eta(L) = Cy_\alpha(0) + Dy_\beta(0)$$

where A, B, C, and D are independent of the starting values (they are complicated functions of ρ, ω^2, and EI). If ω were a natural frequency, it would be possible to choose nonzero values of $y_\alpha(0)$ and $y_\beta(0)$ so that $y_\xi(L)$ and $y_\eta(L)$ both vanished. If ω is not a natural frequency, it is possible to make $y_\eta(L)$ vanish and then use the magnitude of $y_\xi(L)$ as a measure of the error in ω.   Show that, if $y_\alpha(0)$ is arbitrarily taken as unity,

$$y_\xi(L) = \frac{AD - BC}{D}$$

Tabular methods have been set up[1] to carry out approximate solutions to the propagation problem by means of finite differences.   The procedure is similar to that used with the Holzer table[2] for torsional vibrations.

[1] N. O. Myklestad, A New Method of Calculating Natural Modes of Uncoupled Bending Vibration of Airplane Wings and Other Types of Beams, *J. Aeronaut. Sci.*, **2**, 153–163 (1944); M. A. Prohl, A General Method for Calculating Critical Speeds of Rotors, *J. Appl. Mech.*, **12**, A142–A148 (1945).

[2] See Exercise 2-71.

CHAPTER 6

PROPAGATION PROBLEMS IN CONTINUOUS SYSTEMS

This chapter is devoted to a survey of approximate procedures used in studying propagation problems in continuous systems. In some respects the extension of the methods applied to lumped-parameter systems is more difficult here than it was in the case of equilibrium and eigenvalue problems. This is particularly true for the finite-difference methods. We find, for instance, situations where the accuracy of approximation can deteriorate when we use a finer grid. In order to get to the root of such paradoxes, it is necessary to have a good understanding of the nature of the phenomena that are being approximated. In the first section we examine particular problems of diffusion and wave propagation and obtain their governing equations, boundary conditions, and initial conditions. We then turn to a general study of the behavior of these equations, including an introduction to the theory of characteristics of partial differential equations. After these preliminaries we consider approximate numerical procedures, discussing trial solutions with undetermined parameters and finite-difference methods.

6-1. Particular Examples

Five particular propagation problems are analyzed in this section. The governing equations, boundary conditions, and initial conditions, are established and cast into nondimensional form. These will then be drawn upon to illustrate the numerical procedures which follow. The problems are:

6-1. Cooling of a rod.
6-2. Boundary-layer wake.
6-3. Unsteady transverse motion of a beam.
6-4. Transverse motion of a taut string.
6-5. Expansion of a gas behind a piston.

We shall see that the first three problems involve parabolic partial differential equations, while the last two involve hyperbolic partial differential equations. Problems 6-1, 6-3, and 6-4 are linear; the other two are nonlinear.

Problem 6-1.   Cooling of a Rod.   Consider a uniform rod of heat-conducting material of length $2L$.   The rod is taken to be insulated along its length so that all temperature changes are caused by heat transfer at the ends and by heat conduction along the rod.   Under these circumstances if distance along the rod be denoted by x, time by t, and the temperature of the rod by $T(x, t)$, the heat-flux per unit area, $q(x, t)$, must satisfy[1] Fourier's law

$$q = -k \frac{\partial T}{\partial x} \tag{6-1}$$

where k is the thermal conductivity of the rod; also the time rate of decrease of temperature at a point is proportional to the divergence of the heat flux

$$-\rho c_p \frac{\partial T}{\partial t} = \frac{\partial q}{\partial x} \tag{6-2}$$

where ρ is the mass per unit volume and c_p is the specific heat.

Propagation problems for this system consist in determining the subsequent temperature values throughout the rod when given the initial temperature distribution along with the history of the end conditions.   The particular problem we shall consider arises when the rod, initially at the uniform high temperature T_h is cooled by bringing it into an environment of temperature T_c.   A result of practical importance would be the time required for the center of the bar to cool down to some prescribed temperature.

We actually consider two different end conditions.   In (a) we assume that the ends of the bar immediately take on and remain at the temperature T_c of the environment, while in (b) we assume that there is a surface film which resists heat transfer from the ends of the rod to the environment; i.e., we take the heat flux out the ends to be

$$q_{\text{out}} = h(T_{\text{boundary}} - T_c) \tag{6-3}$$

where h is the film coefficient.   In both cases we realize that because of the symmetry there will be no heat transfer across the center of the rod.   These conditions are sketched in Fig. 6-1.

The propagation problems thus consist in finding solutions to (6-1) and (6-2) which satisfy the initial and boundary conditions of Fig. 6-1. Instead of the pair of first-order equations (6-1) and (6-2), it is usually more convenient to consider the single second-order equation

$$\frac{\partial^2 T}{\partial x^2} = \frac{\rho c_p}{k} \frac{\partial T}{\partial t} \tag{6-4}$$

---

[1] See, for example, W. H. McAdams, "Heat Transmission," 3d ed., McGraw-Hill Book Company, Inc., New York, 1954, p. 3.

obtained by eliminating q. Where q enters into the boundary conditions, it must also be eliminated by using (6-1).

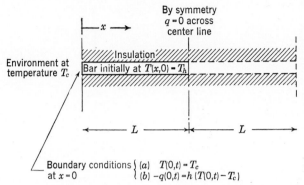

FIG. 6-1. Initial and boundary conditions for a heat-conducting rod being cooled at its ends.

This problem may be brought into nondimensional form by introducing dimensionless lengths, times, and temperatures defined as follows.

$$x' = \frac{x}{L} \qquad t' = \frac{t}{(\rho c_p/k)L^2} \qquad \psi = \frac{T - T_c}{T_h - T_c} \tag{6-5}$$

The governing equation (6-4) becomes

$$\frac{\partial^2 \psi}{\partial x^2} = \frac{\partial \psi}{\partial t} \tag{6-6}$$

after we drop the primes from the dimensionless variables. The initial and boundary conditions for case (a) are

$$\psi(x, 0) = 1 \qquad 0 \le x \le 1 \qquad \left.\begin{array}{c} \psi(0, t) = 0 \\[2mm] \dfrac{\partial \psi}{\partial x}(1, t) = 0 \end{array}\right\} \qquad t > 0 \tag{6-7}$$

These are shown on a diagram of the solution domain in the (x, t) plane in Fig. 6-2a.

In case (b) there is a whole family of problems depending on the relative conductivities of the film and the rod. For numerical treatment we arbitrarily take

$$k = hL \tag{6-8}$$

i.e., the case in which the film has the same resistance to heat flow as half the length of the rod. The boundary condition (6-3) at $x = 0$ then becomes

$$\psi - \frac{\partial \psi}{\partial x} = 0 \qquad t > 0 \tag{6-9}$$

The other conditions are the same as in case (a). The complete formulation is shown in Fig. 6-2b.

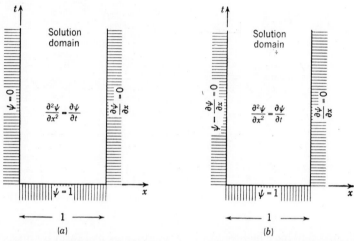

FIG. 6-2.   Dimensionless formulations of Prob. 6-1, cooling of a rod.

Problem 6-2.   Boundary-layer Wake.   Consider a thin flat plate held fixed in a steady uniform stream of incompressible viscous fluid so that the plate is edgewise to the flow.   According to boundary-layer theory[1] the fluid in contact with the plate does not move, and the transition from zero velocity on the plate to very nearly the original free stream velocity is accomplished in a relatively thin boundary layer.   The

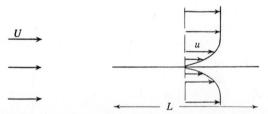

FIG. 6-3.   Velocity profile of laminar flow past a flat plate.

situation is sketched in Fig. 6-3, where a flat plate of length L is fixed in a flow with free stream velocity U.   To study such flows, the momentum equations

$$u \frac{\partial u}{\partial x} + v \frac{\partial u}{\partial y} = -\frac{1}{\rho}\frac{\partial p}{\partial x} + \nu \left(\frac{\partial^2 u}{\partial x^2} + \frac{\partial^2 u}{\partial y^2} \right)$$
$$u \frac{\partial v}{\partial x} + v \frac{\partial v}{\partial y} = -\frac{1}{\rho}\frac{\partial p}{\partial y} + \nu \left(\frac{\partial^2 v}{\partial x^2} + \frac{\partial^2 v}{\partial y^2} \right)$$

(6-10)

for the velocity components u and v in the x and y directions in terms of the pressure p, density ρ, and kinematic viscosity ν and the continuity equation

[1] See S. Goldstein, "Modern Developments in Fluid Dynamics," vol. I, Oxford University Press, New York, 1938, chap. IV, for general background.   The particular problem treated here is described in vol. II of the same work, pp. 571–574.

$$\frac{\partial u}{\partial x} + \frac{\partial v}{\partial y} = 0 \tag{6-11}$$

are approximated in the following way: Taking x along the plate and y normal to the plate, it is assumed that

1. p is constant throughout the flow.
2. $\partial^2 u/\partial x^2$ can be neglected in comparison with $\partial^2 u/\partial y^2$.

The first of (6-10) then reduces to

$$u\frac{\partial u}{\partial x} + v\frac{\partial u}{\partial y} = \nu\frac{\partial^2 u}{\partial y^2} \tag{6-12}$$

which together with (6-11) forms the foundation for the mathematical theory of boundary-layer flow along a flat plate.

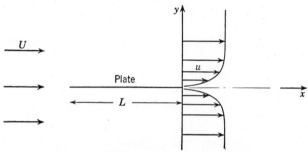

FIG. 6-4. Velocity profile at the beginning of the wake obtained from plate solution of Blasius.

The solution of (6-11) and (6-12) for the flow along the plate was given by Blasius.[1] We here take this solution as known and consider the problem of continuing the solution into the wake of the plate. In Fig. 6-4 we have taken the origin of coordinates at the end of the plate. The propagation problem consists in determining the flow for $x > 0$ from the given flow pattern at $x = 0$. The boundary conditions for the Blasius solution were

$$\left.\begin{matrix} u = 0 \\ v = 0 \end{matrix}\right\} \quad \text{at } y = 0 \qquad \left.\begin{matrix} u = U \\ v = 0 \end{matrix}\right\} \quad \text{at } y = \infty \tag{6-13}$$

In the wake region, $x > 0$, the boundary conditions at $y = \infty$ remain unaltered, but, owing to symmetry, the boundary conditions at $y = 0$ become

$$\frac{\partial u}{\partial y} = 0 \qquad v = 0 \tag{6-14}$$

Thus the particular problem that we shall treat is to find a solution to (6-11) and (6-12) which begins at $x = 0$ with the Blasius pattern and satisfies the altered boundary condition (6-14). In arriving at our final dimensionless formulation of the problem we employ a transformation due to R. von Mises.[2] Define a function $Y(x, y)$ by the relations

[1] H. Blasius, Grenzschichten in Flussigkeiten mit kleinen Reibung, *Z. Math. u. Phys.*, **56**, 4–13 (1908).

[2] R. von Mises, Bemerkung zur Hydrodynamik, *Z. angew. Math. u. Mech.*, **7**, 425–431 (1927).

$$u = \frac{\partial Y}{\partial y} \qquad v = -\frac{\partial Y}{\partial x} \qquad\qquad (6\text{-}15)$$

The continuity equation (6-11) is automatically satisfied by this choice. (Y is actually a stream function.) Then, if in (6-12) we eliminate y and take x and Y as independent variables, we find, after some partial differentiation, the following single equation for $u(x, Y)$:

$$u \frac{\partial u}{\partial x} = \nu u \frac{\partial}{\partial Y}\left(u \frac{\partial u}{\partial Y}\right) \qquad\qquad (6\text{-}16)$$

From the definition of Y we see that, if Y is taken to vanish at the origin, then $Y = 0$ along $y = 0$ and Y increases without limit as y increases. Thus the boundary conditions for (6-16) are

$$\frac{\partial u}{\partial Y} = 0 \quad \text{for } Y = 0 \qquad u = U \quad \text{for } Y = \infty \qquad\qquad (6\text{-}17)$$

for $x > 0$. The initial conditions at $x = 0$ are to be taken from the Blasius solution. Finally, introducing the dimensionless variables

$$x' = \frac{x}{L} \qquad y' = \frac{Y}{UL}\sqrt{\frac{UL}{\nu}} \qquad \psi = 1 - \left(\frac{u}{U}\right)^2 \qquad\qquad (6\text{-}18)$$

our problem becomes that of finding the function ψ which satisfies

$$\frac{\partial \psi}{\partial x} = \sqrt{1 - \psi}\,\frac{\partial^2 \psi}{\partial y^2} \qquad\qquad (6\text{-}19)$$

where we have now dropped the primes from the nondimensional variables. The

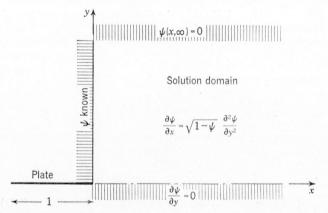

FIG. 6-5.   Dimensionless formulation of Prob. 6-2, boundary-layer wake.

boundary conditions for (6-19) are

$$\frac{\partial \psi}{\partial y} = 0 \quad \text{at } y = 0 \qquad \psi = 0 \quad \text{at } y = \infty \qquad\qquad (6\text{-}20)$$

and the values of ψ at $x = 0$ are to be computed[1] from the Blasius solution. The formulation is sketched in Fig. 6-5.

[1] These values are given in Fig. 6-42.

Problem 6-3. Unsteady Transverse Motion of a Beam. Consider a uniform beam of length L which is clamped at its ends as shown in Fig. 6-6. Propagation problems for such a system involve determining future displacements of the beam from given initial displacement and velocity information. If the beam has flexural rigidity, EI, and mass per unit length, ρ (both taken to be uniform along the beam), the equation of motion[1] is

$$\frac{\partial^4 y}{\partial x^4} + \frac{\rho}{EI}\frac{\partial^2 y}{\partial t^2} = 0 \quad (6\text{-}21)$$

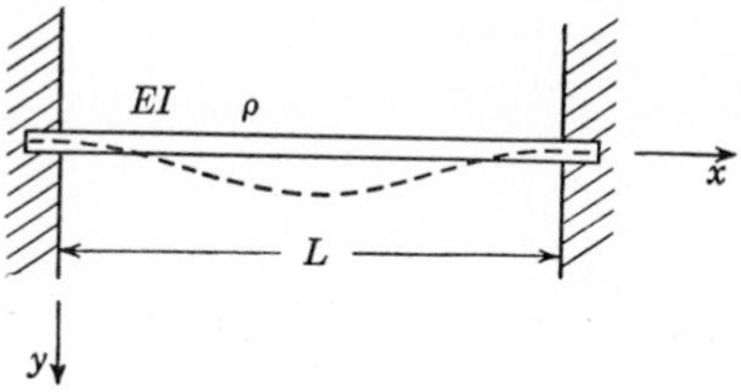

FIG. 6-6. Uniform beam, clamped at the ends, which moves transversely when a load is suddenly removed.

where $y(x, t)$ is the displacement of the neutral axis from the equilibrium position, x is distance along the beam, and t is time. The boundary conditions at the ends of the clamped beam of Fig. 6-6 are

$$\left.\begin{array}{l} y = 0 \\ \dfrac{\partial y}{\partial x} = 0 \end{array}\right\} \quad \text{at } x = 0 \qquad \left.\begin{array}{l} y = 0 \\ \dfrac{\partial y}{\partial x} = 0 \end{array}\right\} \quad \text{at } x = L \qquad (6\text{-}22)$$

for all time t. As a particular initial configuration we take the beam to have the displacement $y(x, 0)$ [due to a static load $w(x) = w_0 \sin (\pi x/L)$],

$$y(x, 0) = \frac{w_0 L^4}{EI\pi^4}\left\{\sin \pi \frac{x}{L} - \pi \frac{x}{L}\left(1 - \frac{x}{L}\right)\right\} \qquad (6\text{-}23)$$

and assume that the beam is without velocity. The propagation problem so defined is thus that of determining the motion of the beam after a sudden removal of load. To arrive at the final nondimensional formulation of the problem, the dimensionless variables

$$x' = \frac{x}{L} \qquad t' = \frac{t}{L^2\sqrt{\rho/EI}} \qquad \psi = \frac{y}{w_0 L^4/EI\pi^4} \qquad (6\text{-}24)$$

are introduced into the differential-equation, boundary, and initial conditions. After dropping the primes from the nondimensional length and time, we obtain the propagation problem for $\psi(x, t)$ defined by the differential equation

$$\frac{\partial^4 \psi}{\partial x^4} + \frac{\partial^2 \psi}{\partial t^2} = 0 \qquad (6\text{-}25)$$

[1] See S. Timoshenko, "Vibration Problems in Engineering," 2d ed., D. Van Nostrand Company, Inc., New York, 1937, p. 332. We neglect the effect of shear and rotatory inertia. Equation (6-25) can be derived from (4-4) by considering $w = -\rho\partial^2 y/\partial t^2$ as an "inertia loading."

to be satisfied for $0 < x < 1$ and $t > 0$ with boundary conditions

$$\left.\begin{array}{c} \psi = 0 \\ \dfrac{\partial \psi}{\partial x} = 0 \end{array}\right\} \quad \text{at } x = 0 \qquad \left.\begin{array}{c} \psi = 0 \\ \dfrac{\partial \psi}{\partial x} = 0 \end{array}\right\} \quad \text{at } x = 1 \qquad (6\text{-}26)$$

to be satisfied for $t > 0$ and the initial conditions

$$\psi = \sin \pi x - \pi x(1 - x)$$
$$\frac{\partial \psi}{\partial t} = 0 \tag{6-27}$$

to be satisfied for $0 < x < 1$ at $t = 0$.

Problem 6-4. Transverse Motion of a Taut String. The following problem is superficially similar to the previous. Instead of a beam we consider in Fig. 6-7 a flexible string of length L, stretched with large tension T, and fixed at the ends. As in the preceding case the propagation problem consists in determining the subsequent motion of the string when given initial displacement and velocity distributions. If

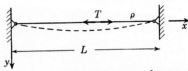

FIG. 6-7. Uniform string under tension T moves transversely when a load is suddenly removed.

x denotes distance along the string, t denotes time, and $y(x, t)$ denotes displacement from the equilibrium position, the equation of motion[1] of the string is the *wave equation*

$$\frac{\partial^2 y}{\partial x^2} = \frac{\rho}{T} \frac{\partial^2 y}{\partial t^2} \tag{6-28}$$

where ρ is the (uniform) mass per unit length of the string. The boundary conditions are simply

$$y = 0 \quad \text{at } x = 0 \qquad y = 0 \quad \text{at } x = L \tag{6-29}$$

for all t. As a particular initial condition we take the case where the string has been supporting a uniform load $w(x) = w_0$ until at $t = 0$ the load is suddenly removed. The corresponding initial conditions are

$$y = \frac{w_0}{2T} x(L - x)$$
$$\frac{\partial y}{\partial t} = 0 \tag{6-30}$$

---

[1] See J. P. Den Hartog, "Mechanical Vibrations," 4th ed., McGraw-Hill Book Company, Inc., New York, 1956, p. 136. Equation (6-28) can be derived from that of Exercise 4-9b by considering $w = -\rho \partial^2 y / \partial t^2$ as an "inertia loading."

at $t = 0$. A dimensionless formulation of the problem is obtained by introducing the nondimensional variables

$$x' = \frac{x}{L} \qquad t' = \frac{t}{L\sqrt{\rho/T}} \qquad \psi = \frac{y}{w_0 L^2/2T} \tag{6-31}$$

into (6-28) to (6-30). After dropping the primes on the nondimensional length and time, the problem is to determine $\psi(x, t)$ which must satisfy

$$\frac{\partial^2\psi}{\partial x^2} = \frac{\partial^2\psi}{\partial t^2} \tag{6-32}$$

for $0 < x < 1$ and $t > 0$ and must satisfy the boundary conditions

$$\psi = 0 \qquad \text{at} \begin{cases} x = 0 \\ x = 1 \end{cases} \tag{6-33}$$

for $t > 0$ and the initial conditions

$$\begin{aligned} \psi &= x(1 - x) \\ \frac{\partial\psi}{\partial t} &= 0 \end{aligned} \tag{6-34}$$

at $t = 0$.

Problem 6-5. Expansion of a Gas behind a Piston. Consider a long tube in which a quantity of gas is contained between a solid end wall and a movable piston. A class of propagation problems for such a system involve determining the future distribution of pressure and velocity within the gas from a given initial distribution and a given piston-displacement history. The particular problem which we shall study is sketched in Fig. 6-8. The gas has been at rest with the piston in the position shown when suddenly at $t = 0$ the piston is withdrawn at a velocity which is one-half the sound velocity of the gas in its initial state.

Assuming we have one-dimensional isentropic flow of a perfect gas, the governing equations[1] are those of momentum, continuity, and the gas laws,

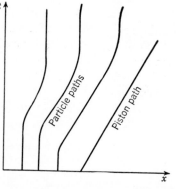

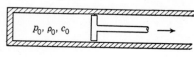

Fig. 6-8. Expansion wave propagates through gas when piston is suddenly moved to the right.

[1] See, for example, A. H. Shapiro, "The Dynamics and Thermodynamics of Compressible Fluid Flow," vol. II, The Ronald Press Company, New York, 1954, chap. 23.

$$\frac{\partial u}{\partial t} + u \frac{\partial u}{\partial x} + \frac{1}{\rho} \frac{\partial p}{\partial x} = 0 \qquad (6\text{-}35)$$

$$\frac{\partial \rho}{\partial t} + \rho \frac{\partial u}{\partial x} + u \frac{\partial \rho}{\partial x} = 0 \qquad (6\text{-}36)$$

$$\frac{p}{\rho^\gamma} = \text{constant} \qquad c^2 = \frac{dp}{d\rho} \qquad (6\text{-}37)$$

where x is displacement, t is time, $u(x, t)$, $p(x, t)$, and $\rho(x, t)$ are the velocity, pressure, and density, respectively of the gas, c is the velocity of sound in the gas, and γ, the ratio of specific heats, is a constant. Initially the gas is at rest, and the pressure, density, and velocity of sound of the gas are p_0, ρ_0, and c_0 throughout. The boundary conditions are that the gas in contact with the fixed wall has zero velocity and the gas in contact with the piston has the piston velocity.

The pressure can be eliminated from (6-35) by introducing (6-37) as follows:

$$\frac{\partial u}{\partial t} + u \frac{\partial u}{\partial x} + \frac{c^2}{\rho} \frac{\partial \rho}{\partial x} = 0 \qquad (6\text{-}38)$$

By combining the two parts of (6-37) we find

$$c^2 = c_0{}^2 \left(\frac{\rho}{\rho_0}\right)^{\gamma-1} \qquad (6\text{-}39)$$

i.e., the sound velocity depends only on the density. Equations (6-36) and (6-38) thus constitute a pair[1] of first-order partial differential equations for u and ρ as unknown functions of x and t.

A dimensionless formulation can be obtained by introducing the following variables:

$$u' = \frac{u}{c_0} \qquad c' = \frac{c}{c_0} \qquad \rho' = \frac{\rho}{\rho_0}$$

$$x' = \frac{x}{L} \qquad t' = \frac{t}{L/c_0} \qquad (6\text{-}40)$$

In terms of these nondimensional variables (we have already dropped the primes) the differential equations of the problem are

$$\frac{\partial u}{\partial t} + u \frac{\partial u}{\partial x} + \frac{c^2}{\rho} \frac{\partial \rho}{\partial x} = 0$$

$$\frac{\partial \rho}{\partial t} + u \frac{\partial \rho}{\partial x} + \rho \frac{\partial u}{\partial x} = 0 \qquad (6\text{-}41)$$

where c is given by

$$c = \rho^{(\gamma-1)/2} \qquad (6\text{-}42)$$

[1] These can be reduced to a single second-order equation if desired (see Exercise 6-6).

In our numerical work we shall take $\gamma = 1.40$. The initial conditions at $t = 0$ are

$$u(x, 0) = 0 \qquad \rho(x, 0) = 1 \tag{6-43}$$

for $0 < x < 1$. The boundary conditions are

$$u(0, t) = 0 \qquad u\left(1 + \frac{t}{2}, t\right) = \frac{1}{2} \tag{6-44}$$

since the piston velocity is $\frac{1}{2}$ in dimensionless units and its position is given by $x = 1 + t/2$ for $t > 0$. This formulation is sketched in Fig. 6-9.

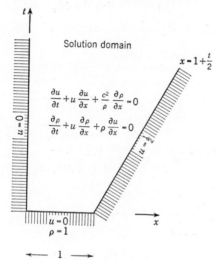

FIG. 6-9. Dimensionless formulation of Prob. 6-5, expansion of a gas behind a piston.

EXERCISES

6-1. An idealized telegraph line is shown in Fig. 6-10. At $t = 0$ the switch is closed, and the battery charges the distributed capacity of the line through the external lumped resistance and the distributed resistance of the line. The voltage $e(x, t)$, and the current $i(x, t)$ in the line satisfy[1] the following relations:

$$\frac{\partial e}{\partial x} + ri = 0$$

$$\frac{\partial i}{\partial x} + c\frac{\partial e}{\partial t} = 0$$

Formulate the propagation problem for $e(x, t)$. Reduce the problem to dimensionless form in the case where $rL = R$. This formulation is almost analogous with that of

[1] See, for example, E. A. Guillemin, "Communication Networks," vol. II, John Wiley & Sons, Inc., New York, 1935, p. 33.

Prob. 6-1, case (b). Verify that, if instead of $e(x, t)$ the quantity $E - e(x, t)$ is taken as the unknown variable, then the analogy becomes perfect.

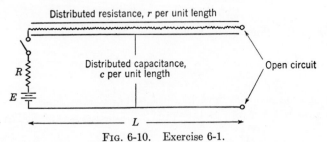

Distributed resistance, r per unit length

Distributed capacitance, c per unit length

Open circuit

R

E

L

FIG. 6-10.   Exercise 6-1.

6-2. Consider the same heat-conducting rod as in Fig. 6-1 except that at $t = 0$ heaters, which supply heat fluxes, Q per unit area, are brought in contact with the ends. Formulate the propagation problem for $T(x, t)$. Show that by suitably defining ψ a dimensionless formulation can be obtained which is the same as those in Fig. 6-2 with the exception that $\psi = 0$ at $t = 0$ and $\partial \psi / \partial x = -1$ at $x = 0$.

$$\textit{Ans. } \psi = (T - T_h)k/QL.$$

6-3. A uniform right-circular cylinder of radius R of heat-conducting material is surrounded by a heating jacket which for $t > 0$ supplies a heat flux, Q per unit area, uniformly to the surface of the cylinder. If the entire cylinder is initially at temperature T_0, formulate the propagation problem for $T(r, t)$.

$$\textit{Ans. } \frac{1}{r} \frac{\partial}{\partial r}\left(r \frac{\partial T}{\partial r}\right) = \frac{\rho c_p}{k} \frac{\partial T}{\partial t} \text{ for } 0 < r < R, \ t > 0, \text{ and } T = T_0 \text{ at } t = 0, \text{ and}$$

$$\frac{\partial T}{\partial r} = \frac{Q}{k} \text{ at } r = R, \text{ and } \frac{\partial T}{\partial r} = 0 \text{ at } r = 0.$$

6-4. Supersonic Nozzle. The governing equations[1] of steady two-dimensional irrotational isentropic flow are

$$u \frac{\partial u}{\partial x} + v \frac{\partial u}{\partial y} + \frac{1}{\rho} \frac{\partial p}{\partial x} = 0$$

$$u \frac{\partial v}{\partial x} + v \frac{\partial v}{\partial y} + \frac{1}{\rho} \frac{\partial p}{\partial y} = 0$$

$$\frac{\partial(\rho u)}{\partial x} + \frac{\partial(\rho v)}{\partial y} = 0$$

$$\frac{\partial v}{\partial x} - \frac{\partial u}{\partial y} = 0$$

$$\frac{p}{\rho^\gamma} = \text{constant} \qquad \frac{dp}{d\rho} = c^2$$

where u and v are velocity components, p is pressure, ρ is density, c is the velocity of sound, and $\gamma = 1.4$ is the ratio of specific heats for air. Show that these can be reduced to the following pair of first-order equations for u and v,

$$(u^2 - c^2) \frac{\partial u}{\partial x} + uv\left(\frac{\partial v}{\partial x} + \frac{\partial u}{\partial y}\right) + (v^2 - c^2) \frac{\partial v}{\partial y} = 0$$

$$\frac{\partial v}{\partial x} - \frac{\partial u}{\partial y} = 0$$

with

$$5c^2 = 6c^{*2} - (u^2 + v^2)$$

[1] See, for example, Shapiro, *op. cit.*, vol. I, chap. 9.

where c^* is the sound velocity when the flow velocity just equals the sound velocity. Formulate the propagation problem for the flow in the two-dimensional nozzle of Fig. 6-11, assuming parallel sonic flow at the throat and that the flow direction is along the walls AB.

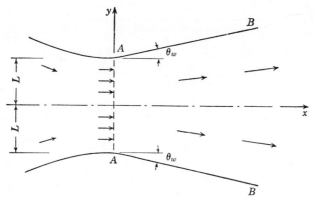

FIG. 6-11. Exercise 6-4.

Ans. In addition to the equations above for u and v, $u = c^*$, $v = 0$ along $x = 0$, $v = 0$ along $y = 0$, and $v = u \tan \theta_w$ along AB.

6-5. Show that by suitably defining a velocity potential φ it is possible to reduce the pair of first-order equations in Exercise 6-4 to a single second-order equation.

$$Ans.\ (u^2 - c^2) \frac{\partial^2 \varphi}{\partial x^2} + 2uv \frac{\partial^2 \varphi}{\partial x\, \partial y} + (v^2 - c^2) \frac{\partial^2 \varphi}{\partial y^2} = 0.$$

6-6. Show that by introducing a velocity potential φ defined by $\partial \varphi / \partial x = u$ it is possible to reduce (6-41) to a single second-order equation.

$$Ans.\ (u^2 - c^2) \frac{\partial^2 \varphi}{\partial x^2} + 2u \frac{\partial^2 \varphi}{\partial x\, \partial t} + \frac{\partial^2 \varphi}{\partial t^2} = 0.$$

6-7. A uniform horizontal elastic bar of length L, Young's modulus E, and mass density ρ per unit volume is fixed at one end and is at rest. At $t = 0$ a constant tensile stress S is applied to the free end. Show that the strain $\epsilon(x, t)$ and velocity $v(x, t)$ must satisfy the compatibility relation $\partial \epsilon / \partial t = \partial v / \partial x$ and that according to Newton's second law $\partial \epsilon / \partial x = (\rho / E)\, \partial v / \partial t$. (a) Formulate the propagation problem for the strains and velocities in the bar in the form of a pair of first-order partial differential equations with appropriate initial and boundary conditions. (b) Reduce these to a single second-order equation, and study the analogy with Prob. 6-4.

Ans. (a) $\epsilon = v = 0$ at $t = 0$; $v = 0$ at $x = 0$; $E\epsilon = S$ at $x = L$ plus the two equations above. (b) $\partial^2 \epsilon / \partial x^2 = (\rho / E)\, \partial^2 \epsilon / \partial t^2$ with $\epsilon = \partial \epsilon / \partial t = 0$ at $t = 0$, $\partial \epsilon / \partial x = 0$ at $x = 0$, and $E\epsilon = S$ at $x = L$.

6-8. Compression of a Plastic Bar. Consider two-dimensional flow of a perfectly plastic bar being squeezed out between rough parallel plates.[1] The material is assumed to suffer no deformation at all if the maximum shear stress is less than k, the yield stress, but to flow plastically when the maximum shear stress reaches k. As

[1] This problem is treated in R. Hill, "Plasticity," Oxford University Press, New York, 1950, p. 226.

shown in Fig. 6-12a there are distinct zones of rigid and plastic behavior. The overhanging ends move horizontally as rigid bodies, the central portions move vertically as rigid extensions of the plates, while in the region $ABCDEF$ there is plastic deformation. Assume the deformation to be so slow that inertia effects can be neglected.

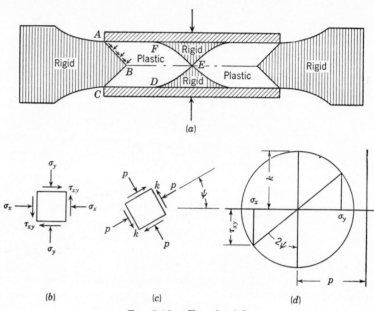

Fig. 6-12.   Exercise 6-8.

With the aid of the Mohr's-circle diagram in Fig. 6-12d show that the conditions of force balance on an element in the plastic region can be reduced to

$$\frac{1}{2k}\frac{\partial p}{\partial x} + \cos 2\psi \frac{\partial \psi}{\partial x} + \sin 2\psi \frac{\partial \psi}{\partial y} = 0$$

$$\frac{1}{2k}\frac{\partial p}{\partial y} + \sin 2\psi \frac{\partial \psi}{\partial x} - \cos 2\psi \frac{\partial \psi}{\partial y} = 0$$

where ψ is the orientation angle of the maximum shear directions at a point and p is the compressive stress in these directions. Formulate the propagation problem for the stresses in the plastic zone $ABEF$. For rigid-body motion of the ends under balanced forces the 45° line AB must be a shear line with the compression p just equal to k. Along AF shearing must take place parallel to the plate, while along BE symmetry requires that no shear stress should act parallel to BE. The location of FE is initially unknown and is actually a part of the solution of the propagation problem.

Ans. In addition to the governing equations for p and ψ given, $\psi = -\pi/4$ and $p = k$ on AB, $\psi = 0$ on AF, and $\psi = -\pi/4$ on BE.

6-9. Show that the pair of first-order equations in Exercise 6-8 can be reduced to a single second-order equation.

Ans.

$$\tan 2\psi \left(\frac{\partial^2 \psi}{\partial y^2} - \frac{\partial^2 \psi}{\partial x^2}\right) + 2\frac{\partial^2 \psi}{\partial x\,\partial y} - 2\left(\frac{\partial \psi}{\partial x}\right)^2 + 2\left(\frac{\partial \psi}{\partial y}\right)^2 - 4\tan 2\psi \left(\frac{\partial \psi}{\partial x}\right)\left(\frac{\partial \psi}{\partial y}\right) = 0.$$

6-10. Let $u = \partial\psi/\partial x$ and $v = \partial\psi/\partial t$ be the dimensionless slope and velocity of the string in Prob. 6-4. Show that in terms of them the problem can be formulated as follows:

$$\left.\begin{aligned}\frac{\partial u}{\partial x} - \frac{\partial v}{\partial t} &= 0\\[4pt]\frac{\partial u}{\partial t} - \frac{\partial v}{\partial x} &= 0\end{aligned}\right\} \quad 0 < x < 1,\, t > 0$$

$$v = 0 \qquad \text{at } \left\{\begin{aligned}x &= 0\\ x &= 1\end{aligned}\right\},\, t > 0$$

$$\left.\begin{aligned}u &= 1 - 2x\\ v &= 0\end{aligned}\right\} \quad 0 < x < 1,\, t = 0$$

What is the physical significance of the second governing equation? Show that a formulation with the same governing equations but with the following initial and boundary conditions

$$\left.\begin{aligned}u &= x(1 - x)\\ v &= 0\end{aligned}\right\} \quad 0 < x < 1,\, t = 0$$

$$u = 0 \qquad \text{at } \left\{\begin{aligned}x &= 0\\ x &= 1\end{aligned}\right\},\, t > 0$$

represents the same problem by setting $u = \psi$. What then is the significance of the variable v?

6-2. Formulation of the General Problem

The five particular problems posed in the preceding section have all been examples of propagation problems in continuous systems. In every case there were two independent variables. The solution domains all had the *open* shape shown in Fig. 6-13a as contrasted[1] with the *closed* shape of Fig. 6-13b, which was characteristic of equilibrium problems. Initial or boundary conditions were specified along the open boundary, and governing partial differential equations were to be satisfied within the solution domain. The forms of the governing equations were somewhat varied, ranging from a pair of first-order equations in two unknowns to a single fourth-order equation for one unknown. Despite this apparent diversity we shall shortly see that there is an underlying unity in that governing equations for propagation problems are of *hyperbolic* or *parabolic* type, whereas in equilibrium problems the equations are of *elliptic* type.

The general propagation problem that we are considering can thus be formulated as follows:

[1] The difference between propagation problems and equilibrium problems was vividly put by L. F. Richardson, How to Solve Differential Equations Approximately by Arithmetic, *Math. Gazette*, **12**, 415–421 (1925), when he called the former *marching* problems and the latter *jury* problems. In propagation problems the solution marches out from initial conditions guided in transit by side boundary conditions. In equilibrium problems the entire solution is judged by a jury demanding simultaneous satisfaction of all the boundary conditions and all the internal requirements.

The problem is to march out the solution of a governing system of partial differential equations of parabolic or hyperbolic type from prescribed conditions on an open boundary.

A more detailed description (e.g., the number and type of boundary conditions required at each point) cannot be given until we have studied the significance of the classification of governing equations and have

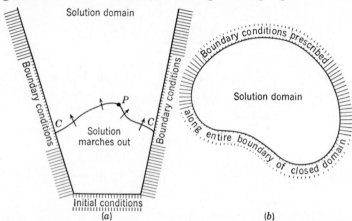

Fig. 6-13. (*a*) Propagation problem is solved in an open domain. (*b*) Equilibrium problem is solved in a closed domain.

become familiar with the *theory of characteristics*. We take up these topics in the following section.

EXERCISES

6-11. In Prob. 6-4, transverse motion of a taut string, would it be physically meaningful to try to specify conditions along the string at some time $T > 0$?

6-12. Consider the equilibrium problem of a thin plate supporting a transverse load. The edge of the plate is maintained in a horizontal plane. Along parts of the boundary the edge is hinged; along other parts it is clamped. Would it be a physically meaningful problem if along a certain portion of the edge there were no specification of the mode of support?

6-3. Mathematical Properties

Let us begin by considering the following pair of simultaneous first-order partial differential equations for the dependent variables u and v,

$$A_1 \frac{\partial u}{\partial x} + B_1 \frac{\partial u}{\partial y} + C_1 \frac{\partial v}{\partial x} + D_1 \frac{\partial v}{\partial y} = 0$$

$$A_2 \frac{\partial u}{\partial x} + B_2 \frac{\partial u}{\partial y} + C_2 \frac{\partial v}{\partial x} + D_2 \frac{\partial v}{\partial y} = 0$$

(6-45)

where the coefficients $A_1, \ldots, D_2$ are functions of u and v but not of x and y. This set of equations is sufficiently general[1] to represent any

---

[1] The case in which the right-hand sides of (6-45) are nonzero is treated in Exercise 6-23.

of the particular examples of Sec. 6-1 except Prob. 6-3, which involves a
fourth-order equation.   Ordinarily (6-45) is nonlinear, but because x and
y do not appear in the coefficients, the system can be reduced to a linear
one by interchanging the roles of the dependent and independent variables, i.e., by considering x and y as unknown functions of u and v.   For
this reason (6-45) is known as a *reducible system.*[1]

Classification of Systems as Hyperbolic, Parabolic, or Elliptic.   Suppose that in solving (6-45) we have reached the state of affairs shown in
Fig. 6-13a.   The solution is known up to the curve CPC.   At P we know
continuously differentiable values of u and v along the curve CPC as well
as all their derivatives in directions that lie below the curve.   We now
pose the question as to whether this is enough information to deduce the
directional derivatives of u and v at P in directions that lie above the
curve.   In other words, is the behavior of the solution just above P
completely determined by the solution below the curve, or is additional
information required from the boundaries at C?

Now the directional derivative of u in a direction along which distance
is measured by s is

$$\frac{\partial u}{\partial s} = \frac{\partial u}{\partial x}\frac{\partial x}{\partial s} + \frac{\partial u}{\partial y}\frac{\partial y}{\partial s} \tag{6-46}$$

so that the directional derivative in any direction is known as soon as
the derivatives $\partial u/\partial x$ and $\partial u/\partial y$ are known.   Our question then may be
restated as follows: "Are the derivatives $\partial u/\partial x$, $\partial u/\partial y$, $\partial v/\partial x$, and $\partial v/\partial y$
of a solution to (6-45) uniquely determined at a point P by the values of
u and v along the curve CPC?"

To answer this question, we write out Eqs. (6-45) at point P.   We also
write out the directional differentials[2] of u and v taken along CPC at P.
These four equations can be assembled into the single matrix equation
below.

$$\begin{bmatrix} A_1 & B_1 & C_1 & D_1 \\ A_2 & B_2 & C_2 & D_2 \\ dx & dy & 0 & 0 \\ 0 & 0 & dx & dy \end{bmatrix} \begin{bmatrix} \dfrac{\partial u}{\partial x} \\ \dfrac{\partial u}{\partial y} \\ \dfrac{\partial v}{\partial x} \\ \dfrac{\partial v}{\partial y} \end{bmatrix} = \begin{bmatrix} 0 \\ 0 \\ du \\ dv \end{bmatrix} \tag{6-47}$$

---

[1] This reduction is actually carried out for Prob. 6-5 in Exercise 6-19.

[2] In (6-47) du is an abbreviation for $\partial u/\partial s\, ds$, where $\partial u/\partial s$ is given by (6-46) and
s now measures distance along the curve CPC.   The symbols dv, dx, and dy stand for
similar abbreviations.

With u and v known at P the coefficients $A_1, \ldots, D_2$ will be known. With the direction of CPC known dx and dy will be known, and if u and v are also known along CPC, then du and dv will be known. The system (6-47) thus constitutes a pair of four simultaneous linear algebraic equations for the four first derivatives. If the determinant of the elements in the square matrix is not zero, these equations have a unique solution; but if the determinant vanishes, we are faced with the exceptional case[1] in which there is an infinity of solutions. In the former case satisfaction of the governing equations requires that the directional derivatives have the same value above and below CPC. In the latter case the governing equations do not determine these derivatives uniquely; i.e., there could be discontinuities as we cross CPC.

Let us see under what circumstances the determinant of the matrix in (6-47) can vanish. Expanding the determinant, we find

$$(A_1C_2 - A_2C_1)\,dy^2 - (A_1D_2 - A_2D_1 + B_1C_2 - B_2C_1)\,dx\,dy \\ + (B_1D_2 - B_2D_1)\,dx^2 = 0 \quad (6\text{-}48)$$

This may be considered as a quadratic equation for the slope dy/dx. If the direction of the curve CPC at P is such that it has a slope satisfying (6-48), then the derivatives of u and v are not uniquely determined by the values of u and v along the curve. Such a direction is called a *characteristic direction*. The quadratic (6-48) gives two real slopes, one real slope, or a pair of complex slopes depending on whether the discriminant

$$(A_1D_2 - A_2D_1 + B_1C_2 - B_2C_1)^2 - 4(A_1C_2 - A_2C_1)(B_1D_2 - B_2D_1) \quad (6\text{-}49)$$

is positive, zero, or negative. This is also the criterion for cataloguing the system (6-45) as *hyperbolic, parabolic,* or *elliptic;* i.e., we call the system hyperbolic at a point if there are two real characteristic directions, parabolic if there is only a single characteristic direction, or elliptic if there are no real characteristic directions at the point.

The foregoing analysis can be repeated for the single second-order *quasi-linear* equation,

$$a\,\frac{\partial^2\psi}{\partial x^2} + b\,\frac{\partial^2\psi}{\partial x\,\partial y} + c\,\frac{\partial^2\psi}{\partial y^2} = f \quad (6\text{-}50)$$

in which a, b, c, and f are functions of x, y, ψ, $\partial\psi/\partial x$, and $\partial\psi/\partial y$. The characteristic directions are determined[2] from the quadratic

$$a\,dy^2 - b\,dx\,dy + c\,dx^2 = 0 \quad (6\text{-}51)$$

[1] See Exercise 6-20.

[2] See Exercises 6-21 and 6-22.

and hence (6-50) is *hyperbolic, parabolic,* or *elliptic* at a point according as the discriminant $b^2 - 4ac$ is positive, zero, or negative.

The Characteristic Curves. We return to the pair of first-order equations (6-45) and suppose that the system is hyperbolic throughout the domain of interest. At every point, then, there are two roots, $(dy/dx)_\alpha$ and $(dy/dx)_\beta$, say, to the quadratic (6-48). A curve which at each of its points has the slope $(dy/dx)_\alpha$ is said to be an α *characteristic.* A curve whose slope is everywhere $(dy/dx)_\beta$ is said to be a β *characteristic.* There are thus two families of characteristic curves filling the domain as shown in Fig. 6-14.

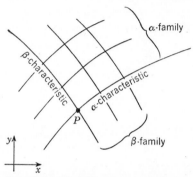

So far we have learned that the characteristics are loci of possible discontinuities in the derivatives of a solution. There is still more to learn from (6-47). If we are in fact considering a characteristic direction so that the determinant in (6-47) is zero, the right-hand column must be compatible with this if there are to be any solutions at all for the first derivatives; i.e., when the right-hand column is substituted for any of the columns on

FIG. 6-14. Slope on an α characteristic is $(dy/dx)_\alpha$; slope on a β characteristic is $(dy/dx)_\beta$.

the left, the resulting determinant must[1] also vanish. Replacing the fourth column on the left with the column on the right and setting the determinant equal to zero lead to

$$(A_1B_2 - A_2B_1)\, du + \left[(A_1C_2 - A_2C_1)\,\frac{dy}{dx} - (B_1C_2 - B_2C_1)\right] dv = 0$$

$$(6\text{-}52)$$

When $(dy/dx)_\alpha$ obtained as a root of (6-48) is inserted in (6-52), it becomes an ordinary differential equation for u and v *along* the α characteristic. A corresponding equation is obtained along a β characteristic.

We thus have outlined a possible method of attack for solving hyperbolic systems: first, locate the characteristic curves, and, second, integrate the ordinary differential equations (6-52) along the characteristic. This procedure is called the *method of characteristics.*

Example. Consider the system

$$\frac{\partial u}{\partial x} - \frac{\partial v}{\partial y} = 0$$

$$\frac{\partial u}{\partial y} - \frac{\partial v}{\partial x} = 0$$

$$(6\text{-}53)$$

[1] See Exercise 6-20.

with the initial and boundary conditions sketched in Fig. 6-15. This is a modified formulation[1] of Prob. 6-4, the taut string. We begin by writing the matrix equation (6-47) for this case.

$$
\begin{bmatrix}
1 & 0 & 0 & -1 \\
0 & 1 & -1 & 0 \\
dx & dy & 0 & 0 \\
0 & 0 & dx & dy
\end{bmatrix}
\begin{bmatrix}
\dfrac{\partial u}{\partial x} \\[2ex]
\dfrac{\partial u}{\partial y} \\[2ex]
\dfrac{\partial v}{\partial x} \\[2ex]
\dfrac{\partial v}{\partial y}
\end{bmatrix}
=
\begin{bmatrix}
0 \\[2ex]
0 \\[2ex]
du \\[2ex]
dv
\end{bmatrix}
\tag{6-54}
$$

The characteristic directions are obtained by equating the determinant of the square matrix on the left to zero. We find as the special case of (6-48)

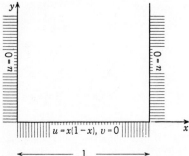

FIG. 6-15. Initial and boundary conditions for the system (6-53).

$$
dy^2 - dx^2 = 0 \qquad \text{or} \qquad \frac{dy}{dx} = \pm 1 \tag{6-55}
$$

which means that the characteristic curves are straight lines making 45° with the coordinate axes. Let us call the characteristics with $dy/dx = +1$ the α family and the characteristics with $dy/dx = -1$ the β family. The ordinary differential equations along the characteristics are obtained by replacing the fourth column of the left-hand matrix by the column on the right side and equating the corresponding determinant to zero. We find as the special case of (6-52)

$$
du - \left(\frac{dy}{dx}\right) dv = 0 \qquad \begin{cases} du - dv = 0 \text{ on } \alpha \text{ characteristics} \\ du + dv = 0 \text{ on } \beta \text{ characteristics} \end{cases} \tag{6-56}
$$

This pair of ordinary differential equations can be integrated to obtain

$$
\begin{aligned}
u - v &= \text{constant} \qquad \text{on } \alpha \text{ characteristics} \\
u + v &= \text{constant} \qquad \text{on } \beta \text{ characteristics}
\end{aligned}
\tag{6-57}
$$

The constants of integration in (6-57) provide a convenient method for distinguishing between the different members of the families of characteristics. Let us designate an α characteristic by the value $\alpha = u - v$, which according to (6-57) can be evaluated at any point along the characteristic. Similarly we designate a β characteristic by the value $\beta = u + v$ carried along it.

We next investigate how the solution of our problem is affected by means of these. In Fig. 6-16 the values of u and v on $y = 0$ have been computed for several points. The values of α and β for the characteristics passing through these points have been obtained from the following relations:

$$
\begin{aligned}
\alpha &= u - v \\
\beta &= u + v
\end{aligned}
\tag{6-58}
$$

[1] See Exercise 6-10. The variable y here plays the role of t.

The solution for u and v at any point within the triangle OAB can now be obtained. All that is necessary is to know the values of α and β on the characteristics passing through the point. We then have

$$u = \frac{\beta + \alpha}{2}$$

$$v = \frac{\beta - \alpha}{2} \tag{6-59}$$

by inverting (6-58). For instance, at the point P $(\frac{13}{24}, \frac{5}{24})$ which is the intersection of the $\alpha = \frac{2}{9}$ line and the $\beta = \frac{3}{16}$ line, as shown, the values of u and v are obtained from (6-59) as $u = \frac{59}{288}$ and $v = -\frac{5}{288}$.

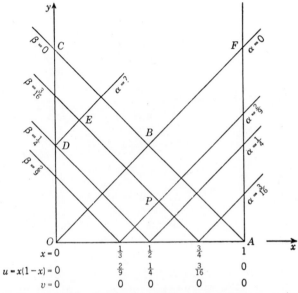

FIG. 6-16. Solution of Prob. 6-4 by the method of characteristics.

In order to continue the solution into the triangle OBC, it is necessary to make use of the boundary condition $u = 0$ along $x = 0$. The way this is done is illustrated at point D. There we already have the characteristic $\beta = \frac{1}{4}$, but the value of α on the α characteristic is unknown until we use the fact that the first of (6-59) must still hold. Substituting our known values of u and β, we find that $\alpha = -\frac{1}{4}$. The α characteristic entering the region at D is often called the *reflection* of the β characteristic, which strikes the boundary at this point. The simple boundary condition $u = 0$ implies that the value of α on the reflected α characteristic is the negative of the value of β on the β characteristic striking the boundary. We are then able to continue the solution to points such as E, where, once more, both α and β are known.

It should now be clear how the solution can be marched out as far as is desired. The given boundary conditions permit us to label successive characteristics, and the solution for u and v is given by (6-59) for any point through which pass two characteristics with known labels.

We note that this has been a mathematically *exact* solution by the method of characteristics. The reasons for the simplicity of the foregoing problem were:

1. The differential equations for the characteristics (6-55) could be integrated and the characteristics completely located prior to studying the solution.

2. The ordinary differential equations along the characteristics (6-56) could be integrated in closed form.

In general neither of these simplifications obtains. It is still possible[1] to use the over-all plan of the method, but it is necessary to advance stepwise, integrating numerically both the ordinary differential equations

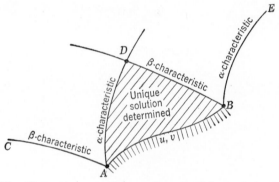

FIG. 6-17.   First uniqueness theorem for pair of first-order equations.

along the characteristics and the differential equations for the characteristics themselves.

Hyperbolic Systems.   We next state some theoretical properties of systems governed by the pair of first-order equations (6-45).   The extent to which initial and boundary conditions determine *unique solutions* can be deduced for a large number of cases from the following three theorems:[2]

In the *first* case, shown in Fig. 6-17, AB is a noncharacteristic curve of continuous slope along which u and v are prescribed as continuously differentiable functions.   A solution to (6-45), assuming these prescribed values, is then uniquely determined in the triangle ABD bounded by AB and the characteristics emanating from A and B.   It is assumed here that the direction of propagation is upward.   The data on AB fix the

---

[1] See Sec. 6-6.

[2] A common method of proof is based on an iteration.   See R. Courant and K. O. Friedrichs, "Supersonic Flow and Shock Waves," Interscience Publishers, New York, 1948, p. 49.   This iteration process may also be used to construct approximate numerical solutions (see Exercise 6-28).   Corresponding theorems for a single second-order equation can be stated by making obvious changes in the wordings (see Exercises 6-21 and 6-22 and Shapiro, *op. cit.*, vol. I, p. 603).

solution equally well in a triangle below AB bounded by the extensions of CA and EB. This theorem applies, for instance, to the triangle OBA in Fig. 6-16. The initial data on OA determine the solution in OBA independently of the boundary conditions at $x = 0$ and $x = 1$.

In the *second* case, sketched in Fig. 6-18, PG and PE are the two characteristics through P. A unique solution is determined in the quadrilateral $PEFG$, where EF and GF are characteristics, when u and v are known[1] at P and continuously differentiable values of either u or v are prescribed along each of the arcs PG and PE. This theorem can be applied, for instance, in Fig. 6-16 after the solution has been marched out to CB and BF.

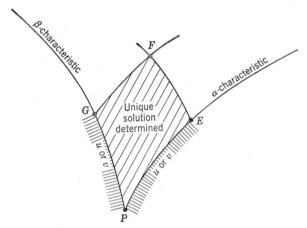

FIG. 6-18.   Second uniqueness theorem.

In the *third* case PA is a noncharacteristic curve of continuous slope as shown in Fig. 6-19. A unique solution is here determined in the quadrilateral $PEFA$, where EF and AF are characteristics, when u and v are known[1] at P and continuously differentiable values of either u or v are prescribed along each of the arcs PE and PA. This theorem applies, for instance, to the triangle COB in Fig. 6-16. When the solution has been marched out to OB, then the boundary condition $u = 0$ on OC fixes the solution within the entire triangle.

In the theorems just given it is assumed that no obstruction or boundary is encountered in the regions of determinancy. In some problems there is a real possibility that unanticipated boundaries may appear within the expected solution domain. Physical examples of such bound-

---

[1] The values at P must be compatible with the characteristic directions as given by (6-48). A unique solution can sometimes be assured in the second and third cases even when the prescribed values are discontinuous at P. For an example see Fig. 6-24.

aries are shock waves[1] and flame fronts in gas flow problems. These are loci of sharp discontinuities of fluid properties which are propagated according to their own special laws. They represent dividing lines between regions where different continuous flow equations must be solved. Generally their locations are unknown in advance and must be determined[2] by marching them out simultaneously with the continuous solutions.

We have seen that the characteristic curves are loci of possible discontinuities in the *derivatives* of u and v; i.e., they are the propagation

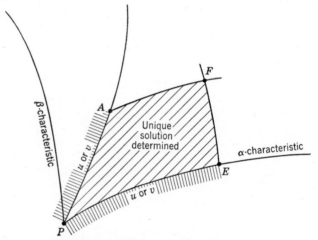

FIG. 6-19. Third uniqueness theorem.

paths for *small* disturbances. The derivatives may, of course, be continuous everywhere, but if a discontinuity exists across a certain characteristic at one point, there will be a discontinuity across that characteristic along its entire length. It is possible, in fact, to obtain[3] differential equations similar to (6-52) governing the propagation of the discontinuity intensity. For instance, in the case of Prob. 6-4, transverse motion of a taut string, it is a simple exercise[4] to show that, if at a point the normal derivative of u across the α characteristic has a jump of a certain magnitude, then there will be a jump in the normal derivative of u *of the same magnitude* at every other point on that α characteristic.

[1] Shapiro, *op. cit.*, vol. I, chap. 16.
[2] For numerical treatment of such problems see H. Geiringer, On Numerical Methods in Wave Interaction Problems, "Advances in Applied Mechanics," vol. I, Academic Press, New York, 1948, pp. 201–248, and H. H. Goldstine and J. Von Neumann, Blast Wave Calculation, *Comm. Pure Appl. Math.*, **8**, 327–353 (1955).
[3] Courant and Friedrichs, *op. cit.*, p. 53.
[4] See Exercise 6-25.

We have also seen that the characteristic curves are the natural border lines as far as determining which parts of a solution domain are controlled by which boundary conditions.   When we come to set up discrete models of continuous systems, we must be sure that our models reflect this fact. If from given boundary information the discrete model can provide a solution in a region not reached by the continuous solution, we must have serious doubts about its value as a model.

Parabolic Systems.   We consider here the governing equation

$$\frac{\partial \psi}{\partial t} = A_2 \frac{\partial^2 \psi}{\partial x^2} + A_1 \frac{\partial \psi}{\partial x} + A_0 \tag{6-60}$$

where the A_j are functions of x, t, and ψ and the solution domain shown in Fig. 6-20.   The initial and boundary conditions are indicated on the

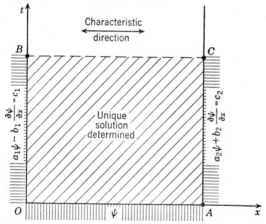

Fig. 6-20.   Uniqueness theorem for second-order parabolic systems.

figure.   The quantities a_1, b_1, c_1, a_2, b_2, c_2 are known functions of t.   This is a sufficiently general formulation to cover most second-order parabolic systems encountered in engineering analysis.   For example, Prob. 6-1, cooling of a rod, and Prob. 6-2, boundary-layer wake, represent special cases of this formulation.   According to (6-51) Eq. (6-60) is parabolic, having only a single family of characteristics: the lines $t = $ constant.

The data shown in Fig. 6-20 determine a solution only within the rectangle bounded by the characteristic BC.   Rigorous proofs for the existence and uniqueness of the solution have apparently[1] been given only when Eq. (6-60) is linear.

If a disturbance is introduced at B in Fig. 6-20, the entire solution above BC will be disturbed; i.e., disturbances are propagated along the

[1] See D. L. Bernstein, "Existence Theorems in Partial Differential Equations," Princeton University Press, Princeton, N.J., 1950, p. 161.

characteristic. The nature of the disturbance may, however, be drastically altered as it travels along the characteristic. A discontinuous change can be smoothed into a continuous change.

For example, if a heater is applied at $t = 0$ to one end of the rod in Prob. 6-1, the temperature at the point of application begins to rise[1] in proportion to $\sqrt{t}$; that is, $\partial\psi/\partial t$ is discontinuous. At an interior point the temperature also begins to rise immediately, but there is *no discontinuity* in $\partial\psi/\partial t$ or *any* of its higher derivatives. Furthermore it is found that, although the temperature begins to rise immediately at all points, the rate of rise falls off very rapidly with distance from the heater.

This behavior is considerably different from that of the taut string of Prob. 6-4. A small disturbance introduced at the left boundary, say, is propagated along an α characteristic in Fig. 6-16 with *constant intensity*. This means that at an interior point there will be no response until that α characteristic arrives, but when it does, the disturbance will be delivered with full strength.

The fourth-order equation (6-25) is also parabolic,[2] and many of its properties are natural extensions of those of the heat-conduction equation. A unique solution is determined within the domain shown in Fig. 6-20 if ψ *and* $\partial\psi/\partial t$ are prescribed at $t = 0$ and if on each of the boundaries OB and AC *two* independent linear relations between ψ and its first three x derivatives are prescribed.

Further Properties of Hyperbolic Systems.   We return to the pair of first-order equations (6-45) with characteristic directions given by (6-48) and with differential equations for u and v along the characteristics given by (6-52). In the example treated in Fig. 6-16 both families of characteristics were *parallel straight* lines. Investigating the reasons for this, we find that in this case all the coefficients $A_1, \ldots, D_2$ in (6-45) were constants, and therefore the characteristic slopes were constant.

The coefficients $A_1, \ldots, D_2$ can be complicated functions of u and v, and still both families of characteristics can be parallel straight lines. This can occur within a region in which both u and v remain constant. Such a region is called a *region of constant state*. The characteristic directions given by (6-48) depend on the coefficients $A_1, \ldots, D_2$, which in turn depend on u and v. If u and v remain constant, the characteristics will be parallel straight lines. This apparently trivial situation is actually of considerable importance. Many practical problems have solutions which consist of several constant-state regions interconnected by regions in which u and v change. Note that in a region of constant state we cannot distinguish between different characteristics within the same

---

[1] See Exercise 6-26.

[2] R. Courant and D. Hilbert, "Methoden der mathematischen Physik," vol. 2, Springer-Verlag OHG, Berlin, 1937, p. 138.

family by using the value of the integration constant for (6-52) as we did in Fig. 6-16.   Under this scheme all the characteristics of one family would bear the same label.

The next step up in complexity is the characteristic configuration shown in Fig. 6-21.   Let u and v remain constant along any one α characteristic, but let their values vary from one α characteristic to another.   The characteristic slopes will be constant on each α characteristic but will differ from one α characteristic to another. The α family thus consists of straight but nonparallel lines.   The characteristic slopes of the β characteristics will vary along their length, and hence the β family consists of curves.   Note that all β curves have the same slope as they cross a given α characteristic.   This configuration is known as a *simple wave*.

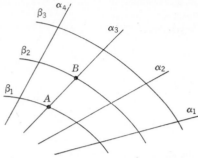

FIG. 6-21.   A simple wave. The α characteristics are straight but non-parallel. The β characteristics are curved.

If we try to distinguish between different characteristics within the same family by using the integration constant for (6-52), we find that the α characteristics will in general carry different labels but that the β characteristics cannot be so distinguished.   To see why, consider the

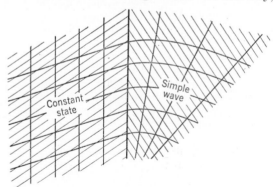

FIG. 6-22.   The common boundary between a constant state and a simple wave is a straight characteristic.

curves marked β_1 and β_2 in Fig. 6-21.   At corresponding points, such as A and B, they carry the same values of u and v: hence the integrated relation between u and v along β_1 must be identical with that along β_2. Thus all the β characteristics bear the same label.

The importance of simple waves arises from the fact that in a solution which contains constant-state regions *the regions adjacent to constant states are always simple waves.*[1]   In Fig. 6-22 a constant state is shown (both families of characteristics are straight lines) bordering on a simple wave (one family of characteristics consists of straight lines).

Example.   To illustrate a case in which constant states and simple waves play an important role, we consider Prob. 6-5, expansion of a gas behind a piston. We begin by determining the equations for the characteristics and the differential equations along the characteristics. To do this, we write the system differential equations (6-41), together with the total differentials for u and ρ in the matrix form of (6-47).

$$
\begin{bmatrix}
u & 1 & \dfrac{c^2}{\rho} & 0 \\
\rho & 0 & u & 1 \\
dx & dt & 0 & 0 \\
0 & 0 & dx & dt
\end{bmatrix}
\begin{bmatrix}
\dfrac{\partial u}{\partial x} \\
\dfrac{\partial u}{\partial t} \\
\dfrac{\partial \rho}{\partial x} \\
\dfrac{\partial \rho}{\partial t}
\end{bmatrix}
=
\begin{bmatrix}
0 \\
0 \\
du \\
d\rho
\end{bmatrix}
\tag{6-61}
$$

The characteristic directions

$$
\frac{dt}{dx} = \frac{1}{u \pm c}
\tag{6-62}
$$

are obtained by setting the determinant of the matrix on the left equal to zero. We call the characteristics on which $dt/dx = 1/(u + c)$ the α family and those on which $dt/dx = 1/(u - c)$ the β family. The equations along the characteristics are obtained by substituting the right-hand column of (6-61) for one of the columns on the left and again equating the determinant to zero. In this way we find

$$
c\frac{d\rho}{\rho} + du = 0 \qquad \text{along } \alpha \text{ characteristics}
$$
$$
c\frac{d\rho}{\rho} - du = 0 \qquad \text{along } \beta \text{ characteristics}
\tag{6-63}
$$

This pair of ordinary differential equations along the characteristics is equivalent to the original system of partial differential equations.

The characteristic curves cannot be obtained in advance in this problem because the characteristic directions (6-62) depend on the solution values of u and ρ. It is possible, however, to integrate (6-63) using (6-42), which gives c as a function of ρ. We use the constants of integration for (6-63) as labels for the individual characteristics in each family. We thus have

$$
\int c\frac{d\rho}{\rho} + u = \frac{2}{\gamma - 1}c + u = 5c + u = \alpha
$$
$$
\int c\frac{d\rho}{\rho} - u = \frac{2}{\gamma - 1}c - u = 5c - u = \beta
\tag{6-64}
$$

on setting $\gamma = 1.40$. When the locations of the characteristics have been found, the relations (6-64) will then yield the complete solution, for by inverting (6-64) we get

$$
u = \frac{\alpha - \beta}{2}
$$
$$
c = \frac{\alpha + \beta}{10}
\tag{6-65}
$$

[1] Courant and Friedrichs, *op. cit.*, p. 59.

which gives the values of u and c at a point in terms of the labels on the two characteristics passing through the point. Our problem thus reduces to the location of the characteristics.

We then turn to a qualitative study of the configuration of characteristic curves within the solution domain. In Fig. 6-23 the initial and boundary conditions of Fig. 6-9 are sketched. Characteristic curves emanating from the initial line OA all bear the labels $\alpha = 5$ or $\beta = 5$ according to (6-64). Their slopes, given by (6-62), are $+1$ for α characteristics and -1 for β characteristics. The boundary condition $u = 0$ on OC implies that the α characteristic reflected from OC should carry the same label as the incident β characteristic. This may be seen from the first of (6-65). The solution within the triangle OAC is thus the constant state $\alpha = \beta = 5$, $u = 0$, $c = 1$. The physical significance of this is simply that the disturbance at A is propagated

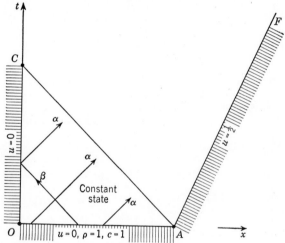

FIG. 6-23. Initial and boundary conditions determine a constant state within the triangle OAC.

with the speed of sound. The region OAC represents points in the gas at a time earlier than the time of arrival of the disturbance wave from A.

We next consider the boundary condition, $u = \frac{1}{2}$, on AF. We know from the previous general discussion that a region adjacent to a region of constant state must be a simple wave and that the common boundary is one of the straight characteristics of the simple wave. We also know that the constant state to the left of CA cannot continue beyond CA because the value of u prescribed on AF is different from that in the constant state. There must then be a simple wave adjacent to CA as shown in Fig. 6-24. The β characteristics are straight lines whose slope and label vary from characteristic to characteristic. The α characteristics are each curved, but all bear the same label. The values of u and c are constant along each β characteristic. Because the prescribed value of $u = \frac{1}{2}$ on AH is not continuous with the value $u = 0$ on CA, the straight characteristics of the simple wave must be centered at A. The extent of the simple wave is determined by the fact that along GA, the last of the fan of β characteristics, the value of u must be equal to $\frac{1}{2}$. Postponing, for the moment, a detailed investigation of the solution in the simple-wave region, we can proceed to the region AGH. Here the prescribed value of u along AH together with the known solution on GA is sufficient to ensure the existence of a unique solution according to the

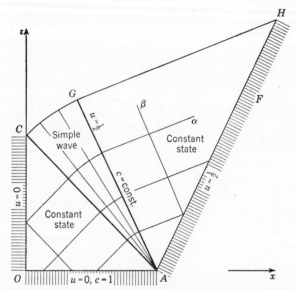

FIG. 6-24.   Centered simple wave between two constant-state regions.

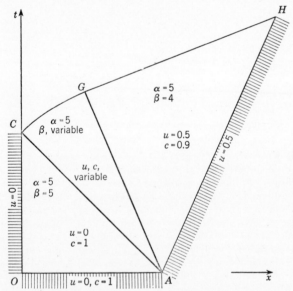

FIG. 6-25.   Solution values within regions of Fig. 6-24.

third uniqueness theorem of Fig. 6-19.   A *constant state* with $u = \frac{1}{2}$ and c equal to its value along GA clearly satisfies these boundary conditions, and therefore, because of the uniqueness theorem, the constant state must be the solution.

Returning to the simple-wave solution, it is possible to give explicit results in the form of analytical expressions.   Inserting $\alpha = 5$ in (6-65), we obtain u and c as functions of β.

$$u = \frac{5 - \beta}{2}$$

$$c = \frac{5 + \beta}{10} \tag{6-66}$$

The slopes of the straight β characteristics are then obtained from (6-62) as

$$\left(\frac{dt}{dx}\right)_\beta = \frac{5}{10 - 3\beta} \tag{6-67}$$

From the fact that $u = \frac{1}{2}$ on GA we find that $\beta = 4.0$, $c = 0.90$, and $(dt/dx)_\beta = -2.50$ along GA. Our solution at this point is sketched in Fig. 6-25, where in the constant-state regions OAC and AGH the complete solution is recorded. The values of u and c at any point in the simple-wave region CAG can be obtained from (6-66) and (6-67). If desired, a more explicit solution may be obtained by eliminating the parameter β from these. Similarly, it is possible to obtain[1] analytical representations of the curved α characteristics. We give the results for the particular α characteristic CG. In terms of the parameter β, points on CG are given by

$$\left.\begin{aligned} x &= 1 - \left(\frac{3\beta}{5} - 2\right)\left(\frac{10}{5 + \beta}\right)^3 \\ t &= \left(\frac{10}{5 + \beta}\right)^3 \end{aligned}\right\} \quad 4 \leqq \beta \leqq 5 \tag{6-68}$$

or, on elimination of the parameter, by the following:

$$x = 1 + 5t - 6t^{\frac{2}{3}} \quad 1 \leq t \leq (\tfrac{10}{9})^3 \tag{6-69}$$

Until now the mathematically exact solution has been obtained without serious difficulty. We have had two regions of constant state and a simple-wave region. To proceed further, we must consider the much more difficult problem in which both families of characteristics are curved. It is in this part of the problem that numerical procedures are usually[2] resorted to. The problem is sketched in Fig. 6-26. We have already located CG and know the solution along it. This together with the boundary condition $u = 0$ on CJ determines a unique solution in the triangle CGJ according to the third uniqueness theorem of Fig. 6-19. As previously noted, the condition $u = 0$ on CJ implies that the value of α on a reflected characteristic is the same as the value of β on the incident characteristic. Therefore, since the labels on the β characteristics crossing CG vary from 4 to 5, the labels on the α characteristics crossing JG must also vary between the same limits in the manner indicated in Fig. 6-26. Thus we have a qualitative picture of the desired solution. The quantitative solution requires the exact location of GJ and the characteristic curves within CGJ. If we try to proceed further, we may recognize that the solution in $KJGM$ is a simple wave but no further quantitative deductions can be made until GJ has been accurately located. Our subsequent treatment of the problem will be limited to the solution within the triangle CGJ.

[1] See Exercise 6-27.
[2] An analytical solution to the present problem in terms of hypergeometric functions was given by B. Riemann, Über die Fortpflanzung ebener Luftwellen von endlicher Schwingungsweite, *Abhandl. Ges. Wiss. Göttingen, Math.-physik. Klasse*, **8**, 43 (1860).

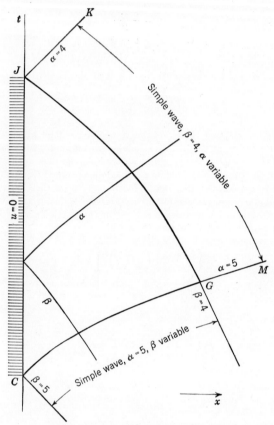

FIG. 6-26.　Both families of characteristics are curved in the triangle CGJ.

EXERCISES

6-13. The formulation of the supersonic-nozzle problem of Exercise 6-4 can be made dimensionless by dividing u, v, and c by c^* and by dividing x and y by L. Obtain the characteristic directions and the differential equations along the characteristics in terms of the dimensionless variables.

$Ans.$ $dy/dx = (uv \pm c \sqrt{u^2 + v^2 - c^2})/(u^2 - c^2);$
$$du(u^2 - c^2) + dv(uv \mp c \sqrt{u^2 + v^2 - c^2}) = 0.$$

6-14. Show that both characteristic directions for the preceding exercise make angles μ with the streamline direction where $\sin \mu = c(u^2 + v^2)^{-\frac{1}{2}}$ and the streamline direction is given by $\tan \theta = v/u$. Reformulate the result of the preceding exercise in terms of q and θ, where $u = q \cos \theta$ and $v = q \sin \theta$.

$Ans.$ $dy/dx = \tan (\theta \pm \mu); q \, d\theta = \pm dq \cot \mu.$

6-15. Show that the equations along the characteristics in the preceding exercise can be integrated to give

$$\left. \begin{array}{l} \theta - \omega = \alpha \\ \theta + \omega = \beta \end{array} \right\} \qquad \omega = \int_1^q \left[\frac{6(q^2 - 1)}{6 - q^2} \right]^{\frac{1}{2}} \frac{dq}{q}$$

where α and β are the constants of integration to be used to distinguish the characteristics. Verify that

$$\omega = \int_\mu^{\pi/2} \frac{5(1 - \sin^2 \mu)}{1 + 5 \sin^2 \mu} \, d\mu$$

Solve for θ and ω in terms of α and β, and make a qualitative study of the location of the characteristics in Fig. 6-11 if $\theta_\omega = 6°$.

Ans. $\theta = \omega = \alpha = \beta = 0$, $\mu = \pi/2$ on AA. There is a fan of β characteristics up to $\beta = 12°$ at A in the upper half nozzle. On the center line $\beta = -\alpha$.

6-16. The formulation of the plasticity problem of Exercise 6-8 can be made dimensionless by setting $p/2k = \omega$ and by dividing all lengths by the half distance between plates. Obtain the characteristic directions and the differential equations along the characteristics.

Ans. $dy/dx = -\cot 2\psi \pm \csc 2\psi$; that is, $(dy/dx)_\alpha = \tan \psi$, $(dy/dx)_\beta = -\cot \psi$; $d\omega + d\psi = 0$ on α characteristics and $d\omega - d\psi = 0$ on β characteristics.

6-17. Integrate the equations along the characteristics in the preceding problem, invert them, and study qualitatively the location of the characteristics.

Ans. There is a simple wave centered at A with α running from $\frac{1}{2} - \pi/4$ to $\frac{1}{2} + \pi/4$. The reflection condition along AF is $\alpha = \beta$, and along the center line it is $\beta = \alpha + \pi/2$.

6-18. The ordinary differential equations for and along the characteristics of a reducible system can be transformed to a single linear partial differential equation if we take the characteristic parameters α and β as new independent variables. Show that in this light (6-62) can be written

$$\frac{\partial t}{\partial \beta} = \frac{1}{u + c} \frac{\partial x}{\partial \beta} = \frac{5}{3\alpha - 2\beta} \frac{\partial x}{\partial \beta}$$

$$\frac{\partial t}{\partial \alpha} = \frac{1}{u - c} \frac{\partial x}{\partial \alpha} = \frac{5}{2\alpha - 3\beta} \frac{\partial x}{\partial \alpha}$$

Eliminate x to obtain a single equation for t.

Ans. $(\alpha + \beta) \, \partial^2 t/\partial\alpha \, \partial\beta = -3(\partial t/\partial\alpha + \partial t/\partial\beta)$.

6-19. With the aid of Exercise 6-18 or by direct transformation of (6-41) show that if x and t are considered dependent variables of u and c the governing equations are

$$5 \frac{\partial x}{\partial u} - \frac{\partial x}{\partial c} - 5(u + c) \frac{\partial t}{\partial u} + (u + c) \frac{\partial t}{\partial c} = 0$$

$$5 \frac{\partial x}{\partial u} + \frac{\partial x}{\partial c} - 5(u - c) \frac{\partial t}{\partial u} - (u - c) \frac{\partial t}{\partial c} = 0$$

6-20. Let $AX = C$ be a matrix equation for the n unknown elements of the column X, where A is a square coefficient matrix and C is a column matrix. Show that, if the determinant of A vanishes, a necessary condition for finite solutions to exist is that when C is substituted for any column of A the resulting determinant must also vanish. Note that in this case if there is a single finite solution then there is an infinity of such solutions. What is the geometric significance of these statements when $n = 2$? See (1-39) and (1-41) for an illustrative case.

6-21. Carry out an analysis similar to that of the text to show that, if $\partial\psi/\partial x$ and $\partial\psi/\partial y$ are known continuously differentiable functions along a curve CPC, then solutions to (6-50) may have discontinuities in the second derivatives if the slope at P is given by (6-51).

Ans. The equation corresponding to (6-47) is

$$\begin{bmatrix} a & b & c \\ dx & dy & 0 \\ 0 & dx & dy \end{bmatrix} \begin{bmatrix} \dfrac{\partial^2 \psi}{\partial x^2} \\ \dfrac{\partial^2 \psi}{\partial x\, \partial y} \\ \dfrac{\partial^2 \psi}{\partial y^2} \end{bmatrix} = \begin{bmatrix} f \\ d\left(\dfrac{\partial \psi}{\partial x}\right) \\ d\left(\dfrac{\partial \psi}{\partial y}\right) \end{bmatrix}$$

6-22. Verify that the differential equations along the characteristics in the previous exercises are

$$a\left(\frac{dy}{dx}\right) d\left(\frac{\partial \psi}{\partial x}\right) + cd\left(\frac{\partial \psi}{\partial y}\right) = f\, dy$$

Show that if $f \equiv 0$

$$\left[\frac{d(\partial \psi/\partial y)}{d(\partial \psi/\partial x)}\right]_\alpha \left(\frac{dy}{dx}\right)_\beta = -1$$

and that this same result holds when α and β are interchanged.

6-23. Repeat the analysis of the text for the case in which Eqs. (6-45) have right-hand sides F_1 and F_2. Show that (6-48) is unchanged but that the following term must be added to the left side of (6-52):

$$\left[(F_1 A_2 - F_2 A_1)\left(\frac{dy}{dx}\right) - (F_1 B_2 - F_2 B_1)\right] dx$$

6-24. Consider the propagation problem in which $\psi = 0$ at $x = 0$ and $x = 1$ and satisfies the equation

$$\frac{\partial^2 \psi}{\partial x^2} = f(t) \qquad \text{in } 0 < x < 1$$

Show that this is a *parabolic* system. It can be interpreted as giving the deflection of a taut, inertialess string subject to a transverse load which is uniformly distributed along the length of the string but whose magnitude is a function of time. Show that if $\partial \psi/\partial t$ is discontinuous at any interior point then $\partial \psi/\partial t$ must be discontinuous for all points at the same instant of time.

6-25. Suppose that in the system of Prob. 6-4 with Eqs. (6-55) to (6-58) $\partial u/\partial \alpha$ has a unit discontinuity across the α characteristic at P. Write the second of (6-56) for a short length of the β characteristic just below P and for a short length just above P, and thereby deduce that $\partial v/\partial \alpha$ has a discontinuity of -1. Then use the first of (6-56) to show that after moving a short distance along the α characteristic the same discontinuity picture prevails; i.e., there is still a unit jump in $\partial u/\partial \alpha$.

6-26. Show that

$$\psi = x \int_{\frac{x}{2\sqrt{t}}}^{\infty} e^{-\xi^2} \frac{d\xi}{\xi^2}$$

satisfies the heat-conduction equation (6-6). It represents the response of a semi-infinite conductor $0 < x < \infty$ to a heater placed at $x = 0$ at $t = 0$. Verify that

$$\frac{\partial \psi}{\partial t} = t^{-\frac{3}{2}} e^{-\frac{x^2}{4t}}$$

and that for $t = 0$ this is infinite at $x = 0$; however, show that, for $x > 0$, $\partial \psi/\partial t$ and all its derivatives vanish at $t = 0$.

6-27. Study the geometry of the simple wave in Fig. 6-25. Show that, if (x, t) are the coordinates of a point on a β characteristic, then

$$\left(\frac{dt}{dx}\right)_\beta = \frac{t}{x-1}$$

Equate this to (6-67), and differentiate with respect to β. Obtain another equation for $dt/d\beta$ and $dx/d\beta$ by writing Eq. (6-62) along an α characteristic. Eliminate x to get the following equation for t along an α characteristic.

$$\frac{dt}{d\beta} + \frac{3t}{5+\beta} = 0$$

Verify the results of (6-68) and (6-69).

6-28. The solution to Prob. 6-5, expansion of a gas behind a piston, within the triangle CGJ of Fig. 6-26 can be obtained by solving the equation given in the answer of Exercise 6-18 within a triangle CGJ in the (α, β) plane, where CG is the line $\alpha = 5$, GJ is the line $\beta = 4$, and CJ is the line $\alpha = \beta$. Show that the boundary condition on CG is $t = 1,000(\alpha + \beta)^{-3}$ and that the boundary condition on CJ is $\partial t/\partial \alpha = \partial t/\partial \beta$. An iterative solution is given by

$$t^{(r+1)} = -\iint \frac{3}{\alpha+\beta}\left(\frac{\partial t^{(r)}}{\partial \alpha} + \frac{\partial t^{(r)}}{\partial \beta}\right) d\alpha\, d\beta \qquad r = 0, 1, 2, \ldots$$

where the arbitrary functions of integration are fixed by the requirement that the two boundary conditions above be satisfied. Starting with $t^{(0)} = 1,000(\alpha + \beta)^{-3}$, which does satisfy these conditions, show that

$$t^{(1)} = \frac{3}{2}\left(\frac{10}{\alpha+\beta}\right)^3 + \frac{1}{2}\left[1 - \left(\frac{10}{\alpha+5}\right)^3 - \left(\frac{10}{5+\beta}\right)^3\right]$$

The t coordinate of J is 2.0579 according to this. The exact[1] answer is 2.1014.

6-4. Trial Solutions with Undetermined Parameters

The procedures here are similar in principle to those described in Chaps. 3 to 5. The basic idea consists in seeking a "best" approximation out of a limited family of functions selected by the analyst. The criteria used are of the weighted-residual types. The trial families may contain undetermined constants[2] or undetermined functions of a single variable. In the latter case the process amounts to approximating a continuous propagation problem by a propagation problem with a finite number of degrees of freedom, i.e., a reduction to a problem of the type treated in Chap. 3. A general treatment[3] is not given here. We present instead an example which illustrates the points in which the treatment of propagation problems by these methods differs from that of equilibrium problems.

---

[1] See Courant and Friedrichs, *op. cit.*, p. 196, Eq. (82-18).

[2] See Exercise 6-33, for an example.

[3] S. Faedo, Un Nuovo metodo per l'analisi esistenzialele quantitativa dei problemi di propagazione, *Ann. scuola norm. sup. Pisa*, (3)**1**, 1947, 1–41 (1949).

Example. Consider case (b) of Prob. 6-1, cooling of a rod, as formulated in Fig. 6-2b. We shall approximate this by a propagation problem for r discrete variables $c_j(t)$ by selecting a trial solution of the form

$$\psi = \sum_{j=1}^{r} c_j(t)\varphi_j(x) \tag{6-70}$$

where the $\varphi_j(x)$ are known functions. Since the boundary conditions on $x = 0$ and $x = 1$ are homogeneous, we can make ψ satisfy them with no restriction on the c_j if we require that each φ_j meet the conditions

$$\varphi_j - \frac{d\varphi_j}{dx} = 0 \qquad \text{at } x = 0$$
$$\frac{d\varphi_j}{dx} = 0 \qquad \text{at } x = 1 \tag{6-71}$$

A simple family of polynomials which meet these conditions is

$$\varphi_j = 1 + x - \frac{x^{j+1}}{j+1} \tag{6-72}$$

A trial (6-70) constructed from (6-72) then satisfies the boundary conditions but does not satisfy *either* the initial conditions *or* the governing equation. By applying weighted-residual criteria we can fix the unknown $c_j(t)$ so that these latter requirements are approximately met. In the work that follows we limit the trial (6-70) to *two* terms.

When $t = 0$, we should have $\psi = 1$, $0 < x < 1$. Let us form the *initial residual*,

$$R[c_1(0), c_2(0), x] = 1 - \left(1 + x - \frac{x^2}{2}\right)c_1(0) - \left(1 + x - \frac{x^3}{3}\right)c_2(0) \tag{6-73}$$

which is a measure of the amount by which this condition is not satisfied. By applying one of the weighted-residual criteria (e.g., collocation or Galerkin's method) we would obtain a pair of constants, $c_1(0)$ and $c_2(0)$, for which (6-73) would be "small."

For $t > 0$ we should have

$$\frac{\partial^2 \psi}{\partial x^2} = \frac{\partial \psi}{\partial t} \qquad 0 < x < 1 \tag{6-74}$$

By forming the *equation residual*

$$R[c_1(t), c_2(t), x] = \left(1 + x - \frac{x^2}{2}\right)\frac{dc_1}{dt} + \left(1 + x - \frac{x^3}{3}\right)\frac{dc_2}{dt} + c_1 + 2xc_2 \tag{6-75}$$

we can apply one of the weighted-residual criteria to obtain a pair of differential equations for $c_1(t)$ and $c_2(t)$. These can be considered as propagation equations for the $c_j(t)$. We then obtain our approximate solution by solving these equations subject to the initial conditions obtained from (6-73).

Any combination of criteria can be applied to the two residuals. There is no obvious way of rating the different criteria[1] as there is in equilibrium problems which can be formulated as minimization problems.

To apply Galerkin's method, we set

$$\int_0^1 R[c_1(t), c_2(t), x]\varphi_j(x)\, dx = 0 \qquad j = 1, 2 \tag{6-76}$$

and obtain

$$\frac{9}{5}\frac{dc_1}{dt} + \frac{691}{360}\frac{dc_2}{dt} + \frac{4}{3}c_1 + \frac{17}{12}c_2 = 0$$
$$\frac{691}{360}\frac{dc_1}{dt} + \frac{1{,}291}{630}\frac{dc_2}{dt} + \frac{17}{12}c_1 + \frac{23}{15}c_2 = 0 \tag{6-77}$$

as governing equations for the c_j. If we also apply Galerkin's method to the initial residual by setting

$$\int_0^1 R[c_1(0), c_2(0), x]\varphi_j(x)\, dx = 0 \qquad j = 1, 2 \tag{6-78}$$

we obtain

$$\frac{4}{3} - \frac{9}{5}c_1(0) - \frac{691}{360}c_2(0) = 0$$
$$\frac{17}{12} - \frac{691}{360}c_1(0) - \frac{1{,}291}{630}c_2(0) = 0 \tag{6-79}$$

for the initial values of c_1 and c_2. The solution to (6-77) which is compatible with the initial conditions (6-79) is found[2] to be

$$c_1 = 0.5862e^{-0.7402t} + 2.4484e^{-11.770t}$$
$$c_2 = 0.1444e^{-0.7402t} - 2.2954e^{-11.770t} \tag{6-80}$$

and with these values the corresponding approximation for ψ is

$$\psi = c_1\left(1 + x - \frac{x^2}{2}\right) + c_2\left(1 + x - \frac{x^3}{3}\right) \tag{6-81}$$

The accuracy of this approximation is illustrated in Fig. 6-27 and Table 6-1. In Fig. 6-27 the approximation to the initial condition $\psi(x, 0) = 1$ is shown. The error in the response $\psi(1, t)$ at the center of the rod is tabulated in Table 6-1.

---

[1] The Galerkin method applied to (6-73) gives the same results as the least-squares criterion (see Exercise 6-32). In this sense then the Galerkin method is an optimum one for fitting *initial* conditions. A comparable result for $t > 0$ is unknown.

[2] Since (6-77) is linear with constant coefficients, the exact solution is easily obtained. In more complicated systems it would be necessary to apply the methods of Chap. 3.

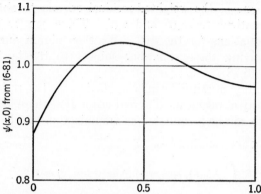

FIG. 6-27.   Approximate fit of the initial condition $\psi = 1$ by the solution (6-81).

Three similar approximate solutions to this problem were obtained by W. G. Bickley[1] by applying different combinations of criteria to the residuals (6-73) and (6-75).   The errors in $\psi(1, t)$ for these are compared in Table 6-1.   A brief description of the methods follows.

A. Collocation of both (6-73) and (6-75) at $x = 0$ and $x = 1$.

B. Collocation at $x = 0$ and $x = 1$ applied to (6-73).   Weighted residual with weighting functions 1 and x applied to (6-75).

C. Weighted residual with weighting functions 1 and x applied to both (6-73) and (6-75).

TABLE 6-1. ABSOLUTE ERROR IN $\psi(1, t)$ FOR VARIOUS CRITERIA

t	A	B	C	(6-81)
0	0.0000	0.0000	−0.0370	−0.0330
0.2	+0.0604	+0.0743	+0.0016	+0.0007
0.4	+0.0390	+0.0673	+0.0020	+0.0005
0.6	+0.0256	+0.0582	+0.0007	+0.0004
0.8	+0.0159	+0.0502	+0.0007	+0.0005
1.0	+0.0083	+0.0432	+0.0005	+0.0004
1.5	−0.0032	+0.0298	+0.0003	+0.0003
2.0	−0.0084	+0.0205	+0.0001	+0.0001
2.5	−0.0091	+0.0141	0.0000	+0.0001
3.0	−0.0096	+0.0097	0.0000	+0.0001
4.0	−0.0071	+0.0045	−0.0001	0.0000

General Remarks.   The procedure just outlined has much in common with that described on page 236 for equilibrium problems.   The difference lies in the fact that the *boundary* conditions there for the unknown functions c_j are selected in advance (or provided automatically by an

[1] W. G. Bickley, Experiments in Approximating to the Solution of a Partial Differential Equation, *Phil. Mag.*, (7)**32**, 50–66 (1941).

extremum principle), whereas here the *initial* conditions are obtained by solving an auxiliary approximation problem. In some cases[1] this auxiliary approximation can be solved exactly by purposely selecting one of the φ_j so that it meets the initial conditions. The initial conditions for the corresponding c_j will then involve unity, and the initial conditions for the other c_j will involve zeros.

The variation between results obtained by applying different criteria to the same trial family (as indicated in Table 6-1) is much less significant than the variations that can result from the choice of different trial families. The selection of the trial family is, of course, the crucial point in any procedure of this type. It is up to the ingenuity of the analyst so to construct the trial solution that a maximum of information can be extracted with a minimum of computation. The more that is known about the expected behavior of a solution (symmetry, etc.), the more intelligently can the trial family be set up. A preliminary study of the qualitative aspects of the expected solution is particularly important in propagation problems. For instance, if a solution has zones of constant state intermixed with zones of varying state, it would be unwise to attempt to approximate the solution by a single continuous analytical expression throughout. The solution domain should be broken up and the separate zones treated individually.

EXERCISES

6-29. Apply weighted residual procedures to the trial $\psi = x(1 - x)c_1(t)$ for Prob. 6-4, transverse motion of a taut string, to obtain the function $c_1(t)$. Note that the initial conditions are satisfied exactly.

Ans. $c_1 = \cos \omega t$ where (*a*) $\omega^2 = 9$ by *collocation* at $x = \frac{1}{3}$; (*b*) $\omega^2 = 12$ by *subdomain* method using $0 < x < 1$; (*c*) $\omega^2 = 10$ by *Galerkin's* method. Compare results with Exercise 5-51.

6-30. Apply Galerkin's method to the trial family

$$\psi = x(1 - x)c_1(t) + x^2(1 - x)^2 c_2(t)$$

for Prob. 6-4, transverse motion of a taut string.

Ans. $c_1 = 0.8035 \cos 3.1416t + 0.1965 \cos 10.1059t$, $c_2 = 0.9105 \cos 3.1416t - 0.9105 \cos 10.1059t$. At $t = \frac{1}{2}$ the exact solution vanishes, while for this trial the maximum amplitude of ψ at $t = \frac{1}{2}$ is 0.0025.

6-31. Consider the trial family (6-70) with $\varphi_j = 1 - \cos 2j\pi x$ for Prob. 6-3, transverse motion of a beam. Set up the equation residual and the initial residual. Actually apply the Galerkin method when there is only one undetermined function $c_1(t)$.

Ans. $c_1 = 0.1107 \cos 22.793t$.

6-32. A function $f(x)$ in $0 < x < 1$ is to be approximated by a linear combination of known functions $\varphi_j(x)$. Define the *residual* as

$$R = f(x) - \Sigma c_j \varphi_j(x)$$

[1] See Exercises 6-29 and 6-30.

Show that if the c_j are determined by applying the Galerkin criterion to R then the integral of R^2 over $0 < x < 1$ is minimized.

6-33. Consider Prob. 6-5, expansion of a gas behind a piston, within the triangle CGJ of the (α, β) plane as formulated in Exercise 6-28. Show that the trial family

$$t = \left(\frac{10}{\alpha + \beta}\right)^3 + c_1 (5 - \alpha)(5 - \beta)$$

satisfies the boundary conditions independently of value of the parameter c_1. Apply the collocation method to evaluate c_1.

Ans. $c_1 = 0.1524$ when $\alpha = \beta = 4.5$ is the selected location. The t coordinate of J is 2.1055 according to this, with error of 0.0041. If Galerkin's criterion is used, $c_1 = 0.2309$ and the error in the t coordinate of J is 0.0826.

6-5. Finite-difference Methods for Parabolic Systems

The essential features of the finite-difference approximations that we have seen in previous chapters extend over to continuous-propagation problems in an obvious way. There are, however, some new features without their counterpart in our earlier discussions. It is quite possible[1] to set up seemingly natural computational programs which produce only gibberish.

A fairly complete discussion of these points is given in this section in terms of the heat conduction equation of Prob. 6-1. The modifications necessary for Probs. 6-2 and 6-3 are then briefly outlined. We begin with a simple illustration in which a straightforward approach is successful.

Example. We consider Prob. 6-1 as formulated in Fig. 6-2. It is natural to adopt a rectangular network as shown in Fig. 6-28. Letting the value of ψ at $P_{j,k}$ be denoted by $\psi_{j,k}$, we can approximate the governing equation as follows:

$$0 = \left(\frac{\partial^2 \psi}{\partial x^2} - \frac{\partial \psi}{\partial t}\right)_{j,k} \approx \frac{\psi_{j-1,k} - 2\psi_{j,k} + \psi_{j+1,k}}{(\Delta x)^2} - \frac{\psi_{j,k+1} - \psi_{j,k}}{\Delta t} \quad (6-82)$$

[1] A historical example is provided in one of the pioneering papers on finite differences, L. F. Richardson, The Approximate Arithmetical Solution by Finite Differences of Physical Problems Involving Differential Equations with an Application to the Stresses in a Masonry Dam, *Trans. Roy. Soc. (London)*, **A210**, 307–357 (1910). In this major contribution equilibrium and eigenvalue problems are handled successfully, but for the cooling of a rod (Prob. 6-1 essentially) a procedure was suggested which only recently was found to be completely unstable. See G. G. O'Brien, M. A. Hyman and S. Kaplan, A Study of the Numerical Solution of Partial Differential Equations, *J. Math. and Phys.*, **29**, 223–251 (1951). See also Exercise 6-46.

Another example is furnished by Prob. 6-3, transverse motion of a beam. In an early paper L. Collatz, Über das Differenzenverfahren partieller Differentialgleichunge, *Z. angew. Math. u. Mech.*, **16**, 239–247 (1936), a "natural" finite-difference process was found to be unstable. The reason for this was given 15 years later: L. Collatz, Zur Stabilität des Differenzenverfahrens bei der Stabschingungsgleichung., *Z. angew. Math. u. Mech.*, **31**, 392–393 (1951).

Let us postpone the question of truncation error. The ratio of the increment Δt to $(\Delta x)^2$ plays an important role in our further discussion.

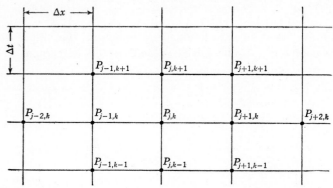

FIG. 6-28. Network of discrete points used to approximate continuous solution domain.

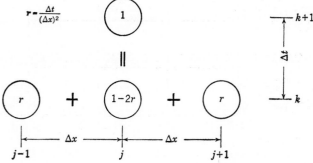

FIG. 6-29. Explicit recurrence formula for the heat-conduction equation $\partial^2 \psi / \partial x^2 = \partial \psi / \partial t$.

We shall call it r; that is, we set

$$r = \frac{\Delta t}{(\Delta x)^2} \qquad (6\text{-}83)$$

Inserting this in (6-82) and solving for $\psi_{j,k+1}$, we get.

$$\psi_{j,k+1} = r\psi_{j-1,k} + (1 - 2r)\psi_{j,k} + r\psi_{j+1,k} \qquad (6\text{-}84)$$

which is indicated schematically in Fig. 6-29. We notice[1] that (6-84) takes an especially simple form when $r = \frac{1}{2}$.

[1] The formula (6-84) was given by E. Schmidt, "Über die Andwendung der Differenzen rechnung auf technische Anheiz-und-Abkühlungsprobleme," A. Föppl Festschrift, Springer-Verlag OHG, Berlin, 1924, pp. 179–189. He showed that for $r = \frac{1}{2}$ it was solvable by a simple graphical construction (see Exercise 6-35). An earlier presentation was given by L. Binder, Über aussere Wärmleitung und Erwärmung elektrischer Maschinen, dissertation, Technische Hochschule Muenchen, W. Knapp Verlag, Halle, Germany, 1911, pp. 20–26.

To illustrate the use of (6-84) and the treatment of initial and boundary conditions, we consider case (a) of Prob. 6-1 in Fig. 6-30, taking $\Delta x = \frac{1}{2}$ and $\Delta t = \frac{1}{8}$ so that we have $r = \frac{1}{2}$. The boundary condition $\psi = 0$ at $x = 0$ has been entered on the diagram. At $x = 1$ the boundary condition $\partial\psi/\partial x = 0$ is handled by reflecting the values $\psi_{1,k}$ to the auxiliary points $P_{3,k}$. The initial value $\psi = 1$ is entered at $x = \frac{1}{2}$ and

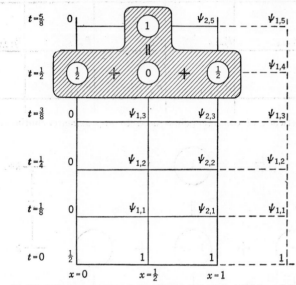

Fig. 6-30. Finite-difference network for Prob. 6-1 with initial and boundary conditions of case (a). Recurrence formula is indicated in shaded region.

$x = 1$. At $P_{0,0}$ there is a unit discontinuity between the boundary value $\psi = 0$ and the initial value $\psi = 1$. We have assigned[1] $\psi_{0,0}$ the mid-value $\frac{1}{2}$.

The solution can now be marched out in Fig. 6-30 by applying the recurrence formula of Fig. 6-29 over and over again. The reader should verify that $\psi_{1,1} = \frac{3}{4}$, $\psi_{2,1} = 1$ and that, in general,

$$\begin{bmatrix} \psi_{1,k+1} \\ \psi_{2,k+1} \end{bmatrix} = \begin{bmatrix} 0 & \frac{1}{2} \\ 1 & 0 \end{bmatrix} \begin{bmatrix} \psi_{1,k} \\ \psi_{2,k} \end{bmatrix} \tag{6-85}$$

for $k \geq 1$.

The values of $\psi_{2,k+1}$ obtained in this fashion have been compared with the exact continuous-solution values of $\psi(1, t)$ and the errors indicated in

---

[1] The accuracy of approximation with crude networks is quite sensitive to the values assigned at such points of discontinuity. See S. H. Crandall, Implicit vs. Explicit Recurrence Formulas for the Linear Diffusion Equation, *J. Assoc. Comp. Mach.*, **2**, 42–49 (1955).

Fig. 6-31. In this figure the corresponding errors for two finer networks are also shown. In each case the ratio r was $\frac{1}{2}$. The convergence of the finite-difference approximation is qualitatively indicated.

Looking closer into the question of convergence, we return to (6-82) and note that the truncation errors are of order $(\Delta x)^2$ and Δt. If we call $\Delta x = h$ and if we keep r fixed (that is, Δt is made a fixed multiple of h^2),

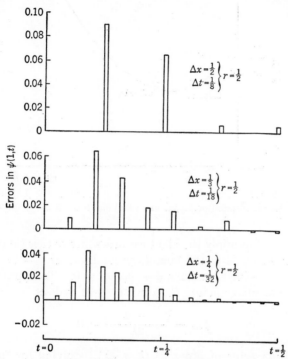

FIG. 6-31. Comparison of errors in approximating $\psi\,(1,\,t)$ by three finite-difference solutions using successively finer networks.

then the total truncation error is $0(h^2)$. We would then expect that under fairly wide circumstances the discretization error of the solution would also be $0(h^2)$ and that h^2 extrapolation would be useful. This is, in fact, borne out by the results of Fig. 6-31. For most values of t, h^2 extrapolation provides a decrease in error. At $t = \frac{1}{2}$, however, we have the interesting situation pictured in Fig. 6-32. Extrapolation based on a straight line through any two of the points A, B, and C does not come close to the exact value E, but a parabola through all three points[1] is in error[2] by only 0.00014. This example does not disprove the princi-

---

[1] See Exercise 3-53.

[2] See Exercise 6-36.

ple of h^2 extrapolation. It merely emphasizes that it is an extrapolation.
The principle states only that when h is small enough for higher-order
terms to be negligible the discretization error is proportional to h^2. From
Fig. 6-32 it is clear that in this case h is not small enough until $h < \frac{1}{4}$.

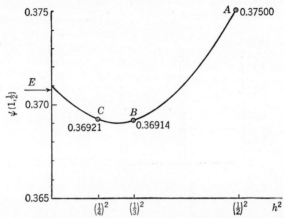

FIG. 6-32. A, B, and C are finite-difference approximations to ψ $(1, \frac{1}{2})$ obtained with
$h = \frac{1}{2}, \frac{1}{3},$ and $\frac{1}{4}$.   Extrapolation based on a parabola through A, B, and C comes close
to the exact solution E.

Stability.   Let us apply the identical procedure to case (b) of Prob. 6-1.
The only difference is in the boundary condition at $x = 0$.   Instead of
having $\psi = 0$ specified, we have $\psi - \partial\psi/\partial x = 0$.   To approximate this,
we introduce auxiliary points $P_{-1,k}$ in Fig. 6-33 and set

$$\psi_{0,k} - \frac{\psi_{1,k} - \psi_{-1,k}}{2h} = 0 \qquad (6\text{-}86)$$

with truncation error of order h^2 $(h = \Delta x)$. Solving for the auxiliary
value, we get

$$\psi_{-1,k} = \psi_{1,k} - \psi_{0,k} \qquad (6\text{-}87)$$

when $h = \frac{1}{2}$.   This relation is entered on the diagram of Fig. 6-33 along
with the reflection condition for $\partial\psi/\partial x = 0$ at $x = 1$ and the initial condition $\psi = 1$ at $t = 0$.   Note that we give $\psi_{0,0}$ the full initial value here.
At each time step there are then three unknowns.   Applying the recurrence formula to each in turn, we can write the result as follows:

$$\begin{bmatrix} \psi_{0,k+1} \\ \psi_{1,k+1} \\ \psi_{2,k+1} \end{bmatrix} = \begin{bmatrix} -\frac{1}{2} & 1 & 0 \\ \frac{1}{2} & 0 & \frac{1}{2} \\ 0 & 1 & 0 \end{bmatrix} \begin{bmatrix} \psi_{0,k} \\ \psi_{1,k} \\ \psi_{2,k} \end{bmatrix} \qquad (6\text{-}88)$$

While this is apparently just as natural a treatment as that given previously for case (a), we find that the solution of (6-88) includes an oscil-

lating component which grows exponentially. The discrepancy between the values of $\psi_{2,k}$ for the first few steps and the exact continuous solution $\psi(1, t)$ are plotted in Fig. 6-34.

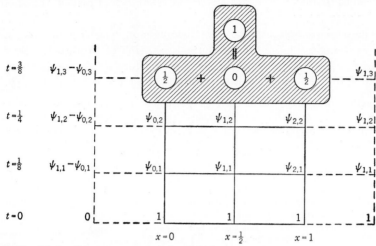

FIG. 6-33.   Finite-difference network for Prob. 6-1 with initial and boundary conditions of case (b).   Recurrence formula is indicated in shaded region.

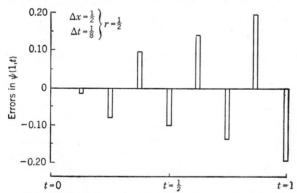

FIG. 6-34.   Growth of oscillating error in unstable finite-difference solution.

We call a finite-difference process within the strip $0 < x < 1$, $t > 0$, *unstable* if for a fixed network and fixed homogeneous boundary conditions there exist initial distributions for which the finite-difference solutions $\psi_{j,k}$ become unbounded[1] as $k \to \infty$. The process (6-88) is unstable

---

[1] This definition is useful only when the corresponding continuous solutions remain bounded. See S. H. Crandall, On a Stability Criterion for Partial Difference Equations, *J. Math. and Phys.*, **32**, 80–81 (1953). A different definition of instability has been given by F. John, On Integration of Parabolic Equations by Difference Methods, *Comm. Pure Appl. Math.*, **5**, 155–211 (1952). Instead of a fixed network a fixed

in this sense. We shall see that the value of the ratio $r = \Delta t/(\Delta x)^2$ is intimately connected with the stability of finite-difference methods for parabolic equations. The value $r = \frac{1}{2}$ used in the two preceding examples happens to be just on the stable side for case (a) and just on the unstable side for case (b).

To gain an understanding of this phenomenon, we shall obtain a closed-form expression for the finite-difference solution of Prob. 6-1. By comparing this with a corresponding expression for the continuous solution we can study the discretization error and the reasons for instability.[1]

Properties of the Explicit Recurrence Formula. The difference equation (6-84) can be solved by assuming that $\psi_{j,k}$ is the product of a function of j alone times a function of k alone. If we set $\psi_{j,k} = X_j T_k$ in (6-84), it is possible to separate the variables.

$$\frac{T_{k+1} - T_k}{T_k} = r\,\frac{X_{j-1} - 2X_j + X_{j+1}}{X_j} \tag{6-89}$$

Since j and k are independent, the common value of the expressions in (6-89) must be independent of both j and k; that is, it must be a constant. Calling the constant $-a$, we can solve separately for X_j and T_k in terms of a. We have

$$X_{j-1} - \left(2 - \frac{a}{r}\right)X_j + X_{j+1} = 0 \tag{6-90}$$
$$T_{k+1} - (1 - a)T_k = 0$$

The first of these is satisfied by $\sin j\beta$ or $\cos j\beta$ if β in turn satisfies[2] the relation $2 - a/r = 2\cos\beta$. The second is satisfied by $(1 - a)^k$. Products of these terms are solutions of (6-84) and because of the linearity sums of such products are also solutions.

To particularize the solution further, we must consider the boundary conditions. For simplicity we take case (a) with the following finite-difference boundary conditions,

$$\psi_{0,k} = 0 \qquad \psi_{S+1,k} = \psi_{S-1,k} \tag{6-91}$$

where $S = 1/h$ is the number of subdivisions in the x direction. These conditions are independent of k and thus represent boundary conditions for X_j. The term $\sin j\beta$ satisfies the condition at $j = 0$ for arbitrary β but satisfies the other condition only when $S\beta = \pi/2, 3\pi/2, \ldots, (2S - 1)\pi/2$. These are the *eigenvalues*, and the $\sin j\beta$ are the *eigenfunctions*, or *space modes*, of the system. If a is restricted to the values which correspond to these β, the solution as a sum of products is

$$\psi_{j,k} = \sum_{n=1}^{S} d_n\left(1 - 4r\sin^2\frac{2n-1}{2S}\frac{\pi}{2}\right)^k \sin\frac{2n-1}{2}\frac{j\pi}{S} \tag{6-92}$$

where the d_n are constants which can be used to fit the initial conditions.

interval $0 < t < T$ is considered with a sequence of finite difference solutions for successively finer networks. If, as $h \to 0$, the finite-difference solutions at $t = T$ can become unbounded, the process is called unstable.

[1] Analyses of this kind were given by C. M. Fowler, Analysis of Numerical Solutions of Transient Heat-flow Problems, *Quart. Appl. Math.*, **3**, 361–376 (1946), and O'Brien, Hyman, and Kaplan, *loc. cit.*

[2] See Exercise 6-37.

In an exactly parallel manner the continuous solution to (6-6) subject to the boundary conditions of (6-7) is[1]

$$\psi(x, t) = \sum_{n=1}^{\infty} c_n \exp\left[-\pi^2 \left(\frac{2n-1}{2}\right)^2 t \right] \sin \frac{2n-1}{2} \pi x \qquad (6\text{-}93)$$

where the c_n are constants to be used to fit the initial conditions. A study of (6-92) and (6-93) can provide[2] a complete picture of the discretization error. Here we consider just a few of the more obvious points.

We note that in each term of (6-93) there is a decaying exponential in t multiplying each space mode. In (6-92) the corresponding element is a geometric progression in k. If the quantity

$$\lambda_n = 1 - 4r \sin^2 \frac{2n-1}{2S} \frac{\pi}{2} \qquad (6\text{-}94)$$

is positive,[3] the progression decays steadily. If $-1 < \lambda_n < 0$, the progression is alternately positive and negative but its amplitude decays. Finally, if $\lambda_n < -1$, the progression oscillates with increasing amplitude.

In the last case (6-92) is unstable. If the space mode corresponding to λ_n is excited by the initial conditions [i.e., if in (6-92) $d_n \neq 0$], the amplitude of that mode will grow without limit as k increases. If d_n is initially zero but is introduced during the working by random round-off errors, the standard deviation of the amplitude will grow without limit. For stability we must have $-1 \leq \lambda_n$ for all n or

$$r \leq \frac{1}{2 \sin^2 \frac{2S-1}{2S} \frac{\pi}{2}} = \frac{1}{1 + \cos \frac{\pi}{2S}} \qquad (6\text{-}95)$$

since the term with $n = S$ tends to become unstable first. The limiting values of r given by (6-95) and the corresponding values obtained[4] for case (b) are tabulated in Table 6-2.

If r is smaller than the values given in Table 6-2, then we are assured that all components of a finite-difference solution will eventually decay. Some of the components may, however, oscillate. To ensure that none of the components oscillate, it is necessary to have $\lambda_n \geq 0$ for all n, and

[1] See Exercise 6-41.
[2] See, for example, M. L. Juncosa and D. Young, On the Convergence of a Solution of a Difference Equation to a Solution of the Equation of Diffusion, *Proc. Am. Math. Soc.*, **5**, 168–174 (1954). See also Exercise 6-45.
[3] Since r is essentially positive, we always have $\lambda < 1$.
[4] See Exercise 6-40. Because of the simplicity of the recurrence formula for $r = \frac{1}{2}$ several alternative boundary conditions have been suggested in place of (6-86) for case (b) which do retain stability up to $r = \frac{1}{2}$. See, for example, P. H. Price and M. R. Slack, Stability and Accuracy of Numerical Solutions of the Heat Flow Equation, *Brit. J. Appl. Phys.*, **3**, 379–384 (1952).

TABLE 6-2. STABILITY LIMITS FOR RECURRENCE FORMULA (6-84)

Number of subdivisions S	Limiting value of $r = \Delta t/(\Delta x)^2$ for stability	
	Case (a)	Case (b)
1	1.0000	0.3821
2	0.5858	0.4606
3	0.5359	0.4812
4	0.5197	0.4892
$S \to \infty$	$\frac{1}{2}(1 + 1.2337S^{-2} - \cdots)$	$\frac{1}{2}(1 - 0.7196S^{-2} + \cdots)$

this is accomplished by taking r *half* as large as the values given by Table 6-2.

We may go a step further and ask that the rate of decay of the geometric progressions in (6-92) should be as close as possible to the corresponding rates of decay in (6-93). When t is increased by Δt, the exponentials in (6-93) are diminished by the factors

$$\Lambda_n = \exp\left[-\pi^2\left(\frac{2n-1}{2}\right)^2 \Delta t\right] = \exp\left[-\frac{\pi^2}{4S^2}(2n-1)^2 r\right] \quad (6\text{-}96)$$

whereas the corresponding decay factors for (6-92) are simply the λ_n of (6-94). When n is small compared with S, we have

$$
\begin{aligned}
\Lambda_n &= 1 - \frac{r(2n-1)^2\pi^2}{4S^2} + \frac{r^2(2n-1)^4\pi^4}{32S^4} - \cdots \\
\lambda_n &= 1 - \frac{r(2n-1)^2\pi^2}{4S^2} + \frac{r(2n-1)^4\pi^4}{(6)32S^4} - \cdots
\end{aligned}
\quad (6\text{-}97)
$$

Thus λ_n and Λ_n agree up to terms in $h^2 = 1/S^2$ for any r, but only for $r = \frac{1}{6}$ do they agree up to terms in h^4. This suggests that $r = \frac{1}{6}$ should provide superior accuracy when applied to case (a).

Milne[1] has illuminated this point somewhat differently by showing that the truncation error of (6-82) while ordinarily $0(h^2)$ becomes[2] $0(h^4)$ when $r = \frac{1}{6}$. The discretization error of the solution, however, remains $0(h^2)$ until the boundary conditions are also approximated with error of order h^4. It happens that the boundary conditions (6-91) of case (a) involve no approximation at all, and thus the use of $r = \frac{1}{6}$ will give $0(h^4)$ discretization error.

[1] W. E. Milne, "Numerical Solution of Differential Equations," John Wiley & Sons, Inc., New York, 1953, p. 122.
[2] See Exercise 6-42.

Example. To illustrate the superiority of $r = \frac{1}{6}$ in case (a) of Prob. (6-1), we take in Fig. 6-35 the crudest possible network with $\Delta x = h = 1$ and $\Delta t = \frac{1}{6}$. The initial and boundary treatment is exactly the same as in Fig. 6-30. The solution values obtained by using Fig. 6-29 with $r = \frac{1}{6}$ are entered directly on the diagram. These values have been compared with the exact solution and the errors shown in Fig. 6-36.

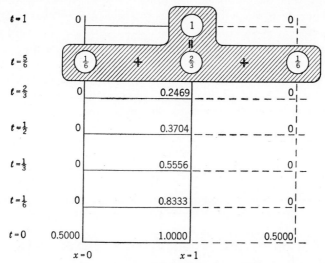

FIG. 6-35.   Solution to case (a) of Prob. 6-1, using $\Delta x = 1$, $\Delta t = \frac{1}{6}$.

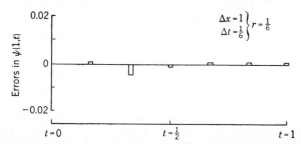

FIG. 6-36.   Errors in the approximate solution obtained in Fig. 6-35.

This should be compared with Fig. 6-31, where much greater errors were obtained from solutions which require $2\frac{2}{3}$, 9, and 21 times as many mesh points to cover the same total time interval. The solution of Fig. 6-35 has no fine-grain detail, but the isolated values obtained are remarkably accurate.

Implicit Recurrence Formulas. We have seen that the explicit recurrence formula of Fig. 6-29 for the heat-conduction equation (6-6) can give useful approximations but that it can be unstable if $r = \Delta t/(\Delta x)^2$ is taken too large. A qualitative argument for expecting trouble is sketched in Fig. 6-37. If $\psi_{j,k}$ is known at every point on AB, then $\psi_{j,k}$ can be obtained at every mesh point within the triangle ABC by successive use of the recurrence formula (shown centered on the point E) *without know-*

ing the values $\psi_{j,k}$ *on* AF *and* BG. The finite-difference model behaves like a *hyperbolic* system. The lines AC and BC play the role of *finite-difference characteristics*. The continuous system on the other hand is *parabolic*: AB and FG are characteristics. The solution at C is not determined until boundary information along AF and BG is known.

An explicit recurrence formula thus provides a somewhat faulty model for a parabolic system. The argument is qualitative only because if r is held constant as $\Delta x \to 0$ the slope of AC (which is $\Delta t/\Delta x$) approaches zero and hence in the limit AC does become the true characteristic.

A truly parabolic finite-difference model would be incapable of producing any values along MN if the boundary conditions at M and N were not known; i.e., the values at each mesh point along MN would be

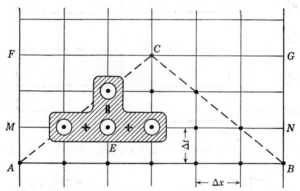

Fig. 6-37. Illustration of "finite-difference characteristics" associated with an explicit recurrence formula.

given as functions of the values along AB *and the boundary conditions at* M *and* N. This kind of behavior is obtained by using an *implicit recurrence formula*.

An implicit recurrence formula is one in which two or more unknown values in the $k + 1$ row are given in terms of known values in the k row (and $k - 1$, $k - 2$, . . . , rows if necessary) by a single application of the formula. If there are S unknown values in the $k + 1$ row, it is necessary to apply the recurrence formula S times across the whole length of the row. The unknown values are then given *implicitly* by the set of S simultaneous equations.

Implicit recurrence formulas for the heat-conduction equation (6-6) can be obtained by approximating the derivative $\partial^2\psi/\partial x^2$ in the $k + 1$ row[1] instead of the k row or by using the average[2] of approximations in

---

[1] This was suggested by O'Brien, Hyman, and Kaplan, *loc. cit.*

[2] This method was used by J. Crank and P. Nicolson, A Practical Method for Numerical Evaluation of Solutions of Partial Differential Equations of Heat-conduction Type, *Proc. Cambridge Phil. Soc.*, **32**, 50–67 (1947).

the k and $k + 1$ rows. More generally one can use

$$\frac{\partial^2 \psi}{\partial x^2} \approx \frac{\theta(\psi_{j-1,k+1} - 2\psi_{j,k+1} + \psi_{j+1,k+1})}{(\Delta x)^2} + \frac{(1 - \theta)(\psi_{j-1,k} - 2\psi_{j,k} + \psi_{j+1,k})}{(\Delta x)^2}$$

$$(6\text{-}98)$$

with $0 \le \theta \le 1$. If $\partial \psi / \partial t$ is still approximated by $(\psi_{j,k+1} - \psi_{j,k})/\Delta t$ and $\Delta t/(\Delta x)^2$ is denoted by r, the resulting recurrence formula is shown schematically in Fig. 6-38. A single application of this equates a linear combination of three unknown values in the $k + 1$ row to a linear combination

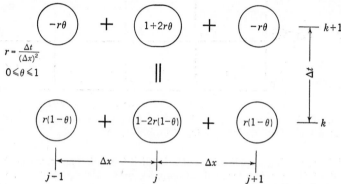

FIG. 6-38. Implicit recurrence formula for the heat-conduction equation $\partial^2 \psi / \partial x^2 = \partial \psi / \partial t$.

of three known values in the k row. Note that if $\theta = 0$ the formula of Fig. 6-38 reduces to that of Fig. 6-29. If $\theta = \frac{1}{2}$, we have the formula of Crank and Nicolson, and if $\theta = 1$, we have the formula of O'Brien, Hyman, and Kaplan. Before examining the advantages and disadvantages of the implicit formula let us work through a simple example to see the method of application.

Example. We consider case (b) of Prob. (6-1) in Fig. 6-39 with $\Delta x = \frac{1}{2}$ and $\Delta t = \frac{1}{6}$ so that $r = \frac{2}{3}$. The treatment of the initial and boundary conditions here is exactly the same as in Fig. 6-33. The recurrence formula, however, is that of Fig. 6-38 with $\theta = \frac{1}{2}$. If the recurrence formula is centered successively at $j = 0, 1$, and 2, the resulting equations can be written as follows:

$$\begin{bmatrix} 2 & -\frac{2}{3} & 0 \\ -\frac{1}{3} & \frac{5}{3} & -\frac{1}{3} \\ 0 & -\frac{2}{3} & \frac{5}{3} \end{bmatrix} \begin{bmatrix} \psi_{0,k+1} \\ \psi_{1,k+1} \\ \psi_{2,k+1} \end{bmatrix} = \begin{bmatrix} 0 & \frac{2}{3} & 0 \\ \frac{1}{3} & \frac{1}{3} & \frac{1}{3} \\ 0 & \frac{2}{3} & \frac{1}{3} \end{bmatrix} \begin{bmatrix} \psi_{0,k} \\ \psi_{1,k} \\ \psi_{2,k} \end{bmatrix} \qquad (6\text{-}99)$$

The reader should verify these and note how the boundary conditions are included.

Equations (6-99) indicate that at each step of the marching process it is necessary to solve a set of simultaneous equations. In systems like this where the coefficients are independent of k we can invert the equations once and for all and thereby obtain an explicit formula. Thus in (6-99) we can solve for the $\psi_{j,k+1}$ in terms of the $\psi_{j,k}$ as follows:

$$\begin{bmatrix} \psi_{0,k+1} \\ \psi_{1,k+1} \\ \psi_{2,k+1} \end{bmatrix} = \frac{1}{64} \begin{bmatrix} 5 & 30 & 6 \\ 15 & 26 & 18 \\ 6 & 36 & 20 \end{bmatrix} \begin{bmatrix} \psi_{0,k} \\ \psi_{1,k} \\ \psi_{2,k} \end{bmatrix} \qquad (6\text{-}100)$$

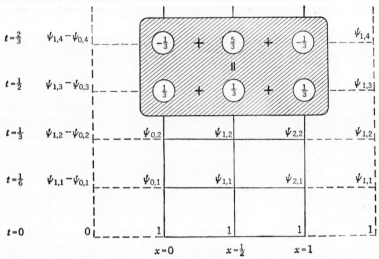

FIG. 6-39. Finite-difference network with initial and boundary conditions and implicit recurrence formula for case (b) of Prob. 6-1.

Once this has been done, application of the implicit formula is not essentially different from that of the explicit formula. Compare (6-88) with (6-100). The solution according to (6-100) has been carried out, and the errors in $\psi(1, t)$ obtained by comparison with the exact solution are shown in Fig. 6-40. We note that the errors are

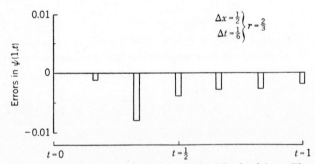

FIG. 6-40. Errors in the approximate solution obtained from Fig. 6-39.

small and well behaved despite the fact r is larger here than it was in Fig. 6-33, which produced the unstable solution depicted in Fig. 6-34.

Properties of the Implicit Recurrence Formula.

The discussion of page 382 regarding the stability, oscillation and truncation error of the explicit recurrence formula of Fig. 6-29 can be readily extended[1] to the

---

[1] S. H. Crandall, An Optimum Implicit Recurrence Formula for the Heat Conduction Equation, *Quart. Appl. Math.*, **13**, 318–320 (1955). See also Exercises 6-42 and 6-44.

implicit recurrence formula of Fig. 6-38. Without repeating the analysis we show the results in Fig. 6-41. Each point with coordinates (r, θ)
represents a different recurrence formula for integrating the heat-conduction equation.

The previous results for the explicit recurrence formula are indicated
on $\theta = 0$. With increasing θ we note that the stability limit increases
and that for $\theta \geq \frac{1}{2}$ the implicit formula is stable for all r. This represents a partial justification of our previous argument concerning the
appropriateness of implicit recurrence formulas for parabolic systems.

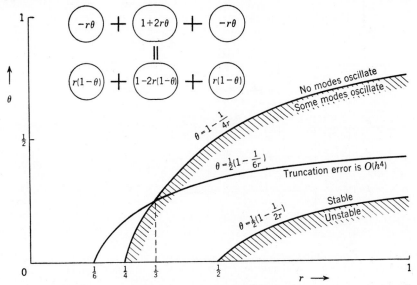

FIG. 6-41. Stability, oscillation, and truncation-error properties of recurrence formulas
for the heat-conduction equation $\partial^2\psi/\partial x^2 = \partial\psi/\partial t$.

As was the case in Table 6-2, the precise values of the stability and
oscillation limits depend on the number of spatial subdivisions S and on
the particular boundary conditions of a problem; however, as S increases,
there is a rapid approach toward fixed limiting values. It is these limiting
values which are shown in Fig. 6-41. Superior accuracy can be expected
from those formulas with $0(h^4)$ truncation error provided the boundary
conditions are also approximated to the same order.

Treatment of Singularities. We consider Prob. 6-2, boundary-layer wake. There
are two interesting points involved in this problem. The governing differential
equation

$$\frac{\partial\psi}{\partial x} = \sqrt{1 - \psi}\,\frac{\partial^2\psi}{\partial y^2} \qquad (6\text{-}101)$$

is nonlinear, and in addition there is an awkward singularity in the solution at the
origin, which corresponds to the point where the flow reunites at the rear of the plate.

The following explicit finite-difference recurrence formula corresponds to (6-101)

$$\psi_{j+1,k} = \psi_{j,k} + \frac{\Delta x}{(\Delta y)^2} \sqrt{1 - \psi_{j,k}} \, (\psi_{j,k-1} - 2\psi_{j,k} + \psi_{j,k+1}) \qquad (6\text{-}102)$$

This differs[1] from (6-84) for the heat-conduction equation only by the inclusion of the square root. Its presence complicates the arithmetical operation required at each step but does not alter the general process of marching out a solution.

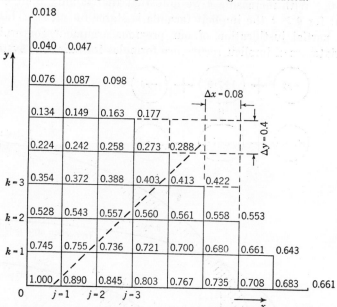

Fig. 6-42. Finite-difference approximation for Prob. 6-2. The solution is marched out to the right from initial data on $x = 0$, using the recurrence formula (6-102).

To begin a finite-difference solution, we select suitable network dimensions Δx and Δy. The similarity of (6-102) to (6-84) suggests[2] that for stability the quantity $(1 - \psi_{j,k})^{\frac{1}{2}} \Delta x/(\Delta y)^2$ should not be larger than one-half. Initially $0 \le \psi_{0,k} \le 1$; so we are probably safe with $\Delta x/(\Delta y)^2 = \frac{1}{2}$. Accordingly we choose $\Delta x = 0.08$ and $\Delta y = 0.4$. The initial conditions (i.e., for $x = 0$) are then taken from the Blasius solution[3] and entered in the first column of Fig. 6-42. The initial condition is known out to $y = \infty$, but we have used only the values out to $y = 3.2$. This permits the most interesting part of the solution domain to be covered and in particular gives the solution on the central streamline for more than half a plate length downstream of the plate. The solution is now to be marched out to the right, using (6-101) and the boundary condition of symmetry at $y = 0$.

[1] The quantities x and t in (6-84) correspond to y and x, respectively, in (6-101). In (6-102) $\psi_{j,k}$ stands for the solution value at the point $(j\,\Delta x,\, k\,\Delta y)$.

[2] A theory of stability for nonlinear difference equations has yet to be given.

[3] These values were obtained from H. Luckert, Über die Integration der Differentialgleichung einer Gleitschicht in Zäher Flüssigkeit, *Schriften Math. Sem. u. Inst. angew. Math. Univ. Berlin,* **1,** 245–274 (1934), where the problem of the text is solved by combining finite-difference treatment in the x direction with graphical integration in the y direction.

There is no difficulty in obtaining that part of the solution which is above the dotted line in Fig. 6-42. The recurrence formula (6-102) is, however, useless for determining $\psi_{1,0}$. Since $\sqrt{1 - \psi_{0,0}} = 0$, blind use of (6-102) would make $\psi_{1,0}$ and successively all the $\psi_{j,0}$ equal to unity. The corresponding solution would then represent not the wake but rather a continuation of the boundary layer. The differential equation (6-101) indicates the nature of the difficulty involved. When $\psi = 1$, $\partial\psi/\partial x$ must vanish unless $\partial^2\psi/\partial y^2 = \infty$. This latter case occurs at the origin. When the flow first reunites, there is a sharp discontinuity in $\partial\psi/\partial y$ and hence $\partial^2\psi/\partial y^2$ is infinite.

Any finite-difference relation based on the assumption that the continuous solution can be expanded in a Taylor's series cannot be expected to provide a good approximation to this sort of behavior. For this reason we have not tried to obtain $\psi_{1,0}$ in Fig. 6-42 by a finite-difference method but have used an available[1] series expansion of the solution in the neighborhood of the origin. When $y = 0$, the leading terms are

$$\psi(x, 0) = 1 - 0.596x^{\frac{2}{3}} + \cdots \tag{6-103}$$

This shows that although ψ is continuous it begins to decrease at an infinite rate along the center line of the wake. The value of $\psi_{1,0}$ in Fig. 6-42 was taken from (6-103).

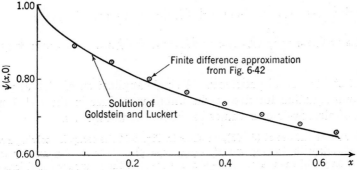

FIG. 6-43. Comparison of the approximate solution of Fig. 6-42 with more accurate solutions.

Once past this point there is no further difficulty in applying (6-102). In Fig. 6-43 the solution along the center line of the wake is compared with considerably more accurate approximations obtained independently[2] by Goldstein and by Luckert. Despite the fairly large mesh used in our finite-difference solution the greatest discrepancy is less than 2 per cent.

Treatment of a Fourth-order System. We consider Prob. 6-3, transverse motion of a beam. An *explicit* recurrence formula can be obtained by approximating $\partial^4\psi/\partial x^4$ in the k row and approximating $\partial^2\psi/\partial t^2$ in the j column as follows:

[1] S. Goldstein, Concerning Some Solutions of the Boundary Layer Equations in Hydrodynamics, *Proc. Cambridge Phil. Soc.*, **26**, 1–30 (1930).

[2] For complete tabulations of these solutions and critical comparisons see L. Rosenhead and J. H. Simpson, Note on the Velocity Distribution in the Wake behind a Flat Plate Placed along a Stream, *Proc. Cambridge Phil. Soc.*, **32**, 385–391 (1936).

$$0 = \left(\frac{\partial^4 \psi}{\partial x^4} + \frac{\partial^2 \psi}{\partial t^2} \right)_{j,k}$$

$$\approx \frac{\psi_{j-2,k} - 4\psi_{j-1,k} + 6\psi_{j,k} - 4\psi_{j+1,k} + \psi_{j+2,k}}{(\Delta x)^4} + \frac{\psi_{j,k-1} - 2\psi_{j,k} + \psi_{j,k+1}}{(\Delta t)^2}$$

$$(6\text{-}104)$$

Setting $\Delta t / (\Delta x)^2 = r$ and solving for $\psi_{j,k+1}$, we get

$$\psi_{j,k+1} = 2\psi_{j,k} - \psi_{j,k-1} - r^2(\psi_{j-2,k} - 4\psi_{j-1,k} + 6\psi_{j,k} - 4\psi_{j+1,k} + \psi_{j+2,k})$$

$$(6\text{-}105)$$

It has been shown[1] that this is stable when $r \le \frac{1}{2}$. A single application of (6-105) gives immediately one unknown value in the $k + 1$ row in terms of already-known values in the $k - 1$ and k rows.

An *implicit* recurrence formula can be obtained by using in (6-104) the average of approximations to $\partial^4 \psi / \partial x^4$ in the $k + 1$ and $k - 1$ rows in place of the approximation obtained from the k row. When this is written out and solved for the values in the $k + 1$ row, we get

$$\psi_{j,k+1} + \frac{r^2}{2} (\psi_{j-2,k+1} - 4\psi_{j-1,k+1} + 6\psi_{j,k+1} - 4\psi_{j+1,k+1} + \psi_{j+2,k+1})$$

$$= 2\psi_{j,k} - \psi_{j,k-1} - \frac{r^2}{2} (\psi_{j-2,k-1} - 4\psi_{j-1,k-1} + 6\psi_{j,k-1} - 4\psi_{j+1,k-1} + \psi_{j+2,k-1})$$

$$(6\text{-}106)$$

This is stable[2] for all positive r. A single application of (6-106) provides only one equation for five of the unknown values in the $k + 1$ row in terms of already-known values in the $k - 1$ and k rows.

To illustrate the use of (6-105), we show in Fig. 6-44 the setup for solving Prob. 6-3 on a network with $\Delta x = \frac{1}{8}$ and $\Delta t = \frac{1}{128}$ ($r = \frac{1}{2}$). Because of symmetry we need consider only half the beam. Note the treatment of the boundary conditions $\psi = 0$, $\partial \psi / \partial x = 0$ at $x = 0$, and the initial condition $\partial \psi / \partial t = 0$ at $t = 0$. The initial ψ values shown in the diagram have been taken from (6-27). When the recurrence formula is applied successively for $j = 1, 2, 3$, and 4, the resulting equations can be written as follows:

$$\begin{bmatrix} \psi_{1,k+1} \\ \psi_{2,k+1} \\ \psi_{3,k+1} \\ \psi_{4,k+1} \end{bmatrix} = \frac{1}{4} \begin{bmatrix} 1 & 4 & -1 & 0 \\ 4 & 2 & 4 & -1 \\ -1 & 4 & 1 & 4 \\ 0 & -2 & 8 & 2 \end{bmatrix} \begin{bmatrix} \psi_{1,k} \\ \psi_{2,k} \\ \psi_{3,k} \\ \psi_{4,k} \end{bmatrix} - \begin{bmatrix} \psi_{1,k-1} \\ \psi_{2,k-1} \\ \psi_{3,k-1} \\ \psi_{4,k-1} \end{bmatrix} \quad (6\text{-}107)$$

The reader should verify these and note how the initial conditions must be used to start the computation at $k = 0$.

The corresponding treatment of the implicit recurrence formula (6-106) is indicated in Fig. 6-45. The network has the same dimensions as in Fig. 6-44, and the treatment of the initial and boundary conditions is identical. The recurrence formula

[1] L. Collatz, Zur Stabilität des Differenzen verfahrens bei der Stabschwingungsgleichung, *Z. angew. Math. u. Mech.*, **31**, 392–394 (1951). See also Exercise 6-47.

[2] S. H. Crandall, Numerical Treatment of a Fourth Order Parabolic Partial Differential Equation, *J. Assoc. Comp. Mach.*, **1**, 111–118 (1954).

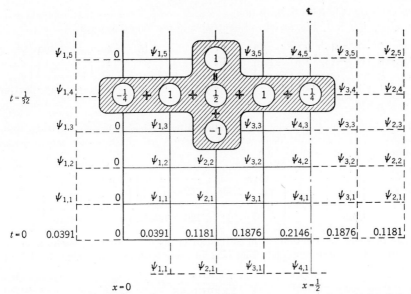

FIG. 6-44. Initial and boundary conditions and explicit recurrence formula for finite-difference approximation to Prob. 6-3.

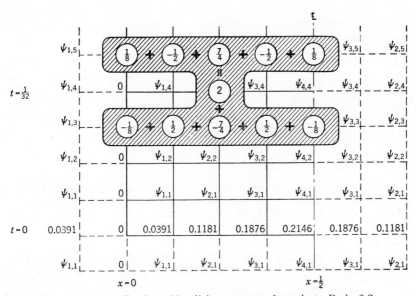

FIG. 6-45. Application of implicit recurrence formula to Prob. 6-3.

shown is that obtained from (6-106) by setting $r = \frac{1}{2}$. The result of applying it successively to $j = 1, 2, 3$, and 4 can be written in the following way:

$$\frac{1}{8}\begin{bmatrix} 15 & -4 & 1 & 0 \\ -4 & 14 & -4 & 1 \\ 1 & -4 & 15 & -4 \\ 0 & 2 & -8 & 14 \end{bmatrix}\begin{bmatrix} \psi_{1,k+1} \\ \psi_{2,k+1} \\ \psi_{3,k+1} \\ \psi_{4,k+1} \end{bmatrix} = 2\begin{bmatrix} \psi_{1,k} \\ \psi_{2,k} \\ \psi_{3,k} \\ \psi_{4,k} \end{bmatrix} + \frac{1}{8}\begin{bmatrix} -15 & 4 & -1 & 0 \\ 4 & -14 & 4 & -1 \\ -1 & 4 & -15 & 4 \\ 0 & -2 & 8 & -14 \end{bmatrix}\begin{bmatrix} \psi_{1,k-1} \\ \psi_{2,k-1} \\ \psi_{3,k-1} \\ \psi_{4,k-1} \end{bmatrix}$$

$$(6\text{-}108)$$

The values in the $k + 1$ row can be obtained by solving these four equations simultaneously. Since the coefficients in the matrices are constant, we can invert the system once and for all to get

$$\begin{bmatrix} \psi_{1,k+1} \\ \psi_{2,k+1} \\ \psi_{3,k+1} \\ \psi_{4,k+1} \end{bmatrix} = \frac{1}{996}\begin{bmatrix} 1{,}151 & 332 & -1 & -24 \\ 332 & 1{,}328 & 332 & 0 \\ -1 & 332 & 1{,}343 & 360 \\ -48 & 0 & 720 & 1{,}344 \end{bmatrix}\begin{bmatrix} \psi_{1,k} \\ \psi_{2,k} \\ \psi_{3,k} \\ \psi_{4,k} \end{bmatrix} - \begin{bmatrix} \psi_{1,k-1} \\ \psi_{2,k-1} \\ \psi_{3,k-1} \\ \psi_{4,k-1} \end{bmatrix}$$

$$(6\text{-}109)$$

Note that this has the same form as the explicit formula (6-107). Useful numerical solutions have been obtained[1] from both of these formulas, on a high-speed digital computer.

EXERCISES

6-34. Show that for a given recurrence formula for a parabolic system, with a fixed value of r, the labor required to march out the solution a given amount is proportional to S^3, where S is the number of spatial subdivisions.

6-35. When $r = \frac{1}{2}$, the recurrence formula (6-84) states that $\psi_{j,k+1}$ is the average of $\psi_{j-1,k}$ and $\psi_{j+1,k}$. Show that if on a graph of ψ vs. x at time $k\,\Delta t$ one draws chords joining points whose abscissas differ by $2\,\Delta x$ the mid-points of these chords give values of ψ at time $(k + 1)\,\Delta t$ according to (6-84). Apply this to Prob. 6-1, case (a), as set up in Fig. 6-30.

6-36. Apply the extrapolation formula of Exercise 3-53 to the points A, B, and C in Fig. 6-32.   *Ans.* 0.37092.

6-37. Show that $A \sin j\beta + B \cos j\beta$ satisfies the first of (6-90) for arbitrary values of A and B if $2 \cos \beta = 2 - a/r$.

6-38. Try a solution of the form $\psi_{j,k} = \lambda^k e^{ij\beta}$, where $i = \sqrt{-1}$ in the difference equation (6-84). Show that this leads readily to the requirement

$$\lambda = 1 - 2r(1 - \cos \beta) = 1 - 4r \sin^2 \frac{\beta}{2}$$

Compare this with (6-94). If we ask that $|\lambda| \leq 1$ for all β, we get $r = \frac{1}{2}$. Compare with Table 6-2. The procedure sketched in this exercise is attributed to J. Von Neumann. For a critical discussion see F. B. Hildebrand, "Methods of Applied Mathematics," Prentice-Hall, Inc., New York, 1952, pages 339 to 344.

6-39. Show that the solution of Exercise 6-37 satisfies the boundary conditions of case (b) [that is, (6-86) at $j = 0$ and (6-91) at $j = S$] only if $A = B \tan S\beta_n$ and β_n is one of the eigenvalues determined by the following transcendental equations:

$$S \sin \beta_n \tan S\beta_n = 1 \qquad n = 1, \ldots, S$$
$$S \sinh \gamma \tanh S\gamma = 1 \qquad \beta_{S+1} = \pi + i\gamma$$

[1] *Ibid.*

6-40. Verify that the $S + 1$ mode in the previous exercise tends to become unstable first and that the stability limit corresponding to (6-95) is given by

$$r = \frac{1}{1 + \cosh \gamma}$$

6-41. Try a solution to (6-6) in the form $\psi = X(x)T(t)$. Verify (6-93) by parallel-ing the text development for the corresponding difference equation. Show that for the boundary conditions of case (b) the space modes are $\tan \alpha_n \sin \alpha_n x + \cos \alpha_n x$ and that the eigenvalues α_n are the roots of the transcendental equation $\alpha_n \tan \alpha_n = 1$.

6-42. By expanding $\psi(x, t)$ in a Taylor's series centered at $P_{j,k}$ show that

$$\frac{\psi_{k+1} - \psi_k}{rh^2} - \frac{1}{h^2} \theta(\psi_{j-1,k+1} - 2\psi_{j,k+1} + \psi_{j+1,k+1}) + 1/h^2(1 - \theta)(\psi_{j-1,k} - 2\psi_{j,k} + \psi_{j+1,k})$$

$$= \left\{ \left(\frac{\partial \psi}{\partial t}\right)_{j,k} - \left(\frac{\partial^2 \psi}{\partial x^2}\right)_{j,k} \right\}$$

$$+ h^2 \left\{ \frac{r}{2}\left(\frac{\partial^2 \psi}{\partial t^2}\right)_{j,k} - r\theta\left(\frac{\partial^3 \psi}{\partial x^2 \, \partial t}\right)_{j,k} - \frac{1}{12}\left(\frac{\partial^4 \psi}{\partial x^4}\right)_{j,k} \right\}$$

$$+ h^4 \left\{ \frac{r^2}{6}\left(\frac{\partial^3 \psi}{\partial t^3}\right)_{j,k} - \frac{r^2\theta}{2}\left(\frac{\partial^4 \psi}{\partial x^2 \, \partial t^2}\right)_{j,k} - \frac{r\theta}{12}\left(\frac{\partial^5 \psi}{\partial x^4 \, \partial t}\right)_{j,k} - \frac{1}{360}\left(\frac{\partial^6 \psi}{\partial x^6}\right)_{j,k} \right\} + \cdots$$

where $\Delta x = h$ and $\Delta t = rh^2$. Verify that if $\psi(x, t)$ is a solution to the heat-conduction equation (6-6) then the partial derivatives within each set of braces are all equal. Hence show that the truncation error of the recurrence formula of Fig. 6-38 is $0(h^4)$ when $\theta = \left(1 - \dfrac{1}{6r}\right) \Big/ 2$ and that the truncation error of (6-84) is $0(h^4)$ when $r = \frac{1}{6}$.

6-43. Show that the truncation error of the recurrence formula of Fig. 6-38 is $0(h^6)$ when

$$r = \frac{\sqrt{5}}{10} = 0.2236$$

$$\theta = \frac{3 - \sqrt{5}}{6} = 0.1273$$

6-44. Obtain the solution to the implicit recurrence formula of Fig. 6-38 in the form of a sum of products of space modes multiplied by geometric progressions in time. Verify that the eigenvalues and space modes are exactly the same as for the explicit recurrence formula but that instead of (6-94) the decay factor is

$$\lambda_n = \frac{1 - 2r(1 - \theta)(1 - \cos \beta_n)}{1 + 2r\theta(1 - \cos \beta_n)}$$

The stability limit for r is obtained by setting $\lambda_n = -1$ and $\beta_S = \pi - \pi/2S$ for case (a) and $\beta_{S+1} = \pi + i\gamma$ for case (b), where γ is defined as in Exercise 6-39. Show that when $S \to \infty$ the stability limit is $r = \frac{1}{2}(1 - 2\theta)^{-1}$.

6-45. Consider Prob. 6-1, case (a), with the exception that the initial condition here is $\psi(x, 0) = \sin \pi x/2$. Then the exact solution is given by (6-93) with $c_1 = 1$ and all other $c_n = 0$. If no round-off errors are made, the finite-difference solution using the explicit recurrence formula (6-84) is given by (6-92) with $d_1 = 1$ and all other $d_n = 0$. Show by using the identity (3-116) or otherwise that if the point (x, t) coincides with a mesh point the difference between the finite-difference solution and the continuous solution (i.e., the discretization error) is

$$e^{-\pi^4 t/4} \sin \frac{\pi x}{2} \left\{ \frac{\pi^4 t}{32 S^2} \left(\frac{1}{6} - r\right) + 0\left(\frac{1}{S^4}\right) \right\}$$

and hence that the discrete solution *converges*[1] to the continuous solution for *arbitrary* values of r with $O(h^2)$ error except when $r = \frac{1}{6}$, in which case the discretization error is $O(h^4)$.

6-46. Show that the recurrence formula

$$\psi_{j,k+1} = \psi_{j,k-1} + 2r(\psi_{j-1,k} - 2\psi_{j,k} + \psi_{j+1,k})$$

for the heat-conduction equation has truncation errors of order $(\Delta x)^2$ and $(\Delta t)^2$ but that it is *unstable* for all values of r.

6-47. Consider Prob. 6-3 but with the boundary condition $\partial^2 \psi / \partial x^2 = 0$ at $x = 0$ and $x = 1$ in place of $\partial \psi / \partial x = 0$. Show that $\psi_{j,k} = \sin j\beta_n \cos k\varphi_n$ satisfies both (6-105) and a finite-difference approximation to the new boundary condition providing

$$\cos \varphi_n = 1 - 2r^2(1 - \cos \beta_n)^2$$

Verify that the stability limit is given by

$$r = \frac{1}{1 + \cos (\pi/S)}$$

6-6. Finite-difference Methods for Hyperbolic Systems

In Sec. 6-3 the importance of characteristic directions and characteristic curves for hyperbolic systems was emphasized. We shall see that finite-difference procedures for hyperbolic systems must also take account of the characteristics if they are to provide useful models.

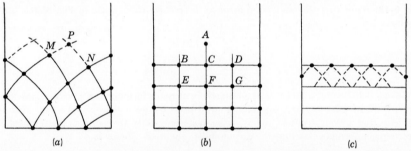

(a)   (b)   (c)

Fig. 6-46. Networks for hyperbolic systems: (a) net consists of arcs of characteristics; (b) fixed rectangular net; (c) compromise in which fixed increment of advance is used.

Probably the finite-difference method most often used is simply the application of step-by-step integration procedures to the method of characteristics illustrated in Sec. 6-3. The solution is marched forward on a grid, or network, made up of arcs of characteristic curves as indicated in Fig. 6-46a. The basic computational operation consists in locating a new mesh point such as P and determining the solution values there.

---

[1] This result is given by F. B. Hildebrand, "Methods of Applied Mathematics," Prentice-Hall, Inc., New York, 1952, p. 331. The convergence is useless in practice since round-off errors *will* occur and the recurrence formula is violently unstable when r is appreciably greater than $\frac{1}{2}$.

Under some circumstances (which depend upon the configuration of the characteristics) it is possible to march out the solution on a fixed rectangular grid as indicated in Fig. 6-46b. The basic computational operation consists in applying recurrence formulas to obtain solution values at points such as A in terms of previously obtained values.

A compromise procedure has been suggested[1] in which the entire solution is marched forward a fixed amount in each step as shown in Fig. 6-46c. The individual values that make up such a step are, however, obtained by tracing back the characteristics to the previous step, using a process very similar to that employed in Fig. 6-46a.

Rectangular Networks. Let us consider Prob. 6-4, transverse motion of a taut string, as formulated in (6-32) to (6-34). Following our usual approach, we select a network of points $P_{j,k}$ with spacings Δx and Δt and approximate the governing equation as follows:

$$0 = \left(\frac{\partial^2 \psi}{\partial t^2} - \frac{\partial^2 \psi}{\partial x^2}\right)_{j,k} \approx \frac{\psi_{j,k-1} - 2\psi_{j,k} + \psi_{j,k+1}}{(\Delta t)^2} - \frac{\psi_{j-1,k} - 2\psi_{j,k} + \psi_{j+1,k}}{(\Delta x)^2}$$
(6-110)

Denoting the ratio $\Delta t/\Delta x$ by m, we solve (6-110) for $\psi_{j,k+1}$ and obtain the recurrence formula shown in Fig. 6-47. This is an explicit recurrence

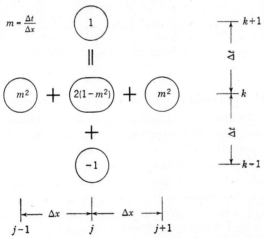

Fig. 6-47.   Recurrence formula for the wave equation $\partial^2 \psi/\partial x^2 = \partial^2 \psi/\partial t^2$.

formula and like that for the heat-conduction equation has the property that, when a solution is known up to AB in Fig. 6-48, then further use of the formula will determine the solution throughout the triangle ADB. The lines AD and BD thus play the role of "finite-difference character-

[1] D. R. Hartree, "Calculating Instruments and Machines," University of Illinois Press, Urbana, Ill., 1949, p. 41.

istics." The slopes of these lines are $\pm m$, that is, determined solely by the selection of Δx and Δt.

We have already seen that for the continuous system the knowledge of a solution up to AB determines the solution in the triangle ACB, where AC and BC are the characteristics of the continuous system. Their slopes (± 1 in this case[1]) are of course independent of the finite-difference model.

We thus have the situation that the choice of network dimensions controls the position of the point D in Fig. 6-48. If $m < 1$, D is beneath C

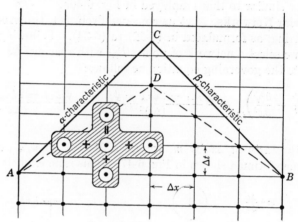

Fig. 6-48. Use of the recurrence formula of Fig. 6-47 establishes "finite-difference characteristics" AD and BD. The true characteristics of the continuous system are shown at AC and BC.

as shown; if $m = 1$, D and C coincide; and if $m > 1$, then D is above C. In the latter case the finite-difference system provides a solution in a region that is not reached by the continuous solution. Such a solution can hardly be correct.

It has in fact been proved[2] that for the finite-difference approximation to be stable and to converge toward the continuous solution as Δt and Δx approach zero with fixed m it is necessary to have $m \leq 1$. In more general systems where the true characteristics are curved the corresponding restriction is that the finite-difference characteristics must never slope more steeply than the continuous characteristics.

It is interesting to study the behavior of the recurrence formula of Fig. 6-47 for different values of m. We find that when $m = 2$ the finite-difference approximation[3] is violently unstable. When $m = 1$, the finite-

[1] See (6-55) and Exercise 6-55.

[2] R. Courant, K. Friedrichs, and H. Lewy, Über die partiellen Differenzengleich-ungen der mathematischen Physik, *Math. Ann.*, **100**, 32–74 (1928).

[3] See Exercise 6-59.

difference solution values are identical[1] with the continuous solution. Finally, when $m < 1$, the discrete approximations are stable but their accuracy deteriorates[2] with decreasing m.

Thus optimum accuracy is obtained when the finite-difference characteristics coincide with the continuous characteristics. The fact that the *exact* solution is obtained in this case is an accident due to the very special properties[3] of the wave equation (6-32). Extrapolating from this result, we can anticipate that optimum accuracy cannot be obtained by using a fixed rectangular network for a system whose true characteristics are curved since necessarily the finite-difference characteristics must have smaller slopes than the true characteristics throughout most of the solution domain.

Characteristic Networks. The finite-difference characteristics will coincide very nearly with curved continuous characteristics if we use a network built up of arcs which are approximations to the continuous characteristics[4] (see Fig. 6-46a). This procedure appears to provide optimum accuracy from coarse networks.

We shall consider here the basic computational step of advancing the solution to the point P in Fig. 6-49 from two points M and N at which the solution is known. We take the case where two unknowns u and v are to be determined from a pair of simultaneous first-order equations. As described in Sec. 6-3, the governing equations are first transformed into four ordinary differential equations: two giving the characteristic directions and two connecting the changes in u and v along the characteristics.

$$\left(\frac{dy}{dx}\right)_\alpha = a \qquad \left(\frac{dy}{dx}\right)_\beta = b$$
$$\left(\frac{dv}{du}\right)_\alpha = A \qquad \left(\frac{dv}{du}\right)_\beta = B \tag{6-111}$$

The quantities a, b, A, and B are functions of u and v in reducible systems. In more general systems they would be functions of x and y as well.

In principle any method of integrating a system of ordinary differential equations could be used to integrate (6-111). In practice either Euler's method or the simple second-category method given by formula (1) in Table 3-6 is usually employed.

---

[1] See Exercise 6-58.
[2] See Exercise 6-60.
[3] See Exercise 6-56.
[4] The method was given by J. Massau, "Mémoire sur l'intégration graphique des équations aux dérivées partielles," Ghent, 1900. See also *Enzykl. Math. Wiss.*, pt. II, sec. 3, vol. 1, Teubner Verlagsgesellschaft, Leipzig, 1909–1921, p. 159.

400 PROPAGATION PROBLEMS IN CONTINUOUS SYSTEMS

Using Euler's method, MP in Fig. 6-49 is taken as a straight line having the slope a_1 (that is, the value taken by the function a at the point M). Similarly NP is taken as a straight line with slope b_2. The intersection of these two locates P and fixes x_3 and y_3. Then along MP and NP, respectively, we set

$$v_3 - v_1 = A_1 (u_3 - u_1)$$
$$v_3 - v_2 = B_2 (u_3 - u_2)$$

(6-112)

These are Euler's approximation to the second pair of (6-111). When they are solved simultaneously for u_3 and v_3, the forward step is complete.

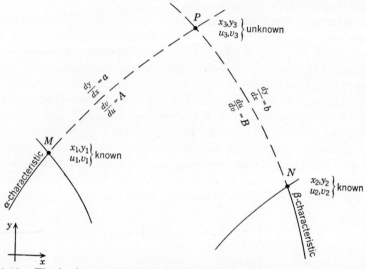

FIG. 6-49. The basic computational step in using a network made up of arcs of characteristic curves consists in determining x_3, y_3, u_3, and v_3 from known values at M and N.

The discretization error can be reduced at the price of increased computational labor by using arcs of parabolas in place of the straight lines of Euler's method. According to formula (1) of Table 3-6 we write

$$\frac{y_3 - y_1}{x_3 - x_1} = \frac{1}{2} (a_1 + a_3) \qquad \frac{y_3 - y_2}{x_3 - x_2} = \frac{1}{2} (b_2 + b_3)$$

$$\frac{v_3 - v_1}{u_3 - u_1} = \frac{1}{2} (A_1 + A_3) \qquad \frac{v_3 - v_2}{u_3 - u_2} = \frac{1}{2} (B_2 + B_3)$$

(6-113)

as an approximation to (6-111). These can be solved[1] for x_3, y_3, u_3, and

[1] See Exercise 6-51.

v_3 by using iteration as described on page 179. Trial estimates of the four unknowns are inserted in the right-hand sides of (6-113) and improved values obtained on the left. This is repeated until the values come to a standstill.

By continually repeating the basic computational step just described the solution can be marched out as indicated in Fig. 6-46a. At a boundary the process requires obvious modifications in order to incorporate the prescribed boundary condition.

Example. Consider Prob. 6-5, expansion of a gas behind a piston. The two unknowns are u and c, the dimensionless gas and sound velocities. This is a reducible system (as are many important physical examples[1]), and hence the second pair of (6-111) can be integrated separately in advance. According to (6-64) we have

$$5c + u = \alpha$$
$$5c - u = \beta \qquad (6\text{-}114)$$

along α and β characteristics, respectively. These can be inverted to give u and c in terms of α and β.

$$u = \frac{\alpha - \beta}{2}$$
$$c = \frac{\alpha + \beta}{10} \qquad (6\text{-}115)$$

The locations of the characteristics are to be obtained by integrating (6-62).

$$\left(\frac{dt}{dx}\right)_\alpha = \frac{1}{u + c} = \frac{5}{3\alpha - 2\beta}$$
$$\left(\frac{dt}{dx}\right)_\beta = \frac{1}{u - c} = \frac{5}{2\alpha - 3\beta} \qquad (6\text{-}116)$$

In Fig. 6-50 the boundary conditions are indicated for the solution within the triangle CGJ of Fig. 6-26. The values of x and t along CG were taken from (6-68). The dashed lines indicate the proposed network of characteristics. Here we can label all our characteristics in advance and also label the solution values for u and c at each intersection, if we desire, by using (6-115). The only thing unknown is the precise location of these intersections.

The solution is started by locating P_1 and continued by locating the succeeding P_j in sequence up to P_{10}. To illustrate the basic step, let us suppose that P_6 has been fixed and we have now to locate P_7. The slope of the α characteristic joining P_6 and P_7 is

$$\left(\frac{dt}{dx}\right)_{\alpha,6} = \frac{5}{(3)(4.50) - (2)(4.25)} = 1.0000 \qquad (6\text{-}117)$$

at P_6 and

$$\left(\frac{dt}{dx}\right)_{\alpha,7} = \frac{5}{(3)(4.50) - (2)(4.00)} = 0.9091 \qquad (6\text{-}118)$$

[1] See the supersonic-nozzle problem of Exercise 6-52 and the plastic-flow problem of Exercise 6-53.

at P_7 according to (6-116). If the arc between P_6 and P_7 were a parabola, the slope of the chord would be the average of these, 0.9545. Similarly we find that

$$\left(\frac{dt}{dx}\right)_{\beta,4} = -2.0000 \qquad \left(\frac{dt}{dx}\right)_{\beta,7} = -1.6667 \tag{6-119}$$

and that, if a parabolic arc had these slopes at its ends, the chord joining the ends would have a slope of -1.83333. The point P_7 can now be located graphically by

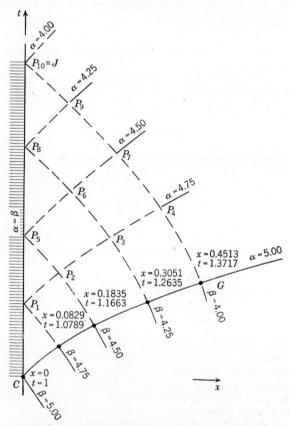

FIG. 6-50.   Characteristic network for Prob. 6-5.

drawing chords with these slopes from P_6 and P_4 or analytically by solving the following equations simultaneously for x_3 and t_3:

$$\frac{t_3 - t_1}{x_3 - x_1} = 0.9545 \qquad \frac{t_3 - t_2}{x_3 - x_2} = -1.83333 \tag{6-120}$$

At a boundary point such as P_8 the process is even simpler. Here we know $x_8 = 0$ so that a chord emanating from P_6 with a slope that is the average of $(dt/dx)_{\beta,6}$ and $(dt/dx)_{\beta,8}$ is all that is needed. The coordinate t_8 is obtained directly from the relation

$$\frac{t_8 - t_6}{0 - x_6} = \frac{1}{2}\left\{\left(\frac{dt}{dx}\right)_{\beta,6} + \left(\frac{dt}{dx}\right)_{\beta,8}\right\} \tag{6-121}$$

When the procedure sketched above is completed, we find the t coordinate of J to be 2.1055. The true value here is 2.1014. Satisfactory results may also be obtained[1] by using two cruder networks together with h^2 extrapolation.

EXERCISES

6-48. Locate the point J in Fig. 6-50 by using the coarsest possible network: a single parabolic arc from G to J.   *Ans.* $t = 2.2179$.

6-49. Locate the point J in Fig. 6-50 by using a network of parabolic arcs in which the only intermediate points are P_5 and P_7.   *Ans.* $t = 2.1280$.

6-50. Apply h^2 extrapolation to the results of the two previous exercises, assuming that the network in the second case can be considered to be twice as fine as that in the first.   *Ans.* $t = 2.0980$.

6-51. Verify that the first pair of (6-113) can be solved as follows:

$$x_3 = \frac{y_2 - y_1 + \frac{1}{2}(a_1 + a_3)x_1 - \frac{1}{2}(b_2 + b_3)x_2}{\frac{1}{2}(a_1 + a_3) - \frac{1}{2}(b_2 + b_3)}$$

$$y_3 = y_1 + \frac{1}{2}(a_1 + a_3)(x_3 - x_1) = y_2 + \frac{1}{2}(b_2 + b_3)(x_3 - x_2)$$

The calculation of both forms of y_3 provides a useful check on the working.

6-52. Organize the computation and check the values given in Fig. 6-51 for the supersonic-nozzle problem of Exercises 6-4 and 6-15. For convenience the dimensionless quantities θ, ω, μ, and β are all considered as angles measured in degrees. The

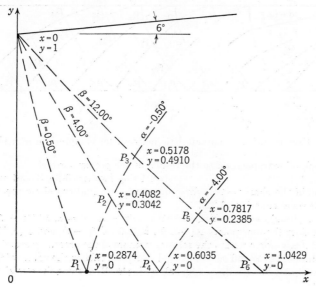

FIG. 6-51. Exercise 6-52. Characteristic network for supersonic nozzle.

points P_1, . . . , P_6 are to be located in sequence. Approximate the characteristics by arcs of parabolas. Verify that the velocity at P_6 is $u = 1.3666$, $v = 0$. A short table[2] of μ vs. ω computed according to the result of Exercise 6-15 is given to simplify the computations.

[1] See Exercise 6-50.

[2] A more extensive table is given by Shapiro, *op. cit.*, vol. I, p. 632.

ω, deg	μ, deg
0.25	75.7395
0.50	72.0988
2.00	61.9969
2.25	60.9340
4.00	55.2048
6.00	50.6186
6.25	50.1385
8.00	47.0818
12.00	41.7007

6-53. Organize the computation for locating the points P_1 to P_{30} in Fig. 6-52 for the plastic-flow problem of Exercises 6-8 and 6-16. Approximate the characteristics by arcs of parabolas. (*a*) Carry out the computation on the coarse network made up only of points P_{16}, P_{20}, and P_{30}. (*b*) Carry out the computation on the intermediate network made up of the points P_6, P_8, P_{10}, P_{16}, P_{18}, P_{20}, P_{25}, P_{27}, P_{30}. (*c*) Carry out the computation on the complete network of Fig. 6-52. Verify that at P_{30} the compression is $p = 7.2832k$.

Ans. The x coordinate of P_{16} is (*a*) 3.4142, (*b*) 3.5391, (*c*) 3.6114, and the x coordinate of P_{30} is (*a*) 5.9142, (*b*) 6.3700, (*c*) 6.6491.

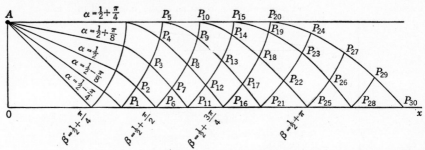

FIG. 6-52. Exercise 6-53. Characteristic network for compression of a plastic bar.

6-54. Apply h^2 extrapolation to the results of the previous exercise.

Ans. Using Exercise 3-54, 3.6391 and 6.7568.

6-55. Obtain the characteristic directions and the equations along the characteristics for the wave equation (6-32).

Ans. $(dt/dx)_\alpha = 1$, $(dt/dx)_\beta = -1$; $\partial\psi/\partial x - \partial\psi/\partial t = \alpha$, $\partial\psi/\partial x + \partial\psi/\partial t = \beta$.

6-56. Consider the four points $A(-h, 0)$, $B(h, 0)$, $C(0, -h)$, and $D(0, h)$ centered about an arbitrary origin within the solution domain of the preceding exercise. Show that the values of $\partial\psi/\partial x$ and $\partial\psi/\partial t$ at C and D depend only on the values of these same quantities at A and B. Set $d\psi = (\partial\psi/\partial x)\,dx + (\partial\psi/\partial t)\,dt$ along the straight line AD, and show that

$$\psi_D - \psi_A = \int_A^D \beta\,dt = \int_C^B \beta\,dt = \psi_B - \psi_C$$

Hence verify that the recurrence formula of Fig. 6-47 gives the exact continuous-solution value when $m = 1$.

6-57. Show that the most general second-order partial differential equation can be approximated by a finite-difference expression utilizing only the seven values at $A, \ldots, G$ in Fig. 6-46*b*.

6-58. Apply the recurrence formula of Fig. 6-47 to Prob. 6-4 as formulated in (6-32) to (6-34), using $\Delta x = \Delta t = \frac{1}{4}$, $m = 1$. Note how the initial condition $\partial \psi / \partial t = 0$ is used to start the computation. Carry the solution up to $t = \frac{1}{2}$.

Ans. $\psi_{1,2} = \psi_{2,2} = \psi_{3,2} = 0$. This agrees with the exact continuous solution.

6-59. Repeat Exercise 6-58, using $\Delta x = \frac{1}{8}$, $\Delta t = \frac{1}{4}$, $m = 2$.

Ans. At $t = \frac{1}{2}$: $\psi_{1,2} = \psi_{7,2} = 0.1091$, $\psi_{2,2} = \psi_{6,2} = -0.0625$, $\psi_{3,2} = \psi_{5,2} = -0.0156$, $\psi_{4,2} = 0$. In the next step $\psi_{1,3} = -0.9219$, $\psi_{2,3} = +0.6250$.

6-60. Repeat Exercise 6-58, using (a) $\Delta x = \frac{1}{4}$, $\Delta t = \frac{1}{8}$, $m = \frac{1}{2}$; (b) $\Delta x = \frac{1}{4}$, $\Delta t = \frac{1}{16}$, $m = \frac{1}{4}$.

Ans. At $t = \frac{1}{2}$: (a) $\psi_{1,4} = \psi_{3,4} = 0.0015$, $\psi_{2,4} = 0.0137$; (b) $\psi_{1,8} = \psi_{3,8} = 0.0054$, $\psi_{2,8} = 0.0178$.

6-61. Show that when $\Delta x = h = 1/S$ the recurrence formula of Fig. 6-47, the boundary conditions (6-33), and the second of the initial conditions (6-34) for Prob. 6-4, transverse motion of a taut string, are satisfied by $\psi_{j,k} = \sin j\beta_n \cos k\varphi_n$ if $\beta_n = \pi/S, 2\pi/S, \ldots, (S-1)\pi/S$ and $\cos \varphi_n = 1 - m^2(1 - \cos \beta_n)$. In order to fit the remaining initial condition, we take a sum of such solutions,

$$\psi_{j,k} = \sum_{n=1}^{S-1} d_n \sin j\beta_n \cos k\varphi_n$$

and adjust the constants d_n.

6-62. Show that a solution to the continuous wave equation (6-32) which corresponds to the discrete approximation of the preceding exercise is

$$\psi(x, t) = \sum_{n=1}^{\infty} c_n \sin n\pi x \cos n\pi t$$

6-63. Verify that the discrete approximation of Exercise 6-61 is *stable* as long as all the φ_n are real but that φ_{S-1} becomes complex when

$$m^2 > 1 + \frac{(1 - \cos \pi/S)}{(1 + \cos \pi/S)}$$

6-64. Suppose that the initial displacement of the string of Prob. 6-4 was simply $\psi(x, 0) = \sin \pi x$. Then the exact solution is given by Exercise 6-62 with $c_1 = 1$ and all the other $c_n = 0$. If no round-off errors are made, the discrete solution using the recurrence formula of Fig. 6-47 is given by Exercise 6-61 with $d_1 = 1$ and all the other $d_n = 0$. Show that, if the point (x, t) coincides with a mesh point, the discretization error is

$$\psi_{j,k} - \psi(x, t) = \sin \pi x \left\{ \cos \left[\pi t \left(1 - \frac{\pi^2(1 - m^2)}{24S^2} + \cdots \right) \right] - \cos \pi t \right\}$$

and hence that the discrete solution *converges* to the continuous solution for *arbitrary* values of m. Moreover the discretization error is $0(h^2)$ in general but $0(h^4)$ (at least) if $m = 1$. See Exercise 6-45 for the corresponding treatment of the heat-conduction equation.

SELECTED REFERENCES

Formulation of Engineering Problems

Johnson, W. C.: "Mathematical and Physical Principles of Engineering Analysis," McGraw-Hill Book Company, Inc., New York, 1944.

Oldenburger, R.: "Mathematical Engineering Analysis," The Macmillan Company, New York, 1950.

Ver Planck, D. W., and B. R. Teare: "Engineering Analysis," John Wiley & Sons, Inc., New York, 1954.

Mathematical Background

Courant, R., and D. Hilbert: "Methoden der mathematischen Physik," 2d ed., Springer-Verlag OHG, Berlin, vol. 1, 1931, vol. 2, 1937.

Hildebrand, F. B.: "Methods of Applied Mathematics," Prentice-Hall, Inc., New York, 1952.

Jeffreys, H., and B. S. Jeffreys: "Methods of Mathematical Physics," 3d ed., Cambridge University Press, New York, 1956.

Numerical Analysis

Collatz, L.: "Numerische Behandlung von Differentialgleichungen," Springer-Verlag OHG, Berlin, 1951.

Forsythe, G. E.: A Numerical Analyst's Fifteen-foot Shelf, *Math. Tables and Other Aids to Comp.*, **7**, 221–228 (1953).

Hartree, D. R.: "Numerical Analysis," Oxford University Press, New York, 1952.

Hildebrand, F. B.: "Introduction to Numerical Analysis," McGraw-Hill Book Company, Inc., New York, 1956.

Kopal, Z.: "Numerical Analysis," John Wiley & Sons, Inc., New York, 1955.

Salvadori, M. G., and M. L. Baron: "Numerical Methods in Engineering," Prentice-Hall, Inc., New York, 1952.

Scarborough, J. B.: "Numerical Mathematical Analysis," 2d ed., John Hopkins Press, Baltimore, 1950.

Computing Machines

Engineering Research Associates, Inc., "High-speed Computing Devices," McGraw-Hill Book Company, Inc., New York, 1950.

Forsythe, G. E.: Selected References on Use of High-speed Computers for Scientific Computation, *Math. Tables and Other Aids to Comp.*, **10**, 25–27 (1956).

Hartree, D. R.: "Calculating Instruments and Machines," University of Illinois Press, Urbana, Ill., 1949.

Soroka, W. W.: "Analog Methods in Computation and Simulation," McGraw-Hill Book Company, Inc., New York, 1954.

NAME INDEX

Adams, J. C., 176
Aitken, A. C., 27, 80, 97, 101, 102
Allen, D. N. deG., 49, 326
Ashley, H., 291

Banachiewicz, T., 27
Barnes, J. L., 134
Baron, M. L., 407
Bashforth, F., 176
Batschelet, E., 271
Beck, M., 289
Benoit, 27
Bernstein, D. L., 213, 361
Bickley, W. G., 245, 374
Biezeno, C. B., 78, 149, 151
Binder, L., 377
Biot, M. A., 75
Birkhoff, G., 16, 54
Bisplinghoff, R. L., 291
Blasius, H., 341
Buckner, H., 308

Cajori, F., 176
Carrier, G. F., 308
Cherry, E. C., 22
Chio, F., 34
Cholesky, A. L., 27
Churchill, R. V., 289
Collar, A. R., 68, 92, 101
Collatz, L., 47, 78, 96, 111, 153, 164, 184,
    192, 218, 262, 269, 271, 287, 289, 291,
    297, 303, 306, 315, 316, 324, 376, 392,
    407
Courant, R., 151, 213, 217, 218, 220, 222,
    224, 253, 270, 287, 291, 358, 360, 362,
    364, 371, 398, 407
Cramer, G., 27
Crandall, S. H., 111, 113, 329, 378, 381,
    388, 392
Crank, J., 386, 387

Cross, H., 58
Crout, P. D., 27, 28, 31

Dawes, C. L., 4, 6
Den Hartog, J. P., 239, 277, 344 .
Diaz, J. B., 16, 54
Dienes, P., 140
Doolittle, M. H., 27
Duffing, G., 239
Duncan, W. J., 68, 92, 101, 149
Dusinberre, G. M., 244
Dwyer, P. S., 27

Engesser, Fr., 21
Engineering Research Associates, 407
Euler, L., 163

Feshbach, H., 145, 218
Forsythe, G. E., 407
Fowler, C. M., 382
Fox, C., 222
Fox, L., 271, 326
Frank, N. H., 280
Frankel, S. P., 253–255, 257
Franklin, P., 147
Frazer, R. A., 68, 148, 153, 232, 314, 318
Freeman, I. M., 290
Friedrichs, K. O., 240, 270, 358, 360, 364,
    371, 398

Galerkin, B. G., 149, 233
Gardner, M. F., 134
Gaunt, J. A., 171
Gauss, C. F., 27, 41, 150
Geiringer, H., 44, 92, 360
Gerling, C. L., 41
Gershgorin, S., 262
Gibson, G. A., 176

409

SUBJECT INDEX